Energy Balance and Cancer

Series Editor:
Nathan A. Berger,
Case Western Reserve University,
Cleveland, OH, USA

For further volumes:
http://www.springer.com/series/8282

Andrew J. Dannenberg • Nathan A. Berger
Editors

Obesity, Inflammation and Cancer

Springer

Editors
Andrew J. Dannenberg
Department of Medicine
Weill Cornell Medical College
New York, NY, USA

Nathan A. Berger
Center for Science, Health & Society
Case Western Reserve University,
School of Medicine
Cleveland, OH, USA

ISBN 978-1-4614-6818-9 ISBN 978-1-4614-6819-6 (eBook)
DOI 10.1007/978-1-4614-6819-6
Springer New York Heidelberg Dordrecht London

Library of Congress Control Number: 2013937294

Printed on acid-free paper

Springer is part of Springer Science+Business Media (www.springer.com)

Preface: What's Old Is New Again and Now It's Red Hot

As the worldwide obesity pandemic expands, obesity has been associated with an increased risk of more and more cancer types. The original malignancies shown to be associated with obesity included esophageal adenocarcinoma, colon cancer, renal cell cancer, postmenopausal breast cancer, endometrial cancer, and advanced prostate cancer. More recently, obesity has been identified as a risk factor for cancers of the pancreas, gall bladder, and ovary and several hematologic malignancies including leukemia, lymphomas, and myeloma, and the list continues to grow.

From a historical viewpoint, while early studies considered the possibility that inflammation initiated the process of carcinogenesis, this was generally considered to be a local effect associated with tissue injury or chronic infection. With elucidation of DNA structure and function and development of the concept of chemical carcinogens as mutagens, attention turned to identification of activated oncogenes and deactivated tumor suppressor genes in the carcinogenic process. Separate studies demonstrated that inflammation extended beyond the local site, mediated by cellular and humoral components. As noted above, independent epidemiologic studies confirmed an association of obesity with cancer incidence, morbidity, and mortality. Studies to identify the mediators of these processes focused on the effects of obesity on growth factors and hormones and the mechanisms of carcinogenesis they commonly affect. More recently, it has become apparent that adipose tissue, in addition to serving as a fat storage depot, is an intensely active metabolic organ. In obesity, low-grade chronic adipose tissue inflammation occurs, resulting in multiple cellular and humoral inflammatory factors. Seminal studies showing that systemic metabolic disorders, such as insulin resistance, could be mediated, in part, by inflammatory cytokines, synthesized and secreted by adipose tissue, resulted in a whole new approach to understanding and attempting to control obesity-associated comorbidities. Moreover, elucidation of the prostaglandin pathway and its role in inflammation, as well as the observations that anti-inflammatory agents, especially the nonsteroidal anti-inflammatory drugs (NSAIDs), could prevent the development and progression of several forms of neoplasia, provided a major stimulus to the field. A major goal of ongoing research is to inhibit inflammation as an approach to cancer prevention and control.

The above brief description traces the complex transdisciplinary evolution of this area of research endeavor. Not only does it illustrate the impact of sometimes divergent disciplines on the evolution of a concept, but it also indicates the potential value of moving forward in this field with a transdisciplinary approach. Accordingly, the goal of this volume of Energy Balance and Cancer, volume 7 in the series, is to highlight the cutting-edge transdisciplinary science linking obesity, inflammation, and cancer. We are grateful to all the authors listed below for their contributions to this volume and look forward to their collective impact in further advancing this rapidly developing field.

This volume first provides information on inflammation as an important link between obesity and insulin resistance, which is in itself linked to promotion of cancer through hyperinsulinemia. The volume then covers some of the most important mechanisms by which obesity leads to inflammation, including the novel inflammasome concept, alterations in chromatin structure, circulating inflammatory factors, unique cellular interactions between adipocytes and macrophages, and the direct link of dietary fat to inflammation and cancer. Subsequently addressed in this volume are a number of target organs and interventional strategies for disrupting inflammation and their effects on cancer prevention and control.

In Chap. 1, Lesley G. Ellies, Andrew Johnson, and Jerrold M. Olefsky (University of California, San Diego) describe the mechanisms by which obesity stimulates low-grade inflammation leading to insulin resistance. Chapter 2, written by Tuo Deng, Christopher J. Lyon, Nan Zhang, Helen Y. Wang, Rong-fu Wang, and Willa A. Hsueh (Weill Cornell Medical College) and Jun Cui (Sun Yat-sen University), reviews the basis for understanding the emerging concept of the inflammasome and its mechanisms of activation and role in obesity. Gerald V. Denis and Deborah J. Bowen (Boston University School of Public Health) describe in Chap. 3 chromatin-based, transcription co-regulatory mechanisms that may link obesity, inflammation, and cancer. Carey Nien-Kai Lumeng (University of Michigan Medical School), in Chap. 4, describes the important role that adipose tissue macrophages play in breast and ovarian cancer. In Chap. 5, Stephanie K. Doerner and Nathan A. Berger (Case Western Reserve University School of Medicine) discuss the impact of different dietary fatty acids on promoting or suppressing colorectal cancer. In Chap. 6, Anamay Sharma, Ahmed Elebiary, Sonia Chowdhury, and Navtej Buttar (Mayo Clinic) describe the contribution of gastric reflux to inflammation in Barrett's esophagus and esophageal adenocarcinoma and potential interventions. In Chap. 7, Stephanie K. Doerner (Case Western Reserve University School of Medicine) and Jason D. Heaney (Baylor College of Medicine) describe the role of obesity-induced intestinal inflammation on colorectal cancer incidence. In Chap. 8, Neil M. Iyengar, Patrick G. Morris and Clifford A. Hudis (Memorial Sloan-Kettering Cancer Center) and Andrew J. Dannenberg (Weill Cornell Medical College) review the emerging evidence supporting the contribution of adipose tissue and chronic breast inflammation to the development of breast cancer. In Chap. 9, the relation of obesity, inflammation, and hepatocellular cancer is discussed by Naim Alkhouri and Arthur McCullough (Cleveland Clinic Lerner College of Medicine at Case Western Reserve University), and in Chap. 10, Jorge Blando, Achinto Saha, Kaoru Kiguchi, and John

DiGiovanni (University of Texas at Austin) describes the role of obesity and inflammation in prostate cancer. Louise R. Howe (Weill Cornell Medical College), in Chap. 11, describes the central role of cyclooxygenase-derived prostaglandins as potential mediators of obesity-related cancer and outlines how targeting this pathway may be protective against obesity-associated carcinogenesis. In Chap. 12, Harmony F. Turk, Jennifer M. Monk, Tim Y. Hou, and Robert S. Chapkin (Texas A&M University) discuss mechanisms through which n-3 polyunsaturated fatty acids interfere with the inflammatory process to suppress carcinogenesis, and in Chap. 13, Gary Stoner and Li-Shu Wang (Medical College of Wisconsin) describe key mechanisms by which naturally occurring dietary compounds reduce the harmful effects of inflammation and the risk for cancer development. In Chap. 14, Stephen D. Hursting, Nikki A. Ford, Sarah M. Dunlap, and Laura M. Lashinger (University of Texas at Austin) and Marcie J. Hursting (Clinical Science Consulting) describe the modification of inflammatory pathways and their impact on cancer by diet and caloric restriction. Ahmad Salameh and Mikhail G. Kolonin, in Chap. 15, describe an innovative approach to adipose tissue control by vascular targeting. In Chap. 16, Michael Gleeson (Loughborough University) describes the anti-inflammatory effects of exercise.

Overall, this volume on Obesity, Inflammation, and Cancer provides an up-to-date status report on the latest developments and state-of-the-art understanding of the role of inflammation in mediating the effects of obesity on cancer and describes possible strategies for targeting inflammation as an approach to cancer prevention and control. The book should be useful for students, researchers, and clinicians, especially those interested in the role of inflammation and its impact on cancer. It is our expectation that this volume will both stimulate research on the role of inflammation in cancer etiology and progression and lead to new approaches and clinical trials for cancer prevention and control by targeting obesity-related inflammation.

New York, NY, USA — Andrew J. Dannenberg
Cleveland, OH, USA — Nathan A. Berger

Contents

Contributors

Naim Alkhouri Digestive Disease Institute, Cleveland Clinic Foundation, Cleveland, OH, USA

Nathan A. Berger Center for Science, Health and Society, Case Western Reserve University, School of Medicine, Cleveland, OH, USA

Jorge Blando Division of Pharmacology and Toxicology, University of Texas at Austin, Austin, TX, USA

Deborah J. Bowen Department of Community Health Sciences, Boston University School of Public Health, Boston, MA, USA

Navtej Buttar Mayo Clinic, Rochester, MN, USA

Robert S. Chapkin Kleberg Center, Texas A&M University, College Station, TX, USA

Sonia Chowdhury Mayo Clinic, Rochester, MN, USA

Jun Cui School of Life Sciences, Sun Yat-sen University, Guangzhou, China

Andrew J. Dannenberg Department of Medicine, Weill Cornell Medical College, New York, NY, USA

Tuo Deng Methodist Hospital Research Institute, Weill Cornell Medical College, Houston, TX, USA

Gerald V. Denis Section of Hematology Oncology, Department of Pharmacology and Experimental Therapeutics and Cancer Research Center, Boston University School of Medicine, Boston, MA, USA

John DiGiovanni Division of Pharmacology and Toxicology, College of Pharmacy, The University of Texas at Austin, Austin, TX, USA

Department of Nutritional Sciences, College of Natural Sciences, The University of Texas at Austin, Austin, TX, USA

Dell Pediatric Research Institute, The University of Texas at Austin, Austin, TX, USA

Stephanie K. Doerner Department of Genetics, Case Western Reserve University School of Medicine, Cleveland, OH, USA

Sarah M. Dunlap Department of Nutritional Sciences, The University of Texas at Austin, Austin, TX, USA

Ahmed Elebiary Mayo Clinic, Rochester, MN, USA

Lesley G. Ellies Department of Pathology, University of California, San Diego, La Jolla, CA, USA

Nikki A. Ford Department of Nutritional Sciences, The University of Texas at Austin, Austin, TX, USA

Michael Gleeson School of Sport, Exercise and Health Sciences, Loughborough University, Loughborough, Leicestershire, UK

Jason D. Heaney Department of Molecular and Human Genetics, Baylor College of Medicine, Houston, TX, USA

Tim Y. Hou Program in Integrative Nutrition and Complex Diseases and the Center for Environmental and Rural Health Texas A&M University, College Station, TX, USA

Louise R. Howe Department of Cell and Developmental Biology, Weill Cornell Medical College, New York, NY, USA

Willa A. Hsueh Methodist Hospital Research Institute, Weill Cornell Medical College, Houston, TX, USA

Clifford A. Hudis Memorial Sloan-Kettering Cancer Center, New York, NY, USA

Marcie J. Hursting Clinical Science Consulting, Austin, TX, USA

Stephen D. Hursting Division of Nutritional Sciences, The University of Texas at Austin, Austin, USA

Department of Molecular Carcinogenesis, The University of Texas-MD Anderson Cancer Center, Smithville, TX, USA

Neil M. Iyengar Memorial Sloan-Kettering Cancer Center, New York, NY, USA

Andrew Johnson Department of Medicine, Division of Endocrinology and Metabolism, University of California, San Diego, La Jolla, CA, USA

Kaoru Kiguchi Division of Pharmacology and Toxicology, University of Texas at Austin, Austin, TX, USA

Mikhail G. Kolonin The University of Texas Health Science Center at Houston, Houston, TX, USA

Laura M. Lashinger Department of Nutritional Sciences, The University of Texas at Austin, Austin, TX, USA

Carey Nien-Kai Lumeng Department of Pediatrics and Communicable Diseases, University of Michigan Medical School, Ann Arbor, MI, USA

Christopher J. Lyon Methodist Hospital Research Institute, Weill Cornell Medical College, Houston, TX, USA

Arthur McCullough Digestive Disease Institute, Cleveland Clinic Foundation, Cleveland, OH, USA

Jennifer M. Monk Program in Integrative Nutrition and Complex Diseases and the Center for Environmental and Rural Health Texas A&M University, College Station, TX, USA

Patrick G. Morris Memorial Sloan-Kettering Cancer Center, New York, NY, USA

Jerrold M. Olefsky Department of Medicine, Division of Endocrinology and Metabolism, University of California, San Diego, La Jolla, CA, USA

Achinto Saha Division of Pharmacology and Toxicology, University of Texas at Austin, Austin, TX, USA

Ahmad Salameh The University of Texas M.D. Anderson Cancer Center, Houston, TX, USA

Anamay Sharma Mayo Clinic, Rochester, MN, USA

Gary Stoner Division of Hematology and Oncology, Department of Medicine, Medical College of Wisconsin, Milwaukee, WI, USA

Harmony F. Turk Program in Integrative Nutrition and Complex Diseases and the Center for Environmental and Rural Health Texas A&M University, College Station, TX, USA

Helen Y. Wang Methodist Hospital Research Institute, Weill Cornell Medical College, Houston, TX, USA

Li-Shu Wang Division of Hematology and Oncology, Department of Medicine, Medical College of Wisconsin, Milwaukee, WI, USA

Rong-fu Wang Methodist Hospital Research Institute, Weill Cornell Medical College, Houston, TX, USA

Nan Zhang Methodist Hospital Research Institute, Weill Cornell Medical College, Houston, TX, USA

Chapter 1
Obesity, Inflammation, and Insulin Resistance

Lesley G. Ellies, Andrew Johnson, and Jerrold M. Olefsky

Abstract Obesity is a pressing public health concern as it leads to a collection of abnormalities often termed the metabolic syndrome. Molecular studies are revealing novel pathways by which obesity-associated hormonal, nutrient, and tissue factors can stimulate the chronic low-grade inflammation that leads to insulin resistance. Signaling interactions between proinflammatory immune cells, particularly macrophages and lymphocytes, and insulin target cells in the liver and adipose tissue are key to this process and provide potential opportunities for the development of targeted therapies to improve insulin sensitivity and correct energy imbalance.

Abbreviations

ATM	Adipose tissue macrophage
DIO	Diet-induced obesity
FFA	Free fatty acid
GPCR	G protein-coupled receptor
HFD	High-fat diet
IL	Interleukin
SAT	Subcutaneous adipose tissue
SFA	Saturated fatty acid

L.G. Ellies, Ph.D.
Department of Pathology, University of California, San Diego, La Jolla, CA 92093, USA
e-mail: lellies@ucsd.edu

A. Johnson • J.M. Olefsky, M.D. (✉)
Department of Medicine, Division of Endocrinology and Metabolism, University of California, San Diego, La Jolla, CA 92093, USA
e-mail: anj004@ucsd.edu; jolefsky@ucsd.edu

A.J. Dannenberg and N.A. Berger (eds.), *Obesity, Inflammation and Cancer*,
Energy Balance and Cancer 7, DOI 10.1007/978-1-4614-6819-6_1,

TNF	Tumor necrosis factor
VAT	Visceral adipose tissue
WAT	White adipose tissue

1.1 Introduction

Overnutrition leads to energy imbalance and obesity, a precursor to the metabolic syndrome and type 2 diabetes, with increased risks of cardiovascular disease and certain types of cancer, resulting in a serious global health issue [1–3]. Traits of efficient nutrient storage and responsive immune cell activation that were advantageous during human evolution have become detrimental to human health in times of food excess [4]. In combination with the reduced physical activity and increasingly sedentary lifestyles associated with improved technology, the sequelae associated with obesity are rapidly increasing. More than one third of adults and almost 17 % of youths in the United States are obese as defined by having a body mass index (BMI, kg/m^2) of at least 30 [5]. Moreover, the most recent data from the National Health and Nutrition Examination Survey indicate that 68.3 % of individuals studied were overweight, having a BMI of at least 25 [4]. Thus, there is a pressing need to understand the molecular mechanisms underpinning obesity so that novel therapies can be developed.

Excess nutrients lead to the expansion of adipose tissues throughout the body and it is not surprising that adipocytes play a major role in obesity-induced insulin resistance [6, 7]. The discovery that low-grade, chronic inflammation in adipose tissue activated proinflammatory pathways critical for the development of insulin resistance has had a major impact on our understanding of the pathophysiology of obesity and type 2 diabetes mellitus. Hotamisligil et al. [8] provided the first evidence that proinflammatory tumor necrosis factor-α (TNFα) produced by adipose tissue could impair insulin signaling and that blocking TNFα activity ameliorated insulin resistance. Another milestone in this field was the work done by Xu et al. and Weisberg et al. [9, 10], who showed that in obesity large numbers of proinflammatory adipose tissue macrophages (ATMs) accumulate in various fat depots. The location of adipose tissue is also important in the development of insulin resistance. Adipose tissue that is deposited centrally, visceral adipose tissue (VAT), is more metabolically detrimental than subcutaneous adipose tissue (SAT) [11], and central obesity is more strongly associated with an increased risk of insulin resistance, the metabolic syndrome, and cardiovascular disease than BMI alone [12, 13]. Gender differences in fat distribution can also affect the incidence of obesity-associated diseases as women have relatively more SAT than VAT compared with men [14]. Men have approximately twice as much VAT as women and this correlates with a higher prevalence of the metabolic syndrome [15]. Estrogen may be a key regulator in mediating these effects as postmenopausal women undergo a redistribution of adipose tissue with increased amounts of VAT and an increased risk of obesity-related metabolic disorders [14]. Interestingly, epidemiologic studies suggest that obesity in premenopausal women is protective against the development of breast cancer, while obesity in postmenopausal women increases the risk for the disease [16]. Regardless of menopausal status, both obesity and type 2 diabetes are associated with breast cancers that are more aggressive at the time of diagnosis and have a poorer prognosis [17, 18].

1.2 Insulin Signaling

Insulin regulates the metabolism of glucose and lipids and insulin signaling is a complex cascade of events downstream of the insulin receptor (IR). There are two main pathways: the phosphatidylinositol 3-kinase (PI3K)-AKT pathway, which mediates glucose uptake and suppresses gluconeogenesis, and the Ras-extracellular signal-related kinase (ERK) pathway, which mediates gene expression and also interacts with the PI3K-AKT pathway to control cell growth and proliferation [19]. Insulin receptor substrates (IRSs) are key mediators of insulin signaling and include four distinct family members, IRS-1–4. IRS-1 and IRS-2 are widely expressed in mammalian tissues, while IRS-3 and IRS-4 have a more restricted distribution. Tyrosine phosphorylation of IRS by the IR generates binding sites for Src homology 2 (SH2) domain proteins, including the p85 regulatory unit of PI3K. The IRS-dependent activation of PI3K is critical for insulin-mediated regulation of metabolism as it leads to the activation of the serine/threonine kinase AKT. AKT phosphorylates key regulatory proteins in multiple tissues, leading to increased glucose transport in muscle and adipocytes, decreased gluconeogenesis and glycogenolysis in hepatocytes, and anti-lipolysis in adipocytes. Shc is another IR substrate, which subsequently engages the Grb21-Sos1-Ras pathway. Although IRS and Shc proteins are the major substrates of the IR and IGFR tyrosine kinases, other substrates such as Grb2-associated binder (GAB) and downstream of kinases (DOKs) can act as tissue- or pathway-specific alternatives to IRS [19].

Insulin resistance can arise when insulin signaling is restricted at any point in this signaling cascade. In inflammation, immune cells release increased levels of obesity-associated inflammatory cytokines which activate serine kinases such as c-Jun amino-terminal kinase (JNK) [20], inhibitor of nuclear factor kappa-B kinase subunit β (IKKβ) [21], and protein kinase Cθ (PKCθ) [22], resulting in serine phosphorylation of IRS-1. This impairs tyrosine phosphorylation and activation of IRS-1, reducing insulin receptor-mediated signaling and leading to insulin resistance.

The importance of immune cells in the etiology of metabolic disease has led to the emergence of a new field termed immunometabolism [23]. While the precise sequence of physiological events initiating inflammation in obesity remains poorly understood, as with other chronic inflammatory conditions, there is a failure in the control mechanisms that rein in overactive immune responses. This review will focus on the immune cells regulating early events in obesity-mediated inflammation and how nutrients and inflammation-related proteins may impact insulin sensitivity.

1.3 Immune Cells

The inflammatory response that occurs during obesity involves multiple immune cell types. Macrophages, neutrophils, $CD4^+$ and $CD8^+$ T cells, B cells, natural killer T (NKT) cells, eosinophils, and mast cells can all be targeted in ways which either positively or negatively regulate inflammation [24]. Thus, the immune response should be considered as a multifaceted, interactive process whereby many cell populations are inter-reliant (Fig. 1.1).

1.3.1 Macrophages

Tissue macrophage populations encompass a heterogeneous group of cells with diverse phenotypes and functions. Canonically, "classically activated" M1 macrophages, expressing the integrin CD11c, are proinflammatory, whereas "alternatively activated" M2 macrophages, which do not express CD11c, are anti-inflammatory [25]. However, such clear distinction into M1 and M2 phenotypes is not always apparent in vivo, and it is likely that a continuum exists between pro- and anti-inflammatory states [25]. Obesity is characterized by a substantial accumulation of $CD11c^+$ macrophages in the VAT and liver and an overall imbalance towards a more proinflammatory phenotype [9, 10, 26, 27].

The central role of macrophages in mediating obesity-associated insulin resistance is best demonstrated by the numerous genetic studies targeting macrophages and macrophage signaling which ameliorate or exacerbate the insulin-resistant state [28]. For example, the depletion of proinflammatory $CD11c^+$ macrophages [29] or the macrophage-intrinsic deletion of the inflammatory mediators JNK1 [30] or IKK-β [31] improves insulin sensitivity, whereas increasing inflammatory macrophage polarization via cell-intrinsic deletion of PPARγ [32] renders mice more insulin resistant [32].

In obesity, adipocytes and endothelial cells secrete chemokines, such as MCP-I and leukotriene B4 (LTB_4), that attract monocytes into the adipose tissue and liver where they differentiate into macrophages [33–36]. Once inflammation is established, the production of chemokines and cytokines by infiltrating immune cells including ATMs provides a feed-forward mechanism exacerbating macrophage recruitment (Fig. 1.1). MCP-1 mediates recruitment by binding to its cognate receptor CCR2 [37, 38]. In diet-induced obesity (DIO) mice, *Ccr2* deficiency reduced ATM accumulation, the inflammatory profile of adipose tissue, and ultimately improved systemic glucose homeostasis and insulin sensitivity [33]. In the same study, short-term treatment of obese mice with a Ccr2 antagonist also reduced ATM content and improved insulin sensitivity. However, these results have not been consistently observed in MCP-I or *Ccr2* knockout (KO) mice [39, 40] suggesting that there is a degree of redundancy in chemoattractant pathways in vivo.

Leukotriene B4 (LTB4) is a proinflammatory lipid mediator generated from arachidonic acid that promotes chemotaxis [41, 42]. Its potent biological actions are

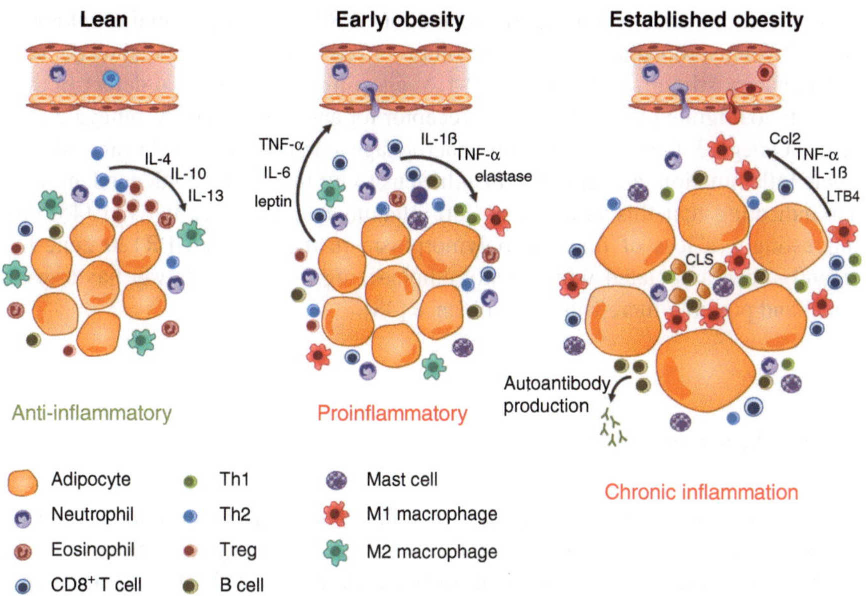

Fig. 1.1 Initiation of adipose tissue inflammation in obesity. In the lean state Th2 T cells, Tregs and eosinophils secrete cytokines such as IL-4, IL-10 and IL-13 that maintain resident macrophages in the M2 state. Early in obesity, there is an influx of neutrophils responding to chemotactic factors released by stressed adipocytes. Neutrophils release neutrophil elastase that promotes proinflammatory cytokine release from $CD8^+$ T cells and Th1 cells that begin to accumulate. This initiates the recruitment of proinflammatory macrophages that are polarized towards the M1 state. In established obesity, M1 macrophages release cyokines that act in a paracrine fashion to maintain the abnormal level of inflammation. Hypoxia and adipocyte cell death fuels the cycle of cytokine and chemokine production that results in a loss of Th2 and Treg cells, while increasing $CD8^+$ T cells, B cells and mast cells. Immune cells congregate in regions of adipocyte lysis forming crown-like structures (CLSs). Cytokines can act locally to cause insulin resistance and can be released into the circulation to promote inflammation in distant tissues

mediated primarily through binding to its G protein-coupled receptor (GPCR), BLT1 [43]. Spite et al. [35] showed that DIO increases the circulating levels of $BLT1^+$ monocytes and genetic ablation of BLT1 reduces ATM accumulation, the expression of proinflammatory cytokines and chemokines and improves insulin sensitivity [35]. Thus, the LTB_4-BLT1 pathway represents a second chemotactic axis promoting macrophage infiltration to adipose tissue and insulin resistance.

Recently, three GPCRs, GPR120, GPR105, and GPR21, have also been implicated in the regulation of macrophage chemotaxis and insulin sensitivity [44–47]. The genetic deletion of GPR105 and GPR21 reduces macrophage chemotaxis to liver or adipose tissue and thus improves insulin sensitivity in DIO mice [44, 46, 47]. The ligand for GPR105 is UDP-glucose, released from injured hepatocytes, and plasma levels of UDP-glucose are elevated in obese mice [44]. Therefore, this is likely to represent a chemotactic pathway whereby macrophages are recruited to sites of liver damage. In contrast, the ligand for GPR21 at present remains unknown. GPR21 KO macrophages are broadly compromised in their cytoskeletal response to inflammatory stimuli, including MCP-1, suggesting a more general defect in their capacity to migrate [46]. GPR120 is a receptor for anti-inflammatory, omega-3 fatty acids. Omega-3 fatty acids inhibit macrophage chemotaxis, ameliorate adipose tissue inflammation, and enhance insulin sensitivity in a GPR120-dependent manner indicating that this is a potent anti-inflammatory pathway in vivo [45, 48]. Interestingly, a loss of function polymorphism in the human GPR120 gene has recently been associated with obesity emphasizing that this pathway has translational and possibly therapeutic significance [48].

1.3.2 Neutrophils

Among the first cells to arrive at sites of inflammation are neutrophils (Fig. 1.1). Neutrophil deployment is normally tightly regulated since they carry potent cargoes of proteases used to dispose of harmful bacteria. Within 3–7 days of initiating a high-fat diet (HFD), an increase in neutrophil recruitment to adipose tissue occurs [49], suggesting that neutrophils could play a role in initiating the inflammatory cascade in response to excess nutrients. Examination of adipose tissue neutrophils (ATNs) over a longer period of HFD showed that ATN numbers remained elevated for up to 90 days [50]. Expression of neutrophil elastase (NE), a potent serine protease known to have proinflammatory effects [51], was also higher in the obese mice. Pharmacologic or genetic inhibition of the NE improved glucose tolerance of obese mice, while administration of recombinant mouse NE to normal chow-fed mice caused glucose intolerance [50]. In concordance with previous studies demonstrating degradation of IRS by NE in tumor cells [52, 53], NE led to decreased IRS1 levels in mouse and human hepatocytes with subsequent impaired insulin signaling, increased hepatic glucose production, and insulin resistance, indicating that neutrophils may play a previously unsuspected role in the initiation of obesity-induced insulin resistance [50].

1.3.3 CD4$^+$ T Cells: T-Helper and T-Regulatory Cell Subsets

CD4$^+$ T cells can be separated into distinct subsets with diverse phenotypes and functions referred to as T-helper (Th1, Th2, and Th17) cells and T-regulatory (Treg) cells, and each of these subsets is present in VAT [24]. Treg and Th2 cells predominate in lean VAT where they play a role in preventing the onset of inflammation [54, 55]. Interestingly, the proportion of Treg cells within the VAT (~40 %) is higher than in any other tissue in the body other than the colon indicating a strong requirement for anti-inflammatory strategies in this tissue [54]. The preferential accumulation of Treg cells in lean VAT is mediated by expression of the transcription factor

PPARγ [56]. Treg cell-specific PPARγ KO mice have reduced Treg cell numbers in VAT and subsequently display enhanced insulin resistance on an HFD [56]. Conversely, treatment of mice with the PPARγ agonist, pioglitazone, enhances Treg cell numbers in obese adipose tissue, and this is part of the mechanism by which pioglitazone increases insulin sensitivity [56]. The precise mechanisms by which Treg and Th2 cells act to limit inflammation in VAT are unknown; however, high expression of the anti-inflammatory cytokine IL-10, especially by Treg cells [54], together with the canonical Th2 cytokines, IL-4 and IL-13, could polarize macrophages towards a less inflammatory state. In contrast, obese adipose tissue is characterized by a specific accumulation of Th1 cells defined by their production of proinflammatory cytokines, such as IFN-γ and unchanged or declining numbers of Th2, Th17, and Treg cells [54, 55, 57]. The accumulation of Th1 cells is subsequent to macrophage and neutrophil infiltration (occurring between 4 and 20 weeks after HFD feeding in mice), and the relatively narrow T cell receptor repertoire of these cells suggests a requirement for antigen presentation [55]. A reversal of this Th1 cell dominance either by anti-CD3 antibody treatment [55] or by pioglitazone-mediated expansion of Treg cells [56] increases insulin sensitivity, and so Th1 cells are thought to contribute to the insulin-resistant state.

1.3.4 CD8⁺ T Cells

CD8⁺ T cells also accumulate in obese adipose tissue and this begins within 2 weeks of HFD feeding and peaks approximately 9 weeks later [58]. Depletion of CD8⁺ T cells either prophylactically or after the onset of inflammation improves insulin sensitivity indicating that CD8⁺ T cells contribute to the insulin-resistant state [58]. CD8⁺ T cells act to enhance inflammatory macrophage recruitment and differentiation emphasizing the integration of different immune cell pathways during obesity.

1.3.5 B Cells

B cells are recruited to adipose tissues shortly after initiation of an HFD [59], and an absence of B cells protects mice from the development of insulin resistance [60]. Transfer of IgG antibodies from obese WT mice to B cell-deficient mice enhances TNFα production, proinflammatory macrophage polarization and decreases glucose tolerance [60]. Therefore, antibody production is one mechanism by which B cells promote insulin resistance in DIO mice. In human studies, islet cell autoantibodies have been identified in ~10 % of type 2 diabetes patients, and the levels of these antibodies correlate with the need for insulin therapy [61]. Furthermore, autoantibodies against glial fibrillary acid protein (GFAP), one of the antigens most strongly associated with insulin resistance, occur in approximately 30 % of people with type 2 diabetes [62]. Interestingly, B cells deficient in the antigen presentation, major histocompatibility molecules (MHC-I and MHC-II), do not promote insulin resistance, indicating that an interaction with T cells is a feature of B cell-mediated insulin resistance [60].

1.3.6 Natural Killer T Cells

NKT cells recognize glycolipid antigens and are present in significant proportions in lean VAT and the liver, although their numbers are depleted after the onset of obesity [63–66]. A similar depletion is also observed in obese humans [65]. Studies in mice have shown little or no effect of NKT cell deficiency on the development of insulin resistance on an HFD [64–66]. However, a recent study found that NKT cell-deficient mice display reduced glucose tolerance in lean settings suggesting that NKT cells might be protective during homeostasis [66]. Furthermore, activation of NKT cells with the model ligand, α-Galactosylceramide, can increase glucose tolerance in obese mice by promoting anti-inflammatory macrophage polarization [65]. However, previous studies in younger NKT cell-deficient mice did not observe the same protective effect [63], and therefore, the role for NKT cells in regulating insulin sensitivity remains to be clarified.

1.3.7 Eosinophils

Eosinophils migrate into adipose tissue by an integrin-dependent process and promote M2 macrophage polarization by secretion of IL-4- or IL-13. Eosinophil-deficient mice fed with an HFD develop increased body fat, increased inflammation, impaired glucose tolerance, and insulin resistance, suggesting a role for eosinophils in protecting from DIO [67]. Interestingly, infection by parasitic helminth worms induces an adipose eosinophilia that enhances glucose tolerance, suggesting that targeting eosinophils to increase their numbers or enhance their function could be a useful therapeutic strategy to increase insulin sensitivity [67].

1.3.8 Mast Cells

Mast cells are typically associated with allergic hyperresponsiveness [68]. Increased numbers of mast cells are found in obese adipose tissues of mice and humans compared with their lean counterparts [69–71], and this is accompanied by increased circulating levels of the mast cell protease tryptase [69]. Genetic depletion of mast cells or pharmacologic inhibition of mast cell function reduces body weight gain; reduces the levels of inflammatory cytokines, chemokines, and proteases in serum and adipose tissue; increases energy expenditure; and improves glucose homeostasis [69]. Therefore, mast cells function to promote obesity-associated metabolic changes.

1.4 Signaling Pathways Linking Inflammation and Insulin Resistance

1.4.1 Cytokine Signaling

The inflammatory state of obese adipose tissue leads to increased local cytokine secretion, which directly causes decreased insulin sensitivity. While a number of immune cell types can produce these factors, the macrophage is the major cell type behind the release of proinflammatory cytokines. TNFα is the most well studied and stimulates serine kinases including IKK [72], JNK [20], S6 kinase (S6K) [73, 74], mammalian target of rapamycin (mTOR) [74], and double-stranded RNA-dependent protein kinase (PKR) [75] that can phosphorylate IRS1 on serine residues attenuating downstream insulin signaling (Fig. 1.2).

A variety of interleukins are released during inflammatory responses, and the two most prominent proinflammatory interleukins upregulated in obesity are IL-1β and IL-6 [76, 77]. IL-1β protein levels are increased in mice on HFD and *Il1r1* KO mice are protected from adipose tissue inflammation [78].

1.4.2 Lipid Signaling

The association between increased circulating free fatty acid (FFA) levels and insulin resistance is well known [79]. In the context of obesity, ATMs and other immune cells are exposed to high local concentrations of FFAs released by adipocyte lipolysis. The effects of saturated fatty acids (SFAs) are mediated in part through activation of the pattern recognition receptors (PRRs) Toll-like receptor 4 (Tlr4) and/or Tlr2 [26, 80, 81]. Tlrs normally recognize pathogen-associated molecular patterns (PAMPs) such as bacterial lipopolysaccharides (LPS). SFA-mediated proinflammatory signaling is attenuated in adipocytes deficient in Tlr4 [82] and *Tlr2* or *Tlr4* KO mice are partially protected against HFD-induced insulin resistance [82–85].

Early binding studies indicated that SFAs do not interact directly with Tlr4 [86]. Recent evidence suggests that fetuin A (FetA), a liver-derived circulating glycoprotein, serves as an adaptor molecule presenting FFAs to Tlr4 and activating the Tlr4 inflammatory pathway [87] (Fig. 1.2). *FetA* knockdown in insulin-resistant DIO mice resulted in reduced Tlr4-mediated inflammatory signaling in adipose tissue, whereas administration of FetA restored inflammatory signaling and induced insulin resistance. In addition, fetuin-deficient mice are protected against aging-associated obesity and insulin resistance [88]. FetA levels are elevated in type 2 diabetes and could therefore serve as a biomarker for the inflammatory state, providing an attractive target to improve glucose homeostasis without affecting immune function [89].

SFAs can also activate the Tlr4 pathway by inducing dimerization and recruitment of Tlr4 into lipid rafts, thereby enhancing its association with adaptor

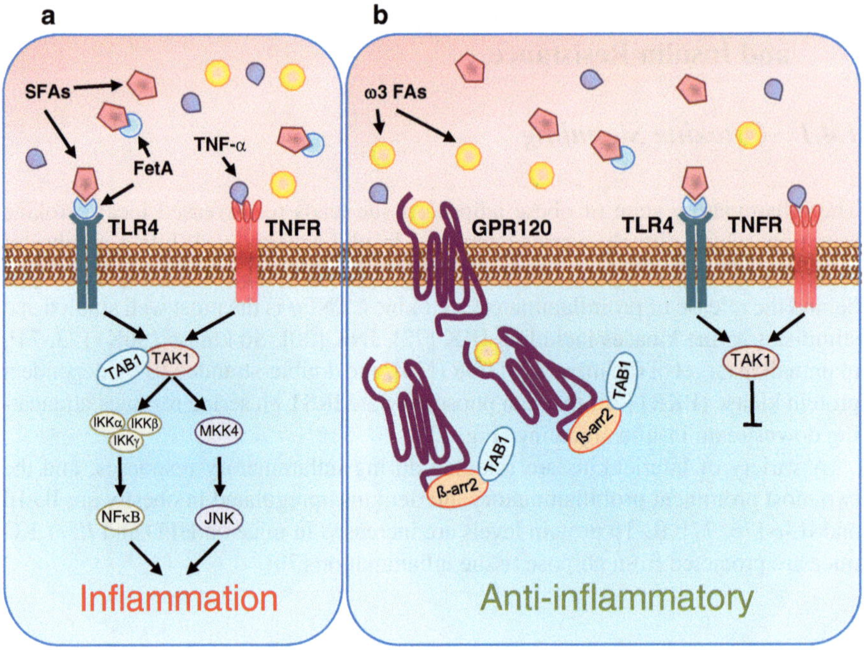

Fig. 1.2 Mechanism for proinflammatory saturated fatty acid (SFA) and anti-inflammatory omega-3 fatty acid (ω3 FA) activity. (**a**) Circulating FetA produced by the liver, functions as an adaptor between SFAs and toll-like receptor 4 (TLR4) signaling. TLR4 activated by SFAs and tumor necrosis factor receptor (TNFR) activated by TNF-α interact with transforming growth factor β (TGF-β) activated kinase 1 (TAK1) with TAK1 binding protein 1(TAB1), initiating a proinflammatory signaling cascade by activating nuclear factor kappa B kinase (NF-κB) and c-Jun N-terminal kinase (JNK). (**b**) Binding of ω3 FAs to GPR120 activates and internalizes the receptor which then binds to β-arrestin 2 (β-arr2) and sequesters TAB1, inhibiting inflammation

molecules TRIF and MyD88 [90]. This process is dependent on the production of reactive oxygen species (ROS) and leads to activation of downstream signaling and increased target gene expression. Other studies have shown that SFAs can alter the membrane distribution of c-Src, partitioning it into lipid rafts where it becomes activated [91]. This leads to signaling via JNK and the transcription of proinflammatory genes. SFAs such as palmitate activate a specialized inflammasome activation pathway in which SFAs inhibit AMP-activated protein kinase (AMPK) leading to defective autophagy, which then activates the inflammasome, resulting in IL-1β cleavage and release [92]. IL-1β stimulates TNFα production and the lack of NLRP3 and its adaptor protein, apoptotic speck protein containing a caspase recruitment domain (ASC), prevents HFD-induced inflammation.

In addition to stimulating inflammatory pathways, metabolites of SFAs, such as ceramide, can directly inhibit insulin signaling by inhibiting Akt. Virtually all stress stimuli increase rates of ceramide synthesis, and numerous studies have demonstrated a strong association between intracellular ceramide levels and the development of insulin resistance [93–96]. Recent studies have revealed an IKKβ-dependent pathway whereby Tpl2 or Tnf2 induce expression of the genes driving ceramide biosynthesis [93], indicating an additional mechanism by which inflammatory cytokines can act to inhibit insulin signaling [93, 97]. Thus, in addition to acting as ceramide precursors, SFAs stimulate ceramide biosynthesis, amplifying the effects of ceramide on insulin signaling [98].

1.4.3 ER Stress

ER stress activates the unfolded protein response (UPR) to restore ER homeostasis by inhibiting protein synthesis, increasing the degradation of proteins from the ER, and increasing the level of chaperone proteins to assist in protein folding [99]. If these adaptive mechanisms are insufficient to restore ER homeostasis, the cell undergoes programmed cell death [100]. Downstream signaling through regulators of the UPR, PKR-like eukaryotic initiation factor 2α kinase (PERK), inositol-requiring enzyme 1 (IRE-1), and activating transcription factor 6 (ATF-6) can activate both JNK and IKK, leading to the expression of inflammatory cytokines [101, 102]. In addition to protein folding, ER stress response genes play an important role in lipid metabolism as indicated by abnormal lipid processing in the absence of these genes [103–105]. Numerous studies have shown increased ER stress in fatty liver tissues from obese mice.

ER chaperones and folding enzymes such as glucose-regulated protein (GRP) 78, GRP94, protein disulphide isomerase (PDI), calnexin (CNX), and calreticulin (CRT) assist in protein folding and prevent aggregation of unfolded or misfolded proteins [106]. Treating obese or diabetic mice with chemical chaperones [107] or overexpressing GRP78 [108] restores systemic insulin sensitivity and improves hepatic steatosis. Mice deficient in X-box-binding protein-1 (XBP-1), a transcription factor, have an aggravated UPR response, develop insulin resistance and glucose intolerance [109], suggesting that ER stress could be a factor in initiating metabolic inflammation.

1.4.4 Hypoxia

In vivo measurements demonstrate that fat depots in obese animal and human subjects contain distinct regions of microvascularity and exist in a hypoxic state [110, 111]. Hypoxic adipose tissue shows a substantial induction of hypoxia-inducible factor (HIF). Evidence suggests that HIF-1α can activate inflammatory pathways and Krishnan et al. [112] have shown in mice that Hif-1α in VAT is critical for development of glucose intolerance and insulin resistance in HFD mice.

1.5 Anti-inflammatory Therapies

Given the important role of chronic tissue inflammation in the etiology of insulin resistance, anti-inflammatory therapy may prove clinically efficacious in type 2 diabetes. Promising studies have already been reported using high-dose salsalate, an inhibitor of the nuclear factor kappa-B kinase (NFκB) pathway in insulin-resistant type 2 diabetic patients [113, 114]. Treatment with this agent leads to an improvement in insulin sensitivity and a modest decrease in hemoglobin A1c levels. Given the modest nature of these effects, more efficacious treatments will be needed in the future, but these studies provide a potential proof of concept that anti-inflammatory therapy can be a beneficial antidiabetic approach.

TNFα plays a role in insulin resistance [115–117] and TNFα blockade can increase insulin sensitivity in mice [8]. However, in man the process is less clear. There is a range of results reported from clinical TNFα neutralization studies on relatively small patient populations. For example, little to no benefit in terms of insulin sensitivity was reported with anti-TNFα therapy used in the treatment of rheumatoid arthritis patients [118] or obese patients with features of metabolic syndrome [119]. In contrast, there are reports of improved insulin sensitivity in patients receiving long-term treatment with TNF inhibitors [120, 121]. Retrospective analysis of a large cohort of rheumatoid arthritis or psoriasis patients showed that individuals treated with TNF inhibitors had a reduced risk for developing type 2 diabetes [122, 123], suggesting a need for further clinical trials in this area that more directly test the hypothesis that long-term anti-TNFα therapy increases insulin sensitivity.

IL1β is a circulating cytokine and antibody-mediated inactivation of IL-1β improves glycemic control in DIO due largely to increased pancreatic beta cell formation [124]. Humans treated with an IL-1 receptor antagonist also show reduced systemic inflammation, but with only modest effects in improving glycemia and no effects on insulin sensitivity [125, 126]. The lowered glucose levels were accompanied by increased insulin secretion, suggesting that IL-1 blockade inhibits damaging effects of IL-1β on pancreatic islets, improving pancreatic beta cell function and increasing insulin secretion [125, 126].

The thiazolidinedione (TZD) group of drugs are currently used in the treatment of type 2 diabetes, and although their precise mechanism of action is not fully understood, they are agonists for nuclear receptor PPARγ [127]. Although adipocytes have been considered the primary target cells for the anti-inflammatory effects of TZDs [128], additional studies indicate an important role for PPARγ in macrophages [32, 129], muscle [130], and the CNS [131, 132]. In addition to the transactivation of genes, PPARγ can also transrepress target genes by antagonizing signal-dependent activation of transcription factors such as NFκB and AP-1, thereby reducing proinflammatory signaling [133]. However, although anti-inflammatory effects are part of the overall mechanism of TZD-induced insulin sensitization, the contributions of this to the beneficial clinical effects of these drugs remain to be defined [134].

The anti-inflammatory properties of omega-3 FAs have long been known and are associated with protection from cardiovascular disease [135, 136]. Fish oil, a major source of omega-3 FAs, leads to decreased production of TNFα IL-1 and IL-6 by macrophages in mice and man [137–139]. The lipid-sensing receptor GRP120, highly expressed in proinflammatory macrophages, can mediate the effects of omega-3 FAs [45]. Thus, when WT and GPR120 KO mice were placed on an HFD with or without omega-3 FA supplementation, the omega-3 FA treatment inhibited inflammation and enhanced systemic insulin sensitivity in WT mice, but was without effect in *Gpr120* KO mice. Signaling studies demonstrated that after ligand stimulation, GPR120 couples to β-arrestin2, which is followed by receptor endocytosis and inhibition of TAB1-mediated activation of TAK1, providing a mechanism for inhibition of both the TLR and TNFα proinflammatory signaling pathways (Fig. 1.2). Importantly, a loss of function mutation in GPR120 in humans increases the risk for obesity and lends support for the therapeutic targeting of this lipid sensor [48].

1.6 Links with Cancer

The chronic low-grade inflammation in obesity is also associated with cancer progression through a variety of common pathways [140–142]. Mechanisms by which inflammation can contribute to carcinogenesis include induction of genomic instability, increased DNA damage, alterations in epigenetic events and subsequent inappropriate gene expression, enhanced proliferation of initiated cells, and resistance to apoptosis. While there is a clear link between obesity, inflammation, and cancer, there are many areas that remain poorly understood. Whether obesity-related inflammation is more important in early events in cancer initiation and how tumor-associated macrophages interact with ATMs during tumor progression are questions that remain to be answered. Inflammation-induced insulin resistance is known to be tissue selective and the resulting hyperinsulinemia may have direct effects on promoting the growth of tumor cells [143]. A detailed discussion of the relationships between obesity, inflammation, insulin resistance, and cancer will be covered in the following chapters.

1.7 Conclusions

Since the discovery that inflammation was a critical component of the pathway from obesity to insulin resistance, research efforts have been focused on understanding the cellular and molecular basis for these changes. It now appears that there are multiple cell types and signaling molecules that offer potential drug discovery efforts to improve insulin resistance.

Given the important role for immune cells in the initiation of obesity-related disease and our knowledge of their protective role in host defense from studies of individuals with genetic abnormalities in immune cell function, finding the balance between therapies that dampen the deleterious aspects of the immune response while maintaining the ability to combat infections and disease will be challenging. For example, targeting of TNFα, a major cytokine released by activated macrophages, has been an effective therapy for the treatment of inflammatory disorders such as rheumatoid arthritis [144]. Concerns regarding long-term effects such as an increase in cancer risk have been largely dispelled by the results of long-term follow-up; however, the risk of exacerbating infection remains [145, 146]. Whether these compounds can be used in the insulin-sensitizing armamentarium is yet to be determined.

Advances in technology and the application of genome-wide association studies (GWAS) are likely to identify novel pathways and linkages between existing pathways that will aid in our understanding of multifactorial inflammatory diseases. A recent study in mice suggests that less than 10 % of transcripts altered in DIO are shared between the classical insulin target tissues, white adipose tissue (WAT), skeletal muscle, liver, and heart [147]. In humans, GWAS have identified novel loci associated with fasting insulin and insulin resistance [148]. These studies and many others provide fruitful avenues for future research.

Acknowledgment This work was supported by NIH grant U54CA155435 and DOD grant BC102147.

References

1. Flegal KM, Carroll MD, Ogden CL, Curtin LR (2010) Prevalence and trends in obesity among US adults, 1999–2008. JAMA 303(3):235–241. doi:2009.2014 [pii]. 10.1001/jama.2009.2014
2. Moller DE, Kaufman KD (2005) Metabolic syndrome: a clinical and molecular perspective. Annu Rev Med 56:45–62. doi:10.1146/annurev.med.56.082103.104751
3. James PT, Leach R, Kalamara E, Shayeghi M (2001) The worldwide obesity epidemic. Obes Res 9(suppl 4):228S–233S. doi:10.1038/oby.2001.123
4. Johnson AR, Justin Milner J, Makowski L (2012) The inflammation highway: metabolism accelerates inflammatory traffic in obesity. Immunol Rev 249(1):218–238. doi:10.1111/j.1600-065X.2012.01151.x
5. Ogden CL, Carroll MD, Kit BK, Flegal KM (2012) Prevalence of obesity in the United States, 2009–2010, vol NCHS data brief, no 82. National Center for Health Statistics, Hyattsville, MD
6. Li P, Fan W, Xu J, Lu M, Yamamoto H, Auwerx J, Sears DD, Talukdar S, Oh D, Chen A, Bandyopadhyay G, Scadeng M, Ofrecio JM, Nalbandian S, Olefsky JM (2011) Adipocyte NCoR knockout decreases PPARgamma phosphorylation and enhances PPARgamma activity and insulin sensitivity. Cell 147(4):815–826. doi:S0092-8674(11)01220-7 [pii]. 10.1016/j.cell.2011.09.050
7. Jacobi D, Stanya KJ, Lee CH (2012) Adipose tissue signaling by nuclear receptors in metabolic complications of obesity. Adipocyte 1(1):4–12. doi:10.4161/adip.19036
8. Hotamisligil GS, Shargill NS, Spiegelman BM (1993) Adipose expression of tumor necrosis factor-alpha: direct role in obesity-linked insulin resistance. Science 259(5091):87–91

9. Xu H, Barnes GT, Yang Q, Tan G, Yang D, Chou CJ, Sole J, Nichols A, Ross JS, Tartaglia LA, Chen H (2003) Chronic inflammation in fat plays a crucial role in the development of obesity-related insulin resistance. J Clin Invest 112(12):1821–1830. doi:10.1172/JCI19451. 112/12/1821 [pii]
10. Weisberg SP, McCann D, Desai M, Rosenbaum M, Leibel RL, Ferrante AW Jr (2003) Obesity is associated with macrophage accumulation in adipose tissue. J Clin Invest 112(12):1796–1808. doi:10.1172/JCI19246
11. Galic S, Oakhill JS, Steinberg GR (2010) Adipose tissue as an endocrine organ. Mol Cell Endocrinol 316(2):129–139. doi:S0303-7207(09)00438-9 [pii]. 10.1016/j.mce.2009.08.018
12. Despres JP, Lemieux I (2006) Abdominal obesity and metabolic syndrome. Nature 444(7121):881–887. doi:10.1038/nature05488
13. Kissebah AH, Vydelingum N, Murray R, Evans DJ, Hartz AJ, Kalkhoff RK, Adams PW (1982) Relation of body fat distribution to metabolic complications of obesity. J Clin Endocrinol Metab 54(2):254–260
14. Nedungadi TP, Clegg DJ (2009) Sexual dimorphism in body fat distribution and risk for cardiovascular diseases. J Cardiovasc Transl Res 2(3):321–327. doi:10.1007/s12265-009-9101-1
15. Ford ES, Li C, Zhao G, Pearson WS, Mokdad AH (2008) Prevalence of the metabolic syndrome among U.S. adolescents using the definition from the International Diabetes Federation. Diabetes Care 31(3):587–589. doi:10.2337/dc07-1030
16. Anderson GL, Neuhouser ML (2012) Obesity and the risk for premenopausal and postmenopausal breast cancer. Cancer Prev Res 5(4):515–521. doi:10.1158/1940-6207.CAPR-12-0091
17. Vona-Davis L, Rose DP (2012) Type 2 diabetes and obesity metabolic interactions: common factors for breast cancer risk and novel approaches to prevention and therapy. Curr Diabetes Rev 8(2):116–130. doi:EPUB-CDR-20120117-002 [pii]
18. Cleary MP, Grossmann ME (2009) Minireview: obesity and breast cancer: the estrogen connection. Endocrinology 150(6):2537–2542. doi:en.2009-0070 [pii]. 10.1210/en.2009-0070
19. Siddle K (2012) Molecular basis of signaling specificity of insulin and IGF receptors: neglected corners and recent advances. Front Endocrinol (Lausanne) 3:34. doi:10.3389/fendo.2012.00034
20. Hirosumi J, Tuncman G, Chang L, Gorgun CZ, Uysal KT, Maeda K, Karin M, Hotamisligil GS (2002) A central role for JNK in obesity and insulin resistance. Nature 420(6913):333–336. doi:10.1038/nature01137
21. Yuan M, Konstantopoulos N, Lee J, Hansen L, Li ZW, Karin M, Shoelson SE (2001) Reversal of obesity- and diet-induced insulin resistance with salicylates or targeted disruption of Ikkbeta. Science 293(5535):1673–1677. doi:10.1126/science.1061620
22. Perseghin G, Petersen K, Shulman GI (2003) Cellular mechanism of insulin resistance: potential links with inflammation. Int J Obes Relat Metab Disord 27(suppl 3):S6–S11. doi:10.1038/sj.ijo.0802491
23. Mathis D, Shoelson SE (2011) Immunometabolism: an emerging frontier. Nat Rev Immunol 11(2):81. doi:10.1038/nri2922
24. Gregor MF, Hotamisligil GS (2011) Inflammatory mechanisms in obesity. Annu Rev Immunol 29:415–445. doi:10.1146/annurev-immunol-031210-101322
25. Liu G, Yang H (2012) Modulation of macrophage activation and programming in immunity. J Cell Physiol 228(3):502–512. doi:10.1002/jcp.24157
26. Nguyen MT, Favelyukis S, Nguyen AK, Reichart D, Scott PA, Jenn A, Liu-Bryan R, Glass CK, Neels JG, Olefsky JM (2007) A subpopulation of macrophages infiltrates hypertrophic adipose tissue and is activated by free fatty acids via toll-like receptors 2 and 4 and JNK-dependent pathways. J Biol Chem 282(48):35279–35292. doi:M706762200 [pii]. 10.1074/jbc.M706762200
27. Lumeng CN, Bodzin JL, Saltiel AR (2007) Obesity induces a phenotypic switch in adipose tissue macrophage polarization. J Clin Invest 117(1):175–184. doi:10.1172/JCI29881

28. Osborn O, Olefsky JM (2012) The cellular and signaling networks linking the immune system and metabolism in disease. Nat Med 18(3):363–374. doi:10.1038/nm.2627. nm.2627 [pii]
29. Patsouris D, Li PP, Thapar D, Chapman J, Olefsky JM, Neels JG (2008) Ablation of CD11c-positive cells normalizes insulin sensitivity in obese insulin resistant animals. Cell Metab 8(4):301–309. doi:S1550-4131(08)00282-9 [pii]. 10.1016/j.cmet.2008.08.015
30. Solinas G, Vilcu C, Neels JG, Bandyopadhyay GK, Luo JL, Naugler W, Grivennikov S, Wynshaw-Boris A, Scadeng M, Olefsky JM, Karin M (2007) JNK1 in hematopoietically derived cells contributes to diet-induced inflammation and insulin resistance without affecting obesity. Cell Metab 6(5):386–397. doi:S1550-4131(07)00292-6 [pii]. 10.1016/j.cmet.2007.09.011
31. Arkan MC, Hevener AL, Greten FR, Maeda S, Li ZW, Long JM, Wynshaw-Boris A, Poli G, Olefsky J, Karin M (2005) IKK-beta links inflammation to obesity-induced insulin resistance. Nat Med 11(2):191–198. doi:nm1185 [pii]. 10.1038/nm1185
32. Odegaard JI, Ricardo-Gonzalez RR, Goforth MH, Morel CR, Subramanian V, Mukundan L, Red Eagle A, Vats D, Brombacher F, Ferrante AW, Chawla A (2007) Macrophage-specific PPARgamma controls alternative activation and improves insulin resistance. Nature 447(7148):1116–1120. doi:nature05894 [pii]. 10.1038/nature05894
33. Weisberg SP, Hunter D, Huber R, Lemieux J, Slaymaker S, Vaddi K, Charo I, Leibel RL, Ferrante AW Jr (2006) CCR2 modulates inflammatory and metabolic effects of high-fat feeding. J Clin Invest 116(1):115–124. doi:10.1172/JCI24335
34. Kanda H, Tateya S, Tamori Y, Kotani K, Hiasa K, Kitazawa R, Kitazawa S, Miyachi H, Maeda S, Egashira K, Kasuga M (2006) MCP-1 contributes to macrophage infiltration into adipose tissue, insulin resistance, and hepatic steatosis in obesity. J Clin Invest 116(6): 1494–1505. doi:10.1172/JCI26498
35. Spite M, Hellmann J, Tang Y, Mathis SP, Kosuri M, Bhatnagar A, Jala VR, Haribabu B (2011) Deficiency of the leukotriene B4 receptor, BLT-1, protects against systemic insulin resistance in diet-induced obesity. J Immunol 187(4):1942–1949. doi:10.4049/jimmunol.1100196
36. Obstfeld AE, Sugaru E, Thearle M, Francisco AM, Gayet C, Ginsberg HN, Ables EV, Ferrante AW Jr (2010) C-C chemokine receptor 2 (CCR2) regulates the hepatic recruitment of myeloid cells that promote obesity-induced hepatic steatosis. Diabetes 59(4):916–925. doi:db09-1403 [pii]. 10.2337/db09-1403
37. Kurihara T, Bravo R (1996) Cloning and functional expression of mCCR2, a murine receptor for the C-C chemokines JE and FIC. J Biol Chem 271(20):11603–11607
38. Charo IF, Myers SJ, Herman A, Franci C, Connolly AJ, Coughlin SR (1994) Molecular cloning and functional expression of two monocyte chemoattractant protein 1 receptors reveals alternative splicing of the carboxyl-terminal tails. Proc Natl Acad Sci U S A 91(7): 2752–2756
39. Inouye KE, Shi H, Howard JK, Daly CH, Lord GM, Rollins BJ, Flier JS (2007) Absence of CC chemokine ligand 2 does not limit obesity-associated infiltration of macrophages into adipose tissue. Diabetes 56(9):2242–2250. doi:db07-0425 [pii]. 10.2337/db07-0425
40. Chen A, Mumick S, Zhang C, Lamb J, Dai H, Weingarth D, Mudgett J, Chen H, MacNeil DJ, Reitman ML, Qian S (2005) Diet induction of monocyte chemoattractant protein-1 and its impact on obesity. Obes Res 13(8):1311–1320. doi:10.1038/oby.2005.159
41. Samuelsson B, Dahlen SE, Lindgren JA, Rouzer CA, Serhan CN (1987) Leukotrienes and lipoxins: structures, biosynthesis, and biological effects. Science 237(4819):1171–1176
42. Haeggstrom JZ (2004) Leukotriene A4 hydrolase/aminopeptidase, the gatekeeper of chemotactic leukotriene B4 biosynthesis. J Biol Chem 279(49):50639–50642. doi:10.1074/jbc.R400027200. R400027200 [pii]
43. Yokomizo T, Izumi T, Chang K, Takuwa Y, Shimizu T (1997) A G-protein-coupled receptor for leukotriene B4 that mediates chemotaxis. Nature 387(6633):620–624. doi:10.1038/42506
44. Xu J, Morinaga H, Oh D, Li P, Chen A, Talukdar S, Lazarowski E, Olefsky JM, Kim JJ (2012) GPR105 ablation prevents inflammation and improves insulin sensitivity in mice with

diet-induced obesity. J Immunol 189(4):1992–1999. doi:jimmunol.1103207 [pii]. 10.4049/jimmunol.1103207
45. Oh DY, Talukdar S, Bae EJ, Imamura T, Morinaga H, Fan W, Li P, Lu WJ, Watkins SM, Olefsky JM (2010) GPR120 is an omega-3 fatty acid receptor mediating potent anti-inflammatory and insulin-sensitizing effects. Cell 142(5):687–698. doi:10.1016/j.cell.2010.07.041
46. Osborn O, da Oh Y, McNelis J, Sanchez-Alavez M, Talukdar S, Lu M, Li P, Thiede L, Morinaga H, Kim JJ, Heinrichsdorff J, Nalbandian S, Ofrecio JM, Scadeng M, Schenk S, Hadcock J, Bartfai T, Olefsky JM (2012) G protein-coupled receptor 21 deletion improves insulin sensitivity in diet-induced obese mice. J Clin Invest 122(7):2444–2453. doi:10.1172/JCI61953. 61953 [pii]
47. Gardner J, Wu S, Ling L, Danao J, Li Y, Yeh WC, Tian H, Baribault H (2012) G-protein-coupled receptor GPR21 knockout mice display improved glucose tolerance and increased insulin response. Biochem Biophys Res Commun 418(1):1–5. doi:10.1016/j.bbrc.2011.11.117
48. Ichimura A, Hirasawa A, Poulain-Godefroy O, Bonnefond A, Hara T, Yengo L, Kimura I, Leloire A, Liu N, Iida K, Choquet H, Besnard P, Lecoeur C, Vivequin S, Ayukawa K, Takeuchi M, Ozawa K, Tauber M, Maffeis C, Morandi A, Buzzetti R, Elliott P, Pouta A, Jarvelin MR, Korner A, Kiess W, Pigeyre M, Caiazzo R, Van Hul W, Van Gaal L, Horber F, Balkau B, Levy-Marchal C, Rouskas K, Kouvatsi A, Hebebrand J, Hinney A, Scherag A, Pattou F, Meyre D, Koshimizu TA, Wolowczuk I, Tsujimoto G, Froguel P (2012) Dysfunction of lipid sensor GPR120 leads to obesity in both mouse and human. Nature 483(7389):350–354. doi:10.1038/nature10798
49. Elgazar-Carmon V, Rudich A, Hadad N, Levy R (2008) Neutrophils transiently infiltrate intra-abdominal fat early in the course of high-fat feeding. J Lipid Res 49(9):1894–1903. doi:M800132-JLR200 [pii]. 10.1194/jlr.M800132-JLR200
50. Talukdar S, Oh DY, Bandyopadhyay G, Li D, Xu J, McNelis J, Lu M, Li P, Yan Q, Zhu Y, Ofrecio J, Lin M, Brenner MB, Olefsky JM (2012) Neutrophils mediate insulin resistance in mice fed a high-fat diet through secreted elastase. Nat Med 18(9):1407–1412. doi:10.1038/nm.2885. nm.2885 [pii]
51. Pham CT (2006) Neutrophil serine proteases: specific regulators of inflammation. Nat Rev Immunol 6(7):541–550. doi:nri1841 [pii]. 10.1038/nri1841
52. Houghton AM (2010) The paradox of tumor-associated neutrophils: fueling tumor growth with cytotoxic substances. Cell Cycle 9(9):1732–1737. doi:11297 [pii]
53. Houghton AM, Rzymkiewicz DM, Ji H, Gregory AD, Egea EE, Metz HE, Stolz DB, Land SR, Marconcini LA, Kliment CR, Jenkins KM, Beaulieu KA, Mouded M, Frank SJ, Wong KK, Shapiro SD (2010) Neutrophil elastase-mediated degradation of IRS-1 accelerates lung tumor growth. Nat Med 16(2):219–223. doi:nm.2084 [pii]. 10.1038/nm.2084
54. Feuerer M, Herrero L, Cipolletta D, Naaz A, Wong J, Nayer A, Lee J, Goldfine AB, Benoist C, Shoelson S, Mathis D (2009) Lean, but not obese, fat is enriched for a unique population of regulatory T cells that affect metabolic parameters. Nat Med 15(8):930–939. doi:nm.2002 [pii]. 10.1038/nm.2002
55. Winer S, Chan Y, Paltser G, Truong D, Tsui H, Bahrami J, Dorfman R, Wang Y, Zielenski J, Mastronardi F, Maezawa Y, Drucker DJ, Engleman E, Winer D, Dosch HM (2009) Normalization of obesity-associated insulin resistance through immunotherapy. Nat Med 15(8):921–929. doi:nm.2001 [pii]. 10.1038/nm.2001
56. Cipolletta D, Feuerer M, Li A, Kamei N, Lee J, Shoelson SE, Benoist C, Mathis D (2012) PPAR-gamma is a major driver of the accumulation and phenotype of adipose tissue Treg cells. Nature 486(7404):549–553. doi:nature11132 [pii]. 10.1038/nature11132
57. Strissel KJ, DeFuria J, Shaul ME, Bennett G, Greenberg AS, Obin MS (2010) T-cell recruitment and Th1 polarization in adipose tissue during diet-induced obesity in C57BL/6 mice. Obesity (Silver Spring) 18(10):1918–1925. doi:oby20101 [pii]. 10.1038/oby.2010.1
58. Nishimura S, Manabe I, Nagasaki M, Eto K, Yamashita H, Ohsugi M, Otsu M, Hara K, Ueki K, Sugiura S, Yoshimura K, Kadowaki T, Nagai R (2009) CD8+ effector T cells contribute to macrophage recruitment and adipose tissue inflammation in obesity. Nat Med 15(8):914–920. doi:10.1038/nm.1964

59. Duffaut C, Galitzky J, Lafontan M, Bouloumie A (2009) Unexpected trafficking of immune cells within the adipose tissue during the onset of obesity. Biochem Biophys Res Commun 384(4):482–485. doi:10.1016/j.bbrc.2009.05.002
60. Winer DA, Winer S, Shen L, Wadia PP, Yantha J, Paltser G, Tsui H, Wu P, Davidson MG, Alonso MN, Leong HX, Glassford A, Caimol M, Kenkel JA, Tedder TF, McLaughlin T, Miklos DB, Dosch HM, Engleman EG (2011) B cells promote insulin resistance through modulation of T cells and production of pathogenic IgG antibodies. Nat Med 17(5):610–617. doi:10.1038/nm.2353
61. Falorni A, Gambelunghe G, Forini F, Kassi G, Cosentino A, Candeloro P, Bolli GB, Brunetti P, Calcinaro F (2000) Autoantibody recognition of COOH-terminal epitopes of GAD65 marks the risk for insulin requirement in adult-onset diabetes mellitus. J Clin Endocrinol Metab 85(1):309–316
62. Gomez-Tourino I, Camina-Darriba F, Otero-Romero I, Rodriguez MA, Hernandez-Fernandez A, Gonzalez-Fernandez A, Pena-Gonzalez E, Rodriguez J, Rodriguez-Segade S, Varela-Calvino R (2010) Autoantibodies to glial fibrillary acid protein and S100beta in diabetic patients. Diabet Med 27(2):246–248. doi:10.1111/j.1464-5491.2009.02911.x
63. Kotas ME, Lee HY, Gillum MP, Annicelli C, Guigni BA, Shulman GI, Medzhitov R (2011) Impact of CD1d deficiency on metabolism. PLoS One 6(9):e25478. doi:10.1371/journal.pone.0025478. PONE-D-11-12529 [pii]
64. Mantell BS, Stefanovic-Racic M, Yang X, Dedousis N, Sipula IJ, O'Doherty RM (2011) Mice lacking NKT cells but with a complete complement of CD8+ T-cells are not protected against the metabolic abnormalities of diet-induced obesity. PLoS One 6(6):e19831. doi:10.1371/journal.pone.0019831. PONE-D-10-04900 [pii]
65. Ji Y, Sun S, Xu A, Bhargava P, Yang L, Lam KS, Gao B, Lee CH, Kersten S, Qi L (2012) Activation of natural killer T cells promotes M2 macrophage polarization in adipose tissue and improves systemic glucose tolerance via interleukin-4 (IL-4)/STAT6 protein signaling axis in obesity. J Biol Chem 287(17):13561–13571. doi:M112.350066 [pii]. 10.1074/jbc.M112.350066
66. Schipper HS, Rakhshandehroo M, van de Graaf SF, Venken K, Koppen A, Stienstra R, Prop S, Meerding J, Hamers N, Besra G, Boon L, Nieuwenhuis EE, Elewaut D, Prakken B, Kersten S, Boes M, Kalkhoven E (2012) Natural killer T cells in adipose tissue prevent insulin resistance. J Clin Invest 122(9):3343–3354. doi:10.1172/JCI62739. 62739 [pii]
67. Wu D, Molofsky AB, Liang HE, Ricardo-Gonzalez RR, Jouihan HA, Bando JK, Chawla A, Locksley RM (2011) Eosinophils sustain adipose alternatively activated macrophages associated with glucose homeostasis. Science 332(6026):243–247. doi:science.1201475 [pii]. 10.1126/science.1201475
68. Zhang J, Shi GP (2012) Mast cells and metabolic syndrome. Biochim Biophys Acta 1822(1):14–20. doi:S0925-4439(10)00290-5 [pii]. 10.1016/j.bbadis.2010.12.012
69. Liu J, Divoux A, Sun J, Zhang J, Clement K, Glickman JN, Sukhova GK, Wolters PJ, Du J, Gorgun CZ, Doria A, Libby P, Blumberg RS, Kahn BB, Hotamisligil GS, Shi GP (2009) Genetic deficiency and pharmacological stabilization of mast cells reduce diet-induced obesity and diabetes in mice. Nat Med 15(8):940–945. doi:nm.1994 [pii]. 10.1038/nm.1994
70. Altintas MM, Azad A, Nayer B, Contreras G, Zaias J, Faul C, Reiser J, Nayer A (2011) Mast cells, macrophages, and crown-like structures distinguish subcutaneous from visceral fat in mice. J Lipid Res 52(3):480–488. doi:jlr.M011338 [pii]. 10.1194/jlr.M011338
71. Divoux A, Moutel S, Poitou C, Lacasa D, Veyrie N, Aissat A, Arock M, Guerre-Millo M, Clement K (2012) Mast cells in human adipose tissue: link with morbid obesity, inflammatory status, and diabetes. J Clin Endocrinol Metab 97(9):E1677–E1685. doi:10.1210/jc.2012-1532
72. Gao Z, Hwang D, Bataille F, Lefevre M, York D, Quon MJ, Ye J (2002) Serine phosphorylation of insulin receptor substrate 1 by inhibitor kappa B kinase complex. J Biol Chem 277(50):48115–48121. doi:10.1074/jbc.M209459200. M209459200 [pii]
73. Gao Z, Zuberi A, Quon MJ, Dong Z, Ye J (2003) Aspirin inhibits serine phosphorylation of insulin receptor substrate 1 in tumor necrosis factor-treated cells through targeting multiple serine kinases. J Biol Chem 278(27):24944–24950. doi:10.1074/jbc.M300423200. M300423200 [pii]

74. Ozes ON, Akca H, Mayo LD, Gustin JA, Maehama T, Dixon JE, Donner DB (2001) A phosphatidylinositol 3-kinase/Akt/mTOR pathway mediates and PTEN antagonizes tumor necrosis factor inhibition of insulin signaling through insulin receptor substrate-1. Proc Natl Acad Sci U S A 98(8):4640–4645. doi:10.1073/pnas.051042298. 051042298 [pii]
75. Nakamura T, Furuhashi M, Li P, Cao H, Tuncman G, Sonenberg N, Gorgun CZ, Hotamisligil GS (2010) Double-stranded RNA-dependent protein kinase links pathogen sensing with stress and metabolic homeostasis. Cell 140(3):338–348. doi:S0092-8674(10)00002-4 [pii]. 10.1016/j.cell.2010.01.001
76. Tack CJ, Stienstra R, Joosten LA, Netea MG (2012) Inflammation links excess fat to insulin resistance: the role of the interleukin-1 family. Immunol Rev 249(1):239–252. doi:10.1111/j.1600-065X.2012.01145.x
77. Pradhan AD, Manson JE, Rifai N, Buring JE, Ridker PM (2001) C-reactive protein, interleukin 6, and risk of developing type 2 diabetes mellitus. JAMA 286(3):327–334
78. McGillicuddy FC, Harford KA, Reynolds CM, Oliver E, Claessens M, Mills KH, Roche HM (2011) Lack of interleukin-1 receptor I (IL-1RI) protects mice from high-fat diet-induced adipose tissue inflammation coincident with improved glucose homeostasis. Diabetes 60(6):1688–1698. doi:10.2337/db10-1278
79. Park E, Wong V, Guan X, Oprescu AI, Giacca A (2007) Salicylate prevents hepatic insulin resistance caused by short-term elevation of free fatty acids in vivo. J Endocrinol 195(2):323–331. doi:195/2/323 [pii]. 10.1677/JOE-07-0005
80. Lee JY, Sohn KH, Rhee SH, Hwang D (2001) Saturated fatty acids, but not unsaturated fatty acids, induce the expression of cyclooxygenase-2 mediated through toll-like receptor 4. J Biol Chem 276(20):16683–16689. doi:10.1074/jbc.M011695200. M011695200 [pii]
81. Senn JJ (2006) Toll-like receptor-2 is essential for the development of palmitate-induced insulin resistance in myotubes. J Biol Chem 281(37):26865–26875. doi:M513304200 [pii]. 10.1074/jbc.M513304200
82. Shi H, Kokoeva MV, Inouye K, Tzameli I, Yin H, Flier JS (2006) TLR4 links innate immunity and fatty acid-induced insulin resistance. J Clin Invest 116(11):3015–3025. doi:10.1172/JCI28898
83. Saberi M, Woods NB, de Luca C, Schenk S, Lu JC, Bandyopadhyay G, Verma IM, Olefsky JM (2009) Hematopoietic cell-specific deletion of toll-like receptor 4 ameliorates hepatic and adipose tissue insulin resistance in high-fat-fed mice. Cell Metab 10(5):419–429. doi:S1550-4131(09)00294-0 [pii]. 10.1016/j.cmet.2009.09.006
84. Himes RW, Smith CW (2010) Tlr2 is critical for diet-induced metabolic syndrome in a murine model. FASEB J 24(3):731–739. doi:fj.09-141929 [pii]. 10.1096/fj.09-141929
85. Ehses JA, Meier DT, Wueest S, Rytka J, Boller S, Wielinga PY, Schraenen A, Lemaire K, Debray S, Van Lommel L, Pospisilik JA, Tschopp O, Schultze SM, Malipiero U, Esterbauer H, Ellingsgaard H, Rutti S, Schuit FC, Lutz TA, Boni-Schnetzler M, Konrad D, Donath MY (2010) Toll-like receptor 2-deficient mice are protected from insulin resistance and beta cell dysfunction induced by a high-fat diet. Diabetologia 53(8):1795–1806. doi:10.1007/s00125-010-1747-3
86. Schaeffler A, Gross P, Buettner R, Bollheimer C, Buechler C, Neumeier M, Kopp A, Schoelmerich J, Falk W (2009) Fatty acid-induced induction of toll-like receptor-4/nuclear factor-kappaB pathway in adipocytes links nutritional signalling with innate immunity. Immunology 126(2):233–245. doi:IMM2892 [pii]. 10.1111/j.1365-2567.2008.02892.x
87. Pal D, Dasgupta S, Kundu R, Maitra S, Das G, Mukhopadhyay S, Ray S, Majumdar SS, Bhattacharya S (2012) Fetuin-A acts as an endogenous ligand of TLR4 to promote lipid-induced insulin resistance. Nat Med 18:1279–1285. doi:10.1038/nm.2851
88. Mathews ST, Rakhade S, Zhou X, Parker GC, Coscina DV, Grunberger G (2006) Fetuin-null mice are protected against obesity and insulin resistance associated with aging. Biochem Biophys Res Commun 350(2):437–443. doi:10.1016/j.bbrc.2006.09.071
89. Heinrichsdorff J, Olefsky JM (2012) Fetuin-A: the missing link in lipid-induced inflammation. Nat Med 18(8):1182–1183. doi:10.1038/nm.2869
90. Wong SW, Kwon MJ, Choi AM, Kim HP, Nakahira K, Hwang DH (2009) Fatty acids modulate toll-like receptor 4 activation through regulation of receptor dimerization and recruitment

into lipid rafts in a reactive oxygen species-dependent manner. J Biol Chem 284(40):27384–27392. doi:M109.044065 [pii]. 10.1074/jbc.M109.044065
91. Holzer RG, Park EJ, Li N, Tran H, Chen M, Choi C, Solinas G, Karin M (2011) Saturated fatty acids induce c-Src clustering within membrane subdomains, leading to JNK activation. Cell 147(1):173–184. doi:S0092-8674(11)01004-X [pii]. 10.1016/j.cell.2011.08.034
92. Wen H, Gris D, Lei Y, Jha S, Zhang L, Huang MT, Brickey WJ, Ting JP (2011) Fatty acid-induced NLRP3-ASC inflammasome activation interferes with insulin signaling. Nat Immunol 12(5):408–415. doi:10.1038/ni.2022
93. Holland WL, Bikman BT, Wang LP, Yuguang G, Sargent KM, Bulchand S, Knotts TA, Shui G, Clegg DJ, Wenk MR, Pagliassotti MJ, Scherer PE, Summers SA (2011) Lipid-induced insulin resistance mediated by the proinflammatory receptor TLR4 requires saturated fatty acid-induced ceramide biosynthesis in mice. J Clin Invest 121(5):1858–1870. doi:10.1172/JCI43378
94. Stratford S, Hoehn KL, Liu F, Summers SA (2004) Regulation of insulin action by ceramide: dual mechanisms linking ceramide accumulation to the inhibition of Akt/protein kinase B. J Biol Chem 279(35):36608–36615. doi:10.1074/jbc.M406499200
95. Ussher JR, Koves TR, Cadete VJ, Zhang L, Jaswal JS, Swyrd SJ, Lopaschuk DG, Proctor SD, Keung W, Muoio DM, Lopaschuk GD (2010) Inhibition of de novo ceramide synthesis reverses diet-induced insulin resistance and enhances whole-body oxygen consumption. Diabetes 59(10):2453–2464. doi:10.2337/db09-1293
96. Frangioudakis G, Garrard J, Raddatz K, Nadler JL, Mitchell TW, Schmitz-Peiffer C (2010) Saturated- and n-6 polyunsaturated-fat diets each induce ceramide accumulation in mouse skeletal muscle: reversal and improvement of glucose tolerance by lipid metabolism inhibitors. Endocrinology 151(9):4187–4196. doi:10.1210/en.2010-0250
97. Dbaibo GS, El-Assaad W, Krikorian A, Liu B, Diab K, Idriss NZ, El-Sabban M, Driscoll TA, Perry DK, Hannun YA (2001) Ceramide generation by two distinct pathways in tumor necrosis factor alpha-induced cell death. FEBS Lett 503(1):7–12
98. Chavez JA, Summers SA (2012) A ceramide-centric view of insulin resistance. Cell Metab 15(5):585–594. doi:10.1016/j.cmet.2012.04.002
99. Hotamisligil GS (2010) Endoplasmic reticulum stress and the inflammatory basis of metabolic disease. Cell 140(6):900–917. doi:S0092-8674(10)00187-X [pii]. 10.1016/j.cell.2010.02.034
100. Engin F, Hotamisligil GS (2010) Restoring endoplasmic reticulum function by chemical chaperones: an emerging therapeutic approach for metabolic diseases. Diabetes Obes Metab 12(suppl 2):108–115. doi:10.1111/j.1463-1326.2010.01282.x
101. Hu P, Han Z, Couvillon AD, Kaufman RJ, Exton JH (2006) Autocrine tumor necrosis factor alpha links endoplasmic reticulum stress to the membrane death receptor pathway through IRE1alpha-mediated NF-kappaB activation and down-regulation of TRAF2 expression. Mol Cell Biol 26(8):3071–3084. doi:26/8/3071 [pii]. 10.1128/MCB.26.8.3071-3084.2006
102. Urano F, Wang X, Bertolotti A, Zhang Y, Chung P, Harding HP, Ron D (2000) Coupling of stress in the ER to activation of JNK protein kinases by transmembrane protein kinase IRE1. Science 287(5453):664–666. doi:8218 [pii]
103. Lee AH, Scapa EF, Cohen DE, Glimcher LH (2008) Regulation of hepatic lipogenesis by the transcription factor XBP1. Science 320(5882):1492–1496. doi:10.1126/science.1158042
104. Yamamoto K, Takahara K, Oyadomari S, Okada T, Sato T, Harada A, Mori K (2010) Induction of liver steatosis and lipid droplet formation in ATF6alpha-knockout mice burdened with pharmacological endoplasmic reticulum stress. Mol Biol Cell 21(17):2975–2986. doi:10.1091/mbc.E09-02-0133
105. Zhang K, Wang S, Malhotra J, Hassler JR, Back SH, Wang G, Chang L, Xu W, Miao H, Leonardi R, Chen YE, Jackowski S, Kaufman RJ (2011) The unfolded protein response transducer IRE1alpha prevents ER stress-induced hepatic steatosis. EMBO J 30(7):1357–1375. doi:10.1038/emboj.2011.52
106. Ni M, Lee AS (2007) ER chaperones in mammalian development and human diseases. FEBS Lett 581(19):3641–3651. doi:10.1016/j.febslet.2007.04.045

107. Ozcan U, Yilmaz E, Ozcan L, Furuhashi M, Vaillancourt E, Smith RO, Gorgun CZ, Hotamisligil GS (2006) Chemical chaperones reduce ER stress and restore glucose homeostasis in a mouse model of type 2 diabetes. Science 313(5790):1137–1140. doi:10.1126/science.1128294
108. Kammoun HL, Chabanon H, Hainault I, Luquet S, Magnan C, Koike T, Ferre P, Foufelle F (2009) GRP78 expression inhibits insulin and ER stress-induced SREBP-1c activation and reduces hepatic steatosis in mice. J Clin Invest 119(5):1201–1215. doi:10.1172/JCI37007
109. Ozcan U, Cao Q, Yilmaz E, Lee AH, Iwakoshi NN, Ozdelen E, Tuncman G, Gorgun C, Glimcher LH, Hotamisligil GS (2004) Endoplasmic reticulum stress links obesity, insulin action, and type 2 diabetes. Science 306(5695):457–461. doi:306/5695/457 [pii]. 10.1126/science.1103160
110. Ye J, Gao Z, Yin J, He Q (2007) Hypoxia is a potential risk factor for chronic inflammation and adiponectin reduction in adipose tissue of ob/ob and dietary obese mice. Am J Physiol Endocrinol Metab 293(4):E1118–E1128. doi:10.1152/ajpendo.00435.2007
111. Pasarica M, Sereda OR, Redman LM, Albarado DC, Hymel DT, Roan LE, Rood JC, Burk DH, Smith SR (2009) Reduced adipose tissue oxygenation in human obesity: evidence for rarefaction, macrophage chemotaxis, and inflammation without an angiogenic response. Diabetes 58(3):718–725. doi:10.2337/db08-1098
112. Krishnan J, Danzer C, Simka T, Ukropec J, Walter KM, Kumpf S, Mirtschink P, Ukropcova B, Gasperikova D, Pedrazzini T, Krek W (2012) Dietary obesity-associated Hif1alpha activation in adipocytes restricts fatty acid oxidation and energy expenditure via suppression of the Sirt2-NAD+ system. Genes Dev 26(3):259–270. doi:10.1101/gad.180406.111
113. Goldfine AB, Fonseca V, Jablonski KA, Pyle L, Staten MA, Shoelson SE (2010) The effects of salsalate on glycemic control in patients with type 2 diabetes: a randomized trial. Ann Intern Med 152(6):346–357. doi:10.1059/0003-4819-152-6-201003160-00004
114. Rumore MM, Kim KS (2010) Potential role of salicylates in type 2 diabetes. Ann Pharmacother 44(7–8):1207–1221. doi:10.1345/aph.1M483
115. Uysal KT, Wiesbrock SM, Marino MW, Hotamisligil GS (1997) Protection from obesity-induced insulin resistance in mice lacking TNF-alpha function. Nature 389(6651):610–614. doi:10.1038/39335
116. Grunfeld C, Feingold KR (1991) The metabolic effects of tumor necrosis factor and other cytokines. Biotherapy 3(2):143–158
117. Hotamisligil GS, Murray DL, Choy LN, Spiegelman BM (1994) Tumor necrosis factor alpha inhibits signaling from the insulin receptor. Proc Natl Acad Sci U S A 91(11):4854–4858
118. Rosenvinge A, Krogh-Madsen R, Baslund B, Pedersen BK (2007) Insulin resistance in patients with rheumatoid arthritis: effect of anti-TNFalpha therapy. Scand J Rheumatol 36(2):91–96. doi:10.1080/03009740601179605
119. Stanley TL, Zanni MV, Johnsen S, Rasheed S, Makimura H, Lee H, Khor VK, Ahima RS, Grinspoon SK (2011) TNF-alpha antagonism with etanercept decreases glucose and increases the proportion of high molecular weight adiponectin in obese subjects with features of the metabolic syndrome. J Clin Endocrinol Metab 96(1):E146–E150. doi:10.1210/jc.2010-1170
120. Gonzalez-Gay MA, Gonzalez-Juanatey C, Vazquez-Rodriguez TR, Miranda-Filloy JA, Llorca J (2010) Insulin resistance in rheumatoid arthritis: the impact of the anti-TNF-alpha therapy. Ann N Y Acad Sci 1193:153–159. doi:NYAS5287 [pii]. 10.1111/j.1749-6632.2009.05287.x
121. Kiortsis DN, Mavridis AK, Vasakos S, Nikas SN, Drosos AA (2005) Effects of infliximab treatment on insulin resistance in patients with rheumatoid arthritis and ankylosing spondylitis. Ann Rheum Dis 64(5):765–766. doi:10.1136/ard.2004.026534. ard.2004.026534 [pii]
122. Yazdani-Biuki B, Stelzl H, Brezinschek HP, Hermann J, Mueller T, Krippl P, Graninger W, Wascher TC (2004) Improvement of insulin sensitivity in insulin resistant subjects during prolonged treatment with the anti-TNF-alpha antibody infliximab. Eur J Clin Invest 34(9):641–642. doi:10.1111/j.1365-2362.2004.01390.x
123. Solomon DH, Massarotti E, Garg R, Liu J, Canning C, Schneeweiss S (2011) Association between disease-modifying antirheumatic drugs and diabetes risk in patients with rheumatoid arthritis and psoriasis. JAMA 305(24):2525–2531. doi:10.1001/jama.2011.878

124. Osborn O, Brownell SE, Sanchez-Alavez M, Salomon D, Gram H, Bartfai T (2008) Treatment with an interleukin 1 beta antibody improves glycemic control in diet-induced obesity. Cytokine 44(1):141–148. doi:10.1016/j.cyto.2008.07.004
125. Larsen CM, Faulenbach M, Vaag A, Volund A, Ehses JA, Seifert B, Mandrup-Poulsen T, Donath MY (2007) Interleukin-1-receptor antagonist in type 2 diabetes mellitus. N Engl J Med 356(15):1517–1526. doi:10.1056/NEJMoa065213
126. Donath MY, Shoelson SE (2011) Type 2 diabetes as an inflammatory disease. Nat Rev Immunol 11(2):98–107. doi:10.1038/nri2925
127. Lehmann JM, Moore LB, Smith-Oliver TA, Wilkison WO, Willson TM, Kliewer SA (1995) An antidiabetic thiazolidinedione is a high affinity ligand for peroxisome proliferator-activated receptor gamma (PPAR gamma). J Biol Chem 270(22):12953–12956
128. Sugii S, Olson P, Sears DD, Saberi M, Atkins AR, Barish GD, Hong SH, Castro GL, Yin YQ, Nelson MC, Hsiao G, Greaves DR, Downes M, Yu RT, Olefsky JM, Evans RM (2009) PPARgamma activation in adipocytes is sufficient for systemic insulin sensitization. Proc Natl Acad Sci U S A 106(52):22504–22509. doi:0912487106 [pii]. 10.1073/pnas.0912487106
129. Hevener AL, Olefsky JM, Reichart D, Nguyen MT, Bandyopadyhay G, Leung HY, Watt MJ, Benner C, Febbraio MA, Nguyen AK, Folian B, Subramaniam S, Gonzalez FJ, Glass CK, Ricote M (2007) Macrophage PPAR gamma is required for normal skeletal muscle and hepatic insulin sensitivity and full antidiabetic effects of thiazolidinediones. J Clin Invest 117(6):1658–1669. doi:10.1172/JCI31561
130. Hevener AL, He W, Barak Y, Le J, Bandyopadhyay G, Olson P, Wilkes J, Evans RM, Olefsky J (2003) Muscle-specific Pparg deletion causes insulin resistance. Nat Med 9(12):1491–1497. doi:10.1038/nm956. nm956 [pii]
131. Ryan KK, Li B, Grayson BE, Matter EK, Woods SC, Seeley RJ (2011) A role for central nervous system PPAR-gamma in the regulation of energy balance. Nat Med 17(5):623–626. doi:nm.2349 [pii]. 10.1038/nm.2349
132. Lu M, Sarruf DA, Talukdar S, Sharma S, Li P, Bandyopadhyay G, Nalbandian S, Fan W, Gayen JR, Mahata SK, Webster NJ, Schwartz MW, Olefsky JM (2011) Brain PPAR-gamma promotes obesity and is required for the insulin-sensitizing effect of thiazolidinediones. Nat Med 17(5):618–622. doi:nm.2332 [pii]. 10.1038/nm.2332
133. Straus DS, Glass CK (2007) Anti-inflammatory actions of PPAR ligands: new insights on cellular and molecular mechanisms. Trends Immunol 28(12):551–558. doi:10.1016/j.it.2007.09.003
134. Cariou B, Zair Y, Staels B, Bruckert E (2011) Effects of the new dual PPAR alpha/delta agonist GFT505 on lipid and glucose homeostasis in abdominally obese patients with combined dyslipidemia or impaired glucose metabolism. Diabetes Care 34(9):2008–2014. doi:10.2337/dc11-0093
135. Calder PC (2006) n-3 polyunsaturated fatty acids, inflammation, and inflammatory diseases. Am J Clin Nutr 83(6 suppl):1505S–1519S
136. Mozaffarian D, Wu JH (2012) (n-3) fatty acids and cardiovascular health: are effects of EPA and DHA shared or complementary? J Nutr 142(3):614S–625S. doi:10.3945/jn.111.149633
137. Renier G, Skamene E, DeSanctis J, Radzioch D (1993) Dietary n-3 polyunsaturated fatty acids prevent the development of atherosclerotic lesions in mice. Modulation of macrophage secretory activities. Arterioscler Thromb 13(10):1515–1524
138. Meydani SN, Endres S, Woods MM, Goldin BR, Soo C, Morrill-Labrode A, Dinarello CA, Gorbach SL (1991) Oral (n-3) fatty acid supplementation suppresses cytokine production and lymphocyte proliferation: comparison between young and older women. J Nutr 121(4):547–555
139. Caughey GE, Mantzioris E, Gibson RA, Cleland LG, James MJ (1996) The effect on human tumor necrosis factor alpha and interleukin 1 beta production of diets enriched in n-3 fatty acids from vegetable oil or fish oil. Am J Clin Nutr 63(1):116–122
140. Ben-Neriah Y, Karin M (2011) Inflammation meets cancer, with NF-kappaB as the matchmaker. Nat Immunol 12(8):715–723. doi:10.1038/ni.2060. ni.2060 [pii]

141. Seki E, Brenner DA, Karin M (2012) A liver full of JNK: signaling in regulation of cell function and disease pathogenesis, and clinical approaches. Gastroenterology 143(2):307–320. doi:S0016-5085(12)00820-7 [pii]. 10.1053/j.gastro.2012.06.004
142. Coussens LM, Werb Z (2002) Inflammation and cancer. Nature 420(6917):860–867
143. Gallagher EJ, LeRoith D (2011) Minireview: IGF, insulin, and cancer. Endocrinology 152(7):2546–2551. doi:10.1210/en.2011-0231
144. Aaltonen KJ, Virkki LM, Malmivaara A, Konttinen YT, Nordstrom DC, Blom M (2012) Systematic review and meta-analysis of the efficacy and safety of existing TNF blocking agents in treatment of rheumatoid arthritis. PLoS One 7(1):e30275. doi:10.1371/journal.pone.0030275
145. van Dartel SA, Fransen J, Kievit W, Flendrie M, den Broeder AA, Visser H, Hartkamp A, van de Laar MA, van Riel PL (2012) Difference in the risk of serious infections in patients with rheumatoid arthritis treated with adalimumab, infliximab and etanercept: results from the Dutch Rheumatoid Arthritis Monitoring (DREAM) registry. Ann Rheum Dis Published online first: 11 August 2012. doi:10.1136/annrheumdis-2012-201338
146. Dreyer L, Mellemkjaer L, Andersen AR, Bennett P, Poulsen UE, Juulsgaard Ellingsen T, Hansen TH, Jensen DV, Linde L, Lindegaard HM, Loft AG, Nordin H, Omerovic E, Rasmussen C, Schlemmer A, Tarp U, Hetland ML (2013) Incidences of overall and site specific cancers in TNFalpha inhibitor treated patients with rheumatoid arthritis and other arthritides—a follow-up study from the DANBIO Registry. Ann Rheum Dis 72(1):79–82. doi:10.1136/annrheumdis-2012-201969
147. Lee RK, Hittel DS, Nyamandi VZ, Kang L, Soh J, Sensen CW, Shearer J (2012) Unconventional microarray design reveals the response to obesity is largely tissue specific: analysis of common and divergent responses to diet-induced obesity in insulin-sensitive tissues. Appl Physiol Nutr Metab 37(2):257–268. doi:10.1139/h11-159
148. Chen G, Bentley A, Adeyemo A, Shriner D, Zhou J, Doumatey A, Huang H, Ramos E, Erdos M, Gerry N, Herbert A, Christman M, Rotimi C (2012) Genome-wide association study identifies novel loci association with fasting insulin and insulin resistance in African Americans. Hum Mol Genet 21(20):4530–4536. doi:10.1093/hmg/dds282

Chapter 2
Inflammasomes and Obesity

Tuo Deng, Jun Cui, Christopher J. Lyon, Nan Zhang, Helen Y. Wang, Rong-fu Wang, and Willa A. Hsueh

Abstract Inflammasomes are a family of protein complexes that recognize diverse microbial and endogenous danger signals to promote innate immune responses, tissue inflammation and injury, or cell death via pyroptosis. Inflammasome activation results in the recruitment and activation of caspase-1, which is required for the production of the proinflammatory cytokines interleukin-1β (IL-1β) and IL-18 that can modulate both adaptive and innate immune responses through effects on leukocyte survival, proliferation, differentiation, and migration. Recent studies suggest that inflammasome activity may also play important roles in several nonmicrobial disease states associated with chronic inflammation. For example, NLRP3 inflammasome expression and IL-1β production are both increased in obesity, and recent work has implicated NLRP3 inflammasome activation in a variety of obesity-linked conditions including gout, type 2 diabetes mellitus, metabolic liver disease, atherosclerosis, Alzheimer's disease, cancer and rheumatoid arthritis. Further, many of the factors associated with these conditions, including elevated plasma glucose, fatty acids, uric acid and cholesterol crystals, and β-amyloid, have been shown to be elevated during obesity and to stimulate NLRP3 inflammasome expression or activation. Since chronic NLRP3 activation appears to play important roles in several common disease states, better understanding of inflammasome regulation and function may lead to better therapeutic approaches. Several agents that attenuate NLRP3 inflammasome activity or inhibit its primary effector, IL-1β, are currently under development or in early clinical trials as therapeutics to treat these common disease

T. Deng, Ph.D. • C.J. Lyon, Ph.D. • N. Zhang, Ph.D. • H.Y. Wang, B.S. • R.-f. Wang, Ph.D. • W.A. Hsueh, M.D. (✉)
Methodist Hospital Research Institute, Weill Cornell Medical College, Houston, TX, USA
e-mail: tdeng@tmhs.org; cjlyon@tmhs.org; nzhang@tmhs.org; Ywang4@tmhs.org; Rwang3@tmhs.org; wahsueh@tmhs.org

J. Cui, Ph.D.
School of Life Sciences, Sun Yat-sen University, Guangzhou 510275, China
e-mail: harrycuijun@hotmail.com

A.J. Dannenberg and N.A. Berger (eds.), *Obesity, Inflammation and Cancer*, Energy Balance and Cancer 7, DOI 10.1007/978-1-4614-6819-6_2,

conditions. This chapter will review new research on inflammasome activation, its role in obesity and other chronic inflammatory states, and the status of approaches to attenuate NLRP3 inflammasome activity.

2.1 Inflammasomes Are Key Mediators of the Innate Immune System

The innate immune system serves as a first line of defense against invading microbes and other harmful agents by sensing pathogen-associated molecular patterns (PAMPs), such as microbial nucleic acid, lipoproteins, and lipopolysaccharides (LPSs), or danger-associated molecular patterns (DAMPs) released during cell stress, including ATP, uric acid, heat shock proteins, and other cellular components [1]. The initiation and regulation of innate immune responses to these factors are orchestrated by several classes of germline-encoded pattern-recognition receptors (PRRs), including toll-like receptors (TLRs), RIG-I-like receptors (RLRs), NOD-like receptors (NLRs), and AIM2 (absent in melanoma 2)-like receptors (ALRs) [2, 3]. NLRs comprise a large protein family of intracellular sensors, the members of which share a conserved central nucleotide-binding, oligomerization domain (NOD), a leucine-rich repeat (LRR) region, and a variable N-terminal effector domain. NLRs can be classified as receptors (Nod1 and Nod2) or by their different central proteins and their diverse functions: negative regulators include NLRX1, NLRC5, and NLRP4 and inflammasome activators include NLRP1, NLRC4, NLRP3, and NLRC5 [4–7]. Several NLRs and AIM2 can form large multi-protein "inflammasome" complexes upon stimulation by PAMP or DAMP signals.

Following activation, NLRs oligomerize and recruit pro-caspases through direct interaction with the NLR caspase recruitment and activation domain (CARD) and through interaction with the pyrin domain (PYD) of the ASC (apoptosis-associated speck-like protein containing a CARD) adaptor protein. ASC resides in the nucleus, but its movement from the nucleus to the cytosol is required for inflammasome assembly [8]. Caspase zymogen recruitment into an inflammasome complex induces a conformational change leading to caspase autoactivation.

IL-1 and IL-18, generated by capase-1 maturation of pro-IL-1 and pro-IL-18 peptides, are the major functional effectors of inflammasome activation, inducing multiple responses to infectious agents. IL-1β induces fever, T cell survival, B cell proliferation, and leukocyte migration and promotes polarization of CD4+ T helper 1 (Th1) cells [9], while IL-18 synergizes with IL-12 to promote Th1 polarization and IFNγ production and facilitates proinflammatory Th17 responses [10]. Caspase-1 activity is also required for pyroptosis, a proinflammatory cell death mechanism for self-clearance of infected macrophages and dendritic cells thought necessary to prevent intracellular replication of pathogenic microorganisms (Fig. 2.1 from Zitvogel et al. [11]). Both IL-1 and IL-18 production and pyroptosis have implications for many diseases including infections, arthritis, metabolic syndrome, cancer, and inflammatory bowel disease. Caspase-1 also contributes to host defense through

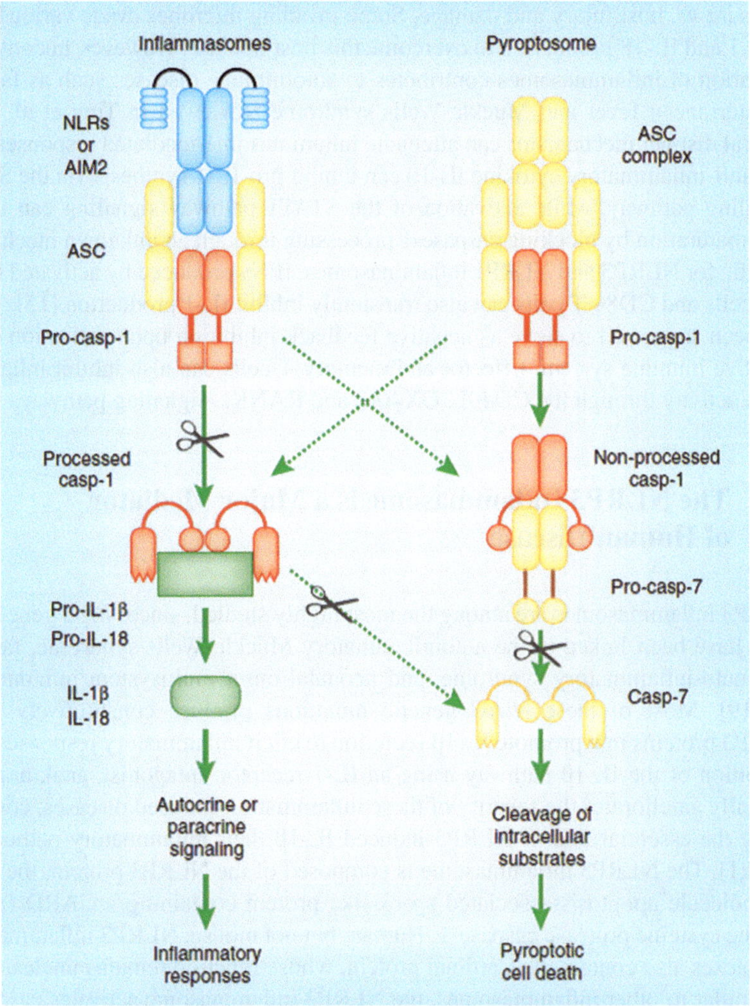

Fig. 2.1 Pyroptosis as a possible outcome of caspase-1 activation. Activation of caspase-1 (casp-1) mediated by the inflammasome or by the supramolecular assembly of ASC dimers known as the "pyroptosome" can derive from the proteolytic cleavage of pro-caspase-1 (pro-casp-1) or follow less well-understood (perhaps conformational) mechanisms. Both examples of caspase-1 activation can lead to the proteolytic maturation of caspase-7, followed by the cleavage of several intracellular substrates and pyroptotic cell death. Nevertheless, cleaved caspase-1 is required for the processing of pro-IL-1β and pro-IL-18, as well as for the release of mature IL-1β and IL-18 into the microenvironment (reprinted with permission from Zitvogel et al. [11])

additional mechanisms, including secretion of leaderless peptides, cleavage of glycolytic pathway enzymes, restriction of bacterial replication, and augmentation of cell repair through control of lipid metabolism (summarized in Lamkanfi and Dixit [12]).

These cytokine- and caspase-1-regulated processes must be tightly controlled balancing the ability to curb infection and develop T and B cell memory to pathogen exposure vs. host injury and damage. Some invading microbes evade various steps in IL-1 and IL-18 production to overcome this host defense. However, uncontrolled activation of inflammasomes contributes to autoimmune diseases such as familial Mediterranean fever and Muckle-Wells syndrome (reviewed in Ting et al. [13]). Several distinct mechanisms can attenuate inflammasome-mediated responses [14]. The anti-inflammatory cytokine IL-10 can inhibit pro-IL-1 synthesis via the STAT3 signaling pathway, while activation of the STAT1 pathway signaling can inhibit IL-1 maturation by blocking caspase-1 processing through an unknown mechanism specific for NLRP3 and NLRP1 inflammasomes. IFNγ produced by activated CD4+ Th1 cells and CD8+ T cells can also transiently inhibit IL-1 production [15], which has been suggested to serve as negative feedback inhibition upon activation of the adaptive immune system. Effector and memory T cells can also inhibit inflammasome activity through the CD40L, OX40L, and RANKL signaling pathways [16].

2.2 The NLRP3 Inflammasome is a Major Mediator of Human Disease

NLRP3 inflammasomes are among the most highly studied, since *Nlrp3* gene mutations have been linked to the autoinflammatory Muckle-Wells syndrome, familial cold auto-inflammatory syndrome, and neonatal-onset multisystem inflammation [17–19]. Most of these *Nlrp3* genetic mutations produce constitutively active NLRP3 proteins that promote IL-1β secretion to elicit inflammatory responses [13]. Inhibition of the IL-1β pathway using an IL-1 receptor antagonist, anakinra, successfully ameliorates the severity of these inflammation-induced diseases, corroborating the essential role of NLRP3-induced IL-1β their inflammatory pathologies [20, 21]. The NLRP3 inflammasome is composed of the NLRP3 protein, the adaptor molecule apoptosis-associated speck-like protein containing a CARD (ASC), and the cysteine protease caspase-1. Human, but not mouse, NLRP3 inflammasome complexes also contain the Cardinal protein, whose function remains unclear [22].

Similar to other inflammasomes, the NLRP3 inflammasome activates caspase-1 to cleave pro-IL-1β and pro-IL-18 into IL-1β and IL-18. NLRP3 inflammasome activation can also induce pyroptosis and modulate immune responses by regulating the secretion of more than 20 leaderless cytokines and growth factors, including HMGB1, IL-1α, fibroblast growth factor 2 (FGF2), galectin-1, and galectin-3 [23]. NLPR3 inflammasome activity thus plays critical roles in inflammation, host defense, and other related activities.

2.2.1 NLRP3 Inflammasome-Priming Signals

IL-1β, unlike most other cytokines, is regulated by both gene transcription and post-translational modification. Pro-IL-1β expression in myeloid cells, such as macrophages and dendritic cells (DCs), is induced upon exposure to LPS, a bacterial-derived

TLR4 ligand, which activates the NF-κB signaling pathway. Many NF-κB pathway activators, including TNF-α, CpG, Pam3CSK4, poly(I:C), R848, imiquimod, and IL-1β itself, have also been shown to increase pro-IL-1β expression [24–28]. However, while NF-κB-activating stimuli are required to induce pro-IL-1β expression in vitro, it is unclear whether such priming agents are also required to increase in vivo pro-IL-1β expression. For example, intraperitoneal injection with alum or monosodium urate (MSU) crystals triggers IL-1β secretion without an exogenous priming step to induce NF-κB activation [29, 30]. Differences between in vitro and in vivo inflammasome responses could be explained, at least in part, by differences in the sensitivity of the responding cell types and their microenvironment.

Most NF-κB-dependent "priming" signals induce pro-IL-1β and increase NLRP3 protein levels, an essential step for inflammasome formation [24]. NLRP3 expression in bone marrow-derived dendritic cells (BMDCs) and macrophages is markedly increased upon TLR stimulation [24, 28], and signaling through the TLR adaptor proteins MyD88 and TRIF increases NLRP3 expression [24, 31]. In contrast, the inflammasome complex proteins caspase-1 and ASC are widely expressed in most cell types and are not further upregulated upon NF-κB activation.

2.2.2 The Assembly of the NLRP3 Inflammasome

NLRP3 is highly expressed in primary mouse neutrophils, peripheral blood mononuclear cells (PBMCs), and BMDCs and moderately expressed in established Th2 and macrophage cell lines [28, 32]. NLRP3 protein is sequestered in an autoinhibitory conformation by heat shock protein 90 (HSP90) and suppressor of G2 allele of SKP1 (SGT1) in resting cells [33], but NLRP3 can recognize stimulatory signals through its LRR domain and undergo a conformational change, leading to ATP-dependent self-oligomerization via its NOD domain. NLRP3 oligomers then recruit the adaptor protein ASC via PYD-PYD domain interactions, which then recruits caspases-1 via CARD-CARD domain interactions to generate the final ~700 kDa NLRP3 inflammasome complex [34, 35]. This multi-protein complex assembly activates caspase-1, which cleaves pro-IL-1β and pro-IL-18 to generate mature IL-1β and IL-18 cytokines for secretion.

It remains unclear how NLRP3 recognizes the diverse range of stimuli that can activate NLRP3 inflammasome assembly and activation. It is plausible that NLRP3 is not a direct sensor of all the external stimuli and cellular "danger signals" it recognizes. Recent results have identified the non-NLR/ALR protein guanylate-binding protein 5 (GBP5) as a sensor for some NLRP3 inflammasome activators, stimulating NLRP3 inflammasome activation in response to pathogenic bacteria and soluble, but not crystalline, inflammasome-priming agents [36]. Interestingly, GBP5-deficient mice demonstrate impairments in both host defense and NLRP3-dependent inflammatory responses, indicating that proteins other than NLRP3, ASC, and caspase-1 are involved in sensing pathogenic "danger signals" that activate the NLRP3 inflammasome. Moreover, the restricted range of NLRP3 inflammasome stimuli recognized by GBP5 suggests that more than one sensor is involved in this process. These results, however, do not fully preclude the possibility of different sensor proteins giving rise to a common signaling intermediate that directly regulates the NLRP3 inflammasome complex. Given the ability of the NLRP3 inflammasome to regulate critical innate and adaptive immune responses to induce both protective and pathogenic responses, future studies are required to characterize signaling small molecules and proteins that regulate NLRP3 inflammasome activation.

2.3 The Diverse Regulation of NLRP3 Inflammasome Activation

Notably, the NLRP3 inflammasome is responsive to a wide array of stimuli, ranging from microorganisms of both endogenous and exogenous origin to inorganic agents such as asbestos, silica, and alum (Table 2.1). Specific activators have been identified for the NLRP1, NLRC4, and AIM2 inflammasomes, but much less is known about what triggers the assembly and activation of the NLRP3 inflammasome [37]. Studies indicate that reactive oxygen species (ROS), potassium efflux, and lysosomal damage are involved in NLRP3 inflammasome activation. Indeed, regulation of NLRP3 inflammasome activity by many stimuli can be explained by these three mechanisms; however, no single mechanism accounts for inflammasome activation by all known stimuli. The following sections summarize and discuss these models and the potential metabolic signals that can induce NLRP3 inflammasome responses.

2.3.1 Reactive Oxygen Species

Many NLRP3 agonists induce ROS formation, and reactive oxygen compounds such as hydrogen peroxide can induce inflammasome formation [38]. NADPH oxidase was initially postulated as the cellular ROS source that activated the NLRP3 inflammasome [39–41], since many factors that induce inflammasome activity also induce NADPH oxidase. However, studies performed with cells lacking various NADPH oxidase components indicate that NADPH oxidase activity is not required for NLRP3 inflammasome activation [42–44].

More recent studies have implicated mitochondrial ROS in inflammasome activation. Mitochondria are the main source of cellular ROS and markedly increase mitochondrial ROS production under various stress conditions [45]. Mitochondria also produce ROS during normal respiration, which can eventually alter respiratory chain function and lead to increased mitochondrial ROS production [46]. Such damaged mitochondria are continuously removed by a specialized form of autophagy, known as mitophagy, to avoid cellular damage from excess ROS exposure. Chemical and genetic manipulations that increase mitochondrial dysfunction and ROS production activate NLRP3 inflammasomes [47, 48], as does pharmacologic or genetic inhibition of mitophagy leading to the accumulation of damaged, ROS-producing mitochondria [46, 47, 49]. Both NLRP3 and ASC relocate to the

Table 2.1 Sensing of microbes, pathogen-associated molecular pattern (PAMPs), bacterial toxins, and "danger signals" by NLRP3

Elicitor	References
Microbial motifs	
MurNAc-L-Ala-D-isoGln (muramyldipeptide)	[221]
Bacterial RNA	[26]
Imidazoquinoline compounds (R837, R848)	[26]
Live bacteria	
Staphylococcus aureus	[27]
Listeria monocytogenes	[27]
Shigella flexneri	[222]
Virus	
Sendai virus	[223]
Influenza virus	[223, 224]
Adenovirus	[225]
Microbial toxins	
Aerolysin (*A. hydrophila*)	[226]
Maitotoxin (*Marine dinoflagellates*)	[27]
Nigericin (*Streptomyces hygroscopicus*)	[27]
Listeriolysin O (*L. monocytogenes*)	[226]
Danger-associated host components	
ATP, NAD+ (P2RX7)	[27, 28]
Mitochondrial membrane potential and generation of ROS	[47, 48]
Crystalline or aggregated substances (asbestos, silica, uric acid (MSU), etc.)	[37]
Necrotic cell components	[227]
Low intracellular potassium	[228]
Cell volume change	[56]

perinuclear space upon inflammasome activation to colocalize with endoplasmic reticulum mitochondria, further implying a role for mitochondrial ROS in inflammasome activation [48].

However, while ROS have been repeatedly implicated in NLRP3 inflammasome activation, it is still not clear precisely how ROS alter the inflammasome complex to activate capase-1 activity. ROS are toxic, but can also serve as signals for various physiologic processes. One recent study found that thioredoxin-interacting protein (TXNIP), a binding partner of the antioxidant protein thioredoxin, can directly bind NLRP3 and regulate inflammasome activation in a ROS-inducible manner [50], and similar results were found in a second study, albeit with minor differences [51]. It is possible that NLRP3 is influenced directly by ROS, although there is currently no experimental evidence to support such a mechanism.

2.3.2 *Cytoplasmic Potassium*

Intracellular potassium is a common initiator of inflammasome activity, as inhibition of potassium efflux by hyperosmotic potassium can block NLRP3 inflammasome activation [39, 52, 53]. Cellular potassium concentration can serve as an indicator of membrane integrity, due to its high cellular and low extracellular levels. NLRP3 inflammasomes are activated by disruption of cell membrane integrity, but extracellular potassium as low as 20 mM is sufficient to block NLRP3 inflammasome formation [54]. NLRP3 inflammasome activity is thus likely regulated at least in part by potassium efflux during loss of membrane integrity, but many stimuli markedly induce inflammasome activity without significantly destabilizing the cell membrane. Inflammasome responses to several such stimuli, including PAMPs, DAMPs, and crystals, can be markedly attenuated by the potassium channel inhibitor glibenclamide [55], however, suggesting that potassium channels mediate many of these responses by increasing intracellular potassium efflux.

Changes in cell volume in response to low environmental osmolarity can also regulate NLRP3 inflammasome activation [56], at least in part by decreasing intracellular potassium concentrations. However, low intracellular potassium is required but not sufficient to induce NLRP3 inflammasome activation during cell swelling. Cell swelling was also found to alter transient receptor potential (TRP) channel activity to regulate an intracellular calcium increase that was found to stimulate transforming growth factor β-activated kinase 1 (TAK1) to activate the NLRP3 inflammasome.

2.3.3 Phagocytosis and Lysosomal Damage

Phagocytosis of crystalline or particulate structures, such as MSU, silica, asbestos, amyloid-β, and alum, can lead to the release of proteolytic lysosomal contents into the cytosol, resulting in NLRP3 inflammasome activation [42, 57]. Leakage of the lysosomal cysteine protease cathepsin B into the cytoplasm has been proposed to result in the cleavage of a putative factor regulating NLRP3 complex activity, since specific inhibition or genetic ablation of cathepsin B activity impairs inflammasome activation induced by phagocytosis of various particulates [42, 58, 59]. However, either cathepsin B or cathepsin L deficiency attenuated NLRP3 inflammasome activation in one of these studies, and repression was less pronounced at higher doses [58], implying functional redundancy or alternative protease activities. Further, not all studies found cathepsin B deficiency attenuated NLRP3 inflammasome activation by crystalline agonists [60].

2.3.4 The NLRP3 Inflammasome is Negatively Regulated by IFNγ and T Cell Responses

Emerging evidence indicates that cytokine signaling events can inhibit inflammasome activation and IL-1β production [15]. For example, IFNγ transiently inhibits LPS-stimulated IL-1β production in mouse macrophages and dendritic cells, although not in human cells [61], but is limited by LPS-induced SOCS1 expression, a negative regulator of the IFNγ signaling cascade. Endogenous IFNγ production in *M. tuberculosis-infected* mice also selectively inhibits IL-1α and IL-1β production by monocyte-macrophages and DCs [15]. IFNγ is a signature cytokine of CD4+ Th1 and CD8+ T cells [62, 63], suggesting that T cell IFNγ secretion during adaptive immune responses may act to attenuate inflammasome responses. Similarly, both effector and memory T cells can directly attenuate NLRP3 and NLRP1 inflammasome activity in macrophages and DCs through a cell-to-cell contact-dependent mechanism apparently mediated by TNF superfamily ligand-receptor interactions [16].

2.4 NLRP3 Inflammasome Is a Metabolic Regulator

Excess weight gain is associated with the dysregulation of multiple metabolic factors that increase inflammasome activation.

2.4.1 *Fatty Acids*

Elevated plasma free fatty acids (FFA) are strongly associated with insulin resistance and type 2 diabetes [64]. Saturated FFAs can activate toll-like receptor 4 (TLR4) signaling to induce proinflammatory responses in adipocytes, macrophages, and pancreatic β-cells to increase insulin resistance in target tissues and reduce pancreatic β-cell function [65, 66]. TLR4-NFκB pathway signaling induces transcription of multiple proinflammatory factors, including IL-1β [67]. Recent work indicates that the saturated FFA palmitate can induce NLRP3-inflammasome activity by suppressing AMP-activated protein kinase activity, resulting in decreased autophagy, accumulation of damaged mitochondria, and increased production of mitochondrial-derived ROS [68]. Thus high FFA levels can provide both the transcription and inflammasome activation signals required for IL-1β production.

Cellular palmitate uptake also induces ceramide accumulation, which can function as a second messenger in several key signaling pathways [69]. Plasma ceramides are increased in obese patients with type 2 diabetes and positively correlate with insulin resistance, implying a potential role for ceramide in obesity-induced insulin resistance [70]. Ceramide has been shown to potently activate NLRP3 inflammasomes in isolated macrophages and adipose tissue explants [71], suggesting a new mechanism for ceramide-mediated insulin resistance in target tissues.

2.4.2 *Hyperglycemia*

Hyperglycemia induces IL-1β production in multiple cell types, including endothelial cells, monocytes, and pancreatic islet β-cells [72–74], and has been shown to activate the PKCα-NFκB signaling pathway to stimulate IL-1β gene transcription [75]. More recent work suggests that TXNIP plays an important role in hyperglycemia-induced pancreatic β-cells IL-1β secretion by directly interacting with the NLRP3-inflammasome in a ROS-dependent manner [50]. TXNIP expression is consistently higher in β-cells of subjects with type 2 diabetes [76], and its expression is robustly induced in response to high glucose [77]. Similar results are found in human adipocytes, where high glucose markedly induces TXNIP expression, caspase-1 activation, and IL-1β secretion [78], and TXNIP knockdown reduces hyperglycemia-induced IL-1β production, albeit primarily through IL-1β transcription [78]. Thus, both FFA and hyperglycemia appear to act through the NF-κB signaling pathway to stimulate IL-1β gene transcription, but use alternate mechanisms to provide the second signal required for inflammasome activation, IL-1β processing, and release.

2.4.3 Uric Acid and MSU Crystals

Multiple studies have indicated that uric acid or MSU crystals can stimulate inflammasome activity. For example, mice intraperitoneally injected with MSU crystals demonstrate NLRP3 inflammasome activation and develop IL-1β-dependent peritonitis [79]. MSU crystals are frequently found in patients with gout, but uric acid has also been implicated in inflammasome activation in other acute and chronic proinflammatory disease conditions. For example, during bleomycin-induced acute lung injury leading to pulmonary inflammation and fibrosis, uric acid released from injured cells can activate the NLRP3 inflammasome to induce IL-1β production [80]. Both adipose tissue uric acid levels [81] and inflammasome activity [82] are elevated in obesity, suggesting that uric acid may also promote NLRP3 inflammasome activation and IL-1β production in adipocytes, although there is no direct evidence for such a mechanism. Similarly, since plasma uric acid concentrations were one of the original diagnostic criteria for metabolic syndrome, it is tempting to speculate on the role of uric acid on systemic complications of obesity. Several recent studies have indicated that elevated plasma uric acid confers increased risk for multiple disease states associated with metabolic syndrome, including obesity, insulin resistance, hypertension, and cardiovascular disease [83].

2.4.4 Cholesterol

Plasma cholesterol levels are strongly linked to atherosclerosis, and cholesterol crystals are a recognized hallmark of advanced atherosclerotic lesions [84]. Cholesterol crystals can induce robust, dose-responsive, and caspase-1-dependent macrophage IL-1β secretion by inducing lysosomal damage [58]. Bone marrow NLRP3, ASC, or IL-1α/β deficiency markedly decreases atherosclerosis and inflammasome-dependent IL-18 levels, suggesting that NLRP3 inflammasomes play major roles in atherosclerosis [58]. Recently, autophagosomes were identified in atherosclerotic plaques, but autophagy was found to be defective [85]. Complete deficiency of autophagy in the apolipoprotein E-deficient mouse led to accelerated atherosclerosis with increased IL-1β levels and ASC protein bodies indicative of inflammasome complex formation and activation. These results are consistent with observations suggesting that lysosomal cholesterol crystals alter autophagic processes [58, 86]. These changes, in addition to increased vascular ROS, contributed to inflammasome activation.

2.4.5 Amyloid

Islet amyloid polypeptide (IAPP), which is co-secreted with insulin, can form amyloid deposits in the pancreas during type 2 diabetes. IAPP aggregates have cytotoxic properties that are believed to play a key role in β-cell loss and type 2 diabetes progression [87]. Recent work now indicates that IAPP crystals can trigger pancreatic NLRP3 inflammasome activation and IL-1β secretion [51].

2.5 Obesity and the NLRP3 Inflammasome

Chronic inflammation and recruitment of macrophages has been recognized as a hallmark of obesity for decades. Recent work indicates that innate and adaptive immune processes in adipose tissue play central roles in the development of obesity-induced inflammatory responses and indicates potential roles for NLRP3 inflammasome activity in these processes.

2.5.1 Chronic Inflammation in Adipose Tissue

Adipose tissue is an important endocrine organ, which can undergo marked changes in both its cellular composition and secretory profile during the onset of obesity. Caloric excess alters adipocyte function and dramatically increases the relative abundance and proinflammatory phenotype of adipose-resident immune cells, resulting in increased serum levels of proinflammatory adipokines [88, 89], while reductions in fat mass strongly correlate with serum adipokine decreases [90], suggesting that adipose tissue plays a central role in the systemic proinflammatory milieu of obesity. Adipocytes are important endocrine cells in this process, and adipocyte secretion of proinflammatory or pathogenic adipokines, such as TNFα, leptin, plasminogen activator inhibitor-1, monocyte chemotactic protein-1, IL-6, resistin, angiotensinogen, and IL-1β, markedly increases with obesity [88, 89]. Obesity also increases the relative adipose abundance of proinflammatory CD4 T cells (Th1) and CD8 T cells, resulting in increased adipose expression of IFNγ [91, 92]. The polarization of adipose tissue macrophages (ATMs) from an anti-inflammatory (M2) to a proinflammatory (M1) phenotype likely results from this increase in adipose IFNγ [93]. However, other immune cells, including B cells [94], mast cells [95], eosinophils [96], and neutrophils [97], have also been reported to regulate obesity-induced inflammation. The trigger signals for immune cell accumulation and polarization are unclear, although adipocyte death, hypoxia, and adipokines secreted in response to metabolic overload may all direct inflammatory cells to adipose tissue [98].

NLRP3 inflammasome components are highly expressed in ATMs, whose abundance can dramatically increase with the onset of obesity and contribute to the metabolic dysregulation and insulin resistance that give rise to obesity-associated complications. IL-1β directly inhibits adipocyte insulin signaling by decreasing insulin receptor substrate-1 [99], and both IL-1β and IL-18 can induce adipose-resident Th1 and CD8+ T cells [71, 100] to further increase adipose tissue inflammation. NLRP3 inflammasome activation in preadipocytes inhibits adipocyte differentiation and fat accumulation by reducing preadipocyte expression of PPARγ, a key regulator of adipogenesis, while caspase-1-deficient adipocytes are more metabolically active, suggesting that obesity-induced changes in NLRP3 inflammasome activity can negatively impact adipocyte function [82]. High-fat diet (HFD)-fed IL-1R knockout mice have less adipose inflammation and better insulin sensitivity than wild-type control mice [101]. Similarly, HFD-fed $Nlrp3^{-/-}$, $Asc^{-/-}$, and $Casp1^{-/-}$ mice demonstrate decreased weight gain and body fat, reduced inflammation, and improved insulin sensitivity [71, 82, 102], although both IL-1R and IL-18 deficiencies were subsequently found to induce hyperphagia resulting in delayed-onset obesity and insulin resistance in chow-fed mice [103–105].

2.5.2 NLRP3 Inflammasome Expression and Activity in Obese Mice

Several studies have investigated IL-1β/IL-18 production and inflammasome activation in mouse models of diet-induced obesity. Visceral adipose IL-1β and Nlrp3 mRNA expression is reported to positively correlate with body weight and adiposity in mice fed with standard chow diet and to markedly decrease upon caloric restriction [71], while adipose tissue IL-1β protein is increased in obese db/db and HFD-fed C57BL/6 mice relative to their lean controls [71, 82]. Similar to IL-1β, adipose tissue and circulating IL-18 protein levels are increased in HFD-fed or genetically obese mice [71, 82, 106], despite no IL-18 mRNA differences, suggestive of increased adipose inflammasome activity. Caspase-1 mRNA and activity are increased in multiple tissues of HFD-fed and genetically obese mice [71, 82]. NLRP3 ablation attenuates obesity-induced caspase-1 activity in adipose tissue and liver, but not kidney [71], implying that kidney caspase-1 activation is either

NLRP3 independent or complemented by another inflammasome. NLRP3 inflammasome-dependent IL-1β secretion by the pancreas also increases during obesity and likely mediates chronic obesity-induced pancreatic damage [107].

2.5.3 NLRP3 Inflammasome Expression and Activity in Obese Patients

Circulating IL-1β concentrations are low in healthy human subjects. Elevated circulating IL-1β levels have been reported in some studies of obese patients [108, 109], but it is questionable whether these concentrations, typically <100 pg/mL, can induce biologically meaningful systemic effects. IL-1β protein concentrations in metabolic tissues, such as adipose tissue and liver, however, are clearly increased in obese human subjects, at levels likely to produce pathological effects, and weight loss reduces adipose and liver IL-1β mRNA expression [71, 110]. IL-1β secretion from human omental fat explants is correlated with the donor's body mass index [111], while LPS-stimulated IL-1β production from primary monocytes of obese alcoholics is correlated with body mass index, percent body fat, abdominal circumference, and total histologic score [112]. Similar to IL-1β, circulating IL-18 concentrations are higher in overweight subjects [113] and are decreased by weight loss [114]. However, unlike IL-1β, weight loss significantly decreases IL-18 mRNA expression in liver, but not in adipose tissue [110].

Few studies have investigated NLRP3 inflammasome genes in obese human subjects. In one study, adipose tissue caspase-1 mRNA expression increased in obese subjects, while NLRP3, IL-1β, and IL-18 mRNA expressions were not significantly different in normal-weight and obese subjects [115]. Results from weight loss studies are contradictory, with weight loss reducing NLRP3 expression in one study [71] but having no effect in another [110], although gender, ethnicity, and adiposity differences between these cohorts may explain this discrepancy.

No comparable data is available for the effect of weight gain and loss on caspase-1 activity in human adipose tissue. However, a recent analysis of paired subcutaneous and visceral adipose tissue biopsies from ten overweight subjects found that IL-1β and IL-18 production as well as caspase-1 activity was dramatically higher in visceral than subcutaneous adipose tissue [100]. Interestingly, caspase-1 activity levels were positively correlated with CD8+ T cell numbers present in both tissues. These findings are consistent with well-established results that visceral adipose tissue is more inflammatory than subcutaneous adipose tissue, and suggest a possible role for caspase-1 activity in the infiltration of CD8+ T cells into human adipose tissue. Further, our recent microarray analysis and RT-PCR analyses of high-purity human and mouse adipocyte fractions from subcutaneous and visceral adipose tissue indicate that the expression of NLR pathway genes, including NLRP3 and ASC, is significantly upregulated during obesity (unpublished data), suggesting that obesity may induce adipocyte inflammasome activity.

2.5.4 Mechanisms of NLRP3 Inflammasome Activation in Adipose Tissue

Adipose tissue TLR4 signaling plays an important role in obesity-associated insulin resistance [66]. In obesity, increased FFA levels stimulate adipose tissue TLR4 signaling by increasing adipose expression of fetuin A, an endogenous TLR4 ligand [116]. Since the proinflammatory TLR4-NFκB signaling cascade regulates both NRLP3 and IL-1β transcription, increased TLR4 signaling may provide the priming signal for the increased NLRP3 inflammasome activity observed in obesity. However, while TLR4 signaling may prime the system, it is unclear what triggers NLRP3 inflammasome activation during obesity. Extracellular ATP can active the NLRP3 inflammasome via a P2X purinoceptor 7 (P2RX7)-dependent mechanism [117], but inflammasome activation and adipose and metabolic phenotypes are similar in HFD-fed P2RX7-deficient and wild-type mice [118], excluding it from a role in this process. Obesity-associated metabolic danger signals, such as elevated FFA and ceramide, hyperglycemia, and mitochondrial dysfunction, may trigger NLRP3 inflammasome activation, as discussed above. Due to the central role adipose inflammation plays in obesity-related complications, identification of the responsible signal(s) regulating adipose tissue inflammasome activity is an important challenge for future investigation.

2.6 The NLRP3 Inflammasome in Metabolic Diseases Associated with Obesity

The prevalence of obesity has led to the increased incidence of obesity-associated metabolic diseases and become a serious threat to public health in the developed world. Overweight and obesity are associated with increased risk of atherosclerosis, type 2 diabetes, nonalcoholic fatty liver disease (NAFLD), neurodegenerative disease, and cancer [119]. Chronic inflammation, a key feature of obesity, plays an important role in the development of many metabolic diseases. As discussed above, the NLRP3 inflammasome has been recognized to mediate inflammatory reactions to several metabolic danger signals that are increased in obesity, implying that these stimuli induce NLRP3-driven inflammation to produce tissue injury in obesity. Indeed, the major effectors of inflammasome activation, IL-1β and IL-18, have been linked to inflammatory responses in various metabolic diseases [9, 120]. However, new research has just begun to explain the specific stimuli and molecular mechanisms acting on the NLRP3 inflammasome for individual metabolic diseases.

2.6.1 Type 1 and Type 2 Diabetes

Diabetes is characterized by uncontrolled blood glucose elevations resulting from inadequate insulin production due to pancreatic β-cell loss (type 1 diabetes) or insulin resistance in skeletal muscle and other metabolic tissues associated that is with a slower decline in pancreatic β-cell function (type 2 diabetes). Mounting evidence now indicates that inflammasome activity plays critical roles in the dysregulation of both insulin production and insulin signaling.

IL-1β production by both pancreatic β-cells and infiltrating immune cells is known to regulate β-cell viability and insulin secretion. IL-1β treatment of both rodent and human pancreatic islets potently inhibits pancreatic β-cell-specific transcription factors and insulin secretion [121], while IL-1β secretion from invading immune cells induces pancreatic β-cell death during the development of autoimmune type 1 diabetes [122], partially through increased β-cell iNOS expression and NO production [123, 124]. Chronic hyperglycemia can also stimulate pancreatic β-cells to produce IL-1β, further impairing their viability and insulin secretion [73], through ROS-mediated activation of the NLRP3 inflammasome [50]. Finally, during type 2 diabetes high insulin secretion can result in the formation of cytotoxic IAPP aggregates that induce NLRP3 inflammasome activation and IL-1β secretion to stimulate β-cell apoptosis [51].

2.6.2 Metabolic Liver Disease

NAFLD is strongly associated with abdominal adiposity, and proinflammatory liver gene expression increases with adiposity [125]. NAFLD affects more than one-third of the Western world population, but only about 25 % of cases develop nonalcoholic steatohepatitis (NASH), which is characterized by chronic inflammation. NASH is a major cause of cirrhosis and liver transplantation, but the mechanism(s) underlying NAFLD progression to NASH remains elusive. Liver expression of NLRP3 inflammasome genes, caspase-1 activity, and mature IL-1β is, however, dramatically increased in mouse models of NASH [126]. NAFLD, the precursor to NASH, is also characterized by a marked increase in hepatocyte lipid accumulation (steatosis), which is attenuated by genetic ablation of NLRP3 inflammasome components [71, 82]. Saturated fatty acid accumulation in NAFLD may play an important role in hepatic inflammation, since the saturated fatty acid palmitate can induce inflammasome activation in cultured hepatocytes [126]. Hepatic inflammasome activation appears to require TLR2 signaling, which is induced by saturated fatty acids, since hepatic caspase-1 activation and serum IL-1β levels are suppressed in $TLR2^{-/-}$ mice fed with a NASH-inducing choline-deficient diet [127]. Finally, a gene regulating the elongation of C12- to C16-length saturated and monounsaturated fatty acids, Elovl6, has also been reported to regulate NASH development, hepatic NLRP3 inflammasome activation, and IL-1β release [128]. Taken together these results suggest that fatty acid accumulation directly or indirectly stimulates NLRP3 inflammasome activity. Genetic or pharmaceutical blockade of IL-1R signaling also attenuates NASH progression in mice [129, 130]. However, NAFLD/NASH phenotypes in these studies could result from non-hepatic inflammasome effects on insulin resistance and inflammation. Liver-specific NLRP3 inflammasome knockouts are required to address this argument.

2.6.3 Atherosclerosis

Chronic inflammation is a well-known contributor to atherosclerosis. Macrophage phagocytosis of modified low-density lipoprotein cholesterol present in the vasculature results in the accumulation of cholesterol-laden foam cells, increased oxidative stress, and vascular inflammation, which can ultimately lead to plaque destabilization and thrombosis. Moreover, IL-1β has been shown to participate in both the development and destabilization of atherosclerotic lesions. For example, genetic ablation of IL-1β or the IL-1 receptor in apolipoprotein E-deficient ($Apoe^{-/-}$) mice, a standard mouse model of atherosclerosis, significantly decreased atherosclerosis development [131, 132], as did treatment with an IL-1 receptor antagonist [133, 134]. Results from low-density lipoprotein receptor-deficient ($Ldlr^{-/-}$) mice fed with a high-cholesterol diet also indicated that macrophage-derived NLRP3 and IL-1 were essential for cholesterol-driven atherosclerosis, since these mice revealed significantly less atherosclerosis after receiving bone marrow transplants from NLRP3- or IL-1α/IL-1β-deficient mice [58]. Surprisingly, however, NLRP3 inflammasome activity was not required for normal atherosclerosis progression in $Apoe^{-/-}$ mice, since mice deficient in NLRP3, ASC, or caspase-1 expression revealed no decrease in atherosclerosis [135].

It is not clear why these results differ from those of $Apoe^{-/-}$ mice lacking IL-1β or IL-1 receptor expression or $Ldlr^{-/-}$ mice with NLRP3- or IL-1α/IL-1β-deficient bone marrow. One possible explanation, however, may be that $Apoe^{-/-}$ mice, unlike $Ldlr^{-/-}$ mice fed with a high-cholesterol diet, represent a nonobese mouse model of atherosclerosis and may therefore lack a metabolic stimulus found in obese mice. Furthermore, while this data indicates that inflammasome activity can play an important role in the pathogenesis of atherosclerosis, the agents that trigger NLRP3 inflammasome activation remain unclear. Duewell et al. have reported that cholesterol crystals, detected in early atherosclerotic lesions, induce NLRP3 inflammasome-dependent IL-1β production in murine macrophages [58], with similar results observed for human macrophages [86]. However, several other factors associated with increased atherosclerotic risk or the atherosclerotic microenvironment (hyperglycemia, ROS, uric acid) may also impact inflammasome activation.

2.6.4 Gout

Gout is the accumulation of MSU crystals in the joints, often leading to an inflammatory arthritis associated with elevated plasma uric acid. MSU crystals have long been identified as the causative agent of gout [136], but only recently has progress been made regarding the mechanism underlying their recognition as a proinflammatory danger signal. MSU crystals induce the production of inflammatory cytokines from monocytes and macrophages, particularly IL-1β [137–139], but do not induce IL-1β production in macrophages with ablations of various NRLP3 inflammasome components [79], indicating that inflammasome activity plays an essential role in MSU recognition. Elevated plasma uric acid was proposed as one of the original criteria for diagnosis of obesity-associated metabolic syndrome [140], and obese patients are more than twice as likely to develop gout [141] and develop it about 11 years earlier than their normal-weight counterparts [142].

2.6.5 *Alzheimer's Disease*

Cerebral accumulation of amyloid-β plaques is one of the main pathologic features of Alzheimer's disease (AD), the most common form of dementia, and this is believed to be a critical stimulus for many proinflammatory components of AD that induce neuronal death and memory loss. Although the role of IL-1β in the pathogenesis of AD is controversial, numerous studies suggest that IL-1β can induce neuronal death and recruit inflammatory cells into the central nervous system [143]. Amyloid fibrils have been reported to trigger IL-1β release from microglia and monocytes [144, 145], suggesting a potential role for inflammasome activity in AD pathology. In keeping with this hypothesis, microglial NLRP3 inflammasome activity and IL-1β secretion are activated by phagocytosis of amyloid-β protein [57], while microglia with genetic ablation of caspase-1 or inflammasome components revealed lack of attenuated or defective inflammatory responses to amyloid-β exposure. Taken together, this data suggests that the inflammasome may represent a novel therapeutic target for the treatment of AD. Small molecule caspase-1 inhibitors such as those currently in clinical trials (see below) appear to be the best candidates for AD therapeutics, due to the difficulty expected in delivering recombinant protein therapeutics across the blood–brain barrier.

2.7 Cancer, Obesity and Inflammation

Mounting evidence suggests that inflammation promotes cancer development. Inflammatory responses are associated with 15–20 % of all cancer deaths worldwide, and inflammatory cells and factors are present in the microenvironment of most tumors, where they are often associated with metastasis and poor prognosis [146]. Proinflammatory cytokines, including IL-1β, IL-6, and TNFα, implicated to link inflammation and increased cancer risk, are expressed by adipose tissue and demonstrate increased systemic expression in overweight and obese human subjects [90]. Body fat is convincingly associated with colorectal, kidney, esophageal, pancreatic, endometrial, and postmenopausal breast cancer, and a recent meta-analysis of prospective cancer studies has expanded this list to include thyroid cancer, leukemia, multiple myeloma, and non-Hodgkin lymphoma in both men and women and increased risk of malignant melanoma in men [147]. Body weight is also associated with increased cancer mortality, since a prospective analysis of more than 900,000

US adults proposed that overweight and obesity may account for 14 % and 20 % of cancer deaths in men and women ≥50 years of age, respectively [148]. Based on the growing worldwide obesity epidemic, weight gain may represent the largest avoidable cause of cancer in nonsmokers.

2.7.1 Cell Death and Inflammasome Activity

An important hallmark of cancer is the ability of cancer cells to escape immunosurveillance [149]. Cancer cells are often resistant to various types of programmed cell death including apoptosis, programmed necrosis, and mitotic catastrophe [150]. Similarly inflammasome-mediated cell death, or pyroptosis, also appears to be involved in oncogenesis, since dysregulation of pyroptosis is tightly associated with tumor progression [151, 152]. Pyroptosis is stimulated by caspase-1 activity and plays key roles in anti-pathogen inflammatory responses, since bacterially infected macrophages and dendritic cells are usually eliminated through pyroptosis [152]. Caspase-1 activity can be activated by several mechanisms during pyroptosis. First, caspase-1 activity can be induced by classical inflammasomes, which usually contain NLRP3 or the cytoplasmic DNA sensor protein AIM2 [153]. Second, ASC can be recruited to a single subcellular location and aggregate into a polymer called an "ASC focus," "ASC speck," or "pyroptosome" that can then recruit and activate caspase-1 [154–156]. Finally, the NLRC4 inflammasome can also activate caspase-1 via a nonclassical, ASC-independent mechanism [157].

In pyroptosis, activated caspase-1 catalyzes the proteolytic activation of caspase-7, but not caspase-3, caspase-8, or caspase-9, to initiate programmed cell death [158]. However, while pyroptosis shares certain features with classical programmed cell death mediated via apoptosis, recent studies have identified characteristics that distinguish pyroptosis from apoptosis. First, apoptotic and pyroptotic cells demonstrate distinct DNA damage patterns. Both apoptotic and pyroptotic DNA damage can be detected by TUNEL staining [159, 160], but pyroptotic cells have a distinct nuclear morphology and usually lack the DNA laddering pattern characteristic of apoptotic DNA damage [159, 161]. Second, these processes have differential effects on cell membrane composition and integrity. For example, apoptosis triggers phosphatidylserine translocation from the inner to the outer plasma membrane, resulting in a cell surface annexin V-staining pattern. By contrast, during pyroptosis the formation of plasma membrane pores allows annexin V to enter the cell to produce an inner cell membrane staining pattern [162–164]. Finally, apoptotic and pyroptotic cell debris demonstrate different fates. During apoptosis, dying cells are cleaved into spherical membrane-bound structures known as apoptotic bodies that are ultimately cleared by phagocytosis [165], while the cytosolic contents of pyroptotic cells are released into extracellular space [159].

Similar to apoptosis, pyroptosis is thought to contribute to cell-autonomous tumor suppression. In support of this hypothesis, caspase-1 is downregulated in most human prostrate cancers, and caspase-1 overexpression in cultured prostate cancer cells enhances their sensitivity to radiation-induced killing [166]. One study has also reported that NLRP3-deficient mice are more susceptible to colitis-associated colon cancer [167], while another has reported that NLRC4-deficient mice, but not NLRP3-deficient mice, have more tumor load than wild-type mice after exposure to azoxymethane-dextran sulfate sodium (AOM-DSS) [168]. Thus, while the function of specific inflammasomes in tumor suppression remains a matter of debate, these results appear to support the hypothesis that pyroptosis plays an important regulatory role in cancer development.

2.7.2 The Inflammasomes and Carcinogenesis

Chronic inflammation is often linked with increased cancer risk, and proinflammatory cytokines, such as IL-1β, TNF-α, and IL-6, are frequently associated with tumor progression [169]. However, the effect of inflammasomes on specific proinflammatory processes associated with increased cancer risk, such as colitis and colitis-associated cancer, remains controversial. For example, capase-1 or NLRP3 deficiency is reported to reduce colitis severity in DSS-treated mice [170, 171]. Consistent with these results, high levels of IL-1 in the tumor microenvironment usually correlate with a poor prognosis [172]. However, other authors have reported that mice with NLRP3, ASC, caspase-1, NLRP6, or NLRP12 deficiencies are more susceptible to DSS-induced colitis and death [167, 173–177]. In one case, an IL-18 deficit following AOM-DSS-induced intestinal damage in caspase-1 and ASC-deficient mice caused a systemic spread of commensal bacteria due to the inability of these mice to repair the mucosal barrier [174]. Colitis-associated cancer has also been reported to be significantly increased upon genetic deletion of components of the NLRP3 inflammasome (ASC, NLRP3, or caspase-1) [167, 168, 177]. These contradictory observations for NLRP3 inflammasome effects on AOM-DSS-induced colitis may be due to animal facility-dependent differences in gut microflora [178].

Despite contradictory reports on inflammasome effects on colitis, in vivo experiments suggest that IL-1β and IL-18 play important roles in promoting gastric, hepatic, and breast cancer progression [179–181]. Further, gastric-specific IL-1β expression in transgenic mice induces gastric tumorigenesis [182], while IL-1β-deficient mice have reduced and retarded subcutaneous tumor development in response to transdermal 3-methylcholanthrene [183], suggesting that local IL-1β-induced inflammation is strongly associated with carcinogenesis.

Based on this link, approaches that inhibit NLRP3, caspase-1, IL-1β, and IL-18 have been used to develop novel therapies to that have been assessed in a variety of experimental cancer systems. Small compounds targeting inflammasomes or NLRP3 have not been successful due to off-target effects [184–187]. However, reagents targeting IL-1β and IL-18, including monoclonal antibodies and recombinant derivatives [188, 189], have been more successful. For example, in patients with smoldering or indolent multiple myeloma, treatment with anakinra, a recombinant non-glycosylated form of the human IL-1 receptor antagonist (IL-1ra), was found to decrease myeloma proliferation rates and high-sensitivity C-reactive protein (hsCRP) levels, leading to a chronic disease state with improved progression-free survival [190]. IL-1 inhibitors have also been shown to reduce the side effects of anticancer therapy [191, 192]. Based on the ability of IL-1R antagonism to reduce tumor burden and metastasis and the relatively low risk of in vivo IL-1 inhibition, future preclinical and clinical trials are needed to examine the effect of IL-1 inhibition on cancer outcomes [193].

2.7.3 Inflammasome-Dependent Anticancer Responses

While the inflammasome has often been linked to innate immunity, chronic inflammation, and carcinogenesis, mounting evidence suggests that the inflammasome is also involved in anticancer responses. Inflammasomes are hypothesized to inhibit tumor progression through several mechanisms, including (1) triggering innate immune reactions against potentially carcinogenic microbiota; (2) inducing the pyroptotic demise of premalignant, infected cells; and (3) facilitating antitumor adaptive immune responses [11].

IL-1β plays an essential role in stimulating adaptive immune responses and facilitating anticancer immunosurveillance [194]. Anticancer chemotherapies primarily eliminate tumors by inducing immunogenic cell death [195]. The immunogenic signals secreted by dying tumor cells not only attract innate immune effector cells into the tumor bed but also stimulate P2RX7 receptor-mediated activation of the NLRP3 inflammasome to produce IL-1β [196]. IL-1β, in conjunction with IL-23, then induces IL-17 secretion by γδ T cells [197, 198] and the polarization of CD8+ αβ T cell responses toward increased IFN-γ secretion [199]. Mice lacking P2RX7, NLRP3, ASC, caspase-1, or IL-1R1 do not respond to chemotherapies that elicit immunogenic cell death signals in wild-type mice [196]. Similarly, blockade of IL-1β signaling significantly impairs the growth-inhibitory effects of tumor chemotherapies in a variety of mouse models [197, 200]. Finally, P2RX7-mediated activation of the NLRP3 inflammasome is also required for the efficacy of anticancer therapies in patients [196, 201–203]. Taken together, these results indicate that inflammasome activity contributes to antitumor adaptive immune responses induced during chemotherapy. The study of the precise molecular mechanisms through which inflammasomes modulate anticancer immune responses may therefore yield insight into better anticancer therapeutics.

2.8 Pharmaceutical Interventions Targeting the Inflammasome

Several treatments designed to attenuate complications of metabolic disorders have now been developed using approaches that target NLRP3 inflammasome activity. Most of these strategies have focused on the development of therapeutic agents to attenuate IL-1β activity, although other agents that directly target inflammasome function and caspase-1 activity are also under development and may provide alternative therapeutic approaches with distinct advantages and disadvantages. One major concern of all these approaches, however, is the potential for reduced tissue repair and immune surveillance, as a result of excessive suppression of NLRP3 inflammasome responses.

2.8.1 Therapeutics Targeting IL-1β Activity

IL-1β can reduce tissue insulin sensitivity and inhibit insulin production by pancreatic β-cell [204, 205], while antagonism of IL-1β signaling, by receptor blockade or cytokine neutralization approaches, has been shown to ameliorate several proinflammatory conditions, including type 2 diabetes. For example, short-term treatment of a small cohort of type 2 diabetic patients with anakinra, a recombinant non-glycosylated form of the human IL-1 receptor antagonist (IL-1ra) that is FDA approved for treatment of rheumatoid arthritis, was found to significantly improve glycemia and β-cell function, while decreasing the plasma level of two surrogate markers of systemic inflammation, hsCRP and IL-6 [206]. These effects occurred within 4 weeks of treatment and were sustained at 13 weeks of treatment. However, no significant improvements were observed in insulin sensitivity, insulin-regulated skeletal muscle gene expression, or serum adipokine levels, indicating anakinra treatment effects were due primarily to improvements in β-cell function rather than enhanced glucose disposal. Similar results were found in a second study performed with a cohort of insulin-resistant but nondiabetic obese human subjects, where anakinra also failed to improve insulin sensitivity [207]. Anakinra-mediated IL-1β antagonism does not, therefore, appear to significantly impact skeletal muscle insulin sensitivity, although the reason for this failure is unclear. Adverse events associated with anakinra usage in these studies were relatively mild, primarily consisting of transient injection-site reactions that most likely result from the solution used to dissolve the recombinant protein, although this is aggravated by a need for daily subcutaneous injections due to the short half-life (<1 h) of anakinra [208].

IL-1 neutralizing agents provide an alternate means to attenuate the deleterious effects of inflammasome activation. Rilonacept, a recombinant therapeutic agent used to neutralize free IL-1α and IL-1β, has been shown to significantly reduce pain scores and plasma hsCRP levels in a small study of ten patients with chronic gouty arthritis, refractive to standard treatment approaches, and to prevent acute gout flares during the initiation of urate-lowering therapy [209, 210]. Rilonacept, however, attenuates both IL-1α and IL-1β activity and thus cannot differentiate between inflammasome-mediated IL-1β effects and those of IL-1α. More recent approaches to specifically inhibit IL-1β bioactivity in the treatment of human inflammatory disorders have resulted in the production of new agents such as canakinumab, a fully human monoclonal antibody that neutralizes IL-1β bioactivity. Canakinumab can be administered intravenously or subcutaneously and is FDA approved for the treatment of cryopyrin-associated periodic syndromes (CAPS), including familial cold auto-inflammatory syndrome and Muckle-Wells syndrome [211], which are associated with mutations in NLRP3 [13]. Canakinumab is also now being used in clinical trials for the treatment of a number of inflammation-related disorders such as type 2 diabetes and its derived complications and chronic obstructive pulmonary disease, with the CANTOS trial (cardiovascular risk reduction in type 2 diabetes) representing the largest trial of any anti-cytokine drug to date [212].

In addition to these three FDA-approved drugs, a number of agents targeting the IL-1 pathway are under preclinical or clinical development. The majority of these agents are neutralizing antibodies, such as gevokizumab (XOMA-052) and LY2189102, that reduce IL-1β bioavailability [212], although IL-1 receptor blocking antibodies, such as MEDI-8968 and AMG-108 (aka MEDI-78998), are also under study as novel therapeutics for treatment of chronic inflammatory diseases. XOMA-052 has been granted orphan drug status by the FDA for the treatment of Behçet's disease, a rare condition where chronic immune-mediated vascular inflammation can result in severe neurological, pulmonary, gastrointestinal, and cardiovascular complications [213]. XOMA-052, LY2189102, MEDI-8968, and AMG-108 are all now under investigation in phase II clinical trials: XOMA-052 and LY2189102 for the treatment of type 2 diabetes, MDI-8968 for the treatment of chronic obstructive pulmonary disease [212], and AMG-108 for the treatment of rheumatoid arthritis. Both MEDI-8968 and AMG-108 are fully human monoclonal antibodies that selectively bind IL-1R to inhibit the binding and subsequent signaling of both IL-1α and IL-1β, resulting in inhibition that is not restricted to inflammasome-mediated IL-1β responses. However, a clinical trial of AMG-108 effects on patients with rheumatoid arthritis or osteoarthritis of the knee found moderate disease improvement coupled with an excellent safety profile [214, 215], suggesting that an "off-target" suppression of IL-1α may not produce severe side effects. Finally, induction of endogenous antibodies by therapeutic vaccines has proven to be a safe and effective means of attenuating other disease conditions [216]. Recently, this approach has also been extended to cytokine-induced disease conditions, and a vaccine targeting IL-1β (CYT-013) is currently in phase I clinical trial in patients with type 2 diabetes [212].

2.8.2 Therapeutic Approaches Acting on the Inflammasome or Caspase-1

Inflammasomes play important roles in immune surveillance and tissue repair. Thus, approaches designed to directly target pathological inflammasome activity need to maintain a careful balance between attenuating deleterious inflammatory activity and maintaining necessary host defense and tissue repair responses. Relatively little is known about the exact mechanisms that regulate inflammasome assembly and caspase-1 activation, however, and these may differ according to the stimulus and NLR subtype of the inflammasome complex. Caspase-1 blockade thus appears to be a far more attractive target for treatment of inflammasome-related disorders due to the greater knowledge base available for caspase-1 inhibition.

Two small molecule caspase-1 inhibitors, VX-765 and VX-740 (pralnacasan), have been tested in clinical trials for the treatment of chronic plaque psoriasis, rheumatoid arthritis, and psoriasis [217]. VX-765 has also been tested on six patients with Muckle-Wells syndrome, resulting from mutations in NLRP3, where it was found to significantly reduce inflammatory markers, recurrent fevers, and arthritis [212]. The rheumatoid arthritis clinical trial using VX-740 was discontinued, however, due to liver abnormalities in animal toxicology studies [217]. No caspase-1 inhibitors are currently approved for clinical use, but other approved drugs with known anti-inflammatory activity may function in part by attenuating inflammasome complex activity. For example, treatment with glyburide, a sulfonylurea drug frequently used for the treatment of type 2 diabetes [51, 55], has been shown to attenuate inflammation under diabetic conditions by blocking IAPP-stimulated NLRP3 inflammasome activation and IL-1β secretion [51]. Glyburide has also been shown to prevent NLRP3 inflammasome activation in response to pathogen-associated and DAMP signals and biological crystals [55]. Glyburide inhibits ATP-sensitive potassium channels (K_{ATP}) in pancreatic β-cells to induce insulin secretion, but NLRP3 activation and glyburide-mediated inhibition were preserved in macrophages lacking K_{ATP} subunits and ATP-binding cassette transporter proteins, indicating that glyburide does not inhibit NLRP3 inflammasome activity through attenuating potassium efflux. Moreover, glyburide had no effect on NLRP3 ATPase activity, strongly suggesting that glyburide acts upstream of the NLRP3 inflammasome [55].

2.8.3 *Other Therapeutic Targets to Attenuate Inflammasome Activity*

Improved knowledge of the signaling networks involved in inflammasome activation has led to the discovery and testing of new therapeutic targets, such as the P2X purinoceptor 7 (P2RX7), an ATP receptor that regulates potassium efflux to induce inflammasome-mediated caspase-1 activation [117]. However, while several studies have now been performed with P2RX7 antagonists, none have shown significant action to reduce inflammation. AZD9056, an oral P2RX7 antagonist, has been evaluated in a phase IIa and subsequent phase IIb clinical trial for treatment of rheumatoid arthritis, but failed to demonstrate any significant efficacy [218]. The P2RX7 antagonist CE-224,535 also failed to reveal efficacy in patients with active rheumatoid arthritis and an inadequate response to methotrexate [219]. Finally, the P2RX7 modulator GSK1482160 was also recently investigated for single-dose safety, tolerability, pharmacokinetics, and pharmacodynamics, in healthy human subjects, but simulations lead to the conclusion that it was not possible to achieve the desired level of pharmacology, resulting in the termination of GSK1482160 development for chronic inflammatory pain [220].

2.9 Summary

Increasing evidence indicates multiple metabolic and cellular factors contribute to NLRP3 inflammasome activation, which plays a key role in adipose tissue inflammation and resultant obesity-associated tissue injury and metabolic derangement. These observations have widespread implications for a variety of diseases that are increased in obesity including diabetes, atherosclerosis, NASH, Alzheimer's disease, and cancer. Better understanding of the roles of NLRP3, IL-1β and IL-18 in mediating specific tissue injuries coupled with new therapeutics that target inflammasome activity may permit the development of novel and more precise interventions to prevent or treat these important disease conditions. Therapeutic control of the inflammasome is in our grasp.

References

1. Strowig T, Henao-Mejia J, Elinav E, Flavell R (2012) Inflammasomes in health and disease. Nature 481:278–286
2. Akira S, Uematsu S, Takeuchi O (2006) Pathogen recognition and innate immunity. Cell 124:783–801
3. Takeuchi O, Akira S (2010) Pattern recognition receptors and inflammation. Cell 140:805–820
4. Cui J, Zhu L, Xia X, Wang HY, Legras X, Hong J, Ji J, Shen P, Zheng S, Chen ZJ et al (2010) NLRC5 negatively regulates the NF-kappaB and type I interferon signaling pathways. Cell 141:483–496
5. Meylan E, Tschopp J, Karin M (2006) Intracellular pattern recognition receptors in the host response. Nature 442:39–44
6. Moore CB, Bergstralh DT, Duncan JA, Lei Y, Morrison TE, Zimmermann AG, Accavitti-Loper MA, Madden VJ, Sun L, Ye Z et al (2008) NLRX1 is a regulator of mitochondrial antiviral immunity. Nature 451:573–577
7. Xia X, Cui J, Wang HY, Zhu L, Matsueda S, Wang Q, Yang X, Hong J, Songyang Z, Chen ZJ et al (2011) NLRX1 negatively regulates TLR-induced NF-kappaB signaling by targeting TRAF6 and IKK. Immunity 34:843–853
8. Bryan NB, Dorfleutner A, Rojanasakul Y, Stehlik C (2009) Activation of inflammasomes requires intracellular redistribution of the apoptotic speck-like protein containing a caspase recruitment domain. J Immunol 182:3173–3182
9. Dinarello CA (2009) Immunological and inflammatory functions of the interleukin-1 family. Annu Rev Immunol 27:519–550
10. Harrington LE, Hatton RD, Mangan PR, Turner H, Murphy TL, Murphy KM, Weaver CT (2005) Interleukin 17-producing CD4+ effector T cells develop via a lineage distinct from the T helper type 1 and 2 lineages. Nat Immunol 6:1123–1132
11. Zitvogel L, Kepp O, Galluzzi L, Kroemer G (2012) Inflammasomes in carcinogenesis and anticancer immune responses. Nat Immunol 13:343–351
12. Lamkanfi M, Dixit VM (2012) Inflammasomes and their roles in health and disease. Annu Rev Cell Dev Biol 28:137–161
13. Ting JP, Kastner DL, Hoffman HM (2006) CATERPILLERs, pyrin and hereditary immunological disorders. Nat Rev 6:183–195
14. Rathinam VA, Vanaja SK, Fitzgerald KA (2012) Regulation of inflammasome signaling. Nat Immunol 13:332–333

15. Mayer-Barber KD, Andrade BB, Barber DL, Hieny S, Feng CG, Caspar P, Oland S, Gordon S, Sher A (2011) Innate and adaptive interferons suppress IL-1alpha and IL-1beta production by distinct pulmonary myeloid subsets during Mycobacterium tuberculosis infection. Immunity 35:1023–1034
16. Guarda G, Dostert C, Staehli F, Cabalzar K, Castillo R, Tardivel A, Schneider P, Tschopp J (2009) T cells dampen innate immune responses through inhibition of NLRP1 and NLRP3 inflammasomes. Nature 460:269–273
17. Dode C, Le Du N, Cuisset L, Letourneur F, Berthelot JM, Vaudour G, Meyrier A, Watts RA, Scott DG, Nicholls A et al (2002) New mutations of CIAS1 that are responsible for Muckle-Wells syndrome and familial cold urticaria: a novel mutation underlies both syndromes. Am J Hum Genet 70:1498–1506
18. Feldmann J, Prieur AM, Quartier P, Berquin P, Certain S, Cortis E, Teillac-Hamel D, Fischer A, de Saint Basile G (2002) Chronic infantile neurological cutaneous and articular syndrome is caused by mutations in CIAS1, a gene highly expressed in polymorphonuclear cells and chondrocytes. Am J Hum Genet 71:198–203
19. Hoffman HM, Mueller JL, Broide DH, Wanderer AA, Kolodner RD (2001) Mutation of a new gene encoding a putative pyrin-like protein causes familial cold autoinflammatory syndrome and Muckle-Wells syndrome. Nat Genet 29:301–305
20. Hawkins PN, Lachmann HJ, McDermott MF (2003) Interleukin-1-receptor antagonist in the Muckle-Wells syndrome. N Engl J Med 348:2583–2584
21. Hoffman HM, Rosengren S, Boyle DL, Cho JY, Nayar J, Mueller JL, Anderson JP, Wanderer AA, Firestein GS (2004) Prevention of cold-associated acute inflammation in familial cold autoinflammatory syndrome by interleukin-1 receptor antagonist. Lancet 364:1779–1785
22. Agostini L, Martinon F, Burns K, McDermott MF, Hawkins PN, Tschopp J (2004) NALP3 forms an IL-1beta-processing inflammasome with increased activity in Muckle-Wells autoinflammatory disorder. Immunity 20:319–325
23. Nickel W, Rabouille C (2009) Mechanisms of regulated unconventional protein secretion. Nat Rev Mol Cell Biol 10:148–155
24. Bauernfeind FG, Horvath G, Stutz A, Alnemri ES, MacDonald K, Speert D, Fernandes-Alnemri T, Wu J, Monks BG, Fitzgerald KA et al (2009) Cutting edge: NF-kappaB activating pattern recognition and cytokine receptors license NLRP3 inflammasome activation by regulating NLRP3 expression. J Immunol 183:787–791
25. Kanneganti TD, Lamkanfi M, Kim YG, Chen G, Park JH, Franchi L, Vandenabeele P, Nunez G (2007) Pannexin-1-mediated recognition of bacterial molecules activates the cryopyrin inflammasome independent of toll-like receptor signaling. Immunity 26:433–443
26. Kanneganti TD, Ozoren N, Body-Malapel M, Amer A, Park JH, Franchi L, Whitfield J, Barchet W, Colonna M, Vandenabeele P et al (2006) Bacterial RNA and small antiviral compounds activate caspase-1 through cryopyrin/Nalp3. Nature 440:233–236
27. Mariathasan S, Weiss DS, Newton K, McBride J, O'Rourke K, Roose-Girma M, Lee WP, Weinrauch Y, Monack DM, Dixit VM (2006) Cryopyrin activates the inflammasome in response to toxins and ATP. Nature 440:228–232
28. Sutterwala FS, Ogura Y, Szczepanik M, Lara-Tejero M, Lichtenberger GS, Grant EP, Bertin J, Coyle AJ, Galan JE, Askenase PW et al (2006) Critical role for NALP3/CIAS1/cryopyrin in innate and adaptive immunity through its regulation of caspase-1. Immunity 24:317–327
29. Chen CJ, Shi Y, Hearn A, Fitzgerald K, Golenbock D, Reed G, Akira S, Rock KL (2006) MyD88-dependent IL-1 receptor signaling is essential for gouty inflammation stimulated by monosodium urate crystals. J Clin Invest 116:2262–2271
30. Guarda G, Braun M, Staehli F, Tardivel A, Mattmann C, Forster I, Farlik M, Decker T, Du Pasquier RA, Romero P et al (2011) Type I interferon inhibits interleukin-1 production and inflammasome activation. Immunity 34:213–223
31. Guarda G, Zenger M, Yazdi AS, Schroder K, Ferrero I, Menu P, Tardivel A, Mattmann C, Tschopp J (2011) Differential expression of NLRP3 among hematopoietic cells. J Immunol 186:2529–2534

32. Anderson JP, Mueller JL, Rosengren S, Boyle DL, Schaner P, Cannon SB, Goodyear CS, Hoffman HM (2004) Structural, expression, and evolutionary analysis of mouse CIAS1. Gene 338:25–34
33. Mayor A, Martinon F, De Smedt T, Petrilli V, Tschopp J (2007) A crucial function of SGT1 and HSP90 in inflammasome activity links mammalian and plant innate immune responses. Nat Immunol 8:497–503
34. Jin C, Flavell RA (2010) Molecular mechanism of NLRP3 inflammasome activation. J Clin Immunol 30:628–631
35. Schroder K, Tschopp J (2010) The inflammasomes. Cell 140:821–832
36. Shenoy AR, Wellington DA, Kumar P, Kassa H, Booth CJ, Cresswell P, MacMicking JD (2012) GBP5 promotes NLRP3 inflammasome assembly and immunity in mammals. Science 336:481–485
37. Latz E (2010) The inflammasomes: mechanisms of activation and function. Curr Opin Immunol 22:28–33
38. Martinon F (2010) Signaling by ROS drives inflammasome activation. Eur J Immunol 40:616–619
39. Dostert C, Petrilli V, Van Bruggen R, Steele C, Mossman BT, Tschopp J (2008) Innate immune activation through Nalp3 inflammasome sensing of asbestos and silica. Science 320:674–677
40. Gross O, Poeck H, Bscheider M, Dostert C, Hannesschlager N, Endres S, Hartmann G, Tardivel A, Schweighoffer E, Tybulewicz V et al (2009) Syk kinase signalling couples to the Nlrp3 inflammasome for anti-fungal host defence. Nature 459:433–436
41. Underhill DM, Rossnagle E, Lowell CA, Simmons RM (2005) Dectin-1 activates Syk tyrosine kinase in a dynamic subset of macrophages for reactive oxygen production. Blood 106:2543–2550
42. Hornung V, Bauernfeind F, Halle A, Samstad EO, Kono H, Rock KL, Fitzgerald KA, Latz E (2008) Silica crystals and aluminum salts activate the NALP3 inflammasome through phagosomal destabilization. Nat Immunol 9:847–856
43. Meissner F, Molawi K, Zychlinsky A (2008) Superoxide dismutase 1 regulates caspase-1 and endotoxic shock. Nat Immunol 9:866–872
44. Meissner F, Seger RA, Moshous D, Fischer A, Reichenbach J, Zychlinsky A (2010) Inflammasome activation in NADPH oxidase defective mononuclear phagocytes from patients with chronic granulomatous disease. Blood 116:1570–1573
45. Brookes PS, Yoon Y, Robotham JL, Anders MW, Sheu SS (2004) Calcium, ATP, and ROS: a mitochondrial love-hate triangle. Am J Physiol 287:C817–C833
46. Zhou L, Aon MA, Almas T, Cortassa S, Winslow RL, O'Rourke B (2010) A reaction–diffusion model of ROS-induced ROS release in a mitochondrial network. PLoS Comput Biol 6:e1000657
47. Nakahira K, Haspel JA, Rathinam VA, Lee SJ, Dolinay T, Lam HC, Englert JA, Rabinovitch M, Cernadas M, Kim HP et al (2011) Autophagy proteins regulate innate immune responses by inhibiting the release of mitochondrial DNA mediated by the NALP3 inflammasome. Nat Immunol 12:222–230
48. Zhou R, Yazdi AS, Menu P, Tschopp J (2011) A role for mitochondria in NLRP3 inflammasome activation. Nature 469:221–225
49. Saitoh T, Fujita N, Jang MH, Uematsu S, Yang BG, Satoh T, Omori H, Noda T, Yamamoto N, Komatsu M et al (2008) Loss of the autophagy protein Atg16L1 enhances endotoxin-induced IL-1beta production. Nature 456:264–268
50. Zhou R, Tardivel A, Thorens B, Choi I, Tschopp J (2010) Thioredoxin-interacting protein links oxidative stress to inflammasome activation. Nat Immunol 11:136–140
51. Masters SL, Dunne A, Subramanian SL, Hull RL, Tannahill GM, Sharp FA, Becker C, Franchi L, Yoshihara E, Chen Z et al (2010) Activation of the NLRP3 inflammasome by islet amyloid polypeptide provides a mechanism for enhanced IL-1beta in type 2 diabetes. Nat Immunol 11:897–904

52. Cassel SL, Eisenbarth SC, Iyer SS, Sadler JJ, Colegio OR, Tephly LA, Carter AB, Rothman PB, Flavell RA, Sutterwala FS (2008) The Nalp3 inflammasome is essential for the development of silicosis. Proc Natl Acad Sci U S A 105:9035–9040
53. Eisenbarth SC, Colegio OR, O'Connor W, Sutterwala FS, Flavell RA (2008) Crucial role for the Nalp3 inflammasome in the immunostimulatory properties of aluminium adjuvants. Nature 453:1122–1126
54. Franchi L, Kanneganti TD, Dubyak GR, Nunez G (2007) Differential requirement of P2X7 receptor and intracellular K+ for caspase-1 activation induced by intracellular and extracellular bacteria. J Biol Chem 282:18810–18818
55. Lamkanfi M, Mueller JL, Vitari AC, Misaghi S, Fedorova A, Deshayes K, Lee WP, Hoffman HM, Dixit VM (2009) Glyburide inhibits the cryopyrin/Nalp3 inflammasome. J Cell Biol 187:61–70
56. Compan V, Baroja-Mazo A, Lopez-Castejon G, Gomez AI, Martinez CM, Angosto D, Montero MT, Herranz AS, Bazan E, Reimers D et al (2012) Cell volume regulation modulates NLRP3 inflammasome activation. Immunity 37:487–500
57. Halle A, Hornung V, Petzold GC, Stewart CR, Monks BG, Reinheckel T, Fitzgerald KA, Latz E, Moore KJ, Golenbock DT (2008) The NALP3 inflammasome is involved in the innate immune response to amyloid-beta. Nat Immunol 9:857–865
58. Duewell P, Kono H, Rayner KJ, Sirois CM, Vladimer G, Bauernfeind FG, Abela GS, Franchi L, Nunez G, Schnurr M et al (2010) NLRP3 inflammasomes are required for atherogenesis and activated by cholesterol crystals. Nature 464:1357–1361
59. Sharp FA, Ruane D, Claass B, Creagh E, Harris J, Malyala P, Singh M, O'Hagan DT, Petrilli V, Tschopp J et al (2009) Uptake of particulate vaccine adjuvants by dendritic cells activates the NALP3 inflammasome. Proc Natl Acad Sci U S A 106:870–875
60. Dostert C, Guarda G, Romero JF, Menu P, Gross O, Tardivel A, Suva ML, Stehle JC, Kopf M, Stamenkovic I et al (2009) Malarial hemozoin is a Nalp3 inflammasome activating danger signal. PLoS One 4:e6510
61. Masters SL, Mielke LA, Cornish AL, Sutton CE, O'Donnell J, Cengia LH, Roberts AW, Wicks IP, Mills KH, Croker BA (2010) Regulation of interleukin-1beta by interferon-gamma is species specific, limited by suppressor of cytokine signalling 1 and influences interleukin-17 production. EMBO Rep 11:640–646
62. O'Rourke RW, White AE, Metcalf MD, Winters BR, Diggs BS, Zhu X, Marks DL (2012) Systemic inflammation and insulin sensitivity in obese IFN-gamma knockout mice. Metabolism 61:1152–1161
63. Rocha VZ, Folco EJ, Sukhova G, Shimizu K, Gotsman I, Vernon AH, Libby P (2008) Interferon-gamma, a Th1 cytokine, regulates fat inflammation: a role for adaptive immunity in obesity. Circ Res 103:467–476
64. Boden G, Shulman GI (2002) Free fatty acids in obesity and type 2 diabetes: defining their role in the development of insulin resistance and beta-cell dysfunction. Eur J Clin Invest 32(suppl 3):14–23
65. Eguchi K, Manabe I, Oishi-Tanaka Y, Ohsugi M, Kono N, Ogata F, Yagi N, Ohto U, Kimoto M, Miyake K et al (2012) Saturated fatty acid and TLR signaling link beta cell dysfunction and islet inflammation. Cell Metab 15:518–533
66. Shi H, Kokoeva MV, Inouye K, Tzameli I, Yin H, Flier JS (2006) TLR4 links innate immunity and fatty acid-induced insulin resistance. J Clin Invest 116:3015–3025
67. Haversen L, Danielsson KN, Fogelstrand L, Wiklund O (2009) Induction of proinflammatory cytokines by long-chain saturated fatty acids in human macrophages. Atherosclerosis 202:382–393
68. Wen H, Gris D, Lei Y, Jha S, Zhang L, Huang MT, Brickey WJ, Ting JP (2011) Fatty acid-induced NLRP3-ASC inflammasome activation interferes with insulin signaling. Nat Immunol 12:408–415
69. Samuel VT, Shulman GI (2012) Mechanisms for insulin resistance: common threads and missing links. Cell 148:852–871

70. Haus JM, Kashyap SR, Kasumov T, Zhang R, Kelly KR, Defronzo RA, Kirwan JP (2009) Plasma ceramides are elevated in obese subjects with type 2 diabetes and correlate with the severity of insulin resistance. Diabetes 58:337–343
71. Vandanmagsar B, Youm YH, Ravussin A, Galgani JE, Stadler K, Mynatt RL, Ravussin E, Stephens JM, Dixit VD (2011) The NLRP3 inflammasome instigates obesity-induced inflammation and insulin resistance. Nat Med 17:179–188
72. Asakawa H, Miyagawa J, Hanafusa T, Kuwajima M, Matsuzawa Y (1997) High glucose and hyperosmolarity increase secretion of interleukin-1 beta in cultured human aortic endothelial cells. J Diabetes Complications 11:176–179
73. Maedler K, Sergeev P, Ris F, Oberholzer J, Joller-Jemelka HI, Spinas GA, Kaiser N, Halban PA, Donath MY (2002) Glucose-induced beta cell production of IL-1beta contributes to glucotoxicity in human pancreatic islets. J Clin Invest 110:851–860
74. Shanmugam N, Reddy MA, Guha M, Natarajan R (2003) High glucose-induced expression of proinflammatory cytokine and chemokine genes in monocytic cells. Diabetes 52:1256–1264
75. Dasu MR, Devaraj S, Jialal I (2007) High glucose induces IL-1beta expression in human monocytes: mechanistic insights. Am J Physiol Endocrinol Metab 293:E337–E346
76. Parikh H, Carlsson E, Chutkow WA, Johansson LE, Storgaard H, Poulsen P, Saxena R, Ladd C, Schulze PC, Mazzini MJ et al (2007) TXNIP regulates peripheral glucose metabolism in humans. PLoS Med 4:e158
77. Stoltzman CA, Peterson CW, Breen KT, Muoio DM, Billin AN, Ayer DE (2008) Glucose sensing by MondoA:Mlx complexes: a role for hexokinases and direct regulation of thioredoxin-interacting protein expression. Proc Natl Acad Sci U S A 105:6912–6917
78. Koenen TB, Stienstra R, van Tits LJ, de Graaf J, Stalenhoef AF, Joosten LA, Tack CJ, Netea MG (2011) Hyperglycemia activates caspase-1 and TXNIP-mediated IL-1beta transcription in human adipose tissue. Diabetes 60:517–524
79. Martinon F, Petrilli V, Mayor A, Tardivel A, Tschopp J (2006) Gout-associated uric acid crystals activate the NALP3 inflammasome. Nature 440:237–241
80. Gasse P, Riteau N, Charron S, Girre S, Fick L, Petrilli V, Tschopp J, Lagente V, Quesniaux VF, Ryffel B et al (2009) Uric acid is a danger signal activating NALP3 inflammasome in lung injury inflammation and fibrosis. Am J Respir Crit Care Med 179:903–913
81. Cheung KJ, Tzameli I, Pissios P, Rovira I, Gavrilova O, Ohtsubo T, Chen Z, Finkel T, Flier JS, Friedman JM (2007) Xanthine oxidoreductase is a regulator of adipogenesis and PPARgamma activity. Cell Metab 5:115–128
82. Stienstra R, Joosten LA, Koenen T, van Tits B, van Diepen JA, van den Berg SA, Rensen PC, Voshol PJ, Fantuzzi G, Hijmans A et al (2010) The inflammasome-mediated caspase-1 activation controls adipocyte differentiation and insulin sensitivity. Cell Metab 12:593–605
83. Lyngdoh T, Marques-Vidal P, Paccaud F, Preisig M, Waeber G, Bochud M, Vollenweider P (2011) Elevated serum uric acid is associated with high circulating inflammatory cytokines in the population-based Colaus study. PLoS One 6:e19901
84. Small DM (1988) George Lyman Duff memorial lecture. Progression and regression of atherosclerotic lesions. Insights from lipid physical biochemistry. Arteriosclerosis 8:103–129
85. Razani B, Feng C, Coleman T, Emanuel R, Wen H, Hwang S, Ting JP, Virgin HW, Kastan MB, Semenkovich CF (2012) Autophagy links inflammasomes to atherosclerotic progression. Cell Metab 15:534–544
86. Rajamaki K, Lappalainen J, Oorni K, Valimaki E, Matikainen S, Kovanen PT, Eklund KK (2010) Cholesterol crystals activate the NLRP3 inflammasome in human macrophages: a novel link between cholesterol metabolism and inflammation. PLoS One 5:e11765
87. Westermark P, Andersson A, Westermark GT (2011) Islet amyloid polypeptide, islet amyloid, and diabetes mellitus. Physiol Rev 91:795–826
88. Clément K (2010) Adipokines, inflammation, and obesity. In: Leff TAG, James G (eds) Adipose tissue in health and disease. Wiley-VCH, Weinheim
89. Trayhurn P, Wood IS (2005) Signalling role of adipose tissue: adipokines and inflammation in obesity. Biochem Soc Trans 33:1078–1081

90. Lyon CJ, Law RE, Hsueh WA (2003) Minireview: adiposity, inflammation, and atherogenesis. Endocrinology 144:2195–2200
91. Nishimura S, Manabe I, Nagasaki M, Eto K, Yamashita H, Ohsugi M, Otsu M, Hara K, Ueki K, Sugiura S et al (2009) CD8+ effector T cells contribute to macrophage recruitment and adipose tissue inflammation in obesity. Nat Med 15:914–920
92. Winer S, Chan Y, Paltser G, Truong D, Tsui H, Bahrami J, Dorfman R, Wang Y, Zielenski J, Mastronardi F et al (2009) Normalization of obesity-associated insulin resistance through immunotherapy. Nat Med 15:921–929
93. Lumeng CN, Bodzin JL, Saltiel AR (2007) Obesity induces a phenotypic switch in adipose tissue macrophage polarization. J Clin Invest 117:175–184
94. Winer DA, Winer S, Shen L, Wadia PP, Yantha J, Paltser G, Tsui H, Wu P, Davidson MG, Alonso MN et al (2011) B cells promote insulin resistance through modulation of T cells and production of pathogenic IgG antibodies. Nat Med 17:610–617
95. Liu J, Divoux A, Sun J, Zhang J, Clement K, Glickman JN, Sukhova GK, Wolters PJ, Du J, Gorgun CZ et al (2009) Genetic deficiency and pharmacological stabilization of mast cells reduce diet-induced obesity and diabetes in mice. Nat Med 15:940–945
96. Wu D, Molofsky AB, Liang HE, Ricardo-Gonzalez RR, Jouihan HA, Bando JK, Chawla A, Locksley RM (2011) Eosinophils sustain adipose alternatively activated macrophages associated with glucose homeostasis. Science 332:243–247
97. Talukdar S, Oh DY, Bandyopadhyay G, Li D, Xu J, McNelis J, Lu M, Li P, Yan Q, Zhu Y et al (2012) Neutrophils mediate insulin resistance in mice fed a high-fat diet through secreted elastase. Nat Med 18(9):1407–1412
98. Lumeng CN, Maillard I, Saltiel AR (2009) T-ing up inflammation in fat. Nat Med 15:846–847
99. Jager J, Gremeaux T, Cormont M, Le Marchand-Brustel Y, Tanti JF (2007) Interleukin-1beta-induced insulin resistance in adipocytes through down-regulation of insulin receptor substrate-1 expression. Endocrinology 148:241–251
100. Koenen TB, Stienstra R, van Tits LJ, Joosten LA, van Velzen JF, Hijmans A, Pol JA, van der Vliet JA, Netea MG, Tack CJ et al (2011) The inflammasome and caspase-1 activation: a new mechanism underlying increased inflammatory activity in human visceral adipose tissue. Endocrinology 152:3769–3778
101. McGillicuddy FC, Harford KA, Reynolds CM, Oliver E, Claessens M, Mills KH, Roche HM (2011) Lack of interleukin-1 receptor I (IL-1RI) protects mice from high-fat diet-induced adipose tissue inflammation coincident with improved glucose homeostasis. Diabetes 60:1688–1698
102. Stienstra R, van Diepen JA, Tack CJ, Zaki MH, van de Veerdonk FL, Perera D, Neale GA, Hooiveld GJ, Hijmans A, Vroegrijk I et al (2011) Inflammasome is a central player in the induction of obesity and insulin resistance. Proc Natl Acad Sci U S A 108:15324–15329
103. Garcia MC, Wernstedt I, Berndtsson A, Enge M, Bell M, Hultgren O, Horn M, Ahren B, Enerback S, Ohlsson C et al (2006) Mature-onset obesity in interleukin-1 receptor I knockout mice. Diabetes 55:1205–1213
104. Netea MG, Joosten LA, Lewis E, Jensen DR, Voshol PJ, Kullberg BJ, Tack CJ, van Krieken H, Kim SH, Stalenhoef AF et al (2006) Deficiency of interleukin-18 in mice leads to hyperphagia, obesity and insulin resistance. Nat Med 12:650–656
105. Zorrilla EP, Sanchez-Alavez M, Sugama S, Brennan M, Fernandez R, Bartfai T, Conti B (2007) Interleukin-18 controls energy homeostasis by suppressing appetite and feed efficiency. Proc Natl Acad Sci U S A 104:11097–11102
106. Membrez M, Ammon-Zufferey C, Philippe D, Aprikian O, Monnard I, Mace K, Darimont C (2009) Interleukin-18 protein level is upregulated in adipose tissue of obese mice. Obesity (Silver Spring) 17:393–395
107. Youm YH, Adijiang A, Vandanmagsar B, Burk D, Ravussin A, Dixit VD (2011) Elimination of the NLRP3-ASC inflammasome protects against chronic obesity-induced pancreatic damage. Endocrinology 152:4039–4045
108. Aygun AD, Gungor S, Ustundag B, Gurgoze MK, Sen Y (2005) Proinflammatory cytokines and leptin are increased in serum of prepubertal obese children. Mediators Inflamm 2005:180–183

109. Garanty-Bogacka B, Syrenicz M, Syrenicz A, Gebala A, Lulka D, Walczak M (2005) Serum markers of inflammation and endothelial activation in children with obesity-related hypertension. Neuro Endocrinol Lett 26:242–246
110. Moschen AR, Molnar C, Enrich B, Geiger S, Ebenbichler CF, Tilg H (2011) Adipose and liver expression of interleukin (IL)-1 family members in morbid obesity and effects of weight loss. Mol Med 17:840–845
111. Nov O, Kohl A, Lewis EC, Bashan N, Dvir I, Ben-Shlomo S, Fishman S, Wueest S, Konrad D, Rudich A (2010) Interleukin-1beta may mediate insulin resistance in liver-derived cells in response to adipocyte inflammation. Endocrinology 151:4247–4256
112. Bunout D, Munoz C, Lopez M, de la Maza MP, Schlesinger L, Hirsch S, Pettermann M (1996) Interleukin 1 and tumor necrosis factor in obese alcoholics compared with normal-weight patients. Am J Clin Nutr 63:373–376
113. Jung C, Gerdes N, Fritzenwanger M, Figulla HR (2009) Circulating levels of interleukin-1 family cytokines in overweight adolescents. Mediators Inflamm 2010:958403
114. Esposito K, Pontillo A, Di Palo C, Giugliano G, Masella M, Marfella R, Giugliano D (2003) Effect of weight loss and lifestyle changes on vascular inflammatory markers in obese women: a randomized trial. JAMA 289:1799–1804
115. Goossens GH, Blaak EE, Theunissen R, Duijvestijn AM, Clement K, Tervaert JW, Thewissen MM (2012) Expression of NLRP3 inflammasome and T cell population markers in adipose tissue are associated with insulin resistance and impaired glucose metabolism in humans. Mol Immunol 50:142–149
116. Pal D, Dasgupta S, Kundu R, Maitra S, Das G, Mukhopadhyay S, Ray S, Majumdar SS, Bhattacharya S (2012) Fetuin-A acts as an endogenous ligand of TLR4 to promote lipid-induced insulin resistance. Nat Med 18(8):1279–1285
117. Di Virgilio F (2007) Liaisons dangereuses: P2X(7) and the inflammasome. Trends Pharmacol Sci 28:465–472
118. Sun S, Xia S, Ji Y, Kersten S, Qi L (2012) The ATP-P2X7 signaling axis is dispensable for obesity-associated inflammasome activation in adipose tissue. Diabetes 61:1471–1478
119. Lumeng CN, Saltiel AR (2011) Inflammatory links between obesity and metabolic disease. J Clin Invest 121:2111–2117
120. Feve B, Bastard JP (2009) The role of interleukins in insulin resistance and type 2 diabetes mellitus. Nat Rev Endocrinol 5:305–311
121. Dinarello CA, Donath MY, Mandrup-Poulsen T (2010) Role of IL-1beta in type 2 diabetes. Curr Opin Endocrinol Diabetes Obes 17:314–321
122. Mandrup-Poulsen T (1996) The role of interleukin-1 in the pathogenesis of IDDM. Diabetologia 39:1005–1029
123. Corbett JA, McDaniel ML (1992) Does nitric oxide mediate autoimmune destruction of beta-cells? Possible therapeutic interventions in IDDM. Diabetes 41:897–903
124. Corbett JA, McDaniel ML (1995) Intraislet release of interleukin 1 inhibits beta cell function by inducing beta cell expression of inducible nitric oxide synthase. J Exp Med 181:559–568
125. Cai D, Yuan M, Frantz DF, Melendez PA, Hansen L, Lee J, Shoelson SE (2005) Local and systemic insulin resistance resulting from hepatic activation of IKK-beta and NF-kappaB. Nat Med 11:183–190
126. Csak T, Ganz M, Pespisa J, Kodys K, Dolganiuc A, Szabo G (2011) Fatty acid and endotoxin activate inflammasomes in mouse hepatocytes that release danger signals to stimulate immune cells. Hepatology 54:133–144
127. Miura K, Yang L, van Rooijen N, Brenner DA, Ohnishi H, Seki E (2013) TLR2 and palmitic acid cooperatively contribute to the development of nonalcoholic steatohepatitis through inflammasome activation. Hepatology 57(2):577–589
128. Matsuzaka T, Atsumi A, Matsumori R, Nie T, Shinozaki H, Suzuki-Kemuriyama N, Kuba M, Nakagawa Y, Ishii K, Shimada M et al (2012) Elovl6 promotes nonalcoholic steatohepatitis in mice and humans. Hepatology 56(6):2199–2208
129. Isoda K, Sawada S, Ayaori M, Matsuki T, Horai R, Kagata Y, Miyazaki K, Kusuhara M, Okazaki M, Matsubara O et al (2005) Deficiency of interleukin-1 receptor antagonist deterio-

rates fatty liver and cholesterol metabolism in hypercholesterolemic mice. J Biol Chem 280:7002–7009
130. Miura K, Kodama Y, Inokuchi S, Schnabl B, Aoyama T, Ohnishi H, Olefsky JM, Brenner DA, Seki E (2010) Toll-like receptor 9 promotes steatohepatitis by induction of interleukin-1beta in mice. Gastroenterology 139(323–334):e327
131. Chi H, Messas E, Levine RA, Graves DT, Amar S (2004) Interleukin-1 receptor signaling mediates atherosclerosis associated with bacterial exposure and/or a high-fat diet in a murine apolipoprotein E heterozygote model: pharmacotherapeutic implications. Circulation 110:1678–1685
132. Kirii H, Niwa T, Yamada Y, Wada H, Saito K, Iwakura Y, Asano M, Moriwaki H, Seishima M (2003) Lack of interleukin-1beta decreases the severity of atherosclerosis in ApoE-deficient mice. Arterioscler Thromb Vasc Biol 23:656–660
133. Elhage R, Maret A, Pieraggi MT, Thiers JC, Arnal JF, Bayard F (1998) Differential effects of interleukin-1 receptor antagonist and tumor necrosis factor binding protein on fatty-streak formation in apolipoprotein E-deficient mice. Circulation 97:242–244
134. Isoda K, Ohsuzu F (2006) The effect of interleukin-1 receptor antagonist on arteries and cholesterol metabolism. J Atheroscler Thromb 13:21–30
135. Menu P, Pellegrin M, Aubert JF, Bouzourene K, Tardivel A, Mazzolai L, Tschopp J (2011) Atherosclerosis in ApoE-deficient mice progresses independently of the NLRP3 inflammasome. Cell Death Dis 2:e137
136. Shi Y, Mucsi AD, Ng G (2010) Monosodium urate crystals in inflammation and immunity. Immunol Rev 233:203–217
137. Di Giovine FS, Malawista SE, Nuki G, Duff GW (1987) Interleukin 1 (IL 1) as a mediator of crystal arthritis. Stimulation of T cell and synovial fibroblast mitogenesis by urate crystal-induced IL1. J Immunol 138:3213–3218
138. Landis RC, Yagnik DR, Florey O, Philippidis P, Emons V, Mason JC, Haskard DO (2002) Safe disposal of inflammatory monosodium urate monohydrate crystals by differentiated macrophages. Arthritis Rheum 46:3026–3033
139. Yagnik DR, Hillyer P, Marshall D, Smythe CD, Krausz T, Haskard DO, Landis RC (2000) Noninflammatory phagocytosis of monosodium urate monohydrate crystals by mouse macrophages. Implications for the control of joint inflammation in gout. Arthritis Rheum 43:1779–1789
140. Reaven GM (1988) Banting lecture 1988. Role of insulin resistance in human disease. Diabetes 37:1595–1607
141. Juraschek SP, Miller ER III, Gelber AC (2013) Body mass index, obesity, and prevalent gout in the United States in 1988–1994 and 2007–2010: body mass index, obesity, and gout. Arthritis Care Res (Hoboken) 65(1):127–132
142. DeMarco MA, Maynard JW, Huizinga MM, Baer AN, Kottgen A, Gelber AC, Coresh J (2011) Obesity and younger age at gout onset in a community-based cohort. Arthritis Care Res (Hoboken) 63:1108–1114
143. Shaftel SS, Griffin WS, O'Banion MK (2008) The role of interleukin-1 in neuroinflammation and Alzheimer disease: an evolving perspective. J Neuroinflammation 5:7
144. Lorton D, Kocsis JM, King L, Madden K, Brunden KR (1996) Beta-amyloid induces increased release of interleukin-1 beta from lipopolysaccharide-activated human monocytes. J Neuroimmunol 67:21–29
145. Yates SL, Burgess LH, Kocsis-Angle J, Antal JM, Dority MD, Embury PB, Piotrkowski AM, Brunden KR (2000) Amyloid beta and amylin fibrils induce increases in proinflammatory cytokine and chemokine production by THP-1 cells and murine microglia. J Neurochem 74:1017–1025
146. Balkwill F, Mantovani A (2001) Inflammation and cancer: back to Virchow? Lancet 357:539–545
147. Renehan AG, Tyson M, Egger M, Heller RF, Zwahlen M (2008) Body-mass index and incidence of cancer: a systematic review and meta-analysis of prospective observational studies. Lancet 371:569–578

148. Calle EE, Rodriguez C, Walker-Thurmond K, Thun MJ (2003) Overweight, obesity, and mortality from cancer in a prospectively studied cohort of U.S. adults. N Engl J Med 348:1625–1638
149. Hanahan D, Weinberg RA (2011) Hallmarks of cancer: the next generation. Cell 144:646–674
150. Galluzzi L, Vitale I, Abrams JM, Alnemri ES, Baehrecke EH, Blagosklonny MV, Dawson TM, Dawson VL, El-Deiry WS, Fulda S et al (2011) Molecular definitions of cell death subroutines: recommendations of the Nomenclature Committee on Cell Death 2012. Cell Death Differ 19:107–120
151. Kepp O, Galluzzi L, Zitvogel L, Kroemer G (2010) Pyroptosis—a cell death modality of its kind? Eur J Immunol 40:627–630
152. Miao EA, Rajan JV, Aderem A (2011) Caspase-1-induced pyroptotic cell death. Immunol Rev 243:206–214
153. Gross O, Thomas CJ, Guarda G, Tschopp J (2011) The inflammasome: an integrated view. Immunol Rev 243:136–151
154. Broz P, Newton K, Lamkanfi M, Mariathasan S, Dixit VM, Monack DM (2010) Redundant roles for inflammasome receptors NLRP3 and NLRC4 in host defense against Salmonella. J Exp Med 207:1745–1755
155. Broz P, von Moltke J, Jones JW, Vance RE, Monack DM (2010) Differential requirement for caspase-1 autoproteolysis in pathogen-induced cell death and cytokine processing. Cell Host Microbe 8:471–483
156. Masumoto J, Taniguchi S, Ayukawa K, Sarvotham H, Kishino T, Niikawa N, Hidaka E, Katsuyama T, Higuchi T, Sagara J (1999) ASC, a novel 22-kDa protein, aggregates during apoptosis of human promyelocytic leukemia HL-60 cells. J Biol Chem 274:33835–33838
157. Mariathasan S, Newton K, Monack DM, Vucic D, French DM, Lee WP, Roose-Girma M, Erickson S, Dixit VM (2004) Differential activation of the inflammasome by caspase-1 adaptors ASC and Ipaf. Nature 430:213–218
158. Lamkanfi M, Kanneganti TD, Van Damme P, Vanden Berghe T, Vanoverberghe I, Vandekerckhove J, Vandenabeele P, Gevaert K, Nunez G (2008) Targeted peptidecentric proteomics reveals caspase-7 as a substrate of the caspase-1 inflammasomes. Mol Cell Proteomics 7:2350–2363
159. Brennan MA, Cookson BT (2000) Salmonella induces macrophage death by caspase-1-dependent necrosis. Mol Microbiol 38:31–40
160. Chen LM, Kaniga K, Galan JE (1996) Salmonella spp. are cytotoxic for cultured macrophages. Mol Microbiol 21:1101–1115
161. Watson PR, Gautier AV, Paulin SM, Bland AP, Jones PW, Wallis TS (2000) Salmonella enterica serovars Typhimurium and Dublin can lyse macrophages by a mechanism distinct from apoptosis. Infect Immun 68:3744–3747
162. Fink SL, Cookson BT (2006) Caspase-1-dependent pore formation during pyroptosis leads to osmotic lysis of infected host macrophages. Cell Microbiol 8:1812–1825
163. Koopman G, Reutelingsperger CP, Kuijten GA, Keehnen RM, Pals ST, van Oers MH (1994) Annexin V for flow cytometric detection of phosphatidylserine expression on B cells undergoing apoptosis. Blood 84:1415–1420
164. Silveira TN, Zamboni DS (2010) Pore formation triggered by Legionella spp. is an Nlrc4 inflammasome-dependent host cell response that precedes pyroptosis. Infect Immun 78:1403–1413
165. Majno G, Joris I (1995) Apoptosis, oncosis, and necrosis. An overview of cell death. Am J Pathol 146:3–15
166. Winter RN, Kramer A, Borkowski A, Kyprianou N (2001) Loss of caspase-1 and caspase-3 protein expression in human prostate cancer. Cancer Res 61:1227–1232
167. Allen IC, TeKippe EM, Woodford RM, Uronis JM, Holl EK, Rogers AB, Herfarth HH, Jobin C, Ting JP (2010) The NLRP3 inflammasome functions as a negative regulator of tumorigenesis during colitis-associated cancer. J Exp Med 207:1045–1056

168. Hu B, Elinav E, Huber S, Booth CJ, Strowig T, Jin C, Eisenbarth SC, Flavell RA (2010) Inflammation-induced tumorigenesis in the colon is regulated by caspase-1 and NLRC4. Proc Natl Acad Sci U S A 107:21635–21640
169. Karin M, Lawrence T, Nizet V (2006) Innate immunity gone awry: linking microbial infections to chronic inflammation and cancer. Cell 124:823–835
170. Bauer C, Duewell P, Mayer C, Lehr HA, Fitzgerald KA, Dauer M, Tschopp J, Endres S, Latz E, Schnurr M (2010) Colitis induced in mice with dextran sulfate sodium (DSS) is mediated by the NLRP3 inflammasome. Gut 59:1192–1199
171. Siegmund B, Lehr HA, Fantuzzi G, Dinarello CA (2001) IL-1 beta-converting enzyme (caspase-1) in intestinal inflammation. Proc Natl Acad Sci U S A 98:13249–13254
172. Rakoff-Nahoum S, Paglino J, Eslami-Varzaneh F, Edberg S, Medzhitov R (2004) Recognition of commensal microflora by toll-like receptors is required for intestinal homeostasis. Cell 118:229–241
173. Chen GY, Liu M, Wang F, Bertin J, Nunez G (2011) A functional role for Nlrp6 in intestinal inflammation and tumorigenesis. J Immunol 186:7187–7194
174. Dupaul-Chicoine J, Yeretssian G, Doiron K, Bergstrom KS, McIntire CR, LeBlanc PM, Meunier C, Turbide C, Gros P, Beauchemin N et al (2010) Control of intestinal homeostasis, colitis, and colitis-associated colorectal cancer by the inflammatory caspases. Immunity 32:367–378
175. Normand S, Delanoye-Crespin A, Bressenot A, Huot L, Grandjean T, Peyrin-Biroulet L, Lemoine Y, Hot D, Chamaillard M (2011) Nod-like receptor pyrin domain-containing protein 6 (NLRP6) controls epithelial self-renewal and colorectal carcinogenesis upon injury. Proc Natl Acad Sci U S A 108:9601–9606
176. Zaki MH, Boyd KL, Vogel P, Kastan MB, Lamkanfi M, Kanneganti TD (2010) The NLRP3 inflammasome protects against loss of epithelial integrity and mortality during experimental colitis. Immunity 32:379–391
177. Zaki MH, Vogel P, Body-Malapel M, Lamkanfi M, Kanneganti TD (2010) IL-18 production downstream of the Nlrp3 inflammasome confers protection against colorectal tumor formation. J Immunol 185:4912–4920
178. Elinav E, Strowig T, Kau AL, Henao-Mejia J, Thaiss CA, Booth CJ, Peaper DR, Bertin J, Eisenbarth SC, Gordon JI et al (2011) NLRP6 inflammasome regulates colonic microbial ecology and risk for colitis. Cell 145:745–757
179. Grivennikov SI, Greten FR, Karin M (2010) Immunity, inflammation, and cancer. Cell 140:883–899
180. Reed JR, Leon RP, Hall MK, Schwertfeger KL (2009) Interleukin-1beta and fibroblast growth factor receptor 1 cooperate to induce cyclooxygenase-2 during early mammary tumourigenesis. Breast Cancer Res 11:R21
181. Schafer M, Werner S (2008) Cancer as an overhealing wound: an old hypothesis revisited. Nat Rev Mol Cell Biol 9:628–638
182. Tu S, Bhagat G, Cui G, Takaishi S, Kurt-Jones EA, Rickman B, Betz KS, Penz-Oesterreicher M, Bjorkdahl O, Fox JG et al (2008) Overexpression of interleukin-1beta induces gastric inflammation and cancer and mobilizes myeloid-derived suppressor cells in mice. Cancer Cell 14:408–419
183. Krelin Y, Voronov E, Dotan S, Elkabets M, Reich E, Fogel M, Huszar M, Iwakura Y, Segal S, Dinarello CA et al (2007) Interleukin-1beta-driven inflammation promotes the development and invasiveness of chemical carcinogen-induced tumors. Cancer Res 67:1062–1071
184. Green DR, Kroemer G (2005) Pharmacological manipulation of cell death: clinical applications in sight? J Clin Invest 115:2610–2617
185. Ito T, Ando H, Suzuki T, Ogura T, Hotta K, Imamura Y, Yamaguchi Y, Handa H (2010) Identification of a primary target of thalidomide teratogenicity. Science 327:1345–1350
186. Juliana C, Fernandes-Alnemri T, Wu J, Datta P, Solorzano L, Yu JW, Meng R, Quong AA, Latz E, Scott CP et al (2010) Anti-inflammatory compounds parthenolide and Bay 11-7082 are direct inhibitors of the inflammasome. J Biol Chem 285:9792–9802

187. Keller M, Sollberger G, Beer HD (2009) Thalidomide inhibits activation of caspase-1. J Immunol 183:5593–5599
188. Dinarello CA (2011) Interleukin-1 in the pathogenesis and treatment of inflammatory diseases. Blood 117:3720–3732
189. Srivastava S, Salim N, Robertson MJ (2010) Interleukin-18: biology and role in the immunotherapy of cancer. Curr Med Chem 17:3353–3357
190. Lust JA, Lacy MQ, Zeldenrust SR, Dispenzieri A, Gertz MA, Witzig TE, Kumar S, Hayman SR, Russell SJ, Buadi FK et al (2009) Induction of a chronic disease state in patients with smoldering or indolent multiple myeloma by targeting interleukin 1{beta}-induced interleukin 6 production and the myeloma proliferative component. Mayo Clin Proc 84:114–122
191. Wu ZQ, Han XD, Wang Y, Yuan KL, Jin ZM, Di JZ, Yan J, Pan Y, Zhang P, Huang XY et al (2011) Interleukin-1 receptor antagonist reduced apoptosis and attenuated intestinal mucositis in a 5-fluorouracil chemotherapy model in mice. Cancer Chemother Pharmacol 68:87–96
192. Zhu J, Zhang J, Xiang D, Zhang Z, Zhang L, Wu M, Zhu S, Zhang R, Han W (2010) Recombinant human interleukin-1 receptor antagonist protects mice against acute doxorubicin-induced cardiotoxicity. Eur J Pharmacol 643:247–253
193. Dinarello CA (2010) Why not treat human cancer with interleukin-1 blockade? Cancer Metastasis Rev 29:317–329
194. Eisenbarth SC, Flavell RA (2009) Innate instruction of adaptive immunity revisited: the inflammasome. EMBO Mol Med 1:92–98
195. Zitvogel L, Kepp O, Kroemer G (2011) Immune parameters affecting the efficacy of chemotherapeutic regimens. Nat Rev Clin Oncol 8:151–160
196. Ghiringhelli F, Apetoh L, Tesniere A, Aymeric L, Ma Y, Ortiz C, Vermaelen K, Panaretakis T, Mignot G, Ullrich E et al (2009) Activation of the NLRP3 inflammasome in dendritic cells induces IL-1beta-dependent adaptive immunity against tumors. Nat Med 15:1170–1178
197. Ma Y, Aymeric L, Locher C, Mattarollo SR, Delahaye NF, Pereira P, Boucontet L, Apetoh L, Ghiringhelli F, Casares N et al (2011) Contribution of IL-17-producing gamma delta T cells to the efficacy of anticancer chemotherapy. J Exp Med 208:491–503
198. Sutton CE, Lalor SJ, Sweeney CM, Brereton CF, Lavelle EC, Mills KH (2009) Interleukin-1 and IL-23 induce innate IL-17 production from gammadelta T cells, amplifying Th17 responses and autoimmunity. Immunity 31:331–341
199. Michaud M, Martins I, Sukkurwala AQ, Adjemian S, Ma Y, Pellegatti P, Shen S, Kepp O, Scoazec M, Mignot G et al (2011) Autophagy-dependent anticancer immune responses induced by chemotherapeutic agents in mice. Science 334:1573–1577
200. Mattarollo SR, Loi S, Duret H, Ma Y, Zitvogel L, Smyth MJ (2011) Pivotal role of innate and adaptive immunity in anthracycline chemotherapy of established tumors. Cancer Res 71:4809–4820
201. Kunzli BM, Bernlochner MI, Rath S, Kaser S, Csizmadia E, Enjyoji K, Cowan P, d'Apice A, Dwyer K, Rosenberg R et al (2011) Impact of CD39 and purinergic signalling on the growth and metastasis of colorectal cancer. Purinergic Signal 7:231–241
202. Sluyter R, Dalitz JG, Wiley JS (2004) P2X7 receptor polymorphism impairs extracellular adenosine 5′-triphosphate-induced interleukin-18 release from human monocytes. Genes Immun 5:588–591
203. Stagg J, Smyth MJ (2010) Extracellular adenosine triphosphate and adenosine in cancer. Oncogene 29:5346–5358
204. Bendtzen K, Mandrup-Poulsen T, Nerup J, Nielsen JH, Dinarello CA, Svenson M (1986) Cytotoxicity of human pI 7 interleukin-1 for pancreatic islets of Langerhans. Science 232:1545–1547
205. Mandrup-Poulsen T, Bendtzen K, Nerup J, Dinarello CA, Svenson M, Nielsen JH (1986) Affinity-purified human interleukin I is cytotoxic to isolated islets of Langerhans. Diabetologia 29:63–67
206. Larsen CM, Faulenbach M, Vaag A, Volund A, Ehses JA, Seifert B, Mandrup-Poulsen T, Donath MY (2007) Interleukin-1-receptor antagonist in type 2 diabetes mellitus. N Engl J Med 356:1517–1526

207. van Asseldonk EJ, Stienstra R, Koenen TB, Joosten LA, Netea MG, Tack CJ (2011) Treatment with Anakinra improves disposition index but not insulin sensitivity in nondiabetic subjects with the metabolic syndrome: a randomized, double-blind, placebo-controlled study. J Clin Endocrinol Metab 96:2119–2126
208. Tack CJ, Stienstra R, Joosten LA, Netea MG (2012) Inflammation links excess fat to insulin resistance: the role of the interleukin-1 family. Immunol Rev 249:239–252
209. Schumacher HR Jr, Sundy JS, Terkeltaub R, Knapp HR, Mellis SJ, Stahl N, Yancopoulos GD, Soo Y, King-Davis S, Weinstein SP et al (2012) Rilonacept (interleukin-1 trap) in the prevention of acute gout flares during initiation of urate-lowering therapy: results of a phase II randomized, double-blind, placebo-controlled trial. Arthritis Rheum 64:876–884
210. Terkeltaub R, Sundy JS, Schumacher HR, Murphy F, Bookbinder S, Biedermann S, Wu R, Mellis S, Radin A (2009) The interleukin 1 inhibitor rilonacept in treatment of chronic gouty arthritis: results of a placebo-controlled, monosequence crossover, non-randomised, single-blind pilot study. Ann Rheum Dis 68:1613–1617
211. Dhimolea E (2010) Canakinumab. MAbs 2:3–13
212. Dinarello CA, Simon A, van der Meer JW (2012) Treating inflammation by blocking interleukin-1 in a broad spectrum of diseases. Nat Rev Drug Discov 11:633–652
213. Gul A, Tugal-Tutkun I, Dinarello CA, Reznikov L, Esen BA, Mirza A, Scannon P, Solinger A (2011) Interleukin-1beta-regulating antibody XOMA 052 (gevokizumab) in the treatment of acute exacerbations of resistant uveitis of Behcet's disease: an open-label pilot study. Ann Rheum Dis 71:563–566
214. Cardiel MH, Tak PP, Bensen W, Burch FX, Forejtova S, Badurski JE, Kakkar T, Bevirt T, Ni L, McCroskery E et al (2010) A phase 2 randomized, double-blind study of AMG 108, a fully human monoclonal antibody to IL-1R, in patients with rheumatoid arthritis. Arthritis Res Ther 12:R192
215. Cohen SB, Proudman S, Kivitz AJ, Burch FX, Donohue JP, Burstein D, Sun YN, Banfield C, Vincent MS, Ni L et al (2011) A randomized, double-blind study of AMG 108 (a fully human monoclonal antibody to IL-1R1) in patients with osteoarthritis of the knee. Arthritis Res Ther 13:R125
216. Weiner LM, Surana R, Murray J (2010) Vaccine prevention of cancer: can endogenous antigens be targeted? Cancer Prev Res (Phila) 3:410–415
217. Leemans JC, Cassel SL, Sutterwala FS (2011) Sensing damage by the NLRP3 inflammasome. Immunol Rev 243:152–162
218. Keystone EC, Wang MM, Layton M, Hollis S, McInnes IB (2012) Clinical evaluation of the efficacy of the P2X7 purinergic receptor antagonist AZD9056 on the signs and symptoms of rheumatoid arthritis in patients with active disease despite treatment with methotrexate or sulphasalazine. Ann Rheum Dis 71:1630–1635
219. Stock TC, Bloom BJ, Wei N, Ishaq S, Park W, Wang X, Gupta P, Mebus CA (2012) Efficacy and safety of CE-224,535, an antagonist of P2X7 receptor, in treatment of patients with rheumatoid arthritis inadequately controlled by methotrexate. J Rheumatol 39:720–727
220. Ali Z, Laurijssens B, Ostenfeld T, McHugh S, Stylianou A, Scott-Stevens P, Hosking L, Dewit O, Richardson JC, Chen C (2013) Pharmacokinetic and pharmacodynamic profiling of P2X7 receptor allosteric modulator GSK1482160 in healthy human subjects. Br J Clin Pharmacol 75(1):197–207
221. Martinon F, Tschopp J (2004) Inflammatory caspases: linking an intracellular innate immune system to autoinflammatory diseases. Cell 117:561–574
222. Willingham SB, Bergstralh DT, O'Connor W, Morrison AC, Taxman DJ, Duncan JA, Barnoy S, Venkatesan MM, Flavell RA, Deshmukh M et al (2007) Microbial pathogen-induced necrotic cell death mediated by the inflammasome components CIAS1/cryopyrin/NLRP3 and ASC. Cell Host Microbe 2:147–159
223. Kanneganti TD, Body-Malapel M, Amer A, Park JH, Whitfield J, Franchi L, Taraporewala ZF, Miller D, Patton JT, Inohara N et al (2006) Critical role for cryopyrin/Nalp3 in activation of caspase-1 in response to viral infection and double-stranded RNA. J Biol Chem 281:36560–36568

224. Ichinohe T, Lee HK, Ogura Y, Flavell R, Iwasaki A (2009) Inflammasome recognition of influenza virus is essential for adaptive immune responses. J Exp Med 206:79–87
225. Muruve DA, Petrilli V, Zaiss AK, White LR, Clark SA, Ross PJ, Parks RJ, Tschopp J (2008) The inflammasome recognizes cytosolic microbial and host DNA and triggers an innate immune response. Nature 452:103–107
226. Gurcel L, Abrami L, Girardin S, Tschopp J, van der Goot FG (2006) Caspase-1 activation of lipid metabolic pathways in response to bacterial pore-forming toxins promotes cell survival. Cell 126:1135–1145
227. Li H, Ambade A, Re F (2009) Cutting edge: necrosis activates the NLRP3 inflammasome. J Immunol 183:1528–1532
228. Petrilli V, Papin S, Dostert C, Mayor A, Martinon F, Tschopp J (2007) Activation of the NALP3 inflammasome is triggered by low intracellular potassium concentration. Cell Death Differ 14:1583–1589

Chapter 3
Uncoupling Obesity from Cancer: Bromodomain Co-regulators That Control Inflammatory Networks

Gerald V. Denis and Deborah J. Bowen

Abstract As the epidemic of overweight and obesity spreads, the number of individuals at risk for metabolic complications of obesity, including cardiovascular disease, type 2 diabetes, and cancer, is expected to increase. Importantly, the risks of complications are not evenly distributed, because not all obesity is biochemically identical. Here we describe "metabolically healthy obese" humans and animal models that show remarkable protection from insulin resistance and glucose intolerance, despite severe obesity. A hallmark of these patients and animals is their reduced inflammatory profile, which we hypothesize confers protection not only from cardiometabolic risk in obesity but also from obesity-associated cancers. Research is urgently required to investigate the basis for this protection, to identify treatment options and prevention strategies for at-risk populations. We explore novel insights into chromatin-based, transcriptional co-regulator mechanisms that link apparently unrelated diseases, with the idea that certain molecularly targeted strategies could moderate multiple risks in obesity. We voice concern that low socioeconomic status citizens are particularly at risk for cardiometabolic disease and obesity-associated cancer, in part because many such individuals live in inflammatory and obesogenic environments. An integrated and hypothesis-driven approach is needed to study and protect these vulnerable and underserved populations from the rising tide of obesity-associated cancer.

G.V. Denis, Ph.D. (✉)
Section of Hematology Oncology, Department of Pharmacology and Experimental Therapeutics and Cancer Research Center, Boston University School of Medicine, 72 East Concord Street, K520, Boston, MA 02118, USA
e-mail: gdenis@bu.edu

D.J. Bowen, Ph.D.
Department of Community Health Sciences, Boston University School of Public Health, 801 Massachusetts Avenue 4th floor, Boston, MA 02118, USA
e-mail: dbowen@bu.edu

A.J. Dannenberg and N.A. Berger (eds.), *Obesity, Inflammation and Cancer*, Energy Balance and Cancer 7, DOI 10.1007/978-1-4614-6819-6_3,

3.1 The Problem of Obesity-Associated Cancer

Diet-induced overnutrition that causes unhealthy weight gain, defined in humans as *overweight* (Body Mass Index [BMI] 25.0–29.9 kg/m^2), *obesity* (BMI ≥30.0–39.9 kg/m^2), and *morbid obesity* (BMI ≥40.0 kg/m^2), has many medical complications. The realization that obesity had become a serious public health concern was initially driven by projected increases in the prevalence of metabolic complications, such as elevated risk for stroke, cardiovascular disease, and type 2 diabetes. The complications of obesity also include dyslipidemia, hypertension, sleep apnea, hepatosteatosis, and glucose intolerance [1]. However, attention has been recently focused on a particularly worrisome complication: obesity-associated malignancy [2, 3]. Recent epidemiological reports have caused serious disquiet that, despite overall declines in cancer rates, particularly the rates for tobacco-associated cancers, the rates of obesity-associated cancers are climbing. Obesity is now thought to be one of the most important preventable causes of several cancers [4]; these include esophageal adenocarcinoma, colorectal cancer, breast cancer in postmenopausal women (but not premenopausal women), and cancers of the endometrium, kidney, pancreas, liver, and gallbladder [3, 5, 6]. The National Cancer Institute, American Cancer Society, and American Association for Cancer Research have been using their influence and expert opinion in recent years to increase the public profile and research portfolio devoted to this problem and its allied risk factors. Obesity-associated malignancies have been estimated to account for 14 % of male and 20 % of female US cancer mortality, notably colorectal cancer and postmenopausal breast cancer [2]. This chapter will present some of the molecular, cellular, and immunological features that link obesity and its complications to cancer.

In view of recent data from the US Centers for Disease Control and Prevention [7], showing that all US states now report at least 20 % prevalence of obesity among adults, as well as the classification of 1.7 billion people worldwide as overweight [8], obesity-associated cancer is positioned to become one of the defining preventable diseases of our time. Diabetes is also a serious complication among the chronic disease burdens of obesity. In fact, overweight and obesity are now well established to be the direct cause of most cases of type 2 diabetes [9]. Three hundred and sixty-six million cases worldwide was a frequently cited early estimate of the incidence of type 2 diabetes by 2030 [10]. Alarmingly, more recent estimates have adjusted upward the anticipated number of diabetic individuals to 439 million by 2030 [11]. Almost all of the incidence will be driven by overweight and obesity. The anticipated further increases in BMI worldwide [12] predict that the seriousness of the problem of obesity-associated cancer will also deepen in coming decades.

3.2 Molecular Features of Insulin-Resistant Obesity

The molecular mechanisms that explain how obesity contributes to cancer risk are still largely unknown in detail. Early epidemiological investigation of obesity and its comorbidities identified an association between the incidence of type 2 diabetes and obesity [13]. Some of the key features of insulin-resistant obesity include elevated concentrations of blood glucose in fasted subjects and impaired glucose tolerance, as well as elevated blood concentrations of insulin in fasted subjects and reduced insulin sensitivity. Leptin, which is a critical regulator of appetite, is produced by adipocytes and is elevated in obesity in proportion to adipose tissue mass [14]. This hyperleptinemia in obesity has been frequently described as "leptin resistance," a term now thought to lack clinical utility [15]. In addition, insulin-resistant obesity frequently features reduced serum concentrations of adiponectin [16, 17], an insulin-sensitizing adipokine [18] that exhibits beneficial, antiatherogenic effects. These characteristics are commonly observed together in obese subjects [19–21] and animal models [22] and reflect the growing inability of peripheral tissues of the obese subject, such as skeletal muscle and fat stores, to transport glucose from blood into cells at normal levels of insulin present in the circulation. This state has been described as peripheral insulin resistance. Commonly, the pancreatic β-cells of such a subject are required to secrete ever higher levels of insulin to compensate for the peripheral insulin resistance. In humans and certain rodent models, this chronic hyperinsulinemia and accompanying β-cell dysfunction are two of the defining characteristics of insulin-resistant obesity and are often associated with increased serum levels and bioavailability of the related mitogenic factor, insulin-like growth factor (IGF)-1 [23].

These clinical presentations provoked questions about which biochemical features of insulin resistance and type 2 diabetes were important for carcinogenesis in obesity. Signal transduction through the insulin receptor and IGF-1 receptor [24] is thought to increase cancer risk in obesity [25, 26]. Fasting insulin concentrations have been used convincingly as a prognostic factor for overall survival among breast cancer patients, with the highest hazard ratio associated with the highest insulin concentrations [27]. Leptin also promotes mitogenic [28–31] and invasive [32] effects in a variety of human cancer cell lines [33–35] and tumor models in animals [36]. For example, leptin-deficient mice appear to be protected from mammary carcinogenesis [37–39]. Adiponectin not only protects insulin-sensitive glucose transport but also appears to be inversely correlated with susceptibility to certain obesity-associated cancers [40–42]. Furthermore, the leptin-adiponectin ratio has been proposed to be a critical predictor of cancer risk [38]. These features have been well summarized elsewhere [43–45]. However, the co-occurrence of these multiple factors in obesity has made it difficult to define the relative importance of each. Overall, rodent models have tended to show that alteration of any single factor in isolation affects mitogenesis, tumor progression, or other relevant parameter of the malignancy. Experimental designs that use rodent models in which multiple variables are manipulated simultaneously to influence cancer risk do not permit straightforward interpretation; thus, the field remains divided about which cellular and molecular factors are of paramount importance for specific obesity-associated cancers.

3.3 Insulin-Resistant Obesity Is Also an Inflammatory Disease

One of the earliest described immunological features of insulin-resistant obesity was subclinical, unresolved, chronic inflammation, which occurs both systemically [47, 67] and in white adipose tissue [48], which is infiltrated with pro-inflammatory macrophages [49–52]. Specifically, such patients demonstrate elevated serum concentrations of acute phase proteins and pro-inflammatory cytokines [53], such as interleukin (IL)-1β, IL-6, and C-reactive protein, that improve over time after intentional weight loss [54, 55, 67] or bariatric surgery [56]. Exposure of glucose-transporting cells to the pro-inflammatory cytokine tumor necrosis factor (TNF)-α was demonstrated as long ago as 1993 to promote insulin resistance directly [57]. Adipose tissue depots, composed of white adipocytes, are typically inflamed, that is, infiltrated with Th1- and Th17-polarized T cells [58] and pro-inflammatory macrophages, both in obese humans [59–63] and animal models [50, 51, 59, 64–66] of diet-induced obesity. In insulin-resistant obesity, the pro-inflammatory macrophages that infiltrate these depots secrete significant amounts of pro-inflammatory cytokines, which, in addition to TNF-α [57], include IL-6, IL-8, and monocyte chemoattractant protein (MCP)-1/chemokine (C–C motif) ligand (CCL)2 [46, 67]. Systemic inflammation is also a feature of insulin-resistant obesity, as indicated by elevated serum levels of C-reactive protein [68] and several of the aforementioned and other cytokines [61]. Furthermore, chemokines such as MCP-1/CCL2 also serve to recruit additional leukocytes, such as peripheral blood monocytes that express the C–C motif chemokine receptor (CCR) 2 [50, 51, 60], to infiltrate the insulin-resistant adipose depot in a deepening cycle of unresolved, chronic inflammation. Thus, a feed-forward loop is established that is difficult for homeostatic forces in the immune system to oppose [58]. Failure of the anti-inflammatory balance may also be an independent, critical factor in the emergence of the many comorbidities of obesity.

Moreover, certain specific, histological features define insulin-resistant adipose tissue. The adipocytes frequently become stressed as their storage limits are exceeded, leading to a large number of apoptotic adipocytes, a process that is thought to recruit additional leukocytes [59]. The dead and dying adipocytes of stressed white adipose tissue appear surrounded with a ring of pro-inflammatory macrophages ($CD68^+$ in humans [69]) that are histologically termed "crown-like structures" [59] and are associated with fibrosis [70] and increased metabolic risk [71–74]. How these structures arise and are resolved by weight loss or drug treatments is not well understood. In mouse models, the macrophages that infiltrate metabolically unhealthy white adipose tissue tend to express a surface phenotype, that is, $CD11b^+$ $CD11c^+$ $F4/80^+$ by flow cytometry, that identifies them as pro-inflammatory. These pro-inflammatory macrophages have been directly implicated in the decline of metabolic health of adipocytes in white adipose tissue in different adipose depots in animal models [75, 76] and humans [101]. Early in the kinetics of diet-induced obesity in rodent models, adipocyte death and the development of whole-body insulin resistance [65] also correlate with a switch in macrophage polarization toward the classically activated, pro-inflammatory (so-called M1) phenotype and away from the alternatively activated, anti-inflammatory ("M2") phenotype [78]. The $CD11c^+$ adipose tissue macrophage populations also transiently remodel the white adipose tissue, which then exhibits activities connected with M2-associated genes, such as increased expression of arginase, IL-10, IL-4, and transforming growth factor (TGF)-β [79]. The net balance of these M1 and M2 inputs defines the profile and magnitude of white adipose tissue inflammation. Relatively high expression of M1 cytokines is associated with metabolic complications of obesity, including insulin resistance. However, in human adipose depots, the molecular details of putative M1 phenotypes and function, and the M1/M2 switch, are less well understood than in animal models.

T cells are also recruited to white adipose tissue in diet-induced obesity through "regulated on normal T cell expressed and secreted" (RANTES/CCL5) and its receptor CCR5 in white adipose tissue [80, 81], where the Th1/Th2 polarization and proliferation of T cells are influenced by macrophage-produced cytokines [82]. T cells play a major role in insulin resistance [62] through macrophage recruitment [83]. T cell polarization between the pro-inflammatory (interleukin-17 producing) subtype (Th17 [84]) and the anti-inflammatory (IL-10 producing) T regulatory subtype ($Foxp3^+$ Treg [85]) also influences metabolism in white adipose tissue. A pro-inflammatory imbalance in $CD4^+$ T cell subsets has been demonstrated both systemically and in adipose depots of type 2 diabetic subjects [58]. The balance of pro-inflammatory and anti-inflammatory cytokines and T cell subsets remains perturbed in insulin-resistant obesity; some investigators hypothesize the adipocyte/T cell cross talk is the critical factor that promotes disease pathogenesis, whereas others hypothesize that macrophages have primary importance. The interpretations have remained controversial. Recent data from human studies also supports a central role for B cells in the pathogenesis of type 2 diabetes in obese subjects [58, 86, 87]. The independent and interdependent roles of T cell subsets, B cell subsets, and monocyte/macrophage polarization, and their specific cross talk with adipocytes that influences risks for obesity-associated cancer and type 2 diabetes, define a central focus of the exciting new field of immunometabolism [58].

Outside the adipose depots, the immune system of the obese and insulin-resistant subject demonstrates systemic, pro-inflammatory changes in T cell, B cell, and myeloid subset differentiation and function that exacerbate the deepening cytokine/chemokine imbalance as metabolism deteriorates in diet-induced obesity. Animal models demonstrate that stoppage of immune cell-mediated inflammatory cascades by any one of several diverse techniques (e.g., genetic, small molecule, or antibody-based) frequently delays or prevents insulin resistance [88, 89]. If metabolic parameters improve through dietary intervention, adipose tissue inflammation also typically improves [90]. The long-established links between chronic, unresolved inflammation and cancer therefore provide a basis to hypothesize that the presence of crown-like structures, for example, or elevation of other local and systemic inflammatory markers, is positively associated with cancer risk for obesity-associated cancers that have an inflammatory component.

3.4 Insight from Unexpected Results

Obesity-associated malignancies are not linked to every type of cancer. Apart from lung cancer, which is associated with cigarette smoking [91, 92] or asbestos inhalation and not with obesity [3], certain other cancers are also clearly not associated with obesity, including but not limited to astrocytoma; glioma; Kaposi's sarcoma; neuroblastoma; head, neck, and oral cancer; pituitary cancer; retinoblastoma; salivary cancer; and testicular cancer. A possibly relevant, shared feature of these cancers is that they do not originate in or near visceral adipose tissue. Compared to subcutaneous depots, the visceral or "central" adipose depot [63, 93, 94] is the most inflamed in obese insulin-resistant patients [95–97] and is independently associated with cardiometabolic risk. In animal models, the epididymal adipose depot of male mice is regarded as a good model for visceral adipose tissue inflammation in diet-induced obesity [59, 65, 66]. Likewise, many (but not all) of the obesity-associated cancers are resident in or likely influenced by inflamed visceral adipose tissue. All female breast carcinomas, for example, are surrounded by significant adipose depots in humans and the mammary fat pad in mice. It is likely that the metabolic and inflammatory properties of this adipose depot are highly relevant to specific aspects of breast cancer progression, invasiveness, metastasis, or recurrence, although this area has not received sufficient attention from investigators.

The observation that insulin-resistant obesity features *systemic* elevations of pro-inflammatory cytokines, as well as systemically elevated glucose, insulin, IGF-1, leptin, and depleted adiponectin, raises a problem. Why are not all cancers obesity-associated? If the argument is made that insulin and IGF-1 signaling cross talk, as well as leptin-promoted, broad-spectrum mitogenesis or diminished protection from adiponectin, are critical factors that explain obesity-associated cancers, why should so many cancers be unrelated to obesity? Presumably these systemic factors affect diverse tissues roughly equally, although different cells of origin of the tumor likely respond differently to the complex endocrine and metabolic microenvironment in each tissue. A recent repeated measures study from the Women's Health Initiative suggests that, at least in the case of colorectal cancer risk in postmenopausal women, the most important association is with elevated glucose, not elevated insulin [98]. It seems likely that additional features of the obese subject influence carcinogenesis and perhaps stratify risk for obesity-associated cancer.

Although insulin-resistant obesity is a chronic inflammatory disease, 20–30 % of adult obese individuals preserve a *reduced* inflammatory profile. The white adipose tissue of these un-inflamed, adult subjects shows lower numbers of infiltrating leukocytes [72], while systemically, serum concentrations of pro-inflammatory cytokine are lower [99, 100] than in insulin-resistant obese adult subjects. These un-inflamed subjects exhibit normal or near-normal glucose tolerance [72] (Fig. 3.1), reduced cardiovascular disease risk [77], and lack metabolic syndrome [72, 102, 169]. They remain relatively "metabolically healthy" with low-inflammatory profiles despite obesity [103, 104, 169] and represent an important off-diagonal population for which cardiovascular risk appears to be uncoupled from obesity [105, 106].

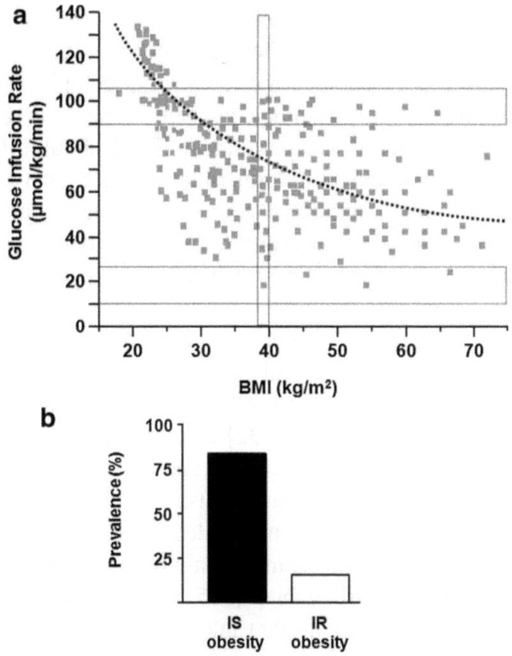

Fig. 3.1 "Metabolically healthy obesity." (**a**) Glucose infusion rates (GIR) in 237 subjects for a broad range of BMI and metabolic health. Insulin sensitivity was determined by GIR during the steady state of a euglycemic-hyperinsulinemic clamp. The highest and lowest quintiles of GIR are marked (*horizontal boxes*) to show that the frequency of insulin resistance is very low in healthy obesity. A regression curve (*dotted line*) of GIR over BMI is based on all available patients. The BMI stratum 39–40 identifies a continuous distribution (*vertical box*) of rates to show that there is no clear separation between insulin-sensitive and insulin-resistant obesity. Note that certain, rare individuals with unusually high BMI (>60) nevertheless display normal, healthy GIR during the clamp. (Subject exclusion criteria were diabetes, hypertension, acute or chronic inflammatory disease with leukocyte count >8,000 Gpt/L, CRP >5.0 mg/dL or clinical signs of infection, and other relevant criteria as detailed in ref. [77].) (**b**) Prevalence of insulin-sensitive (GIR >80 µmol glucose/kg/min) and insulin-resistant (GIR <40 µmol glucose/kg/min) healthy obesity (data from Blüher M (2010). The distinction of metabolically 'healthy' from 'unhealthy' obese individuals. *Curr Opin Lipidol*, 21(1):38–43 are reproduced with permission from M. Blüher and Wolters Kluwer, publishers)

The well-established association between unresolved, chronic inflammation and cancer [107, 108] (e.g., between Crohn's disease and colorectal cancer [109]) suggests that inflamed adipose tissue in insulin-resistant obese adults plays a critical role in obesity-associated carcinogenesis. We have previously hypothesized that the low-inflammatory features and preserved glucose tolerance of these subjects *also protects against risks for obesity-associated cancers* [110].

How do these metabolically healthy obese individuals remain un-inflamed? We have focused on a recently described transcriptional mechanism that may link chronic inflammation, obesity, and cancer [140]. Bromodomain-containing transcriptional co-regulators [112] bind to acetylated histones [113–115] in the nucleosomal chromatin of mammals and target specific genes for transcriptional activation or repression. In this way, they functionally resemble SWItch/Sucrose NonFermentable (SWI/SNF) nucleosome remodeling complexes, which also contain bromodomain subunits and function to activate or silence coordinated networks of genes [116]. Bromodomains are protein motifs of about 110 amino acids in length [117, 118], comprised of four antiparallel α-helices that are linked by connecting loops, which form a binding pocket that is specific for acetylated lysine [119]. These motifs are found in transcription factors, histone acetylases, and related chromatin-directed proteins that are important for gene regulation [117, 120, 121]. Previous work has shown that the double bromodomain-containing proteins Brd2 [122, 123] and Brd4 [124] couple histone acetylation to transcription [125, 126] and are critical for transcriptional control of cell cycle genes [127–131]. Increased or deregulated expression of either protein is oncogenic. In humans, reciprocal chromosomal translocation of *BRD4* [132] creates a dangerous oncoprotein that promotes an aggressive, poorly differentiated, and incurable carcinoma of the midline, called NUT midline carcinoma, that afflicts relatively young people [133, 134]. Remarkable recent studies with small molecule inhibitors of the binding interface between the acetylated lysines and bromodomain have revealed that the chromatin-bromodomain interaction is "druggable" [135], which came as a surprise to the field. Conventional wisdom had held that interactions with such protein motifs in chromatin were unappealing targets for the development of small molecule inhibitors. In the case of midline carcinoma [136] and other human malignancies [137–139], such drugs appear to have significant anticancer benefit [140]. These developments linked Brd2 and Brd4 with cancer and chromatin-controlled networks of gene expression that are coordinated through shared complexes, analogous to the SWI/SNF-regulated array of genes. But the chromatin-based connections between cancer, obesity, and inflammation remained obscure until unexpected results from a bromodomain-manipulated mouse model appeared.

We initially developed a mouse model for Brd2 transgenic expression [141] and showed that upon B cell-restricted expression of Brd2, mice upregulate B cell mitogenic responses through cyclin A transactivation [131] and eventually develop a B cell malignancy [141]. This cancer exhibits a transcriptional fingerprint most similar to the "activated B cell" (ABC) form of diffuse large B cell lymphoma in humans [142], with an inflammatory signature. Surprisingly, whole-body reduction of Brd2 in "Brd2 hypomorph" mice, by *lacZ* gene disruption, caused the development of a glucose-tolerant type of obesity that features elevated serum adiponectin and a remarkably attenuated inflammatory profile [143]. These results suggested that the Brd2 hypomorphic mouse might represent a useful model for human subjects who are metabolically healthy obese (Fig. 3.1) [144]. These humans share with the Brd2 hypomorphic mice a low-inflammatory profile [77, 103, 105] and less reduction in serum adiponectin concentrations despite obesity [72]. The elevated concentrations of adiponectin measured in adiponectin transgenic mice are also metabolically protective [170], although neither adiponectin expression nor any other loci apart from *Brd2* were directly manipulated in Brd2 hypomorphic mice. More significantly, the chromatin-directed networks that these bromodomain-containing co-regulators control likely connect cancer, obesity, and inflammation directly, through coordinated co-activation or co-repression of interacting networks of genes; certain diseases likely share the same or overlapping sets of gene expression co-regulators. This topic has been reviewed [140].

In forthcoming work, we show that small molecule inhibitors of these bromodomain proteins or shRNA ablation are effective as anti-inflammatory strategies, acting as global "uncouplers" of signal transduction that normally activate transcription of diverse cytokine genes [111]. Because many cancers have an inflammatory component, it is reasonable to hypothesize that the inflammatory microenvironment of certain tumors will exacerbate carcinogenesis, tumor progression, invasion, metastasis, or recurrence. Targeted uncoupling of signal transduction from transcription through bromodomain protein-specific small molecule inhibitors [140] may prove to be a novel and efficacious therapeutic or preventive approach to reduce the inflammatory cascades that contribute to obesity-associated cancer. The combined anti-inflammatory [145] and anticancer [136–138] effects of bromodomain protein inhibition are already established. Our previous work has shown that reduced expression of Brd2 protein, equivalent to a haploinsufficient phenotype, de-represses specific genes that are important for metabolism. These reduced levels are sufficient *simultaneously* to stimulate insulin gene transcription in β-cells [143, 146], to increase adipogenesis in pre-adipocytes through stimulation of peroxisome proliferator-activated receptor (PPAR)γ-directed transcriptional programs [143, 147]. The hyperadiponectinemia of obese Brd2 hypomorphic mice [143] suggests that Brd2 is also normally required to corepress transcription of the murine adiponectin gene (*Adipoq*), although this hypothesis has not yet been tested. Conversely, Brd2 reduction also ablates the production of pro-inflammatory cytokines in macrophages such as TNFα, IL-1β, IL-6, and MCP-1 [111, 140, 144]. Taken together with the anticancer properties of Brd2 reduction through attenuated cell cycle progression, as discussed above, these coherent transcriptional and metabolic features (stimulated insulin production, increased adipogenesis, and increased adiponectin production; and reduced production of multiple pro-inflammatory cytokines) lead us to propose a Brd2 mechanism for broad protection against obesity-associated malignancy. Small molecule inhibitors that target this family of transcription co-regulators, or naturally occurring single nucleotide polymorphisms in the human *BRD2* locus that reduce Brd2 expression, may therefore confer multiple forms of metabolic and cancer protection to obese patients.

Several phenotypes of unintended weight loss, such as chronic heart failure [148], share a systemic inflammatory profile [149], with marked elevations in serum levels of IL-1β, IL-6, and TNFα. These observations suggest a mechanistic relationship between the immune system, metabolism, and energy balance, reinforcing the aforementioned argument. Furthermore, it has been noted that "unhealthy aging" [150] is often associated with a pro-inflammatory, pro-senescent phenotype in somatic cells, the local production in skeletal muscle and adipose depots of inflammatory cytokines that are associated with muscle wasting syndromes, and frailty in geriatric patients [151–153]. Investigators have been considering the cause and effect relationships among unresolved, chronic inflammation, energy imbalances associated with weight loss or weight gain, and cancer risk. Many of these relationships may work in both directions. It is reasonable therefore to hypothesize that a chromatin-based therapeutic strategy to treat these connected phenotypes may have broad benefit for more than one type of risk and may be useful for geriatric patients.

3.5 Other Links Between Obesity, Inflammation, and Social Determinants

One social determinant that plays a role in inflammatory disease processes is socioeconomic status (SES). Asthma rates in children are two to three times higher in poor families than in wealthy families; SES shows a dose response relationship with asthma diagnosis and severity [154]. Public housing residents and inner city dwellers, who are among the poorest of urban dwellers in the United States, report higher rates of asthma than do private home and apartment dwellers [155]. This disease arises from allergic reactions to irritants and allergens that are commonly found in public housing, including dust mites, pets, rodents, mold, and cockroaches [156]. Massachusetts public housing has been linked to some of the highest national rates of asthma [157]. The prevalence of asthma is highest among African American families, with overall prevalence of 40 % of adults and 56 % of children [158]. These same populations, that is, poor and low SES individual and public housing residents, report two to three times the obesity rates of other residents who are higher along the SES continuum [159]. According to the American Lung Association, there is no evidence that asthma can cause lung cancer. However, there is evidence that asthma is associated with obesity [155, 160–162]. The risk of asthma has been reported in one study [163] to be up to three times greater for obese subjects compared to lean subjects (odds ratio for obese vs. normal BMI=2.28, 95 % CI: 1.76, 2.96). These observations suggest that socioeconomic factors also influence risk for obesity-associated morbidity, including type 2 diabetes and cancer. Specifically, there may be a rationale to investigate the relationship between asthma, obesity, and obesity-associated cancer. For example, does poverty produce obesity and inflammation, enhancing opportunities for the development of asthma? Are these issues causally related or simply comorbidities of living in high poverty settings? If we are able to reduce one set of comorbid conditions, as is under investigation now in Boston and elsewhere [164–166], will that outcome reduce or alter others?

Several sociological, economic, and behavioral factors have been established to link obesity and type 2 diabetes incidence to cancer incidence. As discussed above, there is strong epidemiological evidence that SES is correlated with both the prevalence of obesity and diabetes and with lung cancer mortality, a malignancy that is associated with smoking (Fig. 3.2a). There is no known molecular association between lung cancer and obesity [3]. However, the use of tobacco in cigarette, cigar, and pipe smoke is strongly associated with lung, tracheal, and oral *cancers* [167]. Furthermore, low SES individuals suffer disproportionately higher health risks due to increased prevalence of smoking [168]. These correlations among chronic diseases that have no downstream molecular connection suggest that the problem of obesity-associated cancer is more complicated in its structural and upstream origins

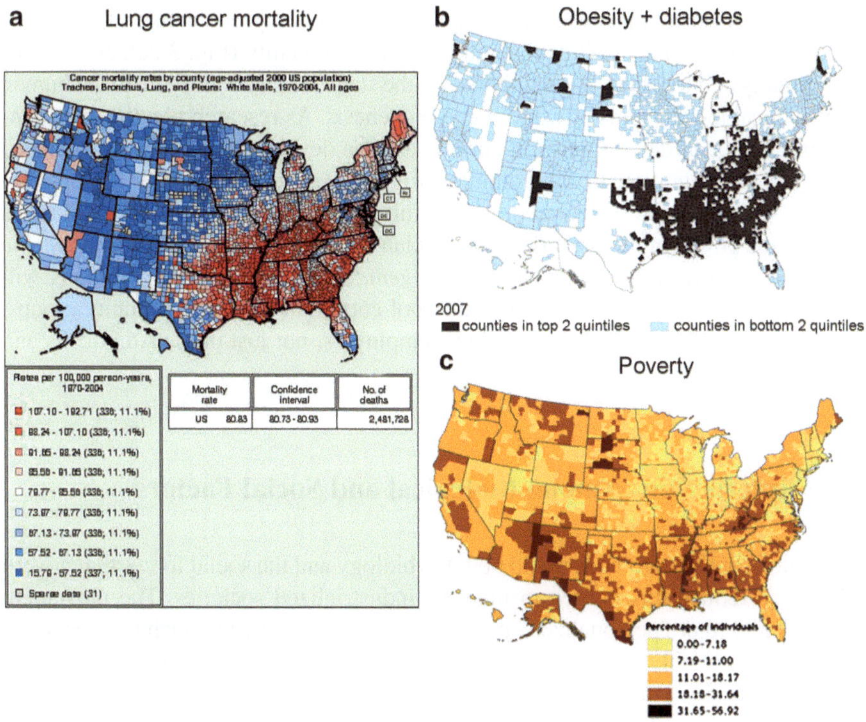

Fig. 3.2 Overlapping morbidities. (**a**) Lung cancer mortality. Mortality among 3,056 United States counties for cancer of the lung, trachea, bronchus, and pleura in white males of all ages, 1970–2004 (age-adjusted 2000 US population). Calculated from *National Cancer Institute data drawn from Atlas of Cancer Mortality in the U.S., 1950–1994*; rates per 100,000 person-years presented here in nine equal intervals with a diverging *red/blue* color scheme. The national rate was 80.83 (CI 80.73–80.93) per 100,000, with the total number of deaths 2,481,728. http://ratecalc.cancer.gov/ratecalc/. (**b**) Diabetes and obesity diagnoses. Estimates among 3,141 United States counties for age-adjusted rates of both diagnosed diabetes and obesity presented together. Estimates were calculated from Census and Behavioral Risk Factor Surveillance System (self-reported) data for 2006–2008. The national proportion of US adults who were obese in 2008 was 26.1 %. In 2007, 8 % of the US population, or 24 million individuals, were diabetic, of which 5.7 million were estimated to be undiagnosed. http://www.cdc.gov/diabetes/pubs/factsheets/countylvlestimates.htm. (**c**) Socioenvironmental map of poverty. County-level data from United States Census Bureau statistics for 2004. Estimated percentage of population living below the poverty threshold as defined by US Census methods is defined by size of family and ages of members and includes information about earnings, unemployment compensation, workers' compensation, Social Security, Supplemental Security Income, public assistance, veterans' payments, survivor benefits, pension or retirement income, interest, dividends, rents, royalties, income from estates, trusts, educational assistance, alimony, child support, assistance from outside the household, and other miscellaneous sources (http://www.cdc.gov/dhdsp/maps/sd_poverty_2004.htm, http://www.census.gov/hhes/www/poverty/about/overview/measure.html)

than a one-to-one correspondence between an obesity exposure and a cancer rate. In other words, the same causal factors that produce increased levels of obesity might also be at work to promote increased rates of lung cancer and lung cancer mortality, as well as type 2 diabetes (Fig. 3.2b). Obesity and cancer are likely to be linked through their upstream causes: downturns in the economy, the nature of work and labor markets, social stratification and economic inequality (Fig. 3.2c), and lack of opportunity and local infrastructure that set the stage for and contribute to the human biochemical mechanisms at work in carcinogenesis. As research identifies the central role of social determinants in chronic disease development, investigators need to pay closer attention to the common origins, even if not directly biologically linked. The social and structural origins of this problem demand structural solutions beyond the power of the prescribing physician's pen: it is clear that *certain specific environments are both obesogenic/diabetogenic and carcinogenic*. Solutions will require focused political will, participation of corporations and community groups, entrepreneurs, school districts, and local employers, not just obese Americans and their physicians.

3.6 Interactions Among Biological and Social Factors

Ultimately, we need to understand both the biology and the social forces that govern obesity to reduce this burden in modern, industrialized societies. The movement toward tailored or "personalized" medicine may be one way in which both perspectives can be not only accommodated but relied upon as translated intervention tools to reduce obesity. For example, identification of an individual's likelihood of being a metabolically unhealthy obese person may provide additional motivation to engage in healthy behaviors. Alteration of the shape of environments for individuals and groups, such that there are clear and accessible food and activity choices, will help families facing both obesity- and asthma-related health issues. Increasing the opportunity for non-obesogenic activities might be a necessary investment for individuals who become inflamed if they become obese or maintain obesity. It may be helpful for the current conditions to consider obesity as a health problem that, for some, causes clear measurable changes related to a variety of chronic diseases, but that is ultimately preventable. Translating the basic research on vulnerability to inflammation with obesity into usable interventions will require new ways of thinking about environment, motivation, and human behavior. From the discussion in this chapter, it is clear that we have begun to consider the broad, powerful mechanistic connections among inflammation, obesity, and cancer and the need to link the cellular, serological, and dietary environments of obese, at-risk individuals to exposures in the built environment, urban infrastructure, and economic policy. Without transdisciplinary, innovative, "out-of-the-box" thinking, the problem of obesity-associated cancer will prove too difficult to address effectively. We therefore call for additional funding and research to investigate these unexpected connections among important variables, with focused conversation among molecular biologists, immunologists, geneticists, cancer and endocrine clinicians, epidemiologists, sociologists, public housing officials, and public health officials. In view of the increasing seriousness of the obesity epidemic, time is running out for this conversation to plan for research and policy priorities.

Acknowledgements The authors thank the National Institutes of Health (DK090455, GVD and DK0704192, DJB), the US Centers for Disease Control and Prevention (U48 DP001922, DJB), the American Cancer Society (RSG-05-072-01, GVD), the Leukemia and Lymphoma Society (6023-09, GVD), the Boston University Clinical and Translational Science Institute (UL1-TR000157, GVD), and the Evans Center for Biomedical Research. The authors are members of an Evans Center-sponsored, multidisciplinary research collaborative, entitled "Obesity, Inflammation and Cancer" based at Boston University. GVD is a former Chair of the Basic Science Section of The Obesity Society and gratefully acknowledges the intellectual and financial support of the Society and its demonstrated and ongoing commitment to address the problem of obesity-associated cancer.

The authors report no conflicts of interest.

References

1. Harris MI, Flegal KM, Cowie CC, Eberhardt MS, Goldstein DE, Little RR, Wiedmeyer HM, Byrd-Holt DD (1998) Prevalence of diabetes, impaired fasting glucose, and impaired glucose tolerance in U.S. adults. The Third National Health and Nutrition Examination Survey, 1988–1994. Diabetes Care 21(4):518–524
2. Calle EE, Rodriguez C, Walker-Thurmond K, Thun MJ (2003) Overweight, obesity, and mortality from cancer in a prospectively studied cohort of U.S. adults. N Engl J Med 348(17):1625–1638
3. Calle EE, Kaaks R (2004) Overweight, obesity and cancer: epidemiological evidence and proposed mechanisms. Nat Rev Cancer 4(8):579–591
4. Wiseman M (2008) The second World Cancer Research Fund/American Institute for Cancer Research expert report. Food, nutrition, physical activity, and the prevention of cancer: a global perspective. Proc Nutr Soc 67(3):253–256
5. Calle EE, Thun MJ (2004) Obesity and cancer. Oncogene 23(38):6365–6378
6. Renehan AG, Tyson M, Egger M, Heller RF, Zwahlen M (2008) Body-mass index and incidence of cancer: a systematic review and meta-analysis of prospective observational studies. Lancet 371(9612):569–578
7. Centers for Disease Control and Prevention (CDC) (2010) Vital signs: state-specific obesity prevalence among adults—United States, 2009. MMWR Morb Mortal Wkly Rep 59(30):951–955. Most recent update: http://www.cdc.gov/obesity/downloads/DNPAO_State_Obesity_Prevalence_Map_2011_508.pdf
8. Haslam DW, James WP (2005) Obesity. Lancet 366(9492):1197–1209
9. Hossain P, Kawar B, El Nahas M (2007) Obesity and diabetes in the developing world—a growing challenge. N Engl J Med 356(9):213–215
10. Wild S, Roglic G, Green A, Sicree R, King H (2004) Global prevalence of diabetes: estimates for the year 2000 and projections for 2030. Diabetes Care 27(5):1047–1053
11. Shaw JE, Sicree RA, Zimmet PZ (2010) Global estimates of the prevalence of diabetes for 2010 and 2030. Diabetes Res Clin Pract 87(1):4–14
12. Finucane MM, Stevens GA, Cowan MJ, Danaei G, Lin JK, Paciorek CJ, Singh GM, Gutierrez HR, Lu Y, Bahalim AN, Farzadfar F, Riley LM, Ezzati M, Global Burden of Metabolic Risk Factors of Chronic Diseases Collaborating Group (Body Mass Index) (2011) National, regional, and global trends in body-mass index since 1980: systematic analysis of health examination surveys and epidemiological studies with 960 country-years and 9.1 million participants. Lancet 377(9765):557–567
13. Mokdad AH, Ford ES, Bowman BA, Dietz WH, Vinicor F, Bales VS, Marks JS (2003) Prevalence of obesity, diabetes, and obesity-related health risk factors, 2001. JAMA 289(1):76–79
14. Frederich RC, Hamann A, Anderson S, Löllmann B, Lowell BB, Flier JS (1995) Leptin levels reflect body lipid content in mice: evidence for diet-induced resistance to leptin action. Nat Med 1(12):1311–1314

15. Myers MG Jr, Heymsfield SB, Haft C, Kahn BB, Laughlin M, Leibel RL, Tschöp MH, Yanovski JA (2012) Challenges and opportunities of defining clinical leptin resistance. Cell Metab 15(2):150–156
16. Kern PA, Di Gregorio GB, Lu T, Rassouli N, Ranganathan G (2003) Adiponectin expression from human adipose tissue: relation to obesity, insulin resistance, and tumor necrosis factor-α expression. Diabetes 52(7):1779–1785
17. Kubota N, Terauchi Y, Yamauchi T, Kubota T, Moroi M, Matsui J et al (2002) Disruption of adiponectin causes insulin resistance and neointimal formation. J Biol Chem 277(29):25863–25866
18. Weyer C, Funahashi T, Tanaka S, Hotta K, Matsuzawa Y, Pratley RE, Tataranni PA (2001) Hypoadiponectinemia in obesity and type 2 diabetes: close association with insulin resistance and hyperinsulinemia. J Clin Endocrinol Metab 86(5):1930–1935
19. Matsubara M, Maruoka S, Katayose S (2002) Inverse relationship between plasma adiponectin and leptin concentrations in normal-weight and obese women. Eur J Endocrinol 147(2):173–180
20. Mojiminiyi OA, Abdella NA, Al Arouj M, Ben Nakhi A (2007) Adiponectin, insulin resistance and clinical expression of the metabolic syndrome in patients with type 2 diabetes. Int J Obes (Lond) 31(2):213–220
21. Snijder MB, Heine RJ, Seidell JC, Bouter LM, Stehouwer CD, Nijpels G, Funahashi T, Matsuzawa Y, Shimomura I, Dekker JM (2006) Associations of adiponectin levels with incident impaired glucose metabolism and type 2 diabetes in older men and women: the Hoorn Study. Diabetes Care 29(11):2498–2503
22. Agil A, Rosado I, Ruiz R, Figueroa A, Zen N, Fernández-Vázquez G (2012) Melatonin improves glucose homeostasis in young Zucker diabetic fatty rats. J Pineal Res 52(2):203–210
23. Renehan AG, Zwahlen M, Minder C, O'Dwyer ST, Shalet SM, Egger M (2004) Insulin-like growth factor (IGF)-1, IGF binding protein-3 and cancer risk: systematic review and meta-regression analysis. Lancet 363(9418):1346–1353
24. Vigneri P, Frasca F, Sciacca L, Pandini G, Vigneri R (2009) Diabetes and cancer. Endocr Relat Cancer 16(4):1103–1123
25. Cohen DH, LeRoith D (2012) Obesity, type 2 diabetes and cancer: the insulin and insulin-like growth factor connection. Endocr Relat Cancer 19(5):F27–F45
26. Lautenbach A, Budde A, Wrann CD, Teichmann B, Vieten G, Karl T, Nave H (2009) Obesity and the associated mediators leptin, estrogen and IGF-1 enhance the cell proliferation and early tumorigenesis of breast cancer cells. Nutr Cancer 61(4):484–491
27. Goodwin PJ, Ennis M, Pritchard KI, Trudeau ME, Koo J, Madarnas Y, Hartwick W, Hoffman B, Hood N (2002) Fasting insulin and outcome in early-stage breast cancer: results of a prospective cohort study. J Clin Oncol 20(1):42–51
28. Aparicio T, Kotelevets L, Tsocas A, Laigneau JP, Sobhani I, Chastre E, Lehy T (2005) Leptin stimulates the proliferation of human colon cancer cells in vitro but does not promote the growth of colon cancer xenografts in nude mice or intestinal tumorigenesis in ApcMin/+ mice. Gut 54(8):1136–1145
29. Frankenberry KA, Skinner H, Somasundar P, McFadden DW, Vona-Davis LC (2006) Leptin receptor expression and cell signaling in breast cancer. Int J Oncol 28(4):985–993
30. Laud K, Gourdou I, Pessemesse L, Peyrat JP, Djiane J (2002) Identification of leptin receptors in human breast cancer: functional activity in the T47-D breast cancer cell line. Mol Cell Endocrinol 188(1–2):219–226
31. Valle A, Sastre-Serra J, Oliver J, Roca P (2011) Chronic leptin treatment sensitizes MCF-7 breast cancer cells to estrogen. Cell Physiol Biochem 28(5):823–832
32. Attoub S, Noe V, Pirola L, Bruyneel E, Chastre E, Mareel M, Wymann MP, Gespach C (2000) Leptin promotes invasiveness of kidney and colonic epithelial cells via phosphoinositide 3-kinase-, rho-, and rac-dependent signaling pathways. FASEB J 14(14):2329–2338
33. Catalano S, Marsico S, Giordano C, Mauro L, Rizza P, Panno ML, Andò S (2003) Leptin enhances, via AP-1, expression of aromatase in the MCF-7 cell line. J Biol Chem 278(31):28668–28676

34. Nkhata KJ, Ray A, Schuster TF, Grossmann ME, Cleary MP (2009) Effects of adiponectin and leptin co-treatment on human breast cancer cell growth. Oncol Rep 21(6):1611–1619
35. Price RS, Cavazos DA, De Angel RE, Hursting SD, deGraffenried LA (2012) Obesity-related systemic factors promote an invasive phenotype in prostate cancer cells. Prostate Cancer Prostatic Dis 15(2):135–143
36. Zheng Q, Hursting SD, Reizes O (2012) Leptin regulates cyclin D1 in luminal epithelial cells of mouse MMTV-Wnt-1 mammary tumors. J Cancer Res Clin Oncol 138(9):1607–1612
37. Cleary MP, Phillips FC, Getzin SC, Jacobson TL, Jacobson MK, Christensen TA, Juneja SC, Grande JP, Maihle NJ (2003) Genetically obese MMTV-TGF-α/LepobLepob female mice do not develop mammary tumors. Breast Cancer Res Treat 77(3):205–215
38. Grossmann ME, Cleary MP (2012) The balance between leptin and adiponectin in the control of carcinogenesis—focus on mammary tumorigenesis. Biochimie 94(10):2164–2171
39. Zheng Q, Dunlap SM, Zhu J, Downs-Kelly E, Rich J, Hursting SD, Berger NA, Reizes O (2011) Leptin deficiency suppresses MMTV-Wnt-1 mammary tumor growth in obese mice and abrogates tumor initiating cell survival. Endocr Relat Cancer 18(4):491–503. Erratum in Endocr Relat Cancer 18(5):X1
40. Cust AE, Kaaks R, Friedenreich C, Bonnet F, Laville M, Lukanova A et al (2007) Plasma adiponectin levels and endometrial cancer risk in pre- and post-menopausal women. J Clin Endocrinol Metab 92(1):255–263
41. Fenton JI, Birmingham JM (2010) Adipokine regulation of colon cancer: adiponectin attenuates interleukin-6-induced colon carcinoma cell proliferation via STAT-3. Mol Carcinog 49(7):700–709
42. Mantzoros C, Petridou E, Dessypris N, Chavelas C, Dalamaga M, Alexe DM, Papadiamantis Y, Markopoulos C, Spanos E, Chrousos G, Trichopoulos D (2004) Adiponectin and breast cancer risk. J Clin Endocrinol Metab 89(3):1102–1107
43. Dalamaga M, Diakopoulos KN, Mantzoros CS (2012) The role of adiponectin in cancer: a review of current evidence. Endocr Rev 33(4):547–594
44. Jardé T, Caldefie-Chézet F, Goncalves-Mendes N, Mishellany F, Buechler C, Penault-Llorca F, Vasson MP (2009) Involvement of adiponectin and leptin in breast cancer: clinical and in vitro studies. Endocr Relat Cancer 16(4):1197–1210
45. Jardé T, Perrier S, Vasson MP, Caldefie-Chézet F (2011) Molecular mechanisms of leptin and adiponectin in breast cancer. Eur J Cancer 47(1):33–43
46. Bastard JP, Jardel C, Bruckert E, Blondy P, Capeau J, Laville M, Vidal H, Hainque B (2000) Elevated levels of interleukin 6 are reduced in serum and subcutaneous adipose tissue of obese women after weight loss. J Clin Endocrinol Metab 85(9):3338–3342
47. Ghanim H, Aljada A, Hofmeyer D, Syed T, Mohanty P, Dandona P (2004) Circulating mononuclear cells in the obese are in a proinflammatory state. Circulation 110(12):1564–1571
48. Wellen KE, Hotamisligil GS (2003) Obesity-induced inflammatory changes in adipose tissue. J Clin Invest 112(12):1785–1788
49. Anderson EK, Gutierrez DA, Hasty AH (2010) Adipose tissue recruitment of leukocytes. Curr Opin Lipidol 21(3):172–177
50. Weisberg SP, McCann D, Desai M, Rosenbaum M, Leibel RL, Ferrante AW Jr (2003) Obesity is associated with macrophage accumulation in adipose tissue. J Clin Invest 112(12):1796–1808
51. Xu H, Barnes GT, Yang Q, Tan G, Yang D, Chou CJ et al (2003) Chronic inflammation in fat plays a crucial role in the development of obesity-related insulin resistance. J Clin Invest 112(12):1821–1830
52. Zeyda M, Gollinger K, Kriehuber E, Kiefer FW, Neuhofer A, Stulnig TM (2010) Newly identified adipose tissue macrophage populations in obesity with distinct chemokine and chemokine receptor expression. Int J Obes (Lond) 34(12):1684–1694
53. Dalmas E, Rouault C, Abdennour M, Rovere C, Rizkalla S et al (2011) Variations in circulating inflammatory factors are related to changes in calorie and carbohydrate intakes early in the course of surgery-induced weight reduction. Am J Clin Nutr 94(2):450–458
54. Cancello R, Clément K (2006) Is obesity an inflammatory illness? Role of low-grade inflammation and macrophage infiltration in human white adipose tissue. BJOG 113(10):1141–1147

55. Clément K, Viguerie N, Poitou C et al (2004) Weight loss regulates inflammation-related genes in white adipose tissue of obese subjects. FASEB J 18(14):1657–1669
56. Illán-Gómez F, Gonzálvez-Ortega M, Orea-Soler I, Alcaraz-Tafalla MS, Aragón-Alonso A, Pascual-Díaz M, Pérez-Paredes M, Lozano-Almela ML (2012) Obesity and inflammation: change in adiponectin, C-reactive protein, tumour necrosis factor-alpha and interleukin-6 after bariatric surgery. Obes Surg 22(6):950–955
57. Hotamisligil GS, Shargill NS, Spiegelman BM (1993) Adipose expression of tumor necrosis factor-α: direct role in obesity-linked insulin resistance. Science 259(5091):87–91
58. Nikolajczyk BS, Jagannathan-Bogdan M, Denis GV (2012) The outliers become a stampede as immunometabolism reaches a tipping point. Immunol Rev 249(1):253–275
59. Cinti SMG, Barbatelli G, Murano I, Ceresi E, Faloia E, Wang S, Fortier M, Greenberg AS, Obin MS (2005) Adipocyte death defines macrophage localization and function in adipose tissue of obese mice and humans. J Lipid Res 46(1):2347–2355
60. Kanda H, Tateya S, Tamori Y, Kotani K, Hiasa K, Kitazawa R, Kitazawa S, Miyachi H, Maeda S, Egashira K, Kasuga M (2006) MCP-1 contributes to macrophage infiltration into adipose tissue, insulin resistance, and hepatic steatosis in obesity. J Clin Invest 116(6):1494–1505
61. Kim CS, Park HS, Kawada T, Kim JH, Lim D, Hubbard NE, Kwon BS, Erickson KL, Yu R (2006) Circulating levels of MCP-1 and IL-8 are elevated in human obese subjects and associated with obesity-related parameters. Int J Obes (Lond) 30(9):1347–1355
62. Kintscher U, Hartge M, Hess K, Foryst-Ludwig A, Clemenz M et al (2008) T-lymphocyte infiltration in visceral adipose tissue: a primary event in adipose tissue inflammation and the development of obesity-mediated insulin resistance. Arterioscler Thromb Vasc Biol 28(7):1304–1310
63. Lê KA, Mahurkar S, Alderete TL, Hasson RE, Adam TC, Kim JS, Beale E, Xie C, Greenberg AS, Allayee H, Goran MI (2011) Subcutaneous adipose tissue macrophage infiltration is associated with hepatic and visceral fat deposition, hyperinsulinemia, and stimulation of NF-κB stress pathway. Diabetes 60(11):2802–2809
64. Rausch ME, Weisberg S, Vardhana P, Tortoriello DV (2008) Obesity in C57BL/6J mice is characterized by adipose tissue hypoxia and cytotoxic T-cell infiltration. Int J Obes (Lond) 32:451–463
65. Strissel KJ, Stancheva Z, Miyoshi H, Perfield JW II, DeFuria J, Jick Z, Greenberg AS, Obin MS (2007) Adipocyte death, adipose tissue remodeling, and obesity complications. Diabetes 56(12):2910–2918
66. Strissel KJ, DeFuria J, Shaul ME, Bennett G, Greenberg AS, Obin MS (2010) T-cell recruitment and Th1 polarization in adipose tissue during diet-induced obesity in C57BL/6 mice. Obesity (Silver Spring) 18(10):1918–1925
67. Bastard JP, Maachi M, Lagathu C, Kim MJ, Caron M, Vidal H, Capeau J, Feve B (2006) Recent advances in the relationship between obesity, inflammation, and insulin resistance. Eur Cytokine Netw 17(1):4–12
68. Kahn SE, Zinman B, Haffner SM, O'Neill MC et al (2006) Obesity is a major determinant of the association of C-reactive protein levels and the metabolic syndrome in type 2 diabetes. Diabetes 55(8):2357–2364
69. Di Gregorio GB, Yao-Borengasser A, Rasouli N, Varma V, Lu T, Miles LM, Ranganathan G, Peterson CA, McGehee RE, Kern PA (2005) Expression of CD68 and macrophage chemoattractant protein-1 genes in human adipose and muscle tissues: association with cytokine expression, insulin resistance, and reduction by pioglitazone. Diabetes 54(8):2305–2313
70. Keophiphath M, Achard V, Henegar C, Rouault C, Clément K, Lacasa D (2009) Macrophage-secreted factors promote a profibrotic phenotype in human preadipocytes. Mol Endocrinol 23(1):11–24
71. Apovian CM, Bigornia S, Mott M, Meyers MR, Ulloor J, Gagua M, McDonnell M, Hess D, Joseph L, Gokce N (2008) Adipose macrophage infiltration is associated with insulin resistance and vascular endothelial dysfunction in obese subjects. Arterioscler Thromb Vasc Biol 28(9):1654–1659

72. Klöting N, Fasshauer M, Dietrich A, Kovacs P, Schön MR, Kern M, Stumvoll M, Blüher M (2010) Insulin-sensitive obesity. Am J Physiol Endocrinol Metab 299(3):E506–E515
73. Permana PA, Menge C, Reaven PD (2006) Macrophage-secreted factors induce adipocyte inflammation and insulin resistance. Biochem Biophys Res Commun 341(2):507–514
74. Wentworth JM, Naselli G, Brown WA, Doyle L, Phipson B, Smyth GK, Wabitsch M, O'Brien PE, Harrison LC (2010) Pro-inflammatory CD11c+CD206+ adipose tissue macrophages are associated with insulin resistance in human obesity. Diabetes 59(7):1648–1656
75. Lumeng CN, Deyoung SM, Saltiel AR (2007) Macrophages block insulin action in adipocytes by altering expression of signaling and glucose transport proteins. Am J Physiol Endocrinol Metab 292(1):E166–E174
76. Westcott DJ, Delproposto JB, Geletka LM, Wang T, Singer K, Saltiel AR, Lumeng CN (2009) MGL1 promotes adipose tissue inflammation and insulin resistance by regulating 7/4hi monocytes in obesity. J Exp Med 206(13):3143–3156
77. Blüher M (2010) The distinction of metabolically 'healthy' from 'unhealthy' obese individuals. Curr Opin Lipidol 21(1):38–43
78. Lumeng CN, Bodzin JL, Saltiel AR (2007) Obesity induces a phenotypic switch in adipose tissue macrophage polarization. J Clin Invest 117(1):175–184
79. Duffield J (2003) The inflammatory macrophage: a story of Jekyll and Hyde. Clin Sci (Lond) 104(1):27–38
80. Surmi BK, Hasty AH (2010) The role of chemokines in recruitment of immune cells to the artery wall and adipose tissue. Vascul Pharmacol 52(1–2):27–36
81. Wu H, Ghosh S, Perrard XD, Feng L, Garcia GE, Perrard JL et al (2007) T-cell accumulation and regulated on activation, normal T cell expressed and secreted upregulation in adipose tissue in obesity. Circulation 115(8):1029–1038
82. Loke P, MacDonald AS, Robb A, Maizels RM, Allen JE (2000) Alternatively activated macrophages induced by nematode infection inhibit proliferation via cell-to-cell contact. Eur J Immunol 30(9):2669–2678
83. Nishimura S, Manabe I, Nagasaki M, Eto K, Yamashita H, Ohsugi M et al (2009) CD8+ effector T cells contribute to macrophage recruitment and adipose tissue inflammation in obesity. Nat Med 15(8):914–920
84. Winer S, Paltser G, Chan Y, Tsui H, Engleman E, Winer D, Dosch HM (2009) Obesity predisposes to Th17 bias. Eur J Immunol 39(9):2629–2635
85. Feuerer M, Herrero L, Cipolletta D, Naaz A, Wong J, Nayer A, Lee J, Goldfine AB, Benoist C, Shoelson S, Mathis D (2009) Lean, but not obese, fat is enriched for a unique population of regulatory T cells that affect metabolic parameters. Nat Med 15(8):930–939
86. Jagannathan M, McDonnell M, Liang Y, Hasturk H, Hetzel J, Rubin D, Kantarci A, Van Dyke TE, Ganley-Leal LM, Nikolajczyk BS (2010) Toll-like receptors regulate B cell cytokine production in patients with diabetes. Diabetologia 53(7):1461–1471
87. Nikolajczyk BS (2010) B cells as under-appreciated mediators of non-auto-immune inflammatory disease. Cytokine 50(3):234–242
88. Perreault M, Marette A (2001) Targeted disruption of inducible nitric oxide synthase protects against obesity-linked insulin resistance in muscle. Nat Med 7(10):1138–1143
89. Saberi M, Woods NB, de Luca C, Schenk S, Lu JC, Bandyopadhyay G, Verma IM, Olefsky JM (2009) Hematopoietic cell-specific deletion of toll-like receptor 4 ameliorates hepatic and adipose tissue insulin resistance in high-fat-fed mice. Cell Metab 10(5):419–429
90. Aron-Wisnewsky J, Tordjman J, Poitou C, Darakhshan F, Hugol D, Basdevant A, Aissat A, Guerre-Millo M, Clément K (2009) Human adipose tissue macrophages: m1 and m2 cell surface markers in subcutaneous and omental depots and after weight loss. J Clin Endocrinol Metab 94(11):4619–4623
91. Centers for Disease Control and Prevention (2004) Cigarette smoking among adults—United States, 2000. MMWR Morb Mortal Wkly Rep 53:427–431
92. U.S. Department of Health and Human Services (2004) The health consequences of smoking: a report of the Surgeon General. U.S. Department of Health and Human Services, Centers for Disease Control and Prevention, Atlanta, GA

93. McLaughlin T, Lamendola C, Liu A, Abbasi F (2011) Preferential fat deposition in subcutaneous versus visceral depots is associated with insulin sensitivity. J Clin Endocrinol Metab 96(11):E1756–E1760
94. Preis SR, Massaro JM, Robins SJ, Hoffmann U, Vasan RS et al (2010) Abdominal subcutaneous and visceral adipose tissue and insulin resistance in the Framingham heart study. Obesity (Silver Spring) 18(11):2191–2198
95. Berg AH, Scherer PE (2005) Adipose tissue, inflammation, and cardiovascular disease. Circ Res 96(9):939–949
96. Després JP, Lemieux I (2006) Abdominal obesity and metabolic syndrome. Nature 444(7121): 881–887
97. Hardy OT, Perugini RA, Nicoloro SM, Gallagher-Dorval K, Puri V, Straubhaar J, Czech MP (2011) Body mass index-independent inflammation in omental adipose tissue associated with insulin resistance in morbid obesity. Surg Obes Relat Dis 7(1):60–67
98. Kabat GC, Kim MY, Strickler HD, Shikany JM, Lane D, Luo J, Ning Y, Gunter MJ, Rohan TE (2012) A longitudinal study of serum insulin and glucose levels in relation to colorectal cancer risk among postmenopausal women. Br J Cancer 106(1):227–232
99. Bonora E, Willeit J, Kiechl S, Oberhollenzer F, Egger G, Bonadonna R, Muggeo M (1998) U-shaped and J-shaped relationships between serum insulin and coronary heart disease in the general population. The Bruneck Study. Diabetes Care 21(2):221–230
100. Wildman RP, Muntner P, Reynolds K, McGinn AP, Rajpathak S, Wylie-Rosett J, Sowers MR (2008) The obese without cardiometabolic risk factor clustering and the normal weight with cardiometabolic risk factor clustering: prevalence and correlates of 2 phenotypes among the US population (NHANES 1999–2004). Arch Intern Med 168(15): 1617–1624
101. Blüher M, Bashan N, Shai I, Harman-Boehm I, Tarnovscki T et al (2009) Activated Ask1-MKK4-p38MAPK/JNK stress signaling pathway in human omental fat tissue may link macrophage infiltration to whole-body insulin sensitivity. J Clin Endocrinol Metab 94(7):2507–2515
102. Ferrannini E, Natali A, Bell P, Cavallo-Perin P, Lalic N, Mingrone G (1997) Insulin resistance and hypersecretion in obesity. European Group for the Study of Insulin Resistance (EGIR). J Clin Invest 100(5):1166–1173
103. Romano M, Guagnano MT, Pacini G, Vigneri S, Falco A, Marinopiccoli M, Manigrasso MR, Basili S, Davì G (2003) Association of inflammation markers with impaired insulin sensitivity and coagulative activation in obese healthy women. J Clin Endocrinol Metab 88(11): 5321–5326
104. Sims EA (2001) Are there persons who are obese, but metabolically healthy? Metabolism 50(12):1499–1504. Erratum in Metabolism 51(4):536 (2002)
105. Karelis AD, Faraj M, Bastard JP, St-Pierre DH, Brochu M, Prud'homme D, Rabasa-Lhoret R (2005) The metabolically healthy but obese individual presents a favorable inflammation profile. J Clin Endocrinol Metab 90(7):4145–4150
106. Succurro E, Marini MA, Frontoni S, Hribal ML, Andreozzi F, Lauro R, Perticone F, Sesti G (2008) Insulin secretion in metabolically obese, but normal weight, and in metabolically healthy but obese individuals. Obesity (Silver Spring) 16(8):1881–1886
107. Karin M, Greten FR (2005) NF-kappaB: linking inflammation and immunity to cancer development and progression. Nat Rev Immunol 5(10):749–759
108. Karin M (2009) NF-κB as a critical link between inflammation and cancer. Cold Spring Harb Perspect Biol 1(5):a000141
109. Kraus S, Arber N (2009) Inflammation and colorectal cancer. Curr Opin Pharmacol 9(4):405–410
110. Denis GV (2010) Bromodomain coactivators in cancer, obesity, type 2 diabetes, and inflammation. Discov Med 10(55):489–499
111. Belkina AC, Nikolajczyk BS, Denis GV (2013) BET protein function is required for inflammation: Brd2 genetic disruption and BET inhibitor JQ1 impair macrophage inflammatory responses. J Immunol. 2013 Feb 18. E-pub ahead of print

112. Denis GV, Green MR (1996) A novel, mitogen-activated nuclear kinase is related to a *Drosophila* developmental regulator. Genes Dev 10(3):261–271
113. Kanno T, Kanno Y, Siegel RM, Jang MK, Lenardo MJ, Ozato K (2004) Selective recognition of acetylated histones by bromodomain proteins visualized in living cells. Mol Cell 13(1):33–43
114. Nakamura Y, Umehara T, Nakano K, Jang MK, Shirouzu M et al (2007) Crystal structure of human *BRD2* bromodomain: Insight into dimerization and recognition of acetylated histone H4. J Biol Chem 282(6):4193–4201
115. Umehara T, Nakamura Y, Jang MK, Nakano K, Tanaka A, Ozato K, Padmanabhan B, Yokoyama S (2010) Structural basis for acetylated histone H4 recognition by the human BRD2 bromodomain. J Biol Chem 285(10):7610–7618
116. Holstege FC, Jennings EG, Wyrick JJ, Lee TI, Hengartner CJ, Green MR, Golub TR, Lander ES, Young RA (1998) Dissecting the regulatory circuitry of a eukaryotic genome. Cell 95(5):717–728
117. Haynes SR, Dollard C, Winston F, Beck S, Trowsdale J, Dawid IB (1992) The bromodomain: a conserved sequence found in human, *Drosophila* and yeast proteins. Nucleic Acids Res 20(10):2603
118. Tamkun JW, Deuring R, Scott MP, Kissenger M, Pattatucci AM, Kaufman TC, Kennison JA (1992) Brahma—a regulator of *Drosophila* homeotic genes structurally related to the yeast transcriptional activator SWI2/SNF2. Cell 68(3):561–572
119. Dhalluin C, Carlson JE, Zeng L, He C, Aggarwal AK, Zhou M-M (1999) Structure and ligand of a histone acetyltransferase bromodomain. Nature 399(6735):491–496
120. Jeanmougin F, Wurtz J-M, Le Douarin B, Chambon P, Losson R (1997) The bromodomain revisited. Trends Biochem Sci 22(5):151–153
121. Sanchez R, Zhou M-M (2009) The role of human bromodomains in chromatin biology and gene transcription. Curr Opin Drug Discov Dev 12(5):659–665
122. Beck S, Hanson I, Kelly A, Pappin DJC, Trowsdale J (1992) A homologue of the *Drosophila female sterile homeotic (fsh)* gene in the class II region of the human MHC. DNA Seq 2(4):203–210
123. Guo N, Faller DV, Denis GV (2000) Activation-induced nuclear translocation of RING3. J Cell Sci 113(pt17):3085–3091
124. Dey A, Chitsaz F, Abbasi A, Misteli T, Ozato K (2003) The double bromodomain protein Brd4 binds to acetylated chromatin during interphase and mitosis. Proc Natl Acad Sci U S A 100(15):8758–8763
125. LeRoy G, Rickards B, Flint SJ (2008) The double bromodomain proteins Brd2 and Brd3 couple histone acetylation to transcription. Mol Cell 30(1):51–60
126. Wu SY, Chiang CM (2007) The double bromodomain-containing chromatin adaptor Brd4 and transcriptional regulation. J Biol Chem 282(18):13141–13145
127. Denis GV, Vaziri C, Guo N, Faller DV (2000) RING3 kinase transactivates promoters of cell cycle regulatory genes through E2F. Cell Growth Differ 11(8):417–424
128. Dey A, Ellenberg J, Farina A, Coleman AE, Maruyama T, Sciortino S, Lippincott-Schwartz J, Ozato K (2000) A bromodomain protein, MCAP, associates with mitotic chromosomes and affects G_2-to-M transition. Mol Cell Biol 20:6537–6549
129. Maruyama T, Farina A, Dey A, Cheong J, Bermudez VP, Tamura T, Sciortino S, Shuman J, Hurwitz J, Ozato K (2002) A Mammalian bromodomain protein, Brd4, interacts with replication factor C and inhibits progression to S phase. Mol Cell Biol 22(18):6509–6520
130. Peng J, Dong W, Chen L, Zou T, Qi Y, Liu Y (2007) Brd2 is a TBP-associated protein and recruits TBP into E2F-1 transcriptional complex in response to serum stimulation. Mol Cell Biochem 294(1–2):45–54
131. Sinha A, Faller DV, Denis GV (2005) Bromodomain analysis of Brd2-dependent transcriptional activation of *cyclin A*. Biochem J 387(pt 1):257–269
132. Kubonishi I, Takehara N, Iwata J, Sonobe H, Ohtsuki Y, Abe T, Miyoshi I (1991) Novel t(15;19)(q15;p13) chromosome abnormality in a thymic carcinoma. Cancer Res 51(12): 3327–3328

133. French CA, Miyoshi I, Kubonishi I, Grier HE, Perez-Atayde AR, Fletcher JA (2003) BRD4-NUT fusion oncogene: a novel mechanism in aggressive carcinoma. Cancer Res 63(2):304–307
134. French CA, Ramirez CL, Kolmakova J, Hickman TT, Cameron MJ et al (2008) BRD-NUT oncoproteins: a family of closely related nuclear proteins that block epithelial differentiation and maintain the growth of carcinoma cells. Oncogene 27(15):2237–2242
135. Muller S, Filippakopoulos P, Knapp S (2011) Bromodomains as therapeutic targets. Expert Rev Mol Med 13:e29
136. Filippakopoulos P, Qi J, Picaud S, Shen Y, Smith WB, Fedorov O, Morse EM et al (2010) Selective inhibition of BET bromodomains. Nature 468(7327):1067–1073
137. Dawson MA, Prinjha RK, Dittmann A, Giotopoulos G, Bantscheff M, Chan WI et al (2011) Inhibition of BET recruitment to chromatin as an effective treatment for MLL-fusion leukaemia. Nature 478(7370):529–533
138. Delmore JE, Issa GC, Lemieux ME, Rahl PB, Shi J, Jacobs HM et al (2011) BET bromodomain inhibition as a therapeutic strategy to target c-Myc. Cell 146(6):904–917
139. Mertz JA, Conery AR, Bryant BM, Sandy P, Balasubramanian S, Mele DA, Bergeron L, Sims RJ III (2011) Targeting MYC dependence in cancer by inhibiting BET bromodomains. Proc Natl Acad Sci U S A 108(40):16669–16674
140. Belkina AC, Denis GV (2012) BET domain co-regulators in obesity, inflammation and cancer. Nat Rev Cancer 12(7):465–477
141. Greenwald R, Tumang JR, Sinha A, Currier N, Cardiff RD, Rothstein TL, Faller DV, Denis GV (2004) Eμ-*BRD2* transgenic mice develop B cell lymphoma and leukemia. Blood 103(4):1475–1484
142. Lenburg ME, Sinha A, Faller DV, Denis GV (2007) Tumor-specific and proliferation-specific gene expression typifies murine transgenic B cell lymphomagenesis. J Biol Chem 282(7):4803–4811
143. Wang F, Liu H, Blanton WP, Belkina A, LeBrasseur NK, Denis GV (2009) *Brd2* disruption in mice causes severe obesity without type 2 diabetes. Biochem J 425(1):7–83
144. Belkina AC, Blanton W, Wang F, Liu H, Denis GV (2010) Whole body Brd2 deficiency protects obese mice from insulin resistance by creating a low inflammatory environment. Obesity 18:S58
145. Nicodeme E, Jeffrey KL, Schaefer U, Beinke S, Dewell S, Chung CW, Chandwani R et al (2010) Suppression of inflammation by a synthetic histone mimic. Nature 468(7327): 1119–1123
146. Wang F, Deeney JT, Denis GV (2013) Brd2 gene disruption causes 'metabolically healthy' obesity: epigenetic and chromatin-based mechanisms that uncouple obesity from type 2 diabetes. Vitam Horm 91:49–75
147. Denis GV, Nikolajczyk BN, Schnitzler GR (2010) An emerging role for bromodomain-containing proteins in chromatin regulation and transcriptional control of adipogenesis. FEBS Lett 584:3260–3268
148. Rauchhaus M, Koloczek V, Volk H, Kemp M, Niebauer J, Francis DP, Coats AJ, Anker SD (2000) Inflammatory cytokines and the possible immunological role for lipoproteins in chronic heart failure. Int J Cardiol 76(2–3):125–133
149. Plata-Salamán CR (2000) Central nervous system mechanisms contributing to the cachexia-anorexia syndrome. Nutrition 16(10):1009–1012
150. Jensen GL (2008) Inflammation: roles in aging and sarcopenia. JPEN J Parenter Enteral Nutr 32(6):656–669
151. Leng S, Chaves P, Koenig K, Walston J (2002) Serum interleukin-6 and hemoglobin as physiological correlates in the geriatric syndrome of frailty: a pilot study. J Am Geriatr Soc 50(7):1268–1271
152. Leng SX, Cappola AR, Andersen RE, Blackman MR, Koenig K, Blair M, Walston JD (2004) Serum levels of insulin-like growth factor-1 (IGF-1) and dehydroepiandrosterone sulfate (DHEA-S), and their relationships with serum interleukin-6, in the geriatric syndrome of frailty. Aging Clin Exp Res 16(2):153–157

153. Leng SX, Xue QL, Tian J, Walston JD, Fried LP (2007) Inflammation and frailty in older women. J Am Geriatr Soc 55(6):864–871
154. Lobstein T, Baur L, Uauy R, IASO International Obesity Task Force (2004) Obesity in children and young people: a crisis in public health. Obes Rev 5(S1):4–104
155. Figueroa-Munoz JI, Chinn S, Rona RJ (2001) Association between obesity and asthma in 4–11 year old children in the UK. Thorax 56(2):133–137
156. Morgan WJ, Crain EF, Gruchalla RS, O'Connor GT, Kattan M, Evans R III, Stout J, Malindzak G, Smartt E, Plaut M, Walter M, Vaughn B, Mitchell H, Inner-City Asthma Study Group (2004) Results of a home-based environmental intervention among urban children with asthma. N Engl J Med 351(11):1068–1080
157. Digenis-Bury EC, Brooks DR, Chen L, Ostrem M, Horsburgh CR (2008) Use of a population-based survey to describe the health of Boston public housing residents. Am J Public Health 98(1):85–91
158. Brugge D, Rice PW, Terry P, Howard L, Best J (2001) Housing conditions and respiratory health in a Boston public housing community. New Solut 11(2):149–164
159. Booth KM, Pinkston MM, Poston WS (2005) Obesity and the built environment. J Am Diet Assoc 105(5 suppl 1):S110–S117
160. Gennuso J, Epstein LH, Paluch RA, Cerny F (1998) The relationship between asthma and obesity in urban minority children and adolescents. Arch Pediatr Adolesc Med 152(12):1197–1200
161. Luder E, Melnik TA, Dimaio M (1998) Association of being overweight with greater asthma symptoms in inner city black and Hispanic children. J Pediatr 132(4):699–703
162. Stenius-Aarniala B, Poussa T, Kvarnstrom J, Gronlund EL, Ylikahri M, Mustajoki P (2000) Immediate and long term effects of weight reduction in obese people with asthma: randomised controlled study. BMJ 320(7238):827–832
163. Ma J, Xiao L, Knowles SB (2010) Obesity, insulin resistance and the prevalence of atopy and asthma in US adults. Allergy 65(11):1455–1463
164. Battaglia TA, Murrell SS, Bhosrekar SG, Caron SE, Bowen DJ, Smith E, Kalish R, Rorie JA (2012) Connecting Boston's public housing developments to community health centers: who's ready for change? Prog Community Health Partnersh 6(3):239–248
165. Conroy K, Sandel M, Zuckerman B (2010) Poverty grown up: how childhood socioeconomic status impacts adult health. J Dev Behav Pediatr 31(2):154–160
166. Trotter LJ, Bowen DJ, Beresford SA (2010) Testing for racial/ethnic differences in the association between childhood socioeconomic position and adult adiposity. Am J Public Health 100(6):1088–1094
167. American Cancer Society (2000) Cancer facts and figures 2000. American Cancer Society, Atlanta, GA
168. Barbeau EM, Kreiger N, Soobader MJ (2004) Working class matters: socioeconomic disadvantage, race/ethnicity, gender, and smoking in NHIS 2000. Am J Public Health 94(2):269–278
169. Blüher M (2012) Are there still healthy obese patients? Curr Opin Endocrinol Diabetes Obes 19(5):341–346
170. Combs TP, Pajvani UB, Berg AH, Lin Y, Jelicks LA et al (2004) A transgenic mouse with a deletion in the collagenous domain of adiponectin displays elevated circulating adiponectin and improved insulin sensitivity. Endocrinology 145(1):367–383

Chapter 4
Adipose Tissue Macrophages in Obesity, Inflammation, and Cancer

Carey Nien-Kai Lumeng

Abstract The chronic low-grade inflammation induced by obesity is a key connection between obesity and disease. It is now understood that the supporting stromal cells in adipose tissue contribute to this inflammatory response in significant ways. Of these, adipose tissue macrophages (ATMs) are major effectors of inflammation in hypertrophic adipose tissue. However, we now know that ATMs are diverse in their phenotypes and have functions that may contribute directly and indirectly to impact cancer risk. This review will summarize our current understanding of the phenotypic diversity in ATMs and how this is altered in obesity. The potential role that ATMs play in breast and ovarian cancer pathogenesis will also be discussed given the close association between adipose tissue and these cancer types.

4.1 Introduction

Adipose tissue has been traditionally thought of as being a static organ composed of only adipocytes, specialized cells designed for the storage of excess nutrients as lipid (primarily triglyceride). By mass, adipocytes make up the bulk of adipose tissue and can drastically enlarge and contract in times of nutrient excess and demand. However, by number, adipocytes may be a minority cell population in adipose tissue as demonstrated in Fig. 4.1. A major advance in the last 5–10 years in obesity research has been the realization that adipose tissue has a complex and dynamic range of cellular components beyond adipocytes that play critical roles in the maintenance of nutrient homeostasis.

Importantly, it is now known that alterations in the non-adipocyte stromal cells in fat form crucial links between obesity and many of its associated diseases. The advancements in understanding obesity-induced inflammation are illustrative of the

C.N.-K. Lumeng, M.D., Ph.D. (✉)
Department of Pediatrics and Communicable Diseases, University of Michigan Medical School, 4520B MSRB I, 1150 West Medical Center Drive, Ann Arbor, MI 48109-0651, USA
e-mail: clumeng@umich.edu

A.J. Dannenberg and N.A. Berger (eds.), *Obesity, Inflammation and Cancer*, Energy Balance and Cancer 7, DOI 10.1007/978-1-4614-6819-6_4,

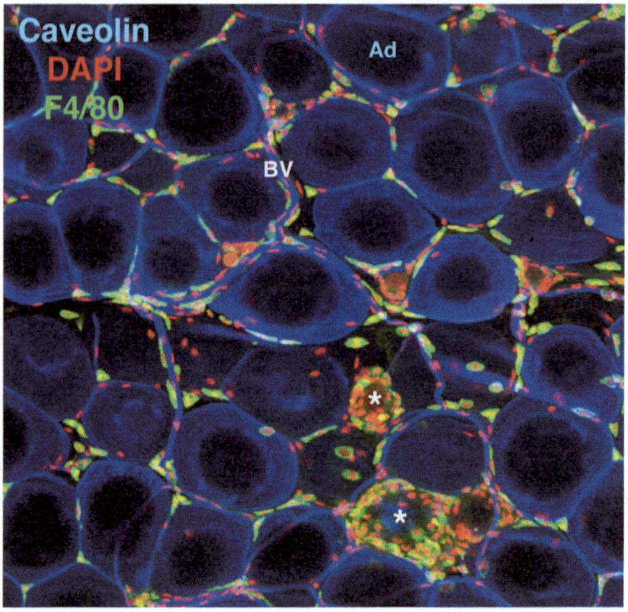

Fig. 4.1 Adipose tissue architecture in obesity. Confocal microscopy image of visceral/gonadal fat depot in C57Bl/6 mice fed with a high-fat diet (60 %) for 16 weeks. Tissue was stained with antibodies against caveolin (*blue*) and F4/80 (macrophages, *green*) and co-stained with DAPI (*red*) to indicate nuclei. Crown-like structure (CLS) of ATMs (*asterisks*) is shown. Adipocytes (Ad) are identified by caveolin staining. Blood vessels (BV) are shown

new directions that are being explored in obesity research. As this field expands, it provides opportunities to identify novel mechanisms by which obesity modifies cancer risk.

This review will summarize the current understanding of adipose tissue macrophages (ATMs), a leukocyte in fat that is fairly well characterized in animal models and in humans. I will review the state of the art view of ATM function and diversity in adipose tissue and how this is altered by obesity. In this framework, I will present the current views of how ATMs promote adipocyte dysfunction and contribute to metabolic dysregulation. This will lead to a better understanding of the ways that ATM activation may both directly and indirectly contribute to the link between obesity and cancer.

4.2 Adipose Tissue Macrophages as Inflammatory Engines in Obesity

When nutrient intake outpaces energy utilization, the only tissue that is designed to efficiently store excess nutrients is adipose tissue. Adipose tissue expands by producing new adipocytes (hyperplasia) and enlarging existing adipocytes (hypertrophy).

For reasons that are still elusive, the chronic and excessive expansion of adipose tissue is associated with the activation of inflammation in fat. Such inflammation was initially identified by the elevated expression of inflammatory cytokines and chemokines such as tumor necrosis factor-α (TNFα) [1], interleukin (IL)-6 [2], monocyte chemoattractant protein 1 (MCP-1) [3, 4], and plasminogen activator inhibitor-1 (PAI-1) [5] from obese adipose tissue. Given the large mass of adipose tissue in obese subjects, such tissue-specific cytokine production likely contributes greatly to the systemic elevations of these cytokines in obesity.

Based on these observations, initial studies focused on how these inflammatory mediators were generated by adipocytes. However, in 2003, two papers demonstrated that most of the inflammatory factors were made not by the adipocytes, but by the non-adipocyte stromal cells which were enriched in macrophages [6, 7]. This led to a series of studies that demonstrated that macrophages accumulate in adipose tissue in obese subjects, are prominent in visceral over subcutaneous fat depots, are associated with measures of metabolic dysfunction, and are decreased with weight loss [8–11].

The evidence for a role of macrophages as a contributor to metabolic disease has been reinforced by many clinical and preclinical studies. In a systems biology study in humans, expression profiling of blood and adipose tissue was combined with genetic dissection of expression quantitative trait loci (eQTL) to find genes associated with obesity and metabolic disease [12, 13]. This unbiased approach identified a gene expression signature in adipose tissue associated with metabolic dysfunction that was enriched for genes typically found in the spleen- and bone marrow-derived macrophages. This macrophage-enriched metabolic network (MEMN) of genes includes *Cd68*, *Emr1* (*F4/80*), *Cd14*, and *C3ar1* and emphasized the link between inflammatory networks in fat and the complex trait of metabolic disease. Follow-up studies validated the importance of many of these genes to metabolic traits using loss of function mouse models [14].

Rodent models of obesity have been instrumental in demonstrating that macrophage activation plays a substantial role in the link between obesity and insulin resistance. Mouse models deficient in receptors critical for inflammatory monocyte/macrophage trafficking have impaired accumulation of ATMs in adipose tissue. These include mice deficient for C-C chemokine receptor type 2 (CCR2) [15], macrophage galactose-type C-type lectin (MGL1) [16], and C-C chemokine receptor type 5 (CCR5) [17]. These models share the similar phenotype when exposed to high-fat diets where adipocyte hypertrophy occurs in the absence of an accumulation of ATMs in crown-like structures (CLS). Despite obesity and increased fat mass in these models, the mice demonstrate an improved metabolic profile (e.g., improved insulin sensitivity and glucose tolerance) and a decrease in hepatic steatosis.

These studies suggest that the ATM activation and inflammation in fat play a major role in impairing the function of fat as an efficient nutrient store (Fig. 4.2). Fat dysfunction promotes lipid and chemokine release as opposed to storage and places pressure on other organs to store the excess nutrient load. Attenuation of ATM accumulation produces "healthier" fat with a sustained capacity to store excess nutrients

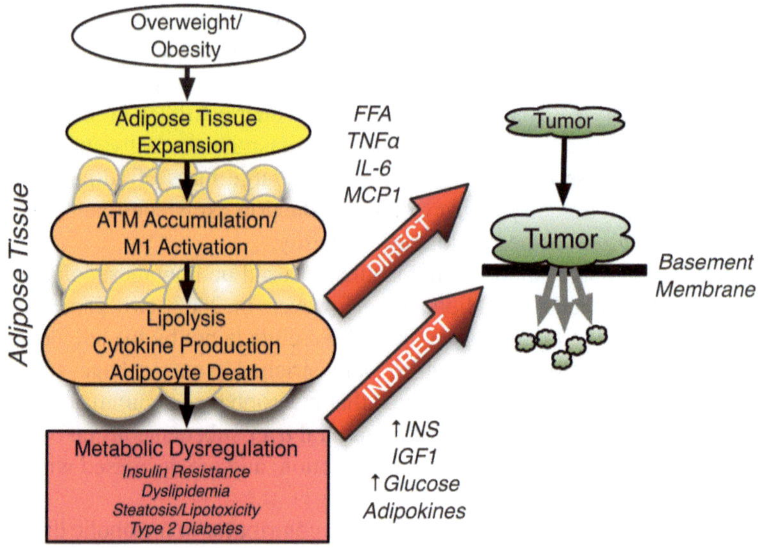

Fig. 4.2 Direct and indirect contributions of ATMs to cancer biology. Overweight and obesity status is associated with expansion of adipose tissue. Triggers concurrent hypertrophy of adipocytes and accumulation of M1-like macrophages in fat. M1 ATMs generate inflammatory cytokines that may have direct effects upon the local tumor environment. ATMs also promote adipocyte dysfunction triggering adipocyte death and release of free fatty acids (FFA). Chronically this process promoted metabolic dysfunction associated with systemic insulin resistance and lipotoxicity. Such metabolic regulation and its associated hormonal changes are downstream of ATM activation and function as a promoter of "sick" adipose tissue

and partition lipids away from other organs such as the liver. A similar phenotype was seen in leptin-deficient mice with transgenic overexpression of adiponectin [18]. Despite morbid obesity, these mice had no evidence of metabolic abnormalities. This was associated with a lack of visceral fat and a preferential expansion of subcutaneous fat depots containing small adipocytes and few ATMs. Overall, these studies have supported the concept that ATMs contribute to the development of insulin resistance and type 2 diabetes with obesity by promoting chronic fat dysfunction.

4.3 Adipose Tissue Macrophage Diversity

A simplified interpretation of the observations above is that all macrophages in fat are deleterious. This generalization is hard to reconcile with the observation that macrophages are found in all fat depots in lean insulin-sensitive subjects [19, 20]. Based on this and other observations, current evidence supports a model where both quantitative and qualitative changes occur in ATMs with obesity. While it is likely that these two events (macrophage accumulation and changes in their activation profile) are coupled together, there is also growing evidence that ATMs are active sensors of the fat environment and can alter their inflammatory output in situ in response to metabolic and inflammatory cues [21, 22].

4.3.1 Macrophage Activation States

Around the time when ATMs were first described, there was a rapid evolution in the understanding of the different activation states that macrophages can assume dependent on the local context. In vitro these have been classified based on the differential responses of macrophages to Th1 and Th2 signals. M1 or classically activated macrophages are generated upon exposure to lipopolysaccharide (a bacterial component) and interferon-γ (IFNγ) and have the capacity to express high levels of pro-inflammatory cytokines/factors (e.g., iNOS, TNFα) that promote killing of foreign pathogens. The M1 program also induces a variety of metabolic changes in macrophages that include an increase in glucose utilization, decreased fatty acid utilization, and an increase in iron uptake to increase their "killing" capacity [23]. In contrast, when macrophages are stimulated with Th2 cytokines such as IL-4 or IL-13, they assume a very different profile. These alternatively activated or M2 macrophages have low production of inflammatory cytokines and instead generate IL-10 and arginase which dampen inflammatory responses and promote repair. Metabolically these cells are different as well and preferentially utilize fatty acids for energy and promote iron release. These metabolic changes and M2 profile appear to be coupled by nuclear transcription factors peroxisome proliferator-activated receptors (PPAR) γ and δ via signal transducer and activator of transcription 6 (STAT6) activation [24–26].

The M1/M2 dichotomy is an in vitro simplification of a complex range of macrophage activation profiles. While the M1/M2 paradigm is useful in many contexts, including that of ATM biology, it is clear that macrophages in vivo assume states along a continuum between these extreme states. In addition, many aspects of the M1/M2 axis differ substantially between human and murine models as gene profiling has shown that there is only 50 % commonality between human and mouse macrophages treated with M1 or M2 stimuli [27].

4.3.2 The M1/M2 Paradigm in Obesity

Despite the limitations noted above, there is evidence to support the concept that the balance between M1 and M2 states is altered by obesity and can explain many of the features of obesity-associated inflammation [28, 29]. In lean animals and humans, the resident population of ATMs is M2-like based on surface marker and gene expression profiles [9, 20, 30, 31]. Animal models with impaired generation of an alternatively activated M2 state demonstrate an exaggerated inflammatory response to obesity and have a more pronounced insulin resistance phenotype. Examples of this include macrophage-specific knockout of *Pparg* and *Ppard* [24–26]. In addition, mice deficient in GPR120, a receptor for omega-3 fatty acids on macrophages, have an impaired M2 gene expression in their ATMs and have an exaggerated pro-inflammatory response to obesity [32, 33].

On the other side of the coin, obesity induces many changes in ATMs consistent with an activation of an M1 profile. At the gene expression level, this includes the induction of *Tnfa* and *Nos2* genes in mice [20, 34]. In both mice and humans, obesity induces the expression of receptors that promote the capacity of ATMs to function as antigen-presenting cells (e.g., CD40, CD11c, and MHC class II), a feature of M1 activation [9, 30]. In addition, mice deficient for critical M1 macrophage activation components have attenuated obesity-induced inflammation. Examples of this include knockout mice for toll-like receptors 2 and 4 (TLR2, TLR4) and mice deficient in IFNγ production [35–38]. Attenuation of IκB kinase β (IKKβ) signaling pathways in macrophages leads to a substantial reduction in inflammation and protection of mice from insulin resistance [39, 40]. Based on this, novel anti-inflammatory treatments for type 2 diabetes are under clinical trial to target pathways that lead to NFκB activation, a key component in the generation of an M1 activation profile [41, 42].

4.3.3 The Nature of the Phenotypic Shift from M2-Like to M1-Like ATMs in Obesity

The mechanisms by which the qualitative and quantitative changes in ATMs are induced are under active investigation. What has become clear is that the changes in the inflammatory profile in fat with obesity are driven by alterations in the balance between at least two distinct types of ATMs (Fig. 4.3). A resident ATM population is found between adipocytes and along blood vessels in fat and is established early in adipose tissue development [43]. These have many features of M2 macrophages and appear to interact with eosinophils and other cells in fat that secrete Th2 cytokines such as IL-4 [44, 45]. The recruitment of the resident M2-like ATMs (type 2 [46]) appears to be independent of traditional chemokine pathways as they are present in normal numbers in *Ccr2*- and *Ccl2*-deficient mice [30].

In obesity, this resident ATM population is retained with similar distribution and surface markers; however, a distinct type of ATM is superimposed on this population. The "phenotypic shift" is generated by an increase in the ratio of these recruited M1-like (type 1) ATMs relative to the resident (type 2) ATMs. In mice and humans, the ATMs induced by obesity preferentially express the surface marker CD11c and have lipid-laden features similar to foam cells in atherosclerosis [47, 48]. The recruitment of $CD11c^+$ ATMs to fat is rapid compared to type 2 ATMs and is concentrated in CLS seen on standard histology preparations [30]. CLS are dense accumulations of ATMs that form around lipid remnants of dead adipocytes and are also sites of fibrosis and accumulation of collagen fibers [49, 50].

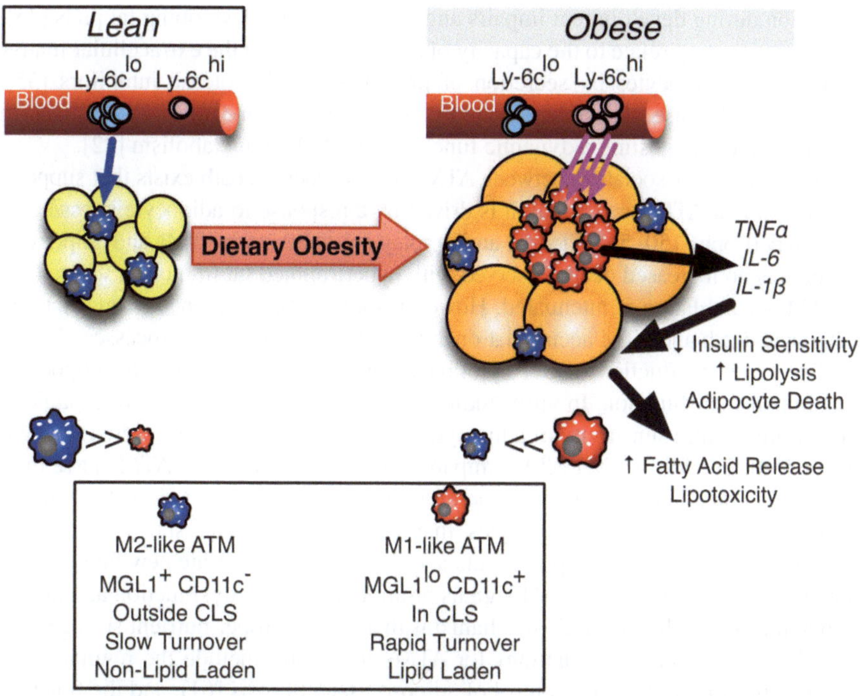

Fig. 4.3 Model of ATM phenotypic changes in obesity. In lean states, M2-like resident ATMs are the dominant ATM population (*blue cells*). These ATMs express M2 surface markers in mice (e.g. MGL1) and are located between adipocytes. With dietary obesity, a distinct type of ATM accumulates with features of M1 macrophages (*red*) that express CD11c and have low MGL1 expression in mice. These ATMs are concentrated in CLS and generate pro-inflammatory cytokines that cause local adipocyte dysfunction which contribute to systemic cytokines. The appearance of M1-like ATMs is coupled to a population of Ly-6c^{hi} monocytes in the circulation that are induced with obesity

The weight of evidence suggests that the accumulation of type 1 CD11c^{+} ATMs is linked to the enhanced trafficking of specific inflammatory monocyte populations from the circulation [51, 52]. In mice, monocytes that express the surface marker Ly6c are induced with obesity and tightly correlate with M1-like ATM accumulation [16, 53]. The specific mediators that promote monocyte trafficking to fat are incompletely understood but include chemokine-dependent (MCP1 [54]) and chemokine-independent (lipolysis) pathways [21].

4.4 ATM Functions

The list of functions of ATMs is incomplete. There is evidence that ATMs contribute to normal physiologic regulation of adipose tissue (e.g., fat development) in addition to their function in pathophysiology (e.g., obesity). In lean subjects, resident ATMs integrate systemic and local fat signals to maintain homeostasis. Macrophage depletion during development impairs angiogenesis in the developing fat pads [43]. This function may relate to the capacity of ATMs to remodel the extracellular matrix surrounding adipocytes by secretion of proteinases and protease inhibitors [55]. Resident ATMs are also able to rapidly sense and respond to acute changes in nutrient status suggesting a dynamic function for ATMs in metabolism [22].

In obesity, an association between ATMs and adipocyte death exists that supports the notion that ATM accumulation is driven by a response to adipocyte stress/death with hypertrophy [50, 56]. Clearly, an interplay exists between ATMs and adipocytes as there is no increased death observed with hypertrophied fat in models where type 1 ATM accumulation is attenuated. However, there is also evidence to suggest that adipocyte death and ATM accumulation are independent unrelated processes [57].

The active production of M1 cytokines from CLS sends signals to adipocytes that impair their function. In vitro studies have shown that M1 ATMs are capable of decreasing insulin sensitivity of adipocytes by direct and indirect (secreted) mechanisms [58]. In addition, M1 ATMs impair adipogenesis while M2 ATMs have little to no effect on adipogenesis [59]. Therefore, the shift toward an M1 profile impairs two crucial functions of adipocytes that can buffer chronic states of nutrient excess—the ability to take up nutrients and the ability to generate new fat cells. By antagonizing both processes, M1 signals promote adipocyte dysfunction and breakdown, the normal homeostatic mechanisms that permit proper nutrient storage.

Additional postulated functions for ATMs in obesity include the regulation of hypoxic responses [60], the control of adipose tissue fibrosis [61], and the function as dendritic cells to activate T cells in fat [62]. All of these may contribute to alter tissue environments in a way that may impact tumor formation, growth, and metastatic disease.

4.5 Links Between Adipose Tissue Macrophages and Cancer

The link between increased adiposity, high BMI, and cancer likely flows through several overlapping mechanisms related to systemic and local factors [63] (Fig. 4.2). These include impaired fatty acid storage in adipocytes, alterations in adipokine secretions (e.g., leptin, adiponectin), shunting of lipids for storage in non-adipose tissues causing lipotoxicity (e.g., hepatic steatosis), systemic elevations in inflammatory cytokines, local production of inflammatory cytokines in tumor-associated adipose tissue, and local activation of leukocytes in adipose tissue near tumor sites (e.g., mammary gland). ATMs have been shown to partially contribute to the development of all of these processes; therefore, one may think of direct and indirect connections between ATM activation and cancer risk.

The relative contributions of these factors to the links between cancer and obesity are still poorly understood and will likely be tissue specific. The systemic alterations in nutrient metabolism (e.g., insulin resistance) will be reviewed in other chapters in this book. I will next discuss the potential for local ATM involvement in cancer progression and metastatic disease by highlighting recent data regarding breast and ovarian cancer biology that intersects with ATM function.

4.5.1 ATMs and Breast Cancer

As in other adipose tissue depots, ATMs are present in normal breast tissue and are critical for the physiologic remodeling that occurs in and around the mammary epithelial cells. Macrophage depletion blocks post-lactation remodeling in the mouse mammary gland [64]. These remodeling changes are associated with an M2-like polarization of the ATMs that contribute to clearance of apoptotic mammary epithelial cells and may modify cancer risk [65]. Similar M2-like changes have been observed in other mouse models where extensive adipose tissue remodeling is induced [66].

The importance of tumor-associated macrophages (TAMs) in breast cancer progression is well described; however, the importance of local ATMs in this process is incompletely understood. Like in adipose tissue, there is significant heterogeneity in the TAMs seen in breast cancer [67]. Macrophages along the invasive front along the tumor periphery in mouse models demonstrate a high degree of motility and migration suggesting that local ATMs may contribute to TAMs [68]. Work in our lab using intravital microscopy corroborates these observations and finds extensive patrolling behavior of ATMs in multiple fat depots (C.N.L., unpublished observation). The function of ATMs at the invasive front may include remodeling of the extracellular matrix, promotion of angiogenesis, direct enhancement of tumor invasion, and blockade of antitumor inflammatory responses by their M2 polarization [65, 69]. Overall, these observations suggest that ATMs from the surrounding adipose tissue stroma may contribute in critical ways to breast cancer induction and metastasis.

These observations also suggest that alterations in ATM phenotypes with obesity may contribute to the association between obesity and breast cancer risk and prognosis [70, 71]. Recent studies have demonstrated a strong association between the density of CLS in the breast and BMI, adipocyte size, inflammation, and aromatase expression [72]. In mouse models, obesity led to an increase in inflammatory cytokine production in mammary ATMs consistent with an M1 profile that was capable of inducing aromatase expression in preadipocytes [73]. Therefore, the induction of inflammatory ATMs in the breast with obesity may be a component of the link between obesity and breast cancer development. This activation may be related to the role that colony stimulating factor 1 (CSF1) appears to play in macrophage activation and the promotion of metastasis [74]. In addition, hormonal regulation and metabolism within ATMs may be important as macrophages express aromatase and are capable of generating estrogen at physiologic levels [75].

Future studies will be required to understand how resident ATMs contribute to the local tumor environment and the leading edge. The idea that ATM accumulation may be a biomarker for disease risk will also deserve further study. Besides ATMs, the possibility that other leukocyte types in obese tumor-associated adipose tissue will need to be studied as potential modifiers of the mammary epithelial environment.

4.5.2 ATMs and Ovarian Cancer

Epithelial ovarian cancer morbidity and mortality is related to the fact that diagnosis of cancer is often delayed until the disease has reached an advanced stage [76]. Primary tumor cells leave the ovarian capsule and are disseminated via the peritoneal fluid. Within this space, the greater omentum, a major visceral adipose tissue depot, is a preferential site of attachment of tumors and a supportive environment for aggressive tumor growth [77]. The interface between the tumors and the omentum is not diffuse but appears to be localized to structures known as milky spots—immune aggregates on the surface of the omental fat pad. Milky spots are collections of immune cells that include large numbers of macrophages and lymphocytes [78]. These regions are highly vascularized and may provide a rich angiogenic environment to support tumor growth. Work from our lab has demonstrated a significant enlargement in milky spots with aging in mice due to an expansion of $CD4^+$ T cells [79] which may play a role in peritoneal immune surveillance and age-related insulin resistance. Milky spots are not limited to the omentum as we have identified them in multiple visceral adipose tissue depots [80].

The nature and function of milky spot ATMs are poorly understood. Evidence suggests that these ATMs play an important role in immune surveillance of the peritoneal cavity as guardians against pathogens as well as tumor cells [81, 82]. Animal models have demonstrated that tumor cells rapidly accumulate in milky spots and then are cleared presumably by the active innate immune system [83]. Cells that escape this clearance can migrate away from the milky spots where they are capable of forming tumors. In vitro studies suggest that communication between ovarian cancer cells may directly influence the function of milky spot macrophages and T cells to promote a tumorigenic environment [84]. This appears to be linked to the ability of ovarian cancer cells to enhance regulatory T cell function and promote macrophage scavenger receptor expression and cytokine production that enhance tissue repair.

The unique nature of milky spots have led some to hypothesize that enhancement of milky spot ATMs (e.g., enhancement of antigen presentation capacity) may be a future tool in cancer treatment and prevention [82]. It has been speculated that milky spots may provide a supportive environment for cancer stem cells and that enhanced ATM function may promote cancer clearance [85]. Given the involvement of the omentum in metastatic colon and stomach cancer, milky spot ATMs may have a broader function outside of ovarian cancer.

4.6 Conclusions

4.6.1 Current Needs in the Field

While this review has centered around the role of ATMs in the inflammatory response to obesity, it is clear that ATMs are one part of a network of leukocytes in adipose tissue that cooperate to initiate and sustain inflammation in fat with nutrient excess.

Essentially all types of leukocytes have been described in fat including regulatory $CD4^+$ T cells, effector/memory $CD4^+$ T cells, $CD8^+$ T cells, activated B cells, eosinophils, mast cells, neutrophils, and NK cells. Given this, it is unclear if ATMs are initiators and/or effectors of adipose tissue inflammation in obesity. In addition, while we know a reasonable amount about the events that sustain fat inflammation in obesity, little is known about the events that lead to a resolution of this inflammation. Understanding how weight loss resolves inflammatory processes may provide insight into new interventions and prevention strategies in cancer biology.

4.6.2 Future Directions

Currently, the descriptive phase of evaluating adipose tissue leukocyte biology is closing. The current need in the field is to integrate the biology of adipose tissue inflammation and understand how the different types of adipose tissue leukocytes communicate as it is unlikely that they are each operating in a vacuum. This challenge is coupled to the need to increase our understanding of human ATMs which have very different activation profiles than those seen in animal models of obesity.

Part of the challenge in human studies is the limitation of most studies to the examination of subcutaneous fat depots and a limited assessment of visceral fat or tumor-associated adipose tissue. Expanding our understanding of tumor-associated adipose tissue and the contribution of leukocytes derived from adipose tissue is a future need. The conceptual shift that adipose tissue is a dynamic inflammatory organ is still young. Given the importance of inflammatory cytokines and pathways in the initiation and maintenance of tumor biology [86, 87], signals from ATMs around tumor sites may have a significant impact upon cancer biology. Advancements in this concept provide new opportunities in human and animal studies to investigate the contribution of ATMs to tumor biology and to identify new treatment approaches.

Acknowledgement Support for this work was derived in part from NIH Grant DK090262 and DK092873.

References

1. Hotamisligil GS, Spiegelman BM (1994) Tumor necrosis factor alpha: a key component of the obesity-diabetes link. Diabetes 43:1271–1278
2. Harkins JM, Moustaid-Moussa N, Chung YJ, Penner KM, Pestka JJ, North CM et al (2004) Expression of interleukin-6 is greater in preadipocytes than in adipocytes of 3T3-L1 cells and C57BL/6J and ob/ob mice. J Nutr 134:2673–2677
3. Christiansen T, Richelsen B, Bruun JM (2005) Monocyte chemoattractant protein-1 is produced in isolated adipocytes, associated with adiposity and reduced after weight loss in morbid obese subjects. Int J Obes Relat Metab Disord 29:146–150
4. Bruun JM, Lihn AS, Pedersen SB, Richelsen B (2005) MCP-1 release is higher in visceral than subcutaneous human adipose tissue. Implication of macrophages resident in the adipose tissue. J Clin Endocrinol Metab 90:2282–2289

5. Eriksson P, Reynisdottir S, Lonnqvist F, Stemme V, Hamsten A, Arner P (1998) Adipose tissue secretion of plasminogen activator inhibitor-1 in non-obese and obese individuals. Diabetologia 41:65–71
6. Weisberg SP, McCann D, Desai M, Rosenbaum M, Leibel RL, Ferrante AW Jr (2003) Obesity is associated with macrophage accumulation in adipose tissue. J Clin Invest 112:1796–1808
7. Xu H, Barnes GT, Yang Q, Tan G, Yang D, Chou CJ et al (2003) Chronic inflammation in fat plays a crucial role in the development of obesity-related insulin resistance. J Clin Invest 112:1821–1830
8. Michaud A, Drolet R, Noel S, Paris G, Tchernof A (2012) Visceral fat accumulation is an indicator of adipose tissue macrophage infiltration in women. Metabolism 61:689–698
9. Aron-Wisnewsky J, Tordjman J, Poitou C, Darakhshan F, Hugol D, Basdevant A et al (2009) Human adipose tissue macrophages: m1 and m2 cell surface markers in subcutaneous and omental depots and after weight loss. J Clin Endocrinol Metab 94:4619–4623
10. Cancello R, Tordjman J, Poitou C, Guilhem G, Bouillot JL, Hugol D et al (2006) Increased infiltration of macrophages in omental adipose tissue is associated with marked hepatic lesions in morbid human obesity. Diabetes 55:1554–1561
11. Harman-Boehm I, Bluher M, Redel H, Sion-Vardy N, Ovadia S, Avinoach E et al (2007) Macrophage infiltration into omental versus subcutaneous fat across different populations: effect of regional adiposity and the comorbidities of obesity. J Clin Endocrinol Metab 92:2240–2247
12. Emilsson V, Thorleifsson G, Zhang B, Leonardson AS, Zink F, Zhu J et al (2008) Genetics of gene expression and its effect on disease. Nature 452:423–428
13. Chen Y, Zhu J, Lum PY, Yang X, Pinto S, MacNeil DJ et al (2008) Variations in DNA elucidate molecular networks that cause disease. Nature 452:429–435
14. Yang X, Deignan JL, Qi H, Zhu J, Qian S, Zhong J et al (2009) Validation of candidate causal genes for obesity that affect shared metabolic pathways and networks. Nat Genet 41:415–423
15. Weisberg SP, Hunter D, Huber R, Lemieux J, Slaymaker S, Vaddi K et al (2005) CCR2 modulates inflammatory and metabolic effects of high-fat feeding. J Clin Invest 116:115–124
16. Westcott DJ, Delproposto JB, Geletka LM, Wang T, Singer K, Saltiel AR et al (2009) MGL1 promotes adipose tissue inflammation and insulin resistance by regulating 7/4hi monocytes in obesity. J Exp Med 206:3143–3156
17. Kitade H, Sawamoto K, Nagashimada M, Inoue H, Yamamoto Y, Sai Y et al (2012) CCR5 plays a critical role in obesity-induced adipose tissue inflammation and insulin resistance by regulating both macrophage recruitment and M1/M2 status. Diabetes 61:1680–1690
18. Kim JY, van de Wall E, Laplante M, Azzara A, Trujillo ME, Hofmann SM et al (2007) Obesity-associated improvements in metabolic profile through expansion of adipose tissue. J Clin Invest 117:2621–2637
19. Lumeng CN, Deyoung SM, Bodzin JL, Saltiel AR (2007) Increased inflammatory properties of adipose tissue macrophages recruited during diet-induced obesity. Diabetes 56:16–23
20. Lumeng CN, Bodzin JL, Saltiel AR (2007) Obesity induces a phenotypic switch in adipose tissue macrophage polarization. J Clin Invest 117:175–184
21. Kosteli A, Sugaru E, Haemmerle G, Martin JF, Lei J, Zechner R et al (2010) Weight loss and lipolysis promote a dynamic immune response in murine adipose tissue. J Clin Invest 120:3466–3479
22. Kishore P, Li W, Tonelli J, Lee DE, Koppaka S, Zhang K et al (2010) Adipocyte-derived factors potentiate nutrient-induced production of plasminogen activator inhibitor-1 by macrophages. Sci Transl Med 2:20ra15
23. Biswas SK, Mantovani A (2012) Orchestration of metabolism by macrophages. Cell Metab 15:432–437
24. Odegaard JI, Ricardo-Gonzalez RR, Red Eagle A, Vats D, Morel CR, Goforth MH et al (2008) Alternative M2 activation of Kupffer cells by PPARdelta ameliorates obesity-induced insulin resistance. Cell Metab 7:496–507

25. Odegaard JI, Ricardo-Gonzalez RR, Goforth MH, Morel CR, Subramanian V, Mukundan L et al (2007) Macrophage-specific PPARgamma controls alternative activation and improves insulin resistance. Nature 447:1116–1120
26. Kang K, Reilly SM, Karabacak V, Gangl MR, Fitzgerald K, Hatano B et al (2008) Adipocyte-derived Th2 cytokines and myeloid PPARdelta regulate macrophage polarization and insulin sensitivity. Cell Metab 7:485–495
27. Martinez FO, Gordon S, Locati M, Mantovani A (2006) Transcriptional profiling of the human monocyte-to-macrophage differentiation and polarization: new molecules and patterns of gene expression. J Immunol 177:7303–7311
28. Odegaard JI, Chawla A (2011) Alternative macrophage activation and metabolism. Annu Rev Pathol 6:275–297
29. Chawla A, Nguyen KD, Goh YP (2011) Macrophage-mediated inflammation in metabolic disease. Nat Rev Immunol 11:738–749
30. Lumeng CN, DelProposto JB, Westcott DJ, Saltiel AR (2008) Phenotypic switching of adipose tissue macrophages with obesity is generated by spatiotemporal differences in macrophage subtypes. Diabetes 57:3239–3246
31. Shaul ME, Bennett G, Strissel KJ, Greenberg AS, Obin MS (2010) Dynamic, M2-like remodeling phenotypes of CD11c+ adipose tissue macrophages during high-fat diet—induced obesity in mice. Diabetes 59:1171–1181
32. Ichimura A, Hirasawa A, Poulain-Godefroy O, Bonnefond A, Hara T, Yengo L et al (2012) Dysfunction of lipid sensor GPR120 leads to obesity in both mouse and human. Nature 483:350–354
33. Oh DY, Talukdar S, Bae EJ, Imamura T, Morinaga H, Fan W et al (2010) GPR120 is an omega-3 fatty acid receptor mediating potent anti-inflammatory and insulin-sensitizing effects. Cell 142:687–698
34. Hotamisligil GS, Shargill NS, Spiegelman BM (1993) Adipose expression of tumor necrosis factor-alpha: direct role in obesity-linked insulin resistance. Science 259:87–91
35. Kim F, Pham M, Luttrell I, Bannerman DD, Tupper J, Thaler J et al (2007) Toll-like receptor-4 mediates vascular inflammation and insulin resistance in diet-induced obesity. Circ Res 100:1589–1596
36. Shi H, Kokoeva MV, Inouye K, Tzameli I, Yin H, Flier JS (2006) TLR4 links innate immunity and fatty acid-induced insulin resistance. J Clin Invest 116:3015–3025
37. Himes RW, Smith CW (2010) Tlr2 is critical for diet-induced metabolic syndrome in a murine model. FASEB J 24:731–739
38. Rocha VZ, Folco EJ, Sukhova G, Shimizu K, Gotsman I, Vernon AH et al (2008) Interferon-gamma, a Th1 cytokine, regulates fat inflammation: a role for adaptive immunity in obesity. Circ Res 103:467–476
39. Yuan M, Konstantopoulos N, Lee J, Hansen L, Li ZW, Karin M et al (2001) Reversal of obesity- and diet-induced insulin resistance with salicylates or targeted disruption of Ikkbeta. Science 293:1673–1677
40. Arkan MC, Hevener AL, Greten FR, Maeda S, Li ZW, Long JM et al (2005) IKK-beta links inflammation to obesity-induced insulin resistance. Nat Med 11:191–198
41. Donath MY, Shoelson SE (2011) Type 2 diabetes as an inflammatory disease. Nat Rev Immunol 11:98–107
42. Goldfine AB, Silver R, Aldhahi W, Cai D, Tatro E, Lee J et al (2008) Use of salsalate to target inflammation in the treatment of insulin resistance and Type 2 diabetes. Clin Transl Sci 1:36–43
43. Han J, Lee JE, Jin J, Lim JS, Oh N, Kim K et al (2011) The spatiotemporal development of adipose tissue. Development 138:5027–5037
44. Wu D, Molofsky AB, Liang HE, Ricardo-Gonzalez RR, Jouihan H, Bando JK et al (2011) Eosinophils sustain adipose alternatively activated macrophages associated with glucose homeostasis. Science 332(6026):243–247

45. Ricardo-Gonzalez RR, Red Eagle A, Odegaard JI, Jouihan H, Morel CR, Heredia JE et al (2010) IL-4/STAT6 immune axis regulates peripheral nutrient metabolism and insulin sensitivity. Proc Natl Acad Sci U S A 107:22617–22622
46. Morris DL, Singer K, Lumeng CN (2011) Adipose tissue macrophages: phenotypic plasticity and diversity in lean and obese states. Curr Opin Clin Nutr Metab Care 14:341–346
47. Wentworth JM, Naselli G, Brown WA, Doyle L, Phipson B, Smyth GK et al (2010) Pro-inflammatory CD11c+CD206+ adipose tissue macrophages are associated with insulin resistance in human obesity. Diabetes 59:1648–1656
48. Li P, Lu M, Nguyen MTA, Bae E, Chapman J, Feng D et al (2010) Functional heterogeneity of CD11C positive adipose tissue macrophages in diet-induced obese mice. J Biol Chem 285:15333–15345
49. Divoux A, Tordjman J, Lacasa D, Veyrie N, Hugol D, Aissat A et al (2010) Fibrosis in human adipose tissue: composition, distribution, and link with lipid metabolism and fat mass loss. Diabetes 59:2817–2825
50. Cinti S, Mitchell G, Barbatelli G, Murano I, Ceresi E, Faloia E et al (2005) Adipocyte death defines macrophage localization and function in adipose tissue of obese mice and humans. J Lipid Res 46:2347–2355
51. Oh DY, Morinaga H, Talukdar S, Bae EJ, Olefsky JM (2012) Increased macrophage migration into adipose tissue in obese mice. Diabetes 61:346–354
52. Nishimura S, Manabe I, Nagasaki M, Seo K, Yamashita H, Hosoya Y et al (2008) In vivo imaging in mice reveals local cell dynamics and inflammation in obese adipose tissue. J Clin Invest 118:710–721
53. Tsou CL, Peters W, Si Y, Slaymaker S, Aslanian AM, Weisberg SP et al (2007) Critical roles for CCR2 and MCP-3 in monocyte mobilization from bone marrow and recruitment to inflammatory sites. J Clin Invest 117:902–909
54. Kanda H, Tateya S, Tamori Y, Kotani K, Hiasa K, Kitazawa R et al (2006) MCP-1 contributes to macrophage infiltration into adipose tissue, insulin resistance, and hepatic steatosis in obesity. J Clin Invest 116:1494–1505
55. Huber J, Loffler M, Bilban M, Reimers M, Kadl A, Todoric J et al (2007) Prevention of high-fat diet-induced adipose tissue remodeling in obese diabetic mice by n-3 polyunsaturated fatty acids. Int J Obes 31:1004–1013
56. Strissel KJ, Stancheva Z, Miyoshi H, Perfield JW II, DeFuria J, Jick Z et al (2007) Adipocyte death, adipose tissue remodeling, and obesity complications. Diabetes 56:2910–2918
57. Feng D, Tang Y, Kwon H, Zong H, Hawkins M, Kitsis RN et al (2011) High-fat diet-induced adipocyte cell death occurs through a cyclophilin d intrinsic signaling pathway independent of adipose tissue inflammation. Diabetes 60:2134–2143
58. Lumeng CN, Deyoung SM, Saltiel AR (2007) Macrophages block insulin action in adipocytes by altering expression of signaling and glucose transport proteins. Am J Physiol Endocrinol Metab 292:E166–E174
59. Bilkovski R, Schulte DM, Oberhauser F, Mauer J, Hampel B, Gutschow C et al (2011) Adipose tissue macrophages inhibit adipogenesis of mesenchymal precursor cells via wnt-5a in humans. Int J Obes 35:1450–1454
60. Trayhurn P, Wang B, Wood IS (2008) Hypoxia in adipose tissue: a basis for the dysregulation of tissue function in obesity? Br J Nutr 100:227–235
61. Bourlier V, Zakaroff-Girard A, Miranville A, De Barros S, Maumus M, Sengenes C et al (2008) Remodeling phenotype of human subcutaneous adipose tissue macrophages. Circulation 117:806–815
62. Stefanovic-Racic M, Yang X, Turner MS, Mantell BS, Stolz DB, Sumpter TL et al (2012) Dendritic cells promote macrophage infiltration and comprise a substantial proportion of obesity-associated increases in CD11c+ cells in adipose tissue and liver. Diabetes 61:2330–2339
63. Khandekar MJ, Cohen P, Spiegelman BM (2011) Molecular mechanisms of cancer development in obesity. Nat Rev Cancer 11:886–895

64. O'Brien J, Martinson H, Durand-Rougely C, Schedin P (2012) Macrophages are crucial for epithelial cell death and adipocyte repopulation during mammary gland involution. Development 139:269–275
65. O'Brien J, Lyons T, Monks J, Lucia MS, Wilson RS, Hines L et al (2010) Alternatively activated macrophages and collagen remodeling characterize the postpartum involuting mammary gland across species. Am J Pathol 176:1241–1255
66. Fischer-Posovszky P, Wang QA, Asterholm IW, Rutkowski JM, Scherer PE (2011) Targeted deletion of adipocytes by apoptosis leads to adipose tissue recruitment of alternatively activated m2 macrophages. Endocrinology 152:3074–3081
67. Laoui D, Movahedi K, Van Overmeire E, Van den Bossche J, Schouppe E, Mommer C et al (2011) Tumor-associated macrophages in breast cancer: distinct subsets, distinct functions. Int J Dev Biol 55:861–867
68. Egeblad M, Ewald AJ, Askautrud HA, Truitt ML, Welm BE, Bainbridge E et al (2008) Visualizing stromal cell dynamics in different tumor microenvironments by spinning disk confocal microscopy. Dis Model Mech 1:155–167; discussion 165
69. Pollard JW (2008) Macrophages define the invasive microenvironment in breast cancer. J Leukoc Biol 84:623–630
70. Cecchini RS, Costantino JP, Cauley JA, Cronin WM, Wickerham DL, Land SR et al (2012) Body mass index and the risk for developing invasive breast cancer among high-risk women in NSABP P-1 and STAR breast cancer prevention trials. Cancer Prev Res 5:583–592
71. McTiernan A, Irwin M, Vongruenigen V (2010) Weight, physical activity, diet, and prognosis in breast and gynecologic cancers. J Clin Oncol 28:4074–4080
72. Morris PG, Hudis CA, Giri D, Morrow M, Falcone DJ, Zhou XK et al (2011) Inflammation and increased aromatase expression occur in the breast tissue of obese women with breast cancer. Cancer Prev Res 4:1021–1029
73. Subbaramaiah K, Howe LR, Bhardwaj P, Du B, Gravaghi C, Yantiss RK et al (2011) Obesity is associated with inflammation and elevated aromatase expression in the mouse mammary gland. Cancer Prev Res (Phila) 4:329–346
74. Lin EY, Nguyen AV, Russell RG, Pollard JW (2001) Colony-stimulating factor 1 promotes progression of mammary tumors to malignancy. J Exp Med 193:727–740
75. Mor G, Yue W, Santen RJ, Gutierrez L, Eliza M, Berstein LM et al (1998) Macrophages, estrogen and the microenvironment of breast cancer. J Steroid Biochem Mol Biol 67:403–411
76. Ozols RF, Bookman MA, Connolly DC, Daly MB, Godwin AK, Schilder RJ et al (2004) Focus on epithelial ovarian cancer. Cancer Cell 5:19–24
77. Gerber SA, Rybalko VY, Bigelow CE, Lugade AA, Foster TH, Frelinger JG et al (2006) Preferential attachment of peritoneal tumor metastases to omental immune aggregates and possible role of a unique vascular microenvironment in metastatic survival and growth. Am J Pathol 169:1739–1752
78. Shimotsuma M, Shields JW, Simpson-Morgan MW, Sakuyama A, Shirasu M, Hagiwara A et al (1993) Morpho-physiological function and role of omental milky spots as omentum-associated lymphoid tissue (OALT) in the peritoneal cavity. Lymphology 26:90–101
79. Lumeng CN, Liu J, Geletka L, Delaney C, Delproposto J, Desai A et al (2011) Aging is associated with an increase in T cells and inflammatory macrophages in visceral adipose tissue. J Immunol 187:6208–6216
80. Morris DL, Oatmen KE, Wang T, DelProposto JL, Lumeng CN (2012) CX3CR1 deficiency does not influence trafficking of adipose tissue macrophages in mice with diet-induced obesity. Obesity 20:1189–1199
81. Cao L, Hu X, Zhang Y, Sun XT (2011) Omental milky spots in screening gastric cancer stem cells. Neoplasma 58:20–26
82. Sorensen EW, Gerber SA, Sedlacek AL, Rybalko VY, Chan WM, Lord EM (2009) Omental immune aggregates and tumor metastasis within the peritoneal cavity. Immunol Res 45:185–194

83. Oosterling SJ, van der Bij GJ, Bogels M, van der Sijp JR, Beelen RH, Meijer S et al (2006) Insufficient ability of omental milky spots to prevent peritoneal tumor outgrowth supports omentectomy in minimal residual disease. Cancer Immunol Immunother 55:1043–1051
84. Alvero AB, Montagna MK, Craveiro V, Liu L, Mor G (2012) Distinct subpopulations of epithelial ovarian cancer cells can differentially induce macrophages and T regulatory cells toward a pro-tumor phenotype. Am J Reprod Immunol 67:256–265
85. Cao L, Hu X, Zhang Y (2009) Omental milky spots—highly efficient "natural filter" for screening gastric cancer stem cells. Med Hypotheses 73:1017–1018
86. Korkaya H, Kim GI, Davis A, Malik F, Henry NL, Ithimakin S et al (2012) Activation of an IL6 inflammatory loop mediates trastuzumab resistance in HER2+ breast cancer by expanding the cancer stem cell population. Mol Cell 47:570–584
87. DiDonato JA, Mercurio F, Karin M (2012) NF-kappaB and the link between inflammation and cancer. Immunol Rev 246:379–400

Chapter 5
Dietary Fats as Mediators of Obesity, Inflammation, and Colon Cancer

Stephanie K. Doerner and Nathan A. Berger

Abstract Obesity and associated low-grade inflammation are clearly risk factors for development of diabetes, cardiovascular disease, and cancer; however, the mechanisms and pathways by which obesity and inflammation lead to these disorders are not clearly defined. Since obesity is largely determined by levels of energy expenditure as well as quantities and composition of consumed nutrients, especially fats and carbohydrates, the question exists as to whether obesity and/or dietary components contribute directly to development of inflammation and/or associated comorbidities including diabetes, cancer, and cardiovascular disease. In this chapter, we examine the evidence supporting a role for dietary fats in the development of inflammation and intestinal tumorigenesis. We also compare different fats and different diets for their ability to promote or prevent intestinal tumorigenesis and explore possible mechanisms of action. These considerations are important for the potential prevention and control of intestinal cancer, since overall diet and specific dietary components are modifiable risk factors and increasing numbers of dietary and pharmacologic interventions are becoming available to control both inflammatory and carcinogenic processes.

S.K. Doerner, Ph.D.
Department of Genetics, Case Western Reserve University School of Medicine, 10900 Euclid Avenue, Cleveland, OH 44106-4915, USA
e-mail: stephanie.doerner@case.edu

N.A. Berger, M.D. (✉)
Center for Science, Health and Society, Case Comprehensive Cancer Center, Case Western Reserve University School of Medicine, 10900 Euclid Avenue, Cleveland, OH 44106-4971, USA
e-mail: nab@case.edu

A.J. Dannenberg and N.A. Berger (eds.), *Obesity, Inflammation and Cancer*, Energy Balance and Cancer 7, DOI 10.1007/978-1-4614-6819-6_5,

5.1 Introduction

Dietary fat was once considered to simply be a source of fatty acids to fuel energy needs and to support development of adipose tissue to serve as an energy reserve, with fatty acid oxidation supplying as much as 2 times the amount of energy that can be derived from equal amounts of carbohydrates or proteins. It is now apparent that dietary fatty acids provide key structural elements for cellular and organelle membrane structure, function and fluidity; support proper growth and development; facilitate absorption of fat-soluble molecules including vitamins A, D, E, and K; and insulate internal organs and neural processes [1, 2]. Fatty acids and their derivatives are critical in regulating and contributing to organogenesis, body temperature, the immune and inflammatory responses, and unique signaling pathways such as those mediated by prostaglandins, phospholipids, sphingolipids, ceramides, and others. Fatty acid consumption and metabolism is regulated by a complex series of ecologic, ethnic and environmental factors, and biobehavioral processes. Disorders of fatty acid intake and metabolism contribute to a wide variety of metabolic, immunologic, and neoplastic disorders including obesity, inflammation, diabetes, and cancer.

Obesity is clearly associated with an increased risk for a number of malignancies including colon cancer, esophageal adenocarcinoma, postmenopausal breast cancer, renal cell carcinoma, uterine cancer, and pancreatic cancer, and the list continues to expand [3, 4].

In addition to its association with metabolic and neoplastic disorders, obesity is now recognized as a chronic low-grade inflammatory disorder, where both dietary fats and the increase in adipose tissue contribute to the inflammatory state and associated comorbidities [5]. Although not consistently confirmed, there is substantial evidence linking dietary fats as contributing factors to a variety of malignancies including breast, pancreatic, hepatic, colon, prostate and others [6–17]. However, because of the complex interactions between diet, obesity, inflammation, and cancer, it is frequently difficult to separate the etiologic contribution of each in the process and their independent role in carcinogenesis. In particular, the fact that high dietary fat leads to both obesity and inflammation makes it difficult to determine which component leads to cancer. Defining these relations is further confounded by the fact that all fats are not equal, in that fatty acids of different lengths and degrees of saturation may have different biologic and pathophysiologic effects, which sometimes may be additive or complimentary, and at others may be competitive or antagonistic. In this regard, it is important to remember that although large studies such as the Women's Health Initiative, in which a total of 48,835 postmenopausal participants were randomized to a diet modification where 19,541 were assigned to consume 20 % lower fat compared to 29,294 women who consumed a usual diet, there was no significant reduction in risk for colorectal cancer (CRC) [18, 19]. In contrast, other studies, outlined below, do show changes in CRC risk in association with different levels of consumption of specific fatty acids.

In this chapter, we focus on the relation of dietary fats to inflammatory processes and to colon cancer. First, we provide a brief outline of fatty acid structure and nomenclature. We next review the epidemiologic and nutritional evidence supporting a role for dietary fatty acids and selected diets in the promotion and prevention of colon cancer. We subsequently review the pro- and anti-inflammatory effects of specific fatty acids and diets and their possible contribution to colon cancer. Finally, we address potential mechanisms and mediators of this relation and how they may serve as targets for cancer prevention and control.

5.2 Fatty Acid Structure and Nomenclature

Fatty acids are hydrocarbon chains with a methyl group at one terminus and a carboxyl at the other. As indicated in Table 5.1, they vary in chain length of carbon atoms connected by bonds that may be saturated or unsaturated [20]. Those with a single unsaturated bond are designated monounsaturated fatty acids (MUFAs), whereas those with more than one double bond are polyunsaturated fatty acids (PUFA). When multiple double bonds exist, they are never adjacent. The hydrogen atoms on either side of the double bond, in naturally occurring fatty acids, are almost always in the *cis* configuration, whereas in synthetic fatty acids, the hydrogen atoms on either side of the double bond may be in the *trans* configuration providing the basis for trans fats. Fatty acids are numbered with the carbon atom in the carboxyl group designated number one, and subsequent numbers progressing sequentially through consecutive carbons to the methyl group at the other end of the molecule which is designated the omega (ω) carbon. The (ω) carbon is sometimes designated as n for its position at the end. In the case of unsaturated fatty acids, the location of the first double bond is counted from the ω carbon atom of the methyl group at the end of the hydrocarbon chain.

Table 5.1 Structure and nomenclature of dietary fatty acids with important impact on colon cancer

Symbol	Common name	Structure	CRC impact
Saturated fatty acids			
C12:0	Lauric	$CH_3(CH_2)_{10}COOH$	↑
C14:0	Myristic	$CH_3(CH_2)_{12}COOH$	↑
C16:0	Palmitic	$CH_3(CH_2)_{14}COOH$	↑
C18:0	Stearic	$CH_3(CH_2)_{16}COOH$	↑
Monounsaturated fatty acids			
C16:1n-7	Palmitoleic	$CH_3(CH_2)_5CH{=}CH(CH_2)_7COOH$	
C18:1n-9	Oleic	$CH_3(CH_2)_7CH{=}CH(CH_2)_7COOH$	↓
Polyunsaturated fatty acids			
C18:2n-6	Linoleic	$CH_3(CH_2)_4(CH{=}CHCH_2)_2\,(CH_2)_6COOH$	↑
C18:3n-3	Linolenic	$CH_3CH_2(CH{=}CHCH_2)_3\,(CH_2)_6COOH$	↓
C20:4n-6	Arachidonic	$CH_3(CH_2)_4(CH{=}CHCH_2)_4\,(CH_2)_2COOH$	↑
C20:5n-3	Eicosapentaenoic	$CH_3CH_2(CH{=}CHCH_2)_5\,(CH_2)_2COOH$	↓
C22:6n-3	Docosahexaenoic	$CH_3CH_2(CH{=}CHCH_2)_6\,CH_2COOH$	↓

Symbol provides number of carbon atoms in fatty acid chain: followed by number of double bonds for unsaturated fatty acids, n-x provides position of first double bond relative to the ω carbon at the terminal methyl end [20]. Impact on colorectal cancer (CRC), ↑ = promoter, ↓ = suppressor

Structure and function of fatty acids are determined by the hydrocarbon chain length and the number, location, and geometric configurations of the double bonds. The saturated fatty acids are the most flexible, with unsaturated fatty acids being more rigid and angled at their double bonds. The long-chain fatty acids of biologic significance usually contain an even number of carbon atoms, between 14 and 22, with 16–18 being most abundant. Lauric and myristic acids are medium-chain saturated fatty acids; palmitic and stearic acids are long-chain saturated fatty acids. Among the unsaturated fatty acids, the most abundant are oleic, linoleic, linolenic and arachidonic acids. The biologically important MUFAs, palmitoleic and oleic acids, can be synthesized in animal tissues from the saturated fatty acids, palmitic and stearic acids by process of fatty acid desaturation, whereas other biologically important MUFAs, linoleic or linolenic, cannot be synthesized by mammalian tissues, must be derived from exogenous sources, and are consequently designated essential fatty acids. Polyunsaturated fatty acids (PUFA) contain more than one double bond. Their composition in the cell membrane can be determined by diet, and they drastically affect membrane function and intracellular signaling [21, 22]. There are two general groups of PUFAs, both are important to general health and disease and specifically to colon cancers. They are the ω3 PUFAs and the ω6 PUFAs. They are not interconvertible and dietary substitution of one results in a competitive reduction in the other [21, 23]. Omega 3 fatty acids, containing unsaturated, double bonds after the third carbon from the methyl end, include eicosapentaenoic acid (EPA) and docosahexaenoic acid (DHA) which are common in marine plants and fish, are generally not made in mammalian tissues, and are also considered essential fatty acids. Another ω3 fatty acid, linolenic acid, is derived from a number of plant seeds and green leafy vegetables. The n6 or ω6 fatty acids, in which the first unsaturated double bond is located after the sixth carbon atom from the methyl end, of biologic importance, include arachidonic acid and linoleic acid, commonly found in corn, soybean, sunflower, sesame, and palm oils [24].

5.3 Epidemiology of Dietary Fat and Colon Cancer

Early observations implicating dietary fats in the etiology of colon cancer were based on differences in the incidence of malignancy where significant anthropological and/or ecological differences existed in diet or occurred in association with diet change, particularly in migrating populations [25–27]. Anthropologic studies indicate that our Paleolithic ancestors consumed a diet high in fiber and carbohydrates, with abundant protein from fruit and vegetable sources, fish, and wild game [28, 29]. In contrast, current diets in affluent Western countries are composed of calorie-dense, nutrient-poor foods that are rich in fat, oil, and sugar [30]. This dietary composition is implicated in the increased risk of colon cancer and other malignancies characteristic of many Western countries [31, 32] and, in particular, the increased cancer risk in populations that migrate from countries that consume low amounts of these same detrimental nutrients [33].

Studying geographic differences, Wynder et al. reported the high incidence of colon cancer in the USA compared to the lower incidence in Japan [27]. They noted also that the increase in colon cancer that occurred in Japan in the 20-year post-World War II period occurred in association with relatively high consumption of eggs, milk, meat, fat, and fruit [27]. They also postulated that the difference might be associated with different intestinal bacteria associated with either the different ethnic groups and/or different diets. Also on the basis of geographic differences in food consumption, especially comparing African to Western countries, Burkitt found that CRC incidence was associated with high dietary fat consumption but low dietary fiber [26]. Based on these observations, he introduced the concept that specific foods could be protective against the carcinogenic effects of others and recommended consumption of diets containing reduced fats and increased fiber to protect against colon cancer, especially in Western countries [26].

Comparison of environmental factors and relation to cancer in 30 countries by Armstrong and Doll [25] showed that colon cancer incidence correlated with per capita consumption of meat, animal protein, and total fat, with the United States and New Zealand showing both the highest per capita consumption and the highest colon cancer incidence compared to Japan and Nigeria with the lowest. These observations were subsequently confirmed and refined to demonstrate that correlation of colon cancer with fat was more specifically associated with animal fat [12].

An important example of emigrational epidemiology involving diet change and cancer risk is shown by the observation that Japanese men born in Japan had a reduced risk of CRC compared with US-born white men, whereas Japanese men born in the USA experienced CRC rates twice as high as foreign-born Japanese men [34–36]. US-born Japanese women had higher rates of CRC than Japanese-born Japanese women or US white women. Colon cancer mortality was also increased in US-born Japanese patients [34, 37]. Evidence suggests that the increase in colon cancer development and mortality is due to an increased consumption of Western dietary components, mainly high dietary fat [35]. Similar observations of increased risk for colon cancer in association with diet change have been made in population groups moving to the United States from South Korea, Vietnam, Cuba, Puerto Rico, and Mexico [38–40].

Another example of the impact of dietary composition on cancer incidence is provided by comparison of Pima Indians who have migrated from Mexico to Arizona [41, 42]. The Pima Indians living in Arizona are exposed to energy-dense foods and a decreased need for physical labor. They have a high prevalence of obesity, metabolic syndrome, diabetes, and multiple malignancies including colon cancer, when compared to the Pima Indians that still live in Mexico, who follow a traditional diet and customs and have a lower prevalence of these diseases [43]. Compared to the general population, Pima Indians have a strong genetic susceptibility to obesity, metabolic syndrome, and significantly higher mortality from heart disease and malignancies. The increase in these disorders associated with migration from Mexico to Arizona demonstrates a powerful impact of environment, especially diet on these conditions.

Another series of studies supporting an association of dietary fat with colon cancer comes from comparing the so-called Mediterranean diet to Western diets. In general, Mediterranean countries have lower rates of colon cancer compared to Western countries [44, 45]. As much as 76 % of the intercountry variation in colon cancer incidence rates have been attributed to three dietary factors, meat, fish, and olive oil, with meat and fish showing a positive association and olive oil showing a negative association with colon cancer [46]. Other studies, including three out of six case–control studies in Mediterranean populations, show an inverse association between monounsaturated and saturated fatty acid ratio with CRC [47].

5.4 Animal Studies Supporting Involvement of Fatty Acids in Colorectal Cancer

As detailed in this section and for each fatty acid, multiple studies in animal models now support the association between high-fat diets and intestinal tumor development. These models usually examine the effects of varying dietary fat quantity and/or quality in rodents treated with primary intestinal carcinogens such as azoxymethane (AOM) or dimethylhydrazine (DMH) or in rodents with hereditary genetic mutations predisposing to intestinal tumorigenesis [48]. A common model used in the latter situation is the APC^{Min} mouse containing a mutation in the FAP gene, the same gene that leads to familial adenomatous polyposis in humans and also shows spontaneous mutations in a high percentage of sporadic cancers [49, 50].

Initial studies to define the specific effects of different fats showed that increased amounts of dietary fat increased intestinal tumorigenesis in rats treated with DMH, and that both corn oil, high in unsaturated fats, and lard, high in saturated fats, each promoted increased intestinal tumorigenesis [51]. Subsequent studies showed differential effects of fat on intestinal tumorigenesis based on source, with corn oil, lard, or beef fat causing an increase in intestinal tumorigenesis in DMH-treated rats, whereas supplementation with olive oil reduced tumors in DMH-treated rats. Similarly, supplementation with fish oil (Menhaden oil) reduced tumors in AOM-treated rats [10, 11]. Dietary fats may also influence carcinogen-induced intestinal tumorigenesis by altering intestinal microbiota. Germ-free rats treated with DMH, which is actually a procarcinogen, showed less tumorigenesis than conventionally housed animals. In contrast, tumorigenesis in germ-free rats treated with the primary carcinogen *N*-methyl-*N*-nitro-*N′*-nitrosoguanidine (MNNG) showed slightly increased tumorigenesis, relative to MNNG-treated conventionally housed animals [52]. Since DMH is a procarcinogen, requiring metabolic activation, whereas MNNG is a direct-acting carcinogen, these studies suggest a role for intestinal microflora in carcinogen activation and/or tumor promotion. These and similar results led to an extensive expansion of studies, summarized below, to define the intestinal tumorigenic effects of specific fatty acids from multiple sources.

5.5 Effects of Specific Fatty Acids

5.5.1 Saturated Fatty Acids

5.5.1.1 Lauric and Myristic Acids

Lauric acid (C12.0) and myristic acid (C14.0) are saturated, medium-chain fatty acids (MCFA) found in high concentration in tropical plants such as coconuts and in dairy fats. They appear to promote colon tumorigenesis and significant increases in lauric, and myristic acids have been identified in subjects at high risk for colon cancer [53]. At least one case–control study noted a significant increase in lauric acid intake with colon cancer risk [54]. These two saturated MCFAs have been shown to induce proinflammatory factors in murine and human macrophages and colon cell lines. Treatment of a murine monocyte macrophage cell line, RAW 264.7, with lauric or myristic acid independently increased NF-κB activation, which serves as a transcription factor for multiple proinflammatory cytokines including interleukin-1β (IL-1β), IL-6, TNF-α, and C-reactive protein (CRP) as well as increased expression of cyclooxygenese-2 (COX-2) [55]. COX-2 is a major enzyme involved in synthesis of prostaglandin E_2 (PGE_2), the latter playing an important role in development of colon cancer [56–58]. Many studies show higher PGE_2 levels in human CRC samples compared to paired tissue samples of normal colon mucosa from the same patient [59]. These latter studies suggest that elevated tumor levels of PGE_2 are determined by specific processes within the tumor that increase PGE_2 and auto-promote tumor growth. PGE_2 levels are regulated both by COX-2-mediated synthesis and degradation by 15-prostaglandin dehydrogenase (15-PGDH). 15-PGDH is an NAD+-linked dehydrogenase that oxidizes the hydroxyl group at the 15 position, reducing the level and biologic consequences of PGE_2. 15-PGDH, the rate-limiting enzyme catalyzing degradation of PGE_2, is commonly inactivated in colon tumors providing a partial explanation for elevated levels of PGE_2 in the tumor tissue [60, 61]. PGE_2 is a proinflammatory mediator that decreases production of IL-2 and IFN-α leading to production of proinflammatory T helper cells. PGE_2 also has direct effects on stimulating growth-promoting pathways, increasing cell proliferation and decreasing apoptosis in intestinal epithelial cells [58, 62].

Treatment of cultures of human nonmalignant colon epithelial cells (HCEC) and multiple human colon cancer cell lines, including HCT 116, SW48, SW480, HT29, and HCA-7, with lauric or myristic acids increases NF-κB, COX-2, and prostaglandin production [63–65].

Further studies suggest that the effects of fatty acids on NF-κB activation and COX-2 expression may result from incorporation of saturated fatty acids into lipopolysaccharide (LPS) followed by signaling through stimulation of Toll-like receptors (TLRs) [66]. These studies show that lauric and myristic acid act on both intestinal epithelial and immune cells to promote increases in inflammatory factors, some of which promote growth in normal and malignant colon tissues.

5.5.1.2 Palmitic and Stearic Acids

Palmitic (C16:0) and stearic acids (C18:0) are long-chain saturated fatty acids found in both animal and plant sources. Epidemiologic studies on the association of palmitic and stearic acids with colon cancer are controversial with many reporting no relation with colon cancer risk [67, 68]. However, in a national prospective case–control study in Scotland that included 1,455 incident cases and 1,455 matched controls, palmitic acid was dose-dependently associated with increased colon cancer risk [69]. In the $Apc^{Min/+}$ murine model, levels of erythrocyte membrane palmitic acid were significantly associated with the development of intestinal tumors [70]. Several studies have demonstrated an increase in stearic acid intake and colon cancer risk, increased stearic acid in the plasma of colon cancer patients compared to controls [67], and increased stearic acid in colon cancer specimens compared to normal tissue [53, 71].

Stearic acid has been shown to increase proliferation of HT-29 human colon cancer cell lines [72] and stearic acid has been shown to promote colon carcinogenesis in male Sprague Dawley rats injected with colon carcinogens [73]. Examination of membrane phospholipids in these rodents showed significantly higher amounts of stearic acid in the tumor-bearing mice compared to controls [73]. From a mechanistic viewpoint, at least one study demonstrated that high levels of membrane stearic acid in tumor cells could inhibit the immune response and subsequent apoptosis [74]. Concentrations of fibrinogen, another marker of inflammation, were shown to be elevated after high dietary consumption of stearic acid [75]. Although not specific to palmitic or stearic acids, examination of the 1999–2000 National Health and Nutrition Examination Study (NHANES 1999–2000) showed a modest association of elevated CRP with consumption of saturated fatty acids [76].

5.5.2 *Monounsaturated Fatty Acids*

5.5.2.1 Palmitoleic Acid

Palmitoleic acid (C16:1n-7), an omega MUFA, commonly found as a minor component of many animal and vegetable oils and in higher concentration in macadamia nuts, has been shown to have anti-inflammatory properties, to reduce hypercholesterolemia, and to have a beneficial effect on muscle insulin resistance; however, its effects on cancer and especially on CRC remain to be investigated [77, 78].

5.5.2.2 Oleic Acid

Oleic acid (C18:1n-9) is a MUFA that is the major component of olive oil and can also be found in meats, nuts, and dairy products. Several studies have suggested that the ratio of oleic acid to stearic acid is important in determining disease outcome. In two human studies, a decreased erythrocyte oleic/stearic acid ratio was reported in patients with CRC, and similar changes were shown in their cancer tissues [71, 79].

To test the ability of oleic acid to interfere with the procarcinogenic effect of stearic acid, mice that were treated with DMH and fed diets high in stearic acid were supplemented with oleic acid. The latter mice showed reduction in intestinal polyp multiplicity and size relative to those without the oleic acid supplement [80]. In human cell culture, it has been shown that oleic acid can inhibit the proliferative effects of stearic acid as well as inhibit stearic acid-induced NF-κB activation and intercellular adhesion molecule-1 (ICAM-1) expression [81]. Studies conducted with human colon cancer cell lines show that treatment of HT-29 and Caco-2 cells with oleic acid results in decreased proliferation and increased apoptosis. Oleic acid has been shown to decrease factors associated with inflammation and tumorigenesis such as COX-2 and Bcl-2 [82]. Raising dietary levels of oleic acid showed a trend to lower circulating inflammatory markers including CRP, IL-6, and E-selectin compared to diets enriched for saturated fatty acids, trans fatty acids, or stearic acids [75, 76]. Oleic acid has also been shown to inhibit calcium influx pathways important for proliferation and inflammation as well as for calcium-induced apoptosis in HT-29 colon cancer cells [83]. Interestingly, olive-oil-fed mice showed reduced production of nitric oxides and other inflammatory mediators [84].

5.5.3 Polyunsaturated Fatty Acids

PUFA composition of the cell membrane can drastically change intercellular signaling and can be modulated by diet [22]. Two general groups of PUFAs that play critical roles in health and disease are the omega-3 and omega-6 PUFA. They are not interconvertible, and dietary substitution of one results in a competitive reduction in concentration of the other in all tissues [21, 23]. Omega-3 (ω-3) PUFAs are essential fatty acids commonly found in plant and marine oils that include α-linolenic, eicosapentaenoic, and DHAs [21]. Omega 6 (ω-6) PUFA are essential fatty acids commonly found in corn, soybean, sunflower, and palm oils and include arachidonic, linoleic, and γ-linoleic acids [24]. Comparing tumor tissue to normal intestinal mucosa in patients undergoing surgery for colon cancer showed that tumor tissue contained elevated PUFA levels, including arachidonic acid and its derivative PGE_2 which serves to promote both inflammation and proliferation. These studies also showed elevated levels of the lipid peroxidation product, malondialdehyde, which serves as a potential DNA-targeted mutagen [85].

5.5.3.1 Linoleic Acid

Linoleic acid (LA) C18:2n-6 is an essential ω6 fatty acid derived mostly from corn oil but also from safflower and sunflower oils; it is the predominant PUFA of the Western diet common in America [86, 87]. Diets high in corn oil have been shown to significantly increase polyp multiplicity in various rodent studies [88–90]. However, data surrounding the effect of LA on intestinal tumorigenesis is complicated by the fact that early studies did not differentiate between LA and an alternative set of isomers, designated conjugated linoleic acid (CLA). Without differentiating between specific LA isomers, numerous studies demonstrated that LA had a beneficial effect on colon cancer outcome [67, 68, 91, 92], whereas more recent studies suggest that different LA isomers could have contrasting effects [93].

Supplementation of AOM-treated rats with linoleic acid resulted in a significantly higher incidence of colon tumors, with histology showing greater malignant differentiation [94]. Further studies of dietary supplementation with LA and/or glucose in Fischer 344 rats treated with the carcinogen AOM showed that both LA and glucose each increased intestinal tumor frequency, multiplicity, and metastasis and the greatest increase occurred in rats treated with the combination of LA and glucose. This increase was associated also with greater weight gain in rats [95]. Several meta-analyses in normal individuals have not shown a relation between linoleic acid intake and levels of inflammation except for an increased level of PGE_2 excretion [96, 97].

Levels of LA have been found to decrease with advanced stages of cancer [22]. In contrast, patients with normal colon exhibited high mucosal LA content compared to those with colon adenomas or colon cancer [91]. Circulating LA concentration have been shown to be decreased in colon cancer patients compared to hospital-based controls [67] and decreased in patients with familial adenomatous polyposis compared to healthy controls [98].

The Netherlands Cohort Study, which comprised 123,852 individuals, of which 531 were diagnosed with colon cancer, detected that LA intake was associated with increased colon tumors. Specifically, this group observed a significant association of LA with tumors that contained KRAS mutations [99, 100]. In contrast, there were no significant associations for CRC with dietary differences in total fat, saturated fats, MUFAs, or linoleic acid [99]. These observations could be explained by lipid peroxidation of ω6 PUFAs with the generation of mutagenic by-products leading to procarcinogenic changes in the K-ras oncogene [99]. Human Caco-2 colon cancer cell lines supplemented with LA showed increases in cell growth and proliferation [101]. Similarly, human SW480 colon cancer cells treated with LA showed significant elevations in cell proliferation and prostaglandins, including PGD_2 and PGE_2, none of which were observed in cells treated with CLA [102].

LA serves as a precursor for the biosynthesis of AA and subsequent proinflammatory prostaglandins, such as PGD_2 and PGE_2. Thus, many of the consequences of dietary supplementation with LA are probably mediated through its conversion to arachidonic acid [103].

5.5.3.2 Conjugated Linoleic Acid

CLAs, C18:2, are a series of 28 different isomers of linoleic acid that vary in the location of their unsaturated double bonds and geometric arrangements [104]. Molecular arrangements around these unsaturated double bonds may be either in the *cis* (c) or *trans* (t) configuration. As such, they provide a source of naturally occurring trans fatty acids derived primarily from dairy products and meat from ruminant animals. The different isomers may have similar and/or opposing effects associated with their unique geometries accounting for conflicting reports of their consequences. Of the various CLA isomers, those most abundant in natural products are the *cis*-9, *trans*-11 isomer, (c9, t11 CLA) and the *trans*-10, *cis*-12 isomer (t10, c12-CLA) [105]. In natural products, the c9, t11 CLA usually exceeds the t10, c12 variety; however, they are commonly present in equal amounts in commercially available CLA supplements used to reduce obesity or build lean body mass [106].

Numerous studies have been conducted in rodent and human cell lines and animal models to examine the outcomes of CLA associated with inflammation and colon cancer development. t10, c12 CLA induces an inflammatory response in cultured adipocytes from Caucasian and African-American women as demonstrated by increased levels of IL-1β, IL-6, IL-8, and COX-2, with the induced inflammatory response being much greater in fully differentiated adipocytes compared to preadipocytes [106]. In contrast, treatment of RAW264.7 murine macrophages with a series of CLAs, both *cis* and *trans*, significantly suppressed IL-1β, TNF-α, INF-γ, COX-2, PGE_2, and iNOS. Many of these effects were eliminated by peroxisome proliferator-activated receptor gamma (PPARγ) mutation, suggesting PPARγ as an important mediator of CLA effect [107].

CLA treatment in tissue culture has also been shown to affect cell growth and other tumor-associated properties. Several studies show that treatment with c9, t11-CLA results in time- and dose-dependent decrease in proliferation in HT-29 and Caco-2 human colon cancer cell lines [101, 108–110]. Treatment of Caco-2 cells with an equal mixture of c9, t11 and t10, c12 reduced their viability, but the effect was blocked by inhibitors of PPARγ, again indicating a role for PPARγ as mediator of the CLA effects [110]. In addition, treatment in tissue culture with c9, t11 CLA but not t10, c12 CLA reduced migration of SW480 human colon cancer cells, and both agents decreased pulmonary metastasis of IV-administered CT-26 mouse colon cancer cells in BalbC mice [111].

Multiple studies show that dietary supplementation with t10, c12 CLA promotes inflammation in murine white adipose tissue [112] whereas c9, t11 CLA reduces inflammation [113] and reduces colonic aberrant crypt formation (ACF) and tumorigenesis in DHM- or AOM-treated murine models [114–116]. The decreased tumor formation in the treated animals was associated with decreased levels of PGE_2, thromboxane A2, arachidonic acid, increased apoptosis, decreased infiltrating macrophages, and increased regulatory T cells in mesenteric lymph nodes [114–116]. CLA supplementation of AOM-treated rats reduced formation of both ACF and tumors and increased PPARγ in tumors and surrounding normal mucosa [117]. Mice, genetically deficient in PPARγ, showed similar polyp numbers and infiltrating immune cells when treated with CLA, demonstrating again that the antitumor effect can be modulated through PPARγ signaling [118].

Similarly, in AOM-treated Sprague Dawley rats, supplementation with CLA containing approximately equal concentrations of the c9, t11 and t10, c12 isomers showed decreased cancer incidence and decreased tumor multiplicity (number of cancers/rat) along with increased apoptosis. The antitumor effect of the CLA supplement was associated with decreased COX-2 and PGE_2. Administration of CLA as free fatty acids was more effective than when they were administered as conjugated triglycerides [119].

Although evidence for beneficial effects of CLA in humans has been hard to establish, given the complications surrounding the LA isomers, the Swedish Mammography Cohort examined 60,708 women and observed that women who consumed high amounts of CLA-containing foods, such as dairy products, showed a significant reduction in colon cancer risk. Each increment of two servings of high-fat dairy foods corresponded to a 13 % reduction in the risk of CRC [120].

5.5.3.3 Arachidonic Acid

Arachidonic acid (AA), 20:4n-6, contained in a variety of dietary food sources and synthesized in the body by desaturation and elongation of linoleic acid, is an ω6 PUFA associated with detrimental effects on disease outcome. Two independent studies demonstrated significantly increased AA levels in plasma and intestinal tissue of colon cancer patients compared to controls, and AA levels were significantly elevated in patients with FAP [98]. In DMH-treated rats, supplementation with AA resulted in a significant increase in colon cell proliferation and tumors [121]. Studies in $APC^{Min/+}$ mice demonstrate that dietary supplementation with AA can increase intestinal AA content, increase PGE_2 formation, but not increase intestinal tumorigenesis. In contrast, EPA supplementation decreased intestinal AA content and PGE_2 production. Supplementation of dietary EPA with AA restored the intestinal AA content, returned capacity for PGE_2 production, and abolished the antitumorigenic effect of EPA [122]. Thus, dietary supplementation with AA did not increase intestinal tumorigenesis in normal-diet-fed mice, but AA was able to overcome the antitumorigenic effect of EPA.

Arachidonic acid is metabolized to PGE_2 and related eicosanoids by the cyclooxygenase system, including COX-1 and COX-2. As noted earlier, COX-2 is upregulated under inflammatory conditions. PGE_2 is degraded by 15-PGDH whose level is decreased in colon neoplasia [56, 57, 60, 61]. Thus, increased COX-2 and decreased PGDH, each contribute to the elevated PGE_2 in colon cancer.

Elevated levels of intestinal PGE_2 lead to increased inflammatory processes, stimulate increased replication of intestinal epithelial cells, prolong their survival due to decreased apoptosis, and increase tumor-associated angiogenesis and neovascularization [123]. The proinflammatory and procarcinogenic effects of arachidonic acid and PGE_2 are further supported by the demonstration that nonsteroidal anti-inflammatory drugs (NSAIDs), which interfere with PGE_2 synthesis, are effective agents in rodents and humans for suppressing development of colorectal adenomas and carcinomas [124] and reducing mortality in patients being treated for CRC [125]. In addition, n-3 PUFA supplementation, especially with EPA and DHA, interfere with the cancer-promoting effects of PGE_2, and knockdown of PGE_2 receptors block development of intestinal tumors in carcinogen-treated rodents.

Although numerous experimental approaches in rodents have been used to demonstrate that AA supplementation and elevated PGE_2 levels promote both local gastrointestinal and systemic proinflammatory states and increased intestinal tumorigenesis, all of these models require either a predisposing genetic mutation, such as the $APC^{Min/+}$ model of familial adenomatous polyposis; a carcinogen-induced mutation, induced by treatment with agents such as AOM or DMH; or spontaneous mutation as expected to occur during free radical generation associated with inflammatory bowel disease [48]. These observations indicate that AA and its growth-promoting derivatives such as PGE_2 function to promote progression of tumors that are initiated by primary mutagenic changes. Overall, we are unaware of rodent studies showing that arachidonic acid initiates tumorigenesis in animals without preexisting genetic alterations. Moreover, while some epidemiologic studies have identified an increase in CRC in association with increased arachidonic acid consumption [126], most studies have not supported this association [54, 69, 127–130]. These observations may account for the lack of any significant increase in reports of adverse effects in populations where diets in the young or the elderly are supplemented with AA [103, 131]. The protumorigenic effects of AA and its use as a dietary supplement in infants and adults are an interesting paradox.

5.5.3.4 Alpha Linolenic Acid

Alpha linolenic acid (ALNA) 18:3n-3 is an essential ω-3 PUFA found in high quantities in vegetable oils, seeds, nuts, and dark green leafy vegetables. Common sources include perilla oil, flaxseeds, kiwi fruit seeds, walnuts, and canola oil. As an n-3 fatty acid, ALNA has similar properties to EPA and DHA and can be converted in the body at limited rates to EPA 20:3n-3 and DHA 22:6n-3.

Dietary supplementation in rodents with perilla oil, walnut oil, or flaxseed oil, all high in ALNA, resulted in increased circulating and tissue ω-3 fatty acids compared to control-fed animals and decreased intestinal ACF, tumor incidence, and tumor load in animals treated with *N*-methyl-*N*-nitrosourea or AOM [132–134]. Similar effects were seen in $APC^{Min/+}$ mice where a flaxseed oil supplemented diet decreased small and large intestine tumor number and size [135].

Further studies in rodents and man showed that diets supplemented with walnuts, flaxseed oil, or purified ALNA decreased circulating and tissue levels of proinflammatory factors including IL-1β, IL-6, TNF-α, myeloperoxidase, CRP, NF-κB, COX-2, and urinary PGEM. ALNA supplementation also decreased oxidative stress as measured by decreased urinary 8 isoprostanes [136–141]. Interestingly, flaxseed-dependent lowering of inflammatory factors in rats was further enhanced by exercise [138]. The ALNA lowering of inflammatory factors is similar to those produced by EPA and DHA [141]. While fish oil supplementation increases circulating EPA and DHA, flaxseed supplementation increases ALNA [142]. Flaxseed oil has also been shown to reduce adipocyte size and monocyte chemoattractant protein 1 in obese rats [143]. The combination of flaxseed oil and fish oil in healthy volunteers produced greater suppression of inflammatory factors than either agent alone [144].

In rodents with inflammatory bowel disease, ALNA decreased inflammatory cytokines and intestinal inflammatory disease. Increased intake of ALNA reduced inflammatory death rate in patients with inflammatory disease, whereas dietary intake of supplementary fish oil was associated with no changes in death from inflammatory diseases [145].

An ALNA-supplemented diet in men and women compared to a linolenic acid-supplemented diet or the average American diet resulted in decreased circulating CRP, decreased intercellular cell adhesion molecule 1 (ICAM-1), and decreased E-selectin which serve as attractants for inflammatory monocytes and macrophages [146]. ALNA caused an increase in apoptosis and dose-dependent decreases in proliferation in human Caco-2 colon cancer cell lines.

5.5.3.5 Eicosapentaenoic Acid and Docosahexaenoic Acid

EPA (20:5n-3) and DHA (22:6n-3) are essential ω-3 long-chain PUFAs found in marine fish and in some seaweeds with multiple health benefits against cardiovascular disease, and with anti-inflammatory, antihypertensive, antiarthritis, antioxidative, and anticancer effects. Numerous studies have demonstrated that EPA and DHA have an inverse relationship with colon cancer [67, 147]. Although EPA and DHA are metabolized differently, nearly all studies which address the association between ω-3 fatty acids and colon cancer in humans use a mixture of EPA and DHA. Because most clinical studies supplement them together, they will be discussed concurrently.

Significantly decreased EPA levels were observed in the mucosa and plasma phospholipid profiles of patients with colon adenomas or colon cancer compared to healthy controls [91]. DHA, but not EPA, was significantly lower in serum from FAP patients compared to controls [98]. Several trials have observed significantly low intake and plasma concentrations of DHA and EPA among colon cancer patients as well as individuals at high risk for developing colon cancer [67, 147]. In a 12-week randomized control trial on the effect of ω-3 fatty acids, specifically EPA, in 20 subjects at risk for colon cancer, a significant reduction was found in the abnormal cell proliferation in the experimental group receiving EPA. EPA supplementation (4.1 g per day for 6 months) increased mucosal composition of this fatty acid and significantly decreased intestinal proliferation in cancer patients [148, 149]. Similar to results with EPA, treatment of colon cancer patients with DHA (1.1 g per day) led to a reduction in cell proliferation in rectal mucosa [148, 149].

The distinct beneficial effects of DHA have been investigated in more detail in animal models and human colon cancer cell lines. DHA decreased aberrant crypt foci (ACF), which are precursors to colon polyps, and decreased polyp number by 40 % in rats treated with DMH and significantly reduced ACF and tumor multiplicity in rats treated with AOM [150, 151].

In human colon cancer cells, DHA was shown to activate many genes involved in apoptosis, such as cytochrome c and caspases 5, 8, 10, and 15, while decreasing expression of anti-apoptotic, Bcl-2 family members as well as components of the prostaglandin synthesis pathway such as COX-2 and PGE_2. In human HT-29 and HCT116 colon cancer cells, DHA reduced the levels of PPARγ, which regulates fatty acid storage and glucose metabolism [152, 153].

Comparing the effect of PUFA treatment on NCM 460 normal colon cells, DHA, but not arachidonic acid, palmitic acid, or oleic acid, selectively stimulated growth, whereas in Caco-2 colon tumor cells, DHA was the only fatty acid that stimulated apoptosis, as demonstrated by cleavage of caspase 3 and poly ADP-ribose polymerase (PARP). Moreover, DHA was the only PUFA tested that reduced viability of Caco-2 cells. This change occurred in association with increased DHA membrane concentration, decreased PI3K, due to decreased concentration of its p85 regulatory subunit, and decreased PTEN phosphorylation, resulting in increasing negative control of PI3K. In addition, DHA and oleic acid significantly decreased Ser^{473} phosphorylation of AkT, thereby interfering with the PI3K/AkT growth-promoting pathway [154]. Additionally, DHA in combination with the saturated fat, butyrate, has been shown to enhance mitochondrial lipid oxidation and calcium-dependent apoptosis in human HCT116 cells [155].

From a mechanistic viewpoint, EPA is a competitive inhibitor of the proinflammatory effects of arachidonic acid mediated through COX-1 and COX-2, yielding PGE 3 which is less inflammatory than PGE_2. EPA and DHA reduce TNF-α, IL-1, IL-6, IL-8, resistin, plasminogen activator inhibitor 1, and monocyte chemoattractant protein 1. The ω-3 PUFAs also increase adiponectin which reduces inflammation [156].

EPA has been independently investigated in more mechanistic detail and in association with its beneficial effects on colon cancer. Prostaglandin synthesis and PPARγ expression were significantly reduced in human HT-29 cells after treatment with EPA. These same studies demonstrated a significant decrease in proliferation and induction of apoptosis in the HT-29 human cancer cell line [157, 158]. Treatment of human HCT116 colon cancer cells with EPA resulted in reduction in DNA polymerase activity and cell cycle arrest at the G1 checkpoint [159]. AOM-treated F344 male rats supplemented with fish oil diets enriched for EPA and DHA decreased intestinal tumor incidence and multiplicity [160].

Studies conducted in $APC^{Min/+}$ mice demonstrated that DHA and EPA supplementation decreased polyp multiplicity and size by 50 % and significantly reduced proinflammatory prostaglandin levels [23, 161, 162]. $Apc^{Min/+}$ mice and AOM-treated rats fed diets supplemented with EPA showed significantly fewer polyp numbers as well as decreased levels of COX-2 and PGE_2 [161, 163, 164]. Interestingly, $APC^{Min/+}$ mice provided with supplemental dietary EPA were protected from the cachexia that usually accompanies intestinal tumorigenesis in these mice [163].

5.5.3.6 Importance of the Omega-3 to Omega-6 Ratio

Many studies have shown that changes in the omega-3/omega-6 (ω-3/ω-6) ratio may contribute to the early development of human colon cancer. Competition occurs between these fatty acids in the COX-signaling pathway which gives rise to a series of prostaglandins, some pro- and others anti-inflammatory. Based on their competitive nature, clinical guidelines recommend a 1:1 ratio between ω-3 and ω-6 fatty acids for optimal health [165]. In contrast to recommendations, it is estimated that Westernized diets are composed of an ω-3/ω-6 ratio of 1:10–25 [166].

A randomized clinical trial examined the plasma phospholipid ω-3/ω-6 ratio as a nutritional marker for the prevention of colonic tumor development and metastasis in 27 patients. A significant increase in the plasma phospholipid content of ω-3 and ω-6 ratio was found in the experimental group which correlated with inhibition of the mucosal neoplastic proliferation, thus associating an increase in an ω-3/ω-6 ratio with decreased cancer risk [167]. In an ecological study in Belgium, ω-3/ω-6 ratios were examined in 11,302 individuals for correlations with mortality associated with colon cancer. A significant inverse relationship was established between colon cancer and high ω-3/ω-6 ratios [168, 169]. Similarly, another study that examined 363 cases of patients with colon adenomas and 498 adenoma-free controls also observed an inverse relationship between an increased ω-3/ω-6 ratio and colon cancer risk [170].

In a prospective study of 73,242 Chinese women participating in the Shanghai Women's Health Study, there was no association of CRC with total ω-6 PUFA or total ω-3 PUFA consumption. There was, however, a dose-dependent association of CRC with arachidonic acid consumption and a strong association of CRC with increasing ratio of ω-6/ω-3 with the highest relative risk of 1.95 comparing highest to lowest quintiles of ω-6 to ω-3 ratios. Interestingly, the risk for rectal cancer was greater than that for colon [126]. In a nested case–control study of 150 cases and 150 controls in women, the ω-6/ω-3 ratio correlated with urinary metabolites of PGE_2 suggesting the possibility that the association between ω-6/ω-3 ratio and CRC is mediated by PGE_2.

5.6 Specific Diets

5.6.1 Animal Fats

Studies supporting an association of CRC with diets containing high intake of red meats and processed meats have been contradictory. A meta-analysis of articles published from 1973 to 1999, examining the relation of meat consumption to CRC, found no significant association for CRC with total meat consumption (red, white, and processed). In contrast, the relative risk of CRC was increased with consumption of red meat and was even greater for processed meat [171, 172]. Results from the Nurses' Health Cohort Study, 121,700 women, provided further support that dietary animal fat, but not vegetable fat consumption, was associated with increased risk for colon cancer. Red meat and processed meats were associated with increased risk, whereas fish and chicken were associated with decreased risk [14]. Some studies show an inverse or no relation between total dietary fat and colon cancer [173, 174], although an increased risk for colon cancer was associated with increased red meat consumption [173].

The presumptive carcinogenic effects associated with meats have been attributed to proteins, fats, pyrolysis products including heterocyclic aromatic amines and polycyclic aromatic hydrocarbons associated with cooking, and *N*-nitroso compounds used as preservatives in processed meats. An interesting example of the effect of diet and genetic interactions on cancer causation is provided by the observation that the rapid acetylator phenotype, associated with NAT2 genetic polymorphisms, contributes to activation of the mutagenic potential of heterocyclic aromatic amines generated by cooking meat at high temperatures, and increases the association of dietary meat consumption with CRC [175].

Rats treated with the carcinogen DMH and fed beef tallow as primary fat source showed significantly increased intestinal tumor incidence and tumor numbers per rat compared to rats fed fish oil diets. The beef tallow-fed rats showed less apoptosis and more PGE_2 in intestinal mucosa compared to rats on the fish oil diet. Interestingly, supplementation of either diet with CLA increased the percentage of cells undergoing apoptosis, decreased PGE_2, and decreased tumorigenesis in DMH-treated animals [176].

In a case–control study in North Carolina, involving 945 cases and 959 controls, oversampled for African Americans, with diet analyzed retrospectively for the past year, the highest intake of red meat was associated with decreased risk of distal CRC (sigmoid, rectosigmoid, or rectal) in both Caucasians and African Americans that only reached significance in Caucasians. There was no statistical difference in distal CRC with total fat, MUFA or PUFA in either race, although a significant trend was noted for decreased risk in African Americans for saturated fat measured as percent energy but not as total fat. The highest level of PUFA intake in African Americans was associated with lower risk, but this did not reach statistical significance [177]. No estimates were provided of absolute amounts or ratios of ω-3/ω-6 PUFAs.

5.6.2 Mediterranean Diet

The Mediterranean diet has high amounts of whole grains, fruits, vegetables, fish, and olive oil, moderate amounts of alcohol and dairy products, low amounts of red or processed meats, and low glycemic loads. As indicated earlier, Mediterranean countries have lower rates of colon cancer compared with other Western countries. For example, colon cancer mortality in Greece is about 40 % lower than that in the United Kingdom [44, 45]. A large meta-analysis of 12 studies examining the health benefits of the Mediterranean diet showed a decreased risk of all cause mortality, cardiovascular disease mortality, and cancer mortality [178]. Another more recent meta-analysis of 19 observational studies, including patients from both Mediterranean and non-Mediterranean countries, showed that the group with the highest consumption of olive oil compared to the lowest had lower odds of having any cancer and specifically lower odds for breast and digestive cancers [179]. More specifically, some, but not all, studies indicate that adherence to the Mediterranean diet is associated with reduced risk of CRC in both men and women [180]. When an altered Mediterranean diet was compared to the DASH diet (Dietary Approach to Stop Hypertension), with higher intake of whole grains, decreased red and preserved meats, and moderate amounts of low-fat dairy products, in two very large cohort studies, The Nurses' Health Study (NHS), 87,256 women, and The Health

Professionals Follow-Up Study (HPFS), 45,490 men, adherence to the altered Mediterranean diet resulted in a reduction in CRC that did not reach statistical significance, whereas adherence to the DASH diet resulted in a statistically significant reduction in CRC in both men and women [181, 182].

Olive oil has been shown to be an important contributor to the cancer preventative effects of the Mediterranean diet [179]. A case–control study conducted in the Marseilles region of southern France examined 399 patients with colon cancer and observed a reduction in olive oil intake compared to age-and sex-matched controls [183]. The relationship between various seasoning fats and colon carcinoma risk was investigated using data from a second case–control study conducted in six Italian areas. Cases were 1,953 patients with histologically confirmed colon carcinomas, while the controls were 4,154 subjects with no history of cancer. This study found that olive oil intake was negatively associated with risk of colon carcinomas [184].

Rats, chemically treated with DMH, fed a higher olive oil-based diet, developed a significantly lower number of ACF than rats fed a low concentration of olive oil. Olive oil dose-dependently downregulated the expression of both Bcl-2 and COX-2 in colonic mucosa and also abrogated the upregulation of Bcl-2 by DMH [185]. Rats treated with AOM and supplemented with 5 % dietary olive oil showed significant decreases in ACF and PGE_2 [186].

5.6.3 Western Diet

The Western-style diet, high in fat, red meat, sugary deserts, low in fiber, fresh fruits, vegetables, whole grains, seafood, and calcium and vitamin D content [30], is associated with increased risk of CRC in both humans and rodents [33, 187]. The Western diet in rodents is both proinflammatory and procarcinogenic; however, supplementation with calcium and cholecalciferol diminishes tumorigenesis [187]. Mice fed a normal lab chow supplemented with 20 % corn oil to simulate a Western diet showed inflammatory changes in their intestinal mucosa including increased F4/80-staining macrophages in the lamina propria, with increased IgA, CRP, fibrinogen, myeloperoxidase, serum amylase, monocyte chemoattractant protein, and macrophage inflammatory protein-1 [188]. Corn oil-supplemented diets clearly promote intestinal tumorigenesis in $APC^{Min/+}$ mice and in AOM-treated F344 rats [10].

5.7 Summary of Dietary Fatty Acid Effects on Colon Cancer

On the basis of epidemiologic studies conducted in multiple countries, with different dietary patterns and with different populations, especially those undergoing significant dietary change, there is a convincing evidence to support the concept that fatty acids and their concentrations in different tissues have important effects on CRC risk, where specific dietary fatty acids can increase the risk of CRC while others can suppress it. Studies in tissue culture and animal models as well as those in patients demonstrate that differences in the effects of fatty acids can be related to variations in their structure, intake, biochemical consequences, and their concentrations in different tissues, especially those obtained from patients with colon cancer compared to individuals free of malignant disease. However, before commenting further on mechanistic and differential effects of fatty acids on CRC, it is important to emphasize that almost all studies in animals require the presence of a carcinogenic or hereditary genetic mutation before the effects of fatty acids can be identified. In general, the fatty acids do not produce colorectal tumors in rodents without preexisting mutations. In contrast, diets supplemented with protumorigenic fatty acids, in rodents treated with AOM or DMH or in mice with the APC^{Min} mutation, result in increased tumorigenesis, including earlier time of onset, higher incidence, greater multiplicity, larger size, and higher tumor load, compared to non-diet-supplemented animals. These studies indicate that protumorigenic fatty acids act primarily as tumor promoters contributing to increased tumor growth but not primary carcinogenesis. Conversely, the reduced tumorigenicity or protective effects of other dietary fats appear to be associated with either a direct or indirect effect on suppression of tumor growth as opposed to any anticarcinogen effect. Since human colorectal adenomas and carcinomas are routinely associated with a series of hereditary and/or spontaneous mutations that serve as initiators of the carcinogenic process, it is expected that, as in rodent studies, fatty acids will serve as tumor promoters or suppressors. Some mutations leading to initiation of CRC in humans include DNA alterations leading to inactivation of tumor suppressors such as FAP, TP53, and TGF-β or activation of oncogene pathways including RAS/RAF, PI3K, and PTEN [58]. As detailed previously and summarized below, fatty acids that function as tumor promoters stimulate cell proliferation through inflammatory pathways. The inflammatory pathways enhance cell proliferation which in turn fixes mutagenic DNA alterations in a population of cells establishing the basis for tumor growth and progression [189].

Figure 5.1 summarizes the results discussed in this section, where medium-chain saturated fatty acids, lauric and myristic acid; long-chain saturated fatty acids, palmitic and stearic acids; and ω6 PUFAs including linoleic and arachidonic acids have detrimental proinflammatory, protumorigenic effects as tumor promoters increasing colon cancer risk, development, and outcome. Conversely, oleic, conjugated linoleic, and ω-3 PUFAs including eicosapentaenoic, docosahexaenoic, and linolenic acids serve as tumor suppressors with beneficial effects, mediated by protecting against inflammation, reducing incidence, and retarding growth in colon cancer.

Overall, it is important to note that the fatty acids that promote inflammation also promote intestinal tumorigenesis, whereas those that suppress inflammation also reduce intestinal tumorigenesis. Moreover, in studies, where tested, the fatty acids that promote or protect against inflammatory bowel disease show similar effects for colon cancer. Ongoing studies in our group are focused on defining the separate contributions and mediators of obesity and inflammation on fatty acid-induced intestinal tumorigenesis and identifying the individual mechanisms [190].

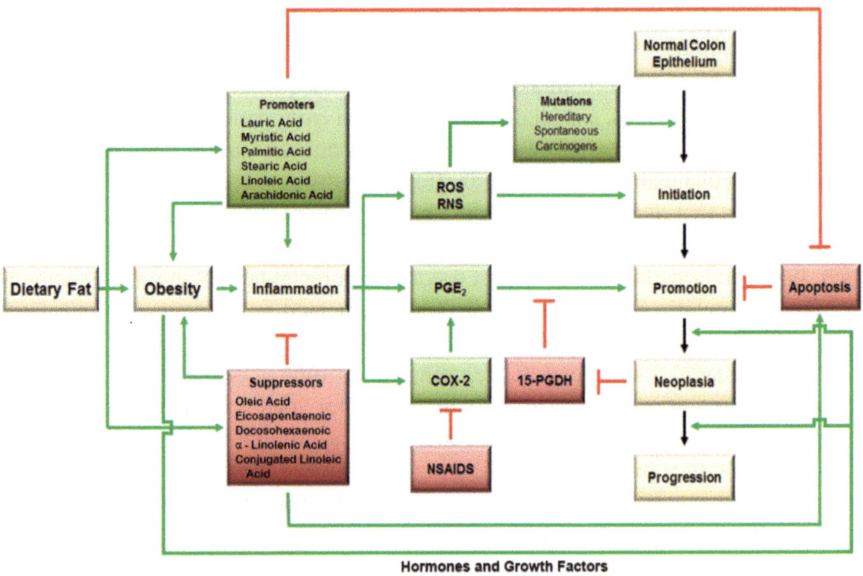

Fig. 5.1 Impact of dietary fatty acids on colorectal cancer. Dietary fatty acids are divided into cancer promoters (*green box*) and cancer suppressors (*red box*). *Green arrows* show positive or stimulatory effects on the indicated process. *Red arrows* show negative or inhibitory effects on the indicated process. *ROS* reactive oxygen species; *RNS* reactive nitrogen species; *PGE_2* prostaglandin E_2; *COX-2* cyclooxygenase 2; *NSAIDs* nonsteroidal anti-inflammatory drugs; *15-PGDH* 15-prostaglandin dehydrogenase

As shown in Fig. 5.1, the specific dietary fatty acids differentially modulate plasma markers and mediators of inflammation in tissue and in circulation. Thus, diets enriched in the fatty acids listed as promoters stimulate synthesis, activation, and release of multiple inflammatory factors including NF-κB, IL-1β, IL-6, TNF-α, E-selectin, fibrinogen, CRP, COX-2, and PGE_2 in both immune system cells and intestinal epithelium. PGE_2 serves as a modulator of inflammation and promotes tumor growth and angiogenesis. PGE_2 levels are determined by its synthetic rate via COX-2 and its catabolic rate determined by 15-PGDH. Elevated levels of PGE_2 in colon cancer are due in part to induction of COX-2 and depression of 15-PGDH in tumor tissue. The increased PGE_2 stimulates proliferation of normal and initiated colon epithelial cells. However, failure of the normal process of apoptosis provides a further basis for unrestricted tumor growth. In addition, fatty acid stimulation of inflammation leads to increased levels of myeloperoxidase indicating increased infiltration by neutrophils and increased synthesis of reactive oxygen species (ROS) and reactive nitrogen species (RNS) which may provide additional opportunities for ongoing mutagenic DNA damage.

The protective fatty acids, including the MUFA, oleic acid, and the ω-3 PUFAs, including EPA, DHA, ALNA, and CLA, inhibit inflammation in tissue culture, murine models, and humans, resulting in decreased COX-2, decreased PGE_2, and, consequently, decreased promotion of cell growth and tumorigenesis. Mechanisms by which ω3 fatty acids downregulate NF-κB and COX-2 include transcriptional regulation and changes in membrane fluidity leading to alterations in growth-factor sensitivity [191]. In addition to preventing increases in COX-2, these fatty acids serve as competitive inhibitors to block metabolism of AA to PGE_2. Moreover, these fatty acids also stimulate apoptosis in normal and transformed intestinal tissues thereby retarding tumor promotion and possibly tumor growth.

While the effects described above are direct consequences of fatty acids, their consumption in high quantities leads also to obesity, which itself is a low-grade inflammatory disorder [5] and forms the basis for a feed-forward cycle that continues to promote inflammation and tumor development and growth. From the viewpoint of cancer prevention, it is noteworthy that weight loss in obese premenopausal women was associated with significant reduction of circulating and rectosigmoid inflammatory markers [192]. Moreover, obesity is responsible for increased circulating growth factors including insulin and leptin, decreased adiponectin, and other changes which further stimulate tumor promotion and growth through cell surface receptors such as the insulin receptor, the leptin receptor, and the IGF-1 receptor as well as their downstream intracellular pathways [193].

5.8 Dietary Fatty Acid-Targeted Interventions in Colon Cancer

5.8.1 Prevention and Control

While physical activity, sleep, and genetics are all important contributors that need to be addressed to maintain healthy body mass, the observations outlined above indicate the importance of not only controlling quantity of fats but also the quality and composition of dietary fats to prevent cancer, especially CRC. Both saturated and unsaturated fats are needed for comprehensive structure and functional development; however, CRC prevention strategies favor dietary consumption of the unsaturated fats relative to saturated fats. As indicated above, there is a significant relationship between increased ratios of ω3 to ω6 fatty acids and reduced colon cancer risk, indicating the importance of maintaining this ratio, either through natural food stuffs or with dietary supplements. Since tumor promotion and progression progresses over many years, it would be useful to adopt these dietary habits early and maintain them on a lifelong basis. Interestingly, dietary supplementation with ω3 fatty acids is more effective in suppressing inflammatory cytokine production in older than younger women [194]. However, since dietary supplements can change circulating and tissue levels of fatty acids over short periods, days to weeks, it seems advisable to change whenever possible. In this regard, it would be helpful to have a single or few easily measured circulating or excreted metabolites that could be used to judge healthy level, balance, and inflammatory status, demonstrated to be predictive of reduced risk of CRC. CRP is an easily measured component of inflammation that has been considered and rejected as a diagnostic and prognostic marker for

CRC, since it is too nonspecific and is reflective of a broad range of disorders encompassing inflammation and tissue damage [195]. Nonetheless, the criteria for such marker(s) have been described [195], and this remains an ongoing goal.

The demonstration that many of the effects of fatty acids are mediated through inflammatory pathways, particularly through COX-2, provides the basis for the use of NSAIDs for prevention of colon cancer. This approach and the associated cardiovascular risk factors that compromise their use have been detailed in previous reviews [56–58, 60–62]. However, the success of NSAIDs in preventing CRC provides proof of principle and points to the need for better understanding of the pathways connecting fatty acids to inflammation as a focus for developing innovative cancer prevention strategies [196]. For example, studies showing that fatty acid-mediated effects are controlled by G protein receptors [197, 198] and fatty acid-binding proteins [199, 200] that regulate cellular uptake suggest these molecules as important targets for preventing CRC. As indicated above, NAT2 gene genetic variations [175] contribute significantly to interindividual variations in fatty acid-induced synthesis and secretions of cytokines. These genetic polymorphisms may govern pro- and anti-inflammatory responses to fatty acids and may serve as the basis for focused neutriceutical-based interventions [201].

TLRs, proteins that act as LPS and nutrient-sensing regulatory proteins, activate T cell inflammation proteins, including NF-κB. Fatty acids may influence this inflammatory pathway by direct effect on TLRs as well as by altering intestinal microbiota so as to alter synthesis and availability of LPS interacting with TLR [202].

5.8.2 Therapy

While ω3 PUFAs have generally been used to prevent colon cancer, they may also have beneficial effects as adjunctive neutriceutical approaches to enhance more traditional therapeutic interventions with chemo- or radiation therapy [203]. Since the ω3 PUFAs reduce levels of PGE_2 and inflammatory factors, they may be expected to reduce colon cancer growth, increase apoptosis, and reduce tumor angiogenesis. In addition, ω3 PUFAs have been shown to potentiate cytotoxicity of chemotherapeutic agents in multiple human colon cancer cell lines in tissue culture and in tumor-bearing animals [204–206]. Interestingly, n-3 PUFAs have been shown to reduce cachexia in mice with colon adenocarcinoma [206]. In addition, pretreatment with ω-3 PUFAs decreases cancer cell survival following both UV and X irradiation [207]. Clearly, clinical trials of n-3 PUFA supplementation of conventional chemoradiation therapy are warranted for CRC.

5.9 Cholesterol Metabolism and Statin Therapy

To this point, we have focused on the fatty acid components of dietary fats and their contribution to CRC. Cholesterol, a polycyclic steroid molecule, is another important component of dietary fat, derived from animal sources. Originating from both exogenous, dietary sources, and endogenous synthesis, overall cholesterol levels have long been a public health concern, especially from a cardiovascular viewpoint [208]. Cholesterol synthesis can be interfered with by a group of agents collectively identified as statins, which are competitive inhibitors of the enzyme 3-hydroxyl-3-methylglutaryl coenzyme A reductase (HMGCoA reductase), a key enzyme in the cholesterol synthesis pathway. A recent study, assessing the effect of statin use over a 12-year period (1995–2007), in the entire population of Denmark, showed a significant decrease among statin users in overall cancer-related mortality [209]. An earlier population-based case–control study in Israel showed statin use to be associated with a significant reduction in CRC [210]. Other population-based studies and meta-analysis have found divergent results, with those showing protective results being associated with longer statin use (≥4 years) [211] whereas those with shorter statin use have tended to not show protective effects [212].

By inhibiting HMGCoA reductase, statins block synthesis of mevalonic acid which serves as a precursor for farnesyl pyrophosphate, the immediate precursor for cholesterol. Farnesyl pyrophosphate is also the precursor for geranylgeranyl pyrophosphate which is required for protein prenylation and activation of proteins including Ras, Rho, and nuclear lamins [213]. Thus, by inhibition of farnesylation, statins could interfere with Ras activation and colon cancer growth. Interestingly, in addition to their effect on cholesterol metabolism, statins have been shown to have anti-inflammatory effects preventing activation of NF-κB; lowering TNF-α, CRP, IL-1β, and IL-6; and acting synergistically with NSAIDs [214–219]. Further research is required to define the role of and potential use of statins in CRC prevention and therapy.

References

1. Campoy C, Escolano-Margarit MV, Anjos T, Szajewska H, Uauy R (2012) Omega 3 fatty acids on child growth, visual acuity and neurodevelopment. Br J Nutr 107(suppl 2):S85–S106
2. Huffman SL, Harika RK, Eilander A, Osendarp SJ (2011) Essential fats: how do they affect growth and development of infants and young children in developing countries? A literature review. Matern Child Nutr 3:44–65
3. Calle EE, Kaaks R (2004) Overweight, obesity and cancer: epidemiological evidence and proposed mechanisms. Nat Rev Cancer 4:579–591
4. Nock N, Berger NA (2010) Obesity and cancer, overview of mechanisms. In: Berger NA (ed) Energy balance and cancer. Springer, New York, pp 129–179
5. Hotamisligil GS, Arner P, Caro JF, Atkinson RL, Spiegelman BM (1995) Increased adipose expression of tumor necrosis factor in human obesity and insulin resistance. J Clin Invest 95:2409–2415
6. Doll R, Peto R (1981) The causes of cancer: quantitative estimates of avoidable risks of cancer in the United States today. J Natl Cancer Inst 66:1191–1308
7. Giovannucci E, Rimm EB, Colditz GA, Stampfer MJ, Ascherio A, Chute CC, Willet WC (1993) A prospective study of dietary fat and risk of prostate cancer. J Natl Cancer Inst 85(19):1571–1579
8. Kolonel LN (1987) Fat and colon cancer: how firm is the epidemiologic evidence? Am J Clin Nutr 45:336–341
9. Kushi L, Giovannucci E (2002) Dietary fat and cancer. Am J Med 113(9B):63S–70S
10. Reddy BS (1986) Effect of dietary fish oil on azoxymethane-induced colon carcinogenesis in male F334 rats. Cancer Res 46:3367–3370

11. Reddy BS (1986) Amount and type of dietary fat and colon cancer: animal model studies. In: Ip C, Birt DF, Rogers AE, Mettlin C (eds) Dietary fat and cancer. Alan R. Liss, New York, pp 295–309
12. Rose DP, Boyar AP, Wynder EL (1986) International comparisons of mortality rates for cancer of the breast, ovary, prostate, and colon, and per capita food consumption. Cancer 58(11):2363–2371
13. Thiebaut ACM, Jiao L, Silverman DT, Cross AJ, Thompson FE, Subar AF, Hollenbeck AR, Schatzkin A, Ztolzenberg-Solomon RZ (2009) Dietary fatty acids and pancreatic cancer in NIH-AARP diet and health study. J Natl Cancer Inst 101:1001–1011
14. Willet WC, Stampfer MJ, Colditz GA, Rosner BA, Speizer FE (1990) Relation of meat, fat, and fiber intake to the risk of colon cancer in a prospective study among women. N Engl J Med 323(24):1664–1672
15. Wynder EL, Cohen LA, Muscat JE, Winters B, Dwyer JT, Blackburn G (1997) Breast cancer: weighing the evidence for a promoting role of dietary fat. J Natl Cancer Inst 89:766–775
16. Rose DP (1997) Dietary fatty acids and cancer. Am J Clin Nutr 66(suppl):998S–1003S
17. Willet WC, Stampfer MJ, Colditz GA, Rosner BA, Hennekens CH, Speizer FE (1987) Dietary fat and the risk of breast cancer. N Engl J Med 316(1):22–28
18. Chlebowski RT, Blackburn GL, Thomson CA, Nixon DW, Shapiro A, Hoy MK, Goodman MT, Giuliano AE, Karanja N, McAndrew P, Hudis C, Butler J, Merkel D, Kristal A, Caan B, Michaelson R, Vinciguerra V, Del Prete S, Winkler M, Hall R, Simon M, Winters BL, Elashoff RM (2006) Dietary fat reduction and breast cancer outcome: interim efficacy results from the Women's Intervention Study. J Natl Cancer Inst 98(24):1767–1776
19. Prentice RL, Thomson CA, Caan B, Hubbell FA, Anderson GL, Beresford SA, Pettinger M, Lane DS, Lessin L, Yasmeen S, Singh B, Khandekar J, Shikany JM, Satterfield S, Chlebowski RT (2007) Low-fat dietary pattern and cancer incidence in the Women's Health Initiative Dietary Modification Randomized Controlled Trial. J Natl Cancer Inst 99(20):1534–1543
20. Voet D, Voet JG, Pratt CW (2006) Fundamentals of biochemistry: life at the molecular level, 2nd edn. Chapter 9, Lipids and biological membranes. Wiley, Hoboken, pp 233–281
21. Li B, Birdwell C, Whelan J (1994) Antithetic relationship of dietary arachidonic acid and eicosapentaenoic acid on eicosanoid production in vivo. J Lipid Res 35:1869–1877
22. Szachowicz-Petelska B, Sulkowski S, Figaszewski ZA (2007) Altered membrane free unsaturated fatty acid composition in human colorectal cancer tissue. Mol Cell Biochem 294:237–242
23. Petrik MB, McEntee MF, Chiu CH, Whelan J (2000) Antagonism of arachidonic acid is linked to the antitumorigenic effect of dietary eicosapentaenoic acid in Apc(Min/+) mice. J Nutr 130(5):1153–1158
24. Cunnane SC (2003) Problems with essential fatty acids: time for a new paradigm? Prog Lipid Res 42:544–568
25. Armstrong B, Doll R (1975) Environmental factors and cancer incidence and mortality in different countries, with special reference to dietary practices. Int J Cancer 15(4):617–631
26. Burkitt DP (1978) Colonic-rectal cancer: fiber and other dietary factors. Am J Clin Nutr 31:S58–S64
27. Wynder EL, Kajitani T, Ishikawa S, Dodo H, Takano A (1969) Environmental factors of cancer of the colon and rectum II. Japanese epidemiological data. Cancer 23(5):1210–1220
28. Carrera-Bastos P, Fontes-Villalba M, O'Keefe JH et al (2011) The western diet and lifestyle and diseases of civilization. Res Rep Clin Cardiol 2011(2):15–35
29. Eaton B, Konner M (1985) Paleolithic nutrition—a consideration of its nature and current implications. N Engl J Med 312:283–289
30. Hu FB, Rimm EB, Stampfer MJ, Ascherio A, Speigelman D, Willet WC (2000) Prospective study of major dietary patterns and risk of coronary heart disease in men. Am J Clin Nutr 72:912–921
31. Giovannucci E (2002) Modifiable risk factors for colon cancer. Gastroenterol Clin North Am 31(4):925–943
32. Slattery ML, Boucher KM, Caan BJ, Potter JD, Ma KN (1998) Eating patterns and risk of colon cancer. Am J Epidemiol 148(1):4–16

33. Marmot M (2007) Food nutrition, physical activity, and the prevention of cancer: a global perspective. World Cancer Research Fund/American Institute for Cancer Research, Washington, DC, pp 4–29
34. Flood DM, Weiss NS, Cook LS, Emerson JC, Schwartz SM, Potter JD (2000) Colorectal cancer incidence in Asian migrants to the United States and their descendants. Cancer Causes Control 11:403–411
35. Potter JD, Slattery ML, Bostick RM, Gapstur SM (1993) Colon cancer: a review of the epidemiology. Epidemiol Rev 15:499–545
36. Shimizu H, Mack TM, Ross RK, Henderson BE (1987) Cancer of the gastrointestinal tract among Japanese and white immigrants in Los Angeles Country. J Natl Cancer Inst 78:223–228
37. Buell P, Dunn JE Jr (1965) Cancer mortality among Japanese Issei and Nissei of California. Cancer 18:656–664
38. Ho GY, Figueroa-Vallés NR, De La Tore-Feliciano T, Tucker KL, Tortolero-Luna G, Rivera WT, Jiménez-Velázquez IZ, Ortiz-Martinez AP, Rohan TE (2009) Cancer disparities between mainland and island Puerto Ricans. Rev Panam Salud Publica 25(5):394–400
39. Lee J, Demissie K, Lu S-E, Ghoads GG (2007) Cancer incidence among Korean-American immigrants in the United States and Native Koreans in South Korea. Cancer Control 14(1):78–85
40. Pinheiro PS, Sherman RL, Trapido EJ, Fleming LE, Huang Y, Gomez-Marin O, Lee D (2009) Cancer incidence in first generation U.S. Hispanics: Cubans, Mexicans, Puerto Ricans, and new Latinos. Cancer Epidemiol Biomarkers Prev 18(8):2162–2169
41. Lillioja S, Bogardus C (1988) Obesity and insulin resistance: lessons learned from the Pima Indians. Diabetes Metab 4:517–540
42. Ravussin E (1993) Energy metabolism in obesity studies in the Pima Indians. Diabetes Care 16:232–238
43. Baier LJ, Hanson RL (2004) Genetic studies of the etiology of type 2 diabetes in Pima Indians: hunting for pieces to a complicated puzzle. Diabetes 53:1181–1186
44. Ferlay J, Bray F, Pisani P et al (2004) GLOBOCAN 2002: cancer incidence, mortality and prevalence worldwide. IARC Press, Lyon
45. Parkin DM, Bray F, Ferlay J, Pisani P (2005) Global cancer statistics, 2002. CA Cancer J Clin 55:74–108
46. Stoneham M, Goldacre M, Seagroatt V, Gill L (2000) Olive oil, diet and colorectal cancer: an ecological study and a hypothesis. J Epidemiol Community Health 54:756–760
47. Trichopoulou A, Lagiou P, Kuper H, Trichopoulos D (2000) Cancer and Mediterranean dietary traditions. Cancer Epidemiol Biomarkers Prev 9:869–873
48. Rogers AE, Nauss KM (1985) Rodent models for carcinoma of the colon. Dig Dis Sci 30(12):87S–102S
49. Cho KR, Vogelstein B (1992) Suppressor gene alterations in the colorectal adenoma-carcinoma sequence. J Cell Biochem Suppl 16G:137–141
50. Cottrell S, Bicknell D, Kaklamanis L, Bodmer WF (1992) Molecular analysis of APC mutations in familial adenomatous polyposis and sporadic colon carcinomas. Lancet 340(8820):626–630
51. Reddy BS, Narisawa T, Vukusich D, Weisburger JH, Wynder EL (1976) Effect of quality and quantity of dietary fat and dimethylhydrazine in colon carcinogenesis in rats. Proc Soc Exp Biol Med 151(2):237–239
52. Reddy BS, Weisburger JH, Narisawa T, Wynder EL (1974) Colon carcinogenesis in germ-free rats with 1, 2-dimethylhydrazine and N-methyl-N′-nitro-N-nitrosoguanidine. Cancer Res 34:2368–2372
53. Neaptolemos JP, Clayton H, Heagerty AM, Nicholson MJ, Johnson B, Mason J, Manson K, James RF, Bell PR (1988) Dietary fat in relation to fatty acid composition of red cells and adipose tissue in colorectal cancer. Br J Cancer 58:575–579
54. Slattery ML, Potter JD, Duncan DM, Berry TD (1997) Dietary fats and colon cancer: assessment of risk associated with specific fatty acids. Int J Cancer 73(5):670–677

55. Lee JY, Sohn KH, Rhee SH, Hwang D (2011) Saturated fatty acids, but not unsaturated fatty acids, induce the expression of cyclooxygenase-2 mediated through Toll-like receptor 4. J Biol Chem 276:16683–16689
56. Backlund MG, Mann JR, DuBois RN (2005) Mechanisms for the prevention of gastrointestinal cancer: the role of prostaglandin E_2. Oncology 69(suppl 1):28–32
57. Backlund MG, Mann JR, Holla VR, Buchanan FG, Tai HH, Musiek ES, Milne GL, Katkuri S, DuBois RN (2005) 15 Hydroxyprostaglandin dehydrogenase is down-regulated in colorectal cancer. J Biol Chem 280(5):3217–3223
58. Markowitz SD, Bertagnolli MM (2009) Molecular origins of cancer: molecular basis of colorectal cancer. N Engl J Med 361(25):2449–2460
59. May V, Colbert LH, Berrigan D, Perkins SN, Pfeiffer R, Lavigne JA, Lanza E, Haines DC, Schatzkin A, Hursting SD (2003) Calorie restriction and diet composition modulate spontaneous intestinal tumorigenesis in Apc(min) mice through different mechanisms. Cancer Res 63:1752–1755
60. Markowitz SD (2007) Aspirin and colon cancer—targeting prevention? N Engl J Med 356(21):2195–2198
61. Yan M, Myung SJ, Fink SP, Lawrence E, Lutterbaugh J, Yang P, Zhou X, Liu D, Rerko RM, Willis J, Dawson D, Tai HH, Barnholtz-Sloan JS, Newman RA, Bertagnolli MM, Markowitz SD (2009) 15-Hydroxyprostaglandin dehydrogenase inactivation as a mechanism of resistance to celecoxib chemoprevention of colon tumors. Proc Natl Acad Sci U S A 106(23): 9409–9413
62. Bertagnolli MM (2012) Cyclooxygenase-2 and chronic inflammation: drivers of colorectal tumorigenesis. In: Markowitz SD, Berger NA (eds) Energy balance and gastrointestinal cancer. Energy balance and cancer, vol 4. Springer, New York, pp 157–182
63. Sharma RA, Gescher A, Plastaras JP, Leuratti C, Singh R, Gallacher-Horley B, Offord E, Marnett LJ, Steward WP, Plummer SM (2001) Cyclooxygenase-2, malondialdehyde and pyrimidopurinone adducts of deoxyguanosine in human colon cells. Carcinogenesis 22:1557–1560
64. Wilson AJ, Byron K, Gibson PR (1999) Interleukin-8 stimulates the migration of human colonic epithelial cells in vitro. Clin Sci (Lond) 97:385–390
65. Zhao L, Kwon MJ, Huang S, Lee JY, Fukase K, Inohara N, Hwang DH (2007) Differential modulation of Nods signaling pathways by fatty acids in human colonic epithelial HCT116 cells. J Biol Chem 282:11618–11628
66. Hwang D (2001) Modulation of the expression of cyclooxygenase-2 by fatty acids mediated through Toll-like receptor 4-derived signaling pathways. FASEB J 15(14):2556–2564
67. Baró L, Hermoso JC, Núñez MC, Jiménez-Rios JA, Gil A (1998) Abnormalities in plasma and red blood cell fatty acid profiles of patients with colorectal cancer. Br J Cancer 77:1978–1983
68. Hietanen E, Bartsch H, Bereziat JC, Camus AM, McClinton S, Eremin O, Davidson L, Boyle P (1994) Diet and oxidative stress in breast, colon and prostate cancer patients: a case-control study. Eur J Clin Nutr 48:575–586
69. Theodoratou E, McNeill G, Cetnarskyj R, Farrington SM, Tenesa A, Barnetson R, Porteous M, Dunlop M, Campbell H (2007) Dietary fatty acids and colorectal cancer: a case-control study. Am J Epidemiol 166(2):181–195
70. Kuriki K, Mutoh M, Tajima K, Wakabayashi K, Tatematsu M (2008) Relationship between intestinal polyp formation and fatty acid levels in plasma, erythrocytes, and intestinal polyps in Min mice. Cancer Sci 99:2410–2416
71. Neaptolemos JP, Husband D, Imray C, Rowley S, Lawson N (1991) Arachidonic acid and docosahexaenoic acid are increased in human colorectal cancer. Gut 32:278–281
72. Awad AB, Hermann T, Fink CS, Horwath PJ (1995) 18:1 n7 fatty acids inhibit growth and decrease inositol phosphate release in HT-29 cells compared to n9 fatty acids. Cancer Lett 91:55–61
73. Habib NA, Hershman MJ, Salem R, Barker W, Apostolov K, Wood CB (1987) Increased erythrocyte stearic acid desaturation in rats with chemically induced colorectal carcinomas. Int J Colorectal Dis 2:12–14

74. Schlager SI, Ohanian SH, Borsos T (1997) Correlation between the ability of tumor cells to incorporate specific fatty acids and their sensitivity to killing by a specific antibody plus guinea pig complement. S Afr Med J 87:152–158
75. Baer DJ, Judd JT, Clevidence BA, Tracy RP (2004) Dietary fatty acids affect plasma markers if inflammation in healthy men fed controlled diets: a randomized crossover study. Am J Clin Nutr 79:969–973
76. King ED, Egan BM, Geesey ME (2003) Relation of dietary fat and fiber to elevation of C-reactive protein. Am J Cardiol 92(11):1335–1339
77. Nestel P, Clifton P, Noakes M (1994) Effects of increasing dietary palmitoleic acid compared with palmitic and oleic acids on plasma lipids of hypercholesterolemic men. J Lipid Res 35(4):656–662
78. Yang ZH, Miyahara H, Hatanaka A (2011) Chronic administration of palmitoleic acid reduces insulin resistance and hepatic lipid accumulation in KK-Ay mice with genetic type 2 diabetes. Lipids Health Dis 10(120):1–8
79. Mosconi C, Agradi E, Gambetta A, Bozzetti F, Galli C (1989) Decrease of polyunsaturated fatty acids and elevation of the oleic/stearic acid ratio in plasma and red blood cell lipids of malnourished cancer patients. JPEN J Parenter Enteral Nutr 13:501–504
80. Takeshita M, Ueda H, Shirabe K, Higuchi Y, Yoshida S (1997) Lack of promotion of colon cancinogenesis by high-oleic safflower oil. Cancer 79:1487–1493
81. Harvey KA, Walker CL, Xu Z, Whitley P, Pavlina TM, Hise M, Zaloga GP, Siddiqui RA (2010) Oleic acid inhibits stearic acid-induced inhibition of cell growth and pro-inflammatory responses in human aortic endothelial cells. J Lipid Res 51(12):3470–3480
82. Llor X, Pons E, Roca A, Alvarez M, Mañé J, Fernández-Bañares F, Gassull MA (2003) The effects of fish oil, olive oil, oleic acid and linoleic acid on colorectal neoplastic processes. Clin Nutr 22:71–79
83. Carrillo C, Cavia Mdel M, Alonso-Torre SR (2012) Oleic acid inhibits store-operated calcium entry in human colorectal adenocarcinoma cells. Eur J Nutr 51(6):677–684
84. de la Puerta R, Marquez-Martin A, Fernandez-Arche A, Ruiz-Gutierrez V (2009) Influence of dietary fat on oxidative stress and inflammation in murine macrophages. Nutrition 25(5):548–554
85. Hendrickse CW, Kelly RW, Radley S, Donovan IA, Keighley MR, Neoptolemos JP (1994) Lipid peroxidation and prostaglandins in colorectal cancer. Br J Surg 81(8):1219–1223
86. Dupont J, White PJ, Carpenter MP, Schaefer EJ, Meydani SN, Elson CE, Woods M, Gorback SL (1990) Food uses and health effects of corn oil. J Am Coll Nutr 9:438–470
87. Kant AK (2000) Consumption of energy-dense, nutrient-poor foods by adult Americans: nutritional and health implications. The third National Health and Nutrition Examination Survey, 1988–1994. Am J Clin Nutr 72:929–936
88. Reddy BS, Maeura Y (1984) Tumor promotion by dietary fat in azoxymethane-induced colon carcinogenesis in female F344 rats: influence of amount and source of dietary fat. J Natl Cancer Inst 72:745–750
89. Reddy BS, Sugie S (1988) Effect of different levels of omega-3 and omega-6 fatty acids on azoxymethane-induced colon carcinogenesis in F344 rats. Cancer Res 48:6642–6647
90. Singh J, Hamid R, Reddy BS (1997) Dietary fat and colon cancer: modulation of cyclooxygenase-2 by types and amount of dietary fat during the postinitiation stage of colon carcinogenesis. Cancer Res 57:3465–3470
91. Fernández-Bañares F, Esteve M, Navarro E, Cabré E, Boix J, Abad-Lacruz A, Klaassen J, Planas R, Humbert P, Pastor C, Gassull MA (1996) Changes of the mucosal n3 and n6 fatty acid status occur early in the colorectal adenoma-carcinoma sequence. Gut 38:254–259
92. Tuyns AJ, Haelterman M, Kaaks R (1987) Colorectal cancer and the intake of nutrients: oligosaccharides are a risk factor, fats are not. A case-control study in Belgium. Nutr Cancer 10:181–196
93. Kelley NS, Hubbard NE, Erickson KL (2007) Conjugated linoleic acid isomers and cancer. J Nutr 137:2599–2607

94. Sakaguchi M, Hiramatsu Y, Takada H, Yamamura M, Hioki K, Saito K, Yamamoto M (1984) Effect of dietary unsaturated and saturated fats on azoxymethane-induced colon carcinogenesis in rats. Cancer Res 44:1472–1477
95. Ohmori H, Lu Y, Fujii K, Sasahira T, Shimomoto T, Denda A, Kuniyasu H (2010) Dietary linoleic acid and glucose enhances azoxymethane-induced colon cancer and metastases via the expression of high-mobility group box 1. Pathobiology 77(4):210–217
96. Johnson GH, Fritsche K (2012) Effect of dietary linoleic acid on markers of inflammation in healthy persons: a systematic review of randomized controlled trials. J Acad Nutr Diet 112(7):1029–1041
97. Zock PL, Katan MB (1998) Linoleic acid intake and cancer risk: a review and meta-analysis. Am J Clin Nutr 68(1):142–153
98. Almendingen K, Høstmark AT, Fausa O, Mosdøl A, Aabakken L, Vatn MH (2007) Familial adenomatous polyposis patients have high levels of arachidonic acid and docosahexaenoic acid and low levels of linoleic acid and alpha-linolenic acid in serum phospholipids. Int J Cancer 120(3):632–637
99. Brink M, Weijenberg MP, De Goeij AF, Schouten LJ, Koedijk FD, Roemen GM, Lentjes MH, De Bruïne AP, Goldbohm RA, Van Den Brandt PA (2004) Fat and K-ras mutations in sporadic colorectal cancer in The Netherlands Cohort Study. Carcinogenesis 25(9):1619–1628
100. Weijenberg MP, Lüchtenborg M, de Goeij AF, Brink M, van Muijen GN, de Bruïne AP, Goldbohm RA, van den Brandt PA (2007) Dietary fat and risk of colon and rectal cancer with aberrant MLH1 expression, APC or KRAS genes. Cancer Causes Control 18(8):865–879
101. Kim EJ, Holthuizen PE, Park HS, Ha YL, Jung KC, Park JH (2002) Trans-10, cis-12-conjugated linoleic acid inhibits Caco-2 colon cancer cell growth. Am J Physiol Gastrointest Liver Physiol 283(2):G357–G367
102. Miller A, Stanton C, Devery R (2001) Modulation of arachidonic acid distribution by conjugated linoleic acid isomers and linoleic acid in MCF-7 and SW480 cancer cells. Lipids 36(10):1161–1168
103. Wielinga PY, Harthoorn LF, Verschuren L, Schoemaker MH, Jouni ZE, van Tol EA, Kleemann R, Kooistra T (2012) Arachidonic acid/docosahexaenoic acid-supplemented diet in early life reduces body weight gain, plasma lipids, and adiposity in later life in ApoE*3Leiden mice. Mol Nutr Food Res 56(7):1081–1089
104. Banni S (2002) Conjugated linoleic acid metabolism. Curr Opin Lipidol 13(3):261–266
105. Nakamora YK, Flintoff-Dye N, Omaye ST (2008) Conjugated linoleic acid modulation of risk factors associated with atherosclerosis. Nutr Metab (Lond) 5:22
106. Martinez K, Kennedy A, West T, Milatovic D, Aschner M, McIntosh M (2010) Trans-10, cis-12-conjugated linoleic acid instigates inflammation in human adipocytes compared with preadipocytes. J Biol Chem 285(23):17701–17712
107. Yu Y, Correll PH, Vanden Heuvel JP (2002) Conjugated linoleic acid decreases production of pro-inflammatory products in macrophages: evidence for a PPARγ-dependent mechanism. Biochim Biophys Acta 1581(3):89–99
108. Huang G, Zhong X, Cao Y, Chen Y (2007) Antiproliferative effects of conjugated linoleic acid on human colon adenocarcinoma cell line Caco-2. Asia Pac J Clin Nutr 16(suppl 1): 432–436
109. Degen C, Ecker J, Piegholdt S, Liebisch G, Schmitz G, Jahreis G (2011) Metabolic and growth inhibitory effects of conjugated fatty acids in the cell line HT-29 with special regard to the conversion of t11, t13-CLA. Biochim Biophys Acta 1811(12):1070–1080
110. Bozzo F, Bocca C, Colombatto S, Miglietta A (2007) Antiproliferative effect of conjugated linoleic acid in caco-2 cells: involvement of PPARγ and APC/β-catenin pathways. Chem Biol Interact 169(2):110–121
111. Soel SM, Choi OS, Bang MH, Park JHY, Kim WK (2007) influence of conjugated linoleic acid isomers on the metastasis of colon cancer cells in vitro and in vivo. J Nutr Biochem 18:650–657
112. Poirier H, Shapiro JS, Kim RJ, Lazar MA (2006) Nutritional supplementation with trans-10, cis-12-conjugated linoleic acid induces inflammation of white adipose tissue. Diabetes 55(6):1634–1641

113. Reynolds CM, Draper E, Keogh B, Rahman A, Moloney AP, Mills KHG, Loscher CE, Roche HM (2009) A conjugated linoleic acid-enriched beef diet attenuates lipopolysaccharide-induced inflammation in mice in part through PPARγ-mediated suppression of Toll-like receptor 4[1-3]. J Nutr 139(12):2351–2357
114. Kohno H, Suzuki R, Noguchi R, Hosokawa M, Miyashita K, Tanaka T (2002) Dietary conjugated linolenic acid inhibits azoxymethane-induced colonic aberrant crypt foci in rats. Jpn J Cancer Res 93:133–142
115. Park J, Euhus DM, Scherer PE (2001) Paracrine and endocrine effects of adipose tissue on cancer development and progression. Endocr Rev 32:550–570
116. Kim EJ, Jun JG, Park HS, Kim SM, Ha YL, Park JH (2002) Conjugated linoleic acid (CLA) inhibits growth of Caco-2 colon cancer cells: possible mediation by oleamide. Anticancer Res 22:2193–2197
117. Kohno H, Suzuki R, Yasui Y, Hosokawa M, Miyashita K, Tanaka T (2004) Pomegranate seed oil rich in conjugated linolenic acid suppresses chemically induced colon carcinogenesis in rats. Cancer Sci 95(6):481–486
118. Evans NP, Misyak SA, Schmelz EM, Guri AJ, Hontecillas R, Bassaganya-Riera J (2010) Conjugated linoleic acid ameliorates inflammation-induced colorectal cancer in mice through activation of PPARgama. J Nutr 140(3):515–521
119. Shiraishi R, Iwakiri R, Takehiro F, Kuroki T, Kakimoto T, Takashima T, Sakata Y, Tsunada S, Nakashima Y, Yanagita T, Jufimoto K (2010) Conjugated linoleic acid suppresses colon carcinogenesis in azoxymethane-pretreated rats with long-term feeding of diet containing beef tallow. J Gastroenterol 45:625–635
120. Larsson SC, Bergkvist L, Wolk A (2005) High-fat dairy food and conjugated linoleic acid intakes in relation to colorectal cancer incidence in the Swedish Mammography Cohort. Am J Clin Nutr 82(4):894–900
121. Petry FM, Tutton PJ, Barkla DH (1984) The influence of arachidonic acid metabolites on cell division in the intestinal epithelium and in colonic tumors. Prostaglandins Leukot Med 15(3):349–358
122. Hansen Petrik MB, McEntee MF, Chiu C-H, Whelan J (2000) Antagonism of arachidonic acid is linked to the antitumorigenic effect of dietary eicosapentaenoic acid in $APC^{Min/+}$ mice. J Nutr 130(5):1153–1158
123. Cathcart MC, Lysaght J, Pidgeon GP (2011) Eicosanoid signalling pathways in the development and progression of colorectal cancer: novel approaches for prevention/intervention. Cancer Metastasis Rev 30(3–4):363–385
124. Nakanishi M, Menoret A, Tanaka T, Miyamoto S, Montrose DC, Vella AT, Rosenberg DW (2011) Selective PGE(2) suppression inhibits colon carcinogenesis and modifies local mucosal immunity. Cancer Prev Res (Phila) 4(8):1198–1208
125. Koehne C-H, Dubois RN (2004) COX-2 inhibition and colorectal cancer. Semin Oncol 31(7):12–21
126. Murff HJ, Shu XO, Li H, Dai Q, Kallianpur A, Yang G, Cai H, Wen W, Gao YT, Zheng W (2009) A prospective study of dietary polyunsaturated fatty acids and colorectal cancer risk in Chinese women. Cancer Epidemiol Biomarkers Prev 18(8):2283–2291
127. Busstra MC, Siezen CL, Grubben MJ, van Kranen HJ, Nagengast FM, van't Veer P (2003) Tissue levels of fish fatty acids and risk of colorectal adenomas: a case-control study (Netherlands). Cancer Causes Control 14(3):269–276
128. Kimura Y, Kono S, Toyomura K, Nagano J, Mizoue T, Moore MA, Mibu R, Tanaka M, Kakeji Y, Maehara Y, Okamura T, Ikejiri K, Futami K, Yasunami Y, Maekawa T, Takenaka K, Ichimiya H, Imaizumi N (2007) Meat, fish and fat intake in relation to subsite-specific risk of colorectal cancer: The Fukuoka Colorectal Cancer Study. Cancer Sci 98(4):590–597
129. Kojima M, Wakai K, Tokudome S, Suzuki K, Tamakoshi K, Watanabe Y, Kawado M, Hashimoto S, Hayakawa N, Ozasa K, Toyoshima H, Suzuki S, Ito Y, Tamakoshi A (2005) JACC Study Group. Serum levels of polyunsaturated fatty acids and risk of colorectal cancer: a prospective study. Am J Epidemiol 161(5):462–471

130. Lin J, Zhang SM, Cook NR, Lee IM, Buring JE (2004) Dietary fat and fatty acids and risk of colorectal cancer in women. Am J Epidemiol 160(10):1011–1022
131. Kakutani S, Ishikura Y, Tateishi N, Horikawa C, Tokuda H, Kontani M, Kawashima H, Sakakibara Y, Kiso Y, Shibata H, Morita I (2011) Supplementation of arachidonic acid-enriched oil increases arachidonic acid contents in plasma phospholipids, but does not increase their metabolites and clinical parameters in Japanese healthy elderly individuals: a randomized controlled study. Lipids Health Dis 10:241
132. Dwivedi C, Natarajan K, Matthees DP (2005) Chemopreventive effects of dietary flaxseed oil on colon tumor development. Nutr Cancer 51(1):52–58
133. Narisawa T, Fukaura Y, Yazawa K, Ishikawa C, Isoda Y, Nishizawa Y (1994) Colon cancer prevention with a small amount of dietary perilla oil high in alpha-linolenic acid in an animal model. Cancer 73(8):2069–2075
134. Onogi N, Okuno M, Komaki C, Moriwaki H, Kawamori T, Tanaka T, Mori H, Muto Y (1996) Suppressing effect of perilla oil on azoxymethane-induced foci of colonic aberrant crypts in rats. Carcinogenesis 17(6):1291–1296
135. Bommareddy A, Zhang X, Schrader D, Kaushik RS, Zeman D, Matthees DP, Dwivedi C (2009) Effects of dietary flaxseed on intestinal tumorigenesis in Apc(Min) mouse. Nutr Cancer 61(2):276–283
136. Chiang YL, Haddad E, Rajaram S, Shavlik D, Sabaté J (2012) The effect of dietary walnuts compared to fatty fish on eicosanoids, cytokines, soluble endothelial adhesion molecules and lymphocyte subsets: a randomized, controlled crossover trial. Prostaglandins Leukot Essent Fatty Acids 87(4–5):111–117
137. Hassan A, Ibrahim A, Mbodji K, Coëffier M, Ziegler F, Bounoure F, Chardigny JM, Skiba M, Savoye G, Déchelotte P, Marion-Letellier R (2010) An α-linolenic acid-rich formula reduces oxidative stress and inflammation by regulating NF-κB in rats with TNBS-induced colitis. J Nutr 140(10):1714–1721
138. Nounou HA, Deif MM, Shalaby MA (2012) Effect of flaxseed supplementation and exercise training on lipid profile, oxidative stress and inflammation in rats with myocardial ischemia. Lipids Health Dis 11(1):129
139. Poudel-Tandukar K, Sato M, Ejima Y, Nanri A, Matsushita Y, Imaizumi K, Mizoue T (2012) Relationship of serum fatty acid composition and desaturase activity to C-reactive protein in Japanese men and women. Atherosclerosis 220(2):520–524
140. Tyagi A, Kumar U, Reddy S, Santosh VS, Mohammed SB, Ehtesham NZ, Ibrahim A (2012) Attenuation of colonic inflammation by partial replacement of dietary linoleic acid with α-linolenic acid in a rat model of inflammatory bowel disease. Br J Nutr 108(9):1612–1622
141. Vanden Heuvel JP (2012) Nutrigenomics and nutrigenetics of ω3 polyunsaturated fatty acids. Prog Mol Biol Transl Sci 108:75–112
142. Park HJ, Park JS, Hayek MG, Reinhart GA, Chew BP (2011) Dietary fish oil and flaxseed oil suppress inflammation and immunity in cats. Vet Immunol Immunopathol 141(3–4):301–306
143. Baranowski M, Enns J, Blewett H, Yakandawala U, Zahradka P, Taylor CG (2012) Dietary flaxseed oil reduces adipocyte size, adipose monocyte chemoattractant protein-1 levels and T-cell infiltration in obese, insulin-resistant rats. Cytokine 59(2):382–391
144. Caughey GE, Mantzioris E, Givson RA, Cleland LG, James MJ (1996) The effect o human tumor necrosis factor α and interleukin 1 production of diets enriched in n-3 fatty acids from vegetable oil or fish oil. Am J Clin Nutr 63:116–122
145. Gopinath B, Buyken AE, Flood VM, Empson M, Rochtchina E, Mitchell P (2011) Consumption of polyunsaturated fatty acids, fish, and nuts and risk of inflammatory disease mortality. Am J Clin Nutr 93(5):1073–1079
146. Zhao G, Etherton TD, Martin KR, West SG, Gillies PJ, Kris-Etherton PM (2004) Dietary alpha-linolenic acid reduces inflammatory and lipid cardiovascular risk factors in hypercholesterolemic men and women. J Nutr 134(11):2991–2997
147. Schloss I, Kidd MS, Tichelaar HY, Young GO, O'Keefe SJ (1997) Dietary factors associated with a low risk of colon cancer in coloured west coast fishermen. S Afr Med J 87:152–158

148. Anti M, Marra G, Armelao F, Bartoli GM, Ficarelli R, Percesepe A, De Vitis I, Maria G, Sofo L, Rapaccini GL et al (1992) Effect of omega-3 fatty acids on rectal mucosal cell proliferation in subjects at risk for colon cancer. Gastroenterology 103:883–891
149. Anti M, Armelao F, Marra G, Percesepe A, Bartoli GM, Palozza P, Parrella P, Canetta C, Gentiloni N, De Vitis I et al (1994) Effects of different doses of fish oil on rectal cell proliferation in patients with sporadic colonic adenomas. Gastroenterology 107:1709–1718
150. Takahashi M, Minamoto T, Yamashita N, Yazawa K, Sugimura T, Esumi H (1993) Reduction in formation and growth of 1,2-dimethylhydrazine-induced aberrant crypt foci in rat colon by docosahexaenoic acid. Cancer Res 53:2786–2789
151. Takahashi M, Fukutake M, Isoi T, Fukuda K, Sato H, Yazawa K, Sugimura T, Wakabayashi K (1997) Suppression of azoxymethane-induced rat colon carcinoma development by a fish oil component, docosahexaenoic acid (DHA). Carcinogenesis 18:1337–1342
152. Narayanan BA, Narayanan NK, Reddy BS (2001) Docosahexaenoic acid regulated genes and transcription factors inducing apoptosis in human colon cancer cells. Int J Oncol 19:1255–1262
153. Calviello G, Di Nicuolo F, Gragnoli S, Piccioni E, Serini S, Maggiano N, Tringali G, Navarra P, Ranalletti FO, Palozza P (2004) n-3 PUFAs reduce VEGF expression in human colon cancer cells modulating the COX-2/PGE2 induced ERK-1 and -2 and HIF-lalpha induction pathway. Carcinogenesis 25:2303–2310
154. Engelbrecht A-M, Toit-Kohn J-L, Ellis B, Thomas M, Nell T, Smith R (2008) Differential induction of apoptosis and inhibition of the PI3-kinase pathway by saturated, monounsaturated and polyunsaturated fatty acids in a colon cancer cell model. Apoptosis 13:1368–1377
155. Kolar SS, Barhoumi R, Callaway ES, Fan YY, Wang N, Lupton JR, Chapkin RS (2007) Synergy between docosahexaenoic acid and butyrate elicits p53-independent apoptosis via mitochondrial Ca(2+) accumulation in colonocytes. Am J Physiol Gastrointest Liver Physiol 293:G935–G943
156. Siriwardhana N, Kalupahana NS, Moustaid-Moussa N (2012) Health benefits of n-3 polyunsaturated fatty acids: eicosapentaenoic acid and docosahexaenoic acid. Adv Food Nutr Res 65:211–222
157. Allred CD, Talbert DR, Southard RC, Wang X, Kilgore MW (2008) PPARgammal as a molecular target of eicosapentaenoic acid in human colon cancer (HT-29) cells. J Nutr 138:250–256
158. Dommels YE, Haring MM, Keestra NG, Alink GM, van Bladeren PJ, van Ommen B (2003) The role of cyclooxygenase in n-6 and n-3 polyunsaturated fatty acid mediated effects on cell proliferation, PGE(2) synthesis and cytotoxicity in human colorectal carcinoma cell lines. Carcinogenesis 24:385–392
159. Kumamoto-Yonezawa Y, Sasaki R, Ota Y, Suzuki Y, Fukushima S, Hada T, Uryu K, Sugimura K, Yoshida H, Mizushina Y (2009) Cell cycle arrest triggered by conjugated eicosapentaenoic acid occurs through several mechanisms including G1 checkpoint activation by induced RPA and ATR expression. Biochim Biophys Acta 1790:339–346
160. Reddy BS, Maruyama H (1986) Effect of dietary fish oil on azoxymethane-induced colon carcinogenesis in male F344 rats. Cancer Res 46(7):3367–3370
161. Whelan J, Petrik MB, McEntee MF, Obukowicz MG (2002) Dietary EPA reduces tumor load in $APC^{Min/+}$ mice by altering arachidonic acid metabolism, but conjugated linoleic acid, gamma-and-alpha-linoleic acids have no effect. Adv Exp Med Biol 507:579–584
162. Petrik MB, McEntee MF, Johnson BT, Obukowicz MB, Whelan J (2000) Highly unsaturated (n-3) fatty acids, but not alpha-linolenic, conjugated linolenic or gamma-linolenic acids, reduce tumorigenesis in Apc(Min/+) mice. J Nutr 130(10):2434–2443
163. Fini L, Piazzi G, Ceccarelli C, Daoud Y, Belluzzi A, Munarini A, Graziani G, Fogliano V, Selgrad M, Garcia M, Gasbarrini A, Genta RM, Boland CR, Ricciardiello L (2010) Highly purified eicosapentaenoic acid as free fatty acids strongly suppresses polyps in Apc(Min/+) mice. Clin Cancer Res 16(23):5703–5711
164. Minoura T, Takata T, Sakaguchi M, Takada H, Yamamura M, Hioki K, Yamamoto M (1988) Effect of dietary eicosapentaenoic acid on azoxymethane-induced colon carcinogenesis in rats. Cancer Res 48:4790–4794

165. Simopoulos AP (2006) Evolutionary aspects of diet, the omega-6/omega-3 ratio and genetic variation: nutritional implications for chronic diseases. Biomed Pharmacother 60(9):502–507
166. Simopoulos AP (2011) Evolutionary aspects of diet: the omega-6/omega-3 ratio and the brain. Mol Neurobiol 44(2):203–215
167. Huang YC, Jessup JM, Forse RA, Flickner S, Pleskow D, Anastopoulos HT, Ritter V, Blackburn GL (1996) N-3 fatty acids decrease colonic epithelial cell proliferation in high-risk bowel mucosa. Lipids 31(suppl):S313–S317
168. Staessen L, De Henauw S, De Bacquer D, De Backer G, Van Peteghem C (1998) Fat sources in the Belgian diet. Ann Nutr Metab 42(3):138–150
169. Staessen L, De Bacquer D, De Henauw S, De Backer G, Van Peteghem C (1998) Fatty acid composition of the Belgian diet: estimates derived from the Belgian Interuniversity Research on Nutrition and Health. Ann Nutr Metab 42(3):151–159
170. Pot GK, Geelen A, van Heijningen EM, Siezen CL, van Kranen HJ, Kampman E (2008) Opposing Associations of serum n-3 and n-6 polyunsaturated fatty acids with colorectal adenoma risk: an endoscopy-based case-control study. Int J Cancer 123(8):1974–1977
171. Larsson SC, Wolk A (2006) Meat consumption and risk of colorectal cancer: a meta-analysis of prospective studies. Int J Cancer 119(11):2657–2664
172. Norat T, Lukanova A, Ferrari P, Riboli E (2002) Meat consumption and colorectal cancer risk: dose-response meta-analysis of epidemiological studies. Int J Cancer 98(2):241–256
173. Giovannucci E, Rimm EB, Stampfer MJ, Colditz GA, Ascherio A, Willet WC (1994) Intake of fat, meat, and fiber in relation to risk of colon cancer in men. Cancer Res 54(9):2390–2397
174. Wakai K, Hirose K, Matsuo K, Ito H, Kuriki K, Suzuki T, Kato T, Hirai T, Kanemitsu Y, Tajima K (2006) Dietary risk factors for colon and rectal cancers: a comparative case-control study. J Epidemiol 16(3):125–135
175. Ognjanovic S, Yamamoto J, Maskarinec G, Le Marchand L (2006) NAT2, meat consumption and colorectal cancer incidence: an ecological study among 27 countries. Cancer Causes Control 17(9):1175–1182
176. Kim KH, Park HS (2003) Dietary supplementation of conjugated linoleic acid reduces colon tumor incidence in DMH-treated rats by increasing apoptosis with modulation of biomarkers. Nutrition 19(9):772–777
177. Williams CD, Satia JA, Adair LS, Stevens J, Galanko J, Keku TO, Sandler RS (2010) Association of red meat, fat, and protein intake with distal colorectal cancer risk. Nutr Cancer 62(6):701–709
178. Sofi F, Cesari F, Abbate R, Gensini GF, Casini A (2008) Adherence to Mediterranean diet and health status: meta-analysis. BMJ 337:a1344
179. Psaltopoulou T, Kosti RI, Haidopoulos D, Dimopoulos M, Panagiotakos DB (2011) Olive oil intake is inversely related to cancer prevalence: a systematic review and a meta-analysis of 13,800 patients and 23,340 controls in 19 observational studies. Lipids Health Dis 10(127):1–16
180. Reddy J, Mitrou PN, Krebs-Smith SM, Wirfalt E, Flood A, Kipnis V, Leitzmann M, Mouw M, Mouw T, Hollenbeck A, Schatzkin A, Subar AF (2008) Index-based dietary patterns and risk of colorectal cancer: the NIH-AARP Diet and Health Study. Am J Epidemiol 168:38–48
181. Dixon LB, Subar AF, Peters U, Weissfeld JL, Bresalier RS, Risch A, Schatzkin A, Hayes RB (2007) Adherence to the USDA Food Guide, DASH Eating Plan, and Mediterranean dietary pattern reduces risk of colorectal adenoma. J Nutr 137(11):2443–2450
182. Fung RR, Hu FB, Chiuve SE, Fuchs CS, Giovannucci E (2010) The Mediterranean and Dietary Approaches to Stop Hypertension (DASH) diets and colorectal cancer. Am J Clin Nutr 92:1429–1435
183. Macquart-Moulin G, Riboli E, Cornée J, Carnay B, Berthezéne P, Day N (1986) Case-control study on colorectal cancer and diet in Marseilles. Int J Cancer 38(2):183–191
184. Braga C, La Vecchia C, Franceschi S, Negri E, Parpinel M, Decarli A, Giacosa A, Trichopoulos D (1998) Olive oil, other seasoning fats and the risk of colorectal carcinoma. Cancer 82:448–453
185. Schwartz B, Birk Y, Raz A, Madar Z (2004) Nutritional-pharmacological combinations—a novel approach to reducing colon cancer incidence. Eur J Nutr 43(4):221–229

186. Bartoli R, Fernández-Bañares F, Navarro E, Castellà E, Mañé J, Alvarez M, Pastor C, Cabré E, Gassull MA (2000) Effect of olive oil on early and late events of colon carcinogenesis in rats: modulation of arachidonic acid metabolism and local prostaglandin E_2 synthesis. Gut 46:191–199
187. Klampfer L, Heerdt BF, Velcich A, Gaffney-Stomberg E, Wang D, Lin E, Augenlicht LH (2012) Dietary modulation of colon cancer: effects on intermediary metabolism, mucosal cell differentiation, and inflammation. In: Markowitz SD, Berger NA (eds) Energy balance and gastrointestinal cancer. Energy balance and cancer, vol 4. Springer, New York, p 4764
188. Erdelyi I, Levelnkova N, Lin EY, Pinto JT, Lipkin M, Quimby FW, Holt PR (2009) Western-style diets induce oxidative stress and dysregulate immune responses in the colon in a mouse model of sporadic colon cancer. J Nutr 139:2072–2078
189. Coussens LM, Werb Z (2002) Inflammation and cancer. Nature 420(6917):860–867
190. Doerner SK, Leung ES, Berger NA, Nadeau JH (2011) High dietary fat promotes inflammation and intestinal neoplasia, independently of diet-induced obesity, in B6.ApcMin/+ congenic-consomic mouse strains. In: Proceedings of the 102nd annual meeting of the American Association for Cancer Research, Orlando, FL, 2–6 Apr 2011. AACR, Philadelphia, PA; 71. Abstract nr 910
191. Larsson SC, Kumlin M, Ingelman-Sundberg M, Wolk A (2004) Dietary long-chain n-3 fatty acids for the prevention of cancer: a review of potential mechanisms. Am J Clin Nutr 79:935–945
192. Pendyala S, Neff LM, Suárez-Fariñas M, Holt PR (2011) Diet-induced weight loss reduces colorectal inflammation: implications for colorectal carcinogenesis. Am J Clin Nutr 93(2):234–242
193. Li L (2012) Obesity, insulin resistance pathway factors and colon cancer. In: Markowitz SD, Berger NA (eds) Energy balance and gastrointestinal cancer. Energy balance and cancer, vol 4. Springer, New York, pp 111–129
194. Meydani SN, Endres S, Woods MM, Goldin BR, Soo C, Morrill-Labrode A, Dinarello CA, Gorbach SL (1991) Oral (n-3) fatty acid supplementation suppresses cytokine production and lymphocyte proliferation: comparison between young and older women. J Nutr 121(4): 547–555
195. Toriola AT, Ulrich CM (2011) Is there a potential use for C-reactive protein as a diagnostic and prognostic marker for colorectal cancer? Future Oncol 7(10):1125–1128
196. Abir F, Alva S, Kaminski DL, Longo WE (2005) The role of arachidonic acid regulatory enzymes in colorectal disease. Dis Colon Rectum 48(7):1471–1483
197. Ichimura A, Hirasawa A, Poulain-Godefroy O, Bonnefond A, Hara T, Yengo L, Kimura I, Leloire A, Liu N, Iida K, Choquet H, Besnard P, Lecoeur C, Vivequin S, Ayukawa K, Takeuchi M, Ozawa K, Tauber M, Maffeis C, Morandi A, Buzzetti R, Elliott P, Pouta A, Jarvelin MR, Körner A, Kiess W, Pigeyre M, Caiazzo R, Van Hul W, Van Gaal L, Horber F, Balkau B, Lévy-Marchal C, Rouskas K, Kouvatsi A, Hebebrand J, Hinney A, Scherag A, Pattou F, Meyre D, Koshimizu TA, Wolowczuk I, Tsujimoto G, Froguel P (2012) Dysfunction of lipid sensor GPR120 leads to obesity in both mouse and human. Nature 483(7389):350–354
198. Oh DY, Talukdar S, Bae EJ, Imamura T, Morinaga H, Fan W, Li P, Lu WJ, Watkins SM, Olefsky JM (2010) GPR120 is an omega-3 fatty acid receptor mediating potent anti-inflammatory and insulin-sensitizing effects. Cell 142(5):687–698
199. Maeda K, Cao H, Kono K, Gorgun CZ, Furuhashi M, Uysal KT, Cao Q, Atsumi G, Malone H, Krishnan B, Minokoshi Y, Kahn BB, Parker RA, Hotamisligil GS (2005) Adipocyte/macrophage fatty acid binding proteins control integrated metabolic responses in obesity and diabetes. Cell Metab 1(2):107–119
200. Makowski L, G khan SH (2005) The role of fatty acid binding proteins in metabolic syndrome and atherosclerosis. Curr Opin Lipidol 16:543–548
201. Stryjecki C, Mutch DM (2011) Fatty acid-gene interactions, adipokines and obesity. Eur J Clin Nutr 65(3):285–297
202. Sun S, Ji Y, Kersten S, Qi L (2012) Mechanisms of inflammatory responses in obese adipose tissue. Annu Rev Nutr 32:261–286

203. Dupertuis YM, Meguid MM, Pichard C (2007) Colon cancer therapy: new perspectives of nutritional manipulations using polyunsaturated fatty acids. Curr Opin Clin Nutr Metab Care 10(4):427–432
204. Calviello G, Di Nicuolo F, Serini S, Piccioni E, Boninsegna A, Maggiano N, Ranelletti FO, Palozza P (2005) Docosahexaenoic acid enhances the susceptibility of human colorectal cancer cells to 5-fluorouracil. Cancer Chemother Pharmacol 55(1):12–20
205. Hardman WE (2004) (n-3) Fatty acids and cancer therapy. J Nutr 134:3427S–3430S
206. Wynter MP, Russell ST, Tisdale MJ (2004) Effect of n-3 fatty acids on the antitumour effects of cytotoxic drugs. In Vivo 18(5):543–547
207. Benais-Pont G, Dupertuis YM, Kossovsky MP, Nouet P, Allal AS, Buchegger F, Pichard C (2006) Omega-3 polyunsaturated fatty acids and ionizing radiation: combined cytotoxicity on human colorectal adenocarcinoma cells. Nutrition 22(9):931–939
208. Graveline D (2012) Statins, inflammation and heart disease. spacedoc.com; [cited 12/13/2012]. Available from http://www.spacedoc.com/statins_inflammation_heart_disease.html
209. Nielsen SF, Nordestgaard BG, Bojesen SE (2012) Statin use and reduced cancer-related mortality. N Engl J Med 367(19):1792–1802
210. Poynter JN, Gruber SB, Higgins PD, Almog R, Bonner JD, Rennert HS, Low M, Greenson JK, Rennert G (2005) Statins and the risk of colorectal cancer. N Engl J Med 352(21): 2184–2192
211. Graaf MR, Beiderbeck AB, Egberts ACG, Richel DJ, Guchelaar H-J (2004) The risk of cancer in users of statins. J Clin Oncol 22(12):2388–2394
212. Dale KM, Coleman CI, Henyan NN, Kluger J, White CM (2006) Statins and cancer risk. JAMA 295(1):74–80
213. Dechend R, Muller D, Park JK, Fiebeler A, Haller H, Luft FC (2002) Statins and angiotensin II-induced vascular injury. Nephrol Dial Transplant 17:349–353
214. Cimino M, Banfi C, Brioschi M, Gelosa P, Nobili E, Gianella A, Tremoli E, Sironi L (2005) Experimental ischemia: therapy. poster session. J Cereb Blood Flow Metab 25:S39
215. Hernández-Presa MA, Ortego M, Tuñón J, Martín-Ventura JL, Mas S, Blanco-Colio LM, Aparicio C, Ortega L, Gómez-Gerique J, Vivanco F, Egido J (2003) Simvastatin reduces NF-kappaB activity in peripheral mononuclear and in plaque cells of rabbit atheroma more markedly than lipid lowering diet. Cardiovasc Res 57(1):168–177
216. Quist-Paulsen P (2010) Statins and inflammation: an update. Curr Opin Cardiol 25:399–405
217. Schonbeck U, Libby P (2004) Inflammation, immunity and HMG-CoA reductase inhibitors: statins as anti-inflammatory agents? Circulation 109(21 suppl 1):II18–II26
218. Shyamsundar M, McKeown ST, O'Kane CM, Craig TR, Brown V, Thickett DR, Matthay MA, Taggart CC, Backman JT, Elborn JS, McAuley DF (2009) Simvastatin decreases lipopolysaccharide-induced pulmonary inflammation in healthy volunteers. Am J Respir Crit Care Med 179(12):1107–1114
219. Swamy MV, Cooma I, Reddy BS, Rao CV, Lamin B (2002) caspase-3 activity, and apoptosis induction by a combination of HMG-CoA reductase inhibitor and COX-2 inhibitors: a novel approach in developing effective chemopreventive regimens. Int J Oncol 20(4):753–759

Chapter 6
Inflammation, Obesity, Barrett's Esophagus, and Esophageal Adenocarcinoma

Anamay Sharma, Ahmed Elebiary, Sonia Chowdhury, and Navtej Buttar

Abstract Esophageal adenocarcinoma arising in metaplastic Barrett's esophagus is one of the most rapidly increasing cancers in Western countries. Accumulating epidemiological evidence provides support that both chronic reflux injury and being overweight are strongly associated with the risk of esophageal adenocarcinoma. It is proposed that being overweight could contribute to increased predisposition to reflux by mechanically disrupting the physiological mechanisms that prevent reflux injury to the esophagus. Furthermore, mechanistic investigations also provide a link between being overweight to the risk of esophageal adenocarcinoma through increased loco-systemic injury response and metabolic syndrome. Together these observations provide the basis for the hypothesis that being overweight could be a key early trigger for the initiation and an ongoing stimulus for the progression of esophageal adenocarcinoma. In this chapter we will summarize the existing data that supports this hypothesis and discuss ongoing and future investigations to address this hypothesis that links obesity to risk of esophageal adenocarcinoma.

6.1 Introduction, Obesity, and Esophageal Adenocarcinoma: Guilt by Association

The anthropometry of population of Western countries, particularly in the United States, is rapidly changing with an alarming increase in the incidence of obesity or being overweight. Of late there has been a sharp increase in the prevalence of obesity, almost to epidemic proportions, and the rate of increase in esophageal adenocarcinoma has essentially mirrored this rising rate of obesity [1, 2]. Several recent studies found close associations between the risk of esophageal adenocarcinoma

A. Sharma, M.D. • A. Elebiary, M.D. • S. Chowdhury, M.D. • N. Buttar, M.D. (✉)
Mayo Clinic, 200 First Street SW, Rochester, MN 55905, USA
e-mail: Buttar.Navtej@mayo.edu

A.J. Dannenberg and N.A. Berger (eds.), *Obesity, Inflammation and Cancer*, Energy Balance and Cancer 7, DOI 10.1007/978-1-4614-6819-6_6,

and being overweight [3–7]. However, the conclusion drawn by these studies does differ depending upon how the obesity is defined or characterized. Interestingly, a study examining the relationship between obesity and Barrett's esophagus, a precursor of esophageal adenocarcinoma, noted that when BMI is used as a continuous variable and adjusted for central adiposity, the BMI did not appear to be associated with increased risk of Barrett's esophagus [8]. According to WHO definition, obesity is defined as BMI of 30 or more but this does not take into account variation in body fat distribution [9]. Several parameters like waist circumference, waist to hip ratio, or imaging techniques like CT and DEXA have been used to assess the body fat distribution [10]. Male preponderance of central obesity as well as esophageal adenocarcinoma suggests that the association between being obese and esophageal adenocarcinoma could be through central adiposity [11, 12]. Indeed, several studies have found stronger relationships between abdominal obesity and esophageal disorders like gastroesophageal reflux disease (GERD) and Barrett's esophagus [3, 4, 10, 13]. Further studies examined and suggested that instead of BMI, there exists a closer association between the risk of esophageal adenocarcinoma and how the fat is distributed in the body [4, 14]. Patients with truncal accumulation of fat, as determined by hip to waist ratio, are more likely to develop esophageal adenocarcinoma [4]. Recent, carefully undertaken studies using CT of the abdomen to examine the relative contribution of visceral vs. subcutaneous fat suggest that the excess visceral fat, compared to subcutaneous fat, more likely predict the risk of esophageal adenocarcinoma [3]. In summary, it appears that being overweight is one of the important pathogenic processes that predispose patients to the development of esophageal adenocarcinoma. The development of esophageal adenocarcinoma is a long protracted process, and its risk can be influenced at several steps. Being obese could facilitate the degree of reflux or the consequence of the reflux by altering local or systemic inflammatory response that could facilitate the development of premalignant Barrett's esophagus. Finally, being overweight could also activate pro-oncogenic pathways or interfere with tumor-suppressive mechanisms in Barrett's mucosa. In the following paragraphs we will discuss these issues and the putative mechanisms.

6.2 Obesity and Degree of Reflux: Disruption of Mechanical Barrier

Long-standing reflux injury is associated with replacement of normal squamous mucosa of the esophagus by premalignant metaplastic Barrett's columnar mucosa that, in a subset of patients, progress to esophageal adenocarcinoma. It has been postulated that the pressure effect of abdominal obesity could change the relationship between the gastroesophageal junction and diaphragmatic anti-reflux mechanisms. These changes could decrease in LES (lower esophageal sphincter) pressure to below 10 mmHg from the usual 10–35 mmHg that leads to increase in reflux, eventually causing Barrett's esophagus and esophageal adenocarcinoma [10, 15].

This possibility along with the findings that obese patients are known to have an increased incidence of asymptomatic reflux might explain how the neoplastic progression can go undetected in obese patients [16]. Several studies have demonstrated increased association between obesity and drop in LES pressure and increased incidence of hiatal hernia eventually leading to more reflux. Kuper et al. conducted manometric studies on 47 obese patients without reflux symptoms and 15 normal-weight individuals and found that obesity is associated with a significant drop in LES pressure as well as disruption of normal esophageal motility [17]. Similarly, in a large study conducted by Ayazi et al., where 24-h pH monitoring as well as esophageal manometry was performed, obesity was correlated with low esophageal pH and lower LES pressure. This suggests that obesity likely increase the reflux of acidic gastroduodenal contents by altering LES relaxation leading to a drop in LES pressure [18]. Schneider et al. did a manometric and pH monitoring study in patients where they had groups with obesity, GERD without obesity, diffuse esophageal spasm, and normal controls and found that obese patients and GERD patients without obesity showed significantly increased transient LES relaxation [19]. In a novel manometric study Pandolfino et al. found that obesity is positively correlated with the disruption of esophageal gastric junction and higher value of gastroesophageal pressure gradient. This study supports the mechanical basis of increased reflux and hiatal hernia in patients with obesity [20]. Therefore it appears that augmented relaxation of the LES is one of the underlying mechanisms for reflux in the obese. Together these findings support the idea that by increasing the degree of reflux, truncal obesity could play a role in the development of esophageal adenocarcinoma. In addition, being obese could also make the esophageal mucosa more vulnerable to injury as discussed below.

6.3 Obesity and Altered Composition of Reflux: Facilitating Injury and Carcinogenesis

Contents of reflux, particularly bile acids in a pH-dependent manner inflict mucosal injury as well as facilitate neoplastic progression in Barrett's esophagus. Bile acids are derived from hepatic metabolism of cholesterol. Dietary and systemic derangements that are associated with being obese could potentially modify composition and the amount of bile salts in reflux. In a systematic review McQuaid et al. found that obese patients with GERD had higher concentration of bile acid in their esophageal aspirate [21]. Development of Barrett's esophagus is considered a protective mechanism against reflux injury as intestinal columnar cells are more resistant to toxic insults compared to squamous cells. Continued injury leads to release of increased concentration of pro-inflammatory cytokines, prostaglandins, reactive oxygen species (ROS) which through various signaling pathways induce cell proliferation and mutagenesis and inhibit apoptosis that likely facilitate the neoplastic transformation [22]. In various in vivo/in vitro studies it is noted that bile acids stimulate metaplastic and squamous esophageal cells to produce pro-carcinogenic inflammatory mediators like IL-8/COX-2 and cause oxidative stress and DNA damage [43]. Obesity-associated changes in bile acid composition and subsequent generation of free radicals have been advocated to play an important role both in perpetuation of inflammation as well as induction of various signaling pathways which eventually promote growth, angiogenesis, and tumor development. These free radicals include ROS like superoxide anion, hydrogen peroxide, and hydroxyl radical and reactive nitrogen species like peroxynitrite [22]. In experimental studies, bile acid-induced intracellular ROS mediate proliferation, DNA damage, and apoptosis through activation of tyrosine kinase, ERK1/2, and AKT as well as through protein kinase C signaling pathways [23, 24]. Song et al. in a study on Barrett's and EAC cell lines found that unconjugated bile acids induced COX-2 through ROS-mediated activation of PI3K/AKT and ERK1/2 and their downstream effectors CREB and AP-1 to facilitate neoplastic transformation [45]. These experiments were replicated in vivo in rats where it was found that bile acid-associated COX-2, p-AKT, and p-CREB overexpression induced metaplastic and neoplastic changes in the esophagus [45]. In an in vitro study performed by our group on Barrett's cell lines, we found that COX-2 inhibition significantly decreases proliferation of Barrett's esophageal cells, whereas treatment of these cells with PGE2 (product of COX-2 activity) restored proliferation of the cell lines, suggesting a possible role of COX-2 inhibitors for chemoprevention in Barrett's esophagus [25]. We further tested the role of COX-2 inhibitors in an animal model, where reflux injury was induced by performing esophagojejunostomy, and the animals were then randomized to groups receiving a COX-2 inhibitor vs. placebo [26]. After 28 weeks it was found that COX-2 inhibition significantly decreased the degree of inflammation and PGE2 biosynthesis with a parallel decline in the risk of development of esophageal cancer giving direct evidence for the chemopreventive role of anti-inflammatory agents in Barrett's esophagus [26]. Based on these findings and epidemiological evidence supporting the role of anti-inflammatory agents with reduced risk of esophageal adenocarcinoma, we conducted a multicenter double blind, randomized, placebo-controlled, phase 2 clinical trial where 114 patients with Barrett's esophagus were randomized in three groups [27]. Patients received 40 mg esmoprazole twice daily with either placebo or low-dose aspirin (81 mg) or 325 mg aspirin daily for 28 days, and esophageal endoscopic biopsy was performed pre- and post-intervention to measure PGE2 levels as a biochemical surrogate marker of neoplastic progression [27]. We noted that the higher dose of aspirin in combination with esmoprazole significantly reduced tissue concentrations of PGE2 while low-dose aspirin and placebo did not have a significant effect [27]. No effect with low-dose aspirin might be due to small sample size and a short trial period [27]. Further trials will be needed to establish the role of aspirin and other NSAIDS for chemoprevention of Barrett's esophagus. Epigenetic changes like DNA methylation and histone modifications also occur secondary to pro-inflammatory signals that obesity induces and could facilitate neoplastic transformation [52]. In summary, altered bile acid composition in response to being obese appears to be an attractive hypothesis; however, this premise needs further investigation.

6.4 Obesity and Mucosal Response to Injury: Role of Altered Systemic and Local Cytokine Levels

Inflammation plays a very important role in mounting an immune response against various environmental triggers and when such triggers persists, such as reflux injury, a state of chronic inflammation develops [23]. Virchow proposed the association of inflammation with cancer in 1850s as cancer being a local inflammation resulting from humoral stasis and injury [28]. Inflammation has been linked to various cancers including breast cancer, endometrial cancer, colon cancer, gastric cancer, and esophageal cancer [29]. Inflammation is increasingly being recognized as an important link between obesity, insulin resistance, tissue injury, and carcinogenesis [10].

Two important aspects of esophageal inflammation include the injury-inducing reflux and mucosal response to this injury. Obesity results in systemic and local pro-inflammatory state where adipocytes release high circulating concentrations of inflammatory cytokines like TNF-α, IL6, IL1B, IL10, CRP, etc. [30]. These cytokines create a pro-inflammatory state that has been proposed as an important connecting link between obesity and various cancers including esophageal adenocarcinoma [30]. Several other pro-inflammatory factors like interferon-γ, monocyte chemotactic protein (MCP-1), PAI-1, and fibrinogen have also been found to be associated with adiposity [31]. In a translational study on patients with esophageal adenocarcinoma, Lysaght et al. found higher concentrations of activated inflammatory cells particularly CD8+ cells in the visceral adipose tissue compared to the subcutaneous adipose tissue [32]. Specifically, IFN-γ was the dominant cytokine produced by the activated T cells resulting in upregulation of adipose tissue inflammation and TNF-α production [32]. It is proposed that systemic increase as well as increased esophageal stromal TNF-α promotes esophageal inflammation [32]. We have shown that neutralizing antibodies against TNF-α in the setting of ongoing reflux injury nearly completely mitigate esophageal inflammation [33]. We induced reflux in Sprague–Dawley rats ($n=25$) and randomized animals at 2 months receive to either specific rat anti-TNF-α Ab (1× week, intraperitoneal, monoclonal mouse IgG2a,K Ab 15 mg/kg) or PBS control injection for 8 weeks (time to develop Barrett's). We found that the inhibition of TNF-α with anti-TNF-α Ab for 2 months in rats with chronic reflux significantly reduced PGE2 (4,907 ± 1,200 vs. 116 ± 18 pg/mg, $p<0.001$) and cPLA2α (0.87 ± 0.2 vs. 0.32 ± 0.02 OD450 $p=0.02$) in Barrett's mucosa compared to controls. The histology showed marked reduction in stromal response to injury. Commonly seen squamous hyperplasia and hyperkeratinization was also absent in anti-TNF-α-treated rats. We further performed in vitro assays that showed that compared to control, exogenous TNF-α treatment increased the number of viable cells by 20 % ($p<0.05$) in primary Barrett's cells. These findings suggest a chemopreventive role of anti-TNF-α Ab on esophageal carcinogenesis. We also show that TNF-α increases prostaglandin biosynthesis in Barrett's mucosa, which increases cell growth and facilitate development of esophageal adenocarcinoma [24].

Emerging evidence supports the hypothesis that TNF-α switches stromal macrophage phenotype to modulate inflammation in target tissues including esophagus. These observations along with increasing expression of both the ligand and the receptor of TNF-α during carcinogenesis in Barrett's esophagus [34] and an increased expression of TNF-α in adipose tissue of obese rats compared to controls [35] suggest that in overweight patients, TNF-α could be one of the mechanistic links where adipocytes in a paracrine manner influence systemic and local inflammatory response and the facilitation of neoplastic transformation. The repertoire of pro-inflammatory cytokines that link obesity-inflammation-carcinogenesis is rapidly expanding. Dvorakova et al. did the tissue analysis of Barrett's esophagus, duodenum, and squamous epithelium distant from metaplastic area and found an increased concentration of IL-6 in metaplastic tissue [36]. Chronic inflammation creates and maintains a tumor microenvironment where there is migration and relative abundance of immune inflammatory cells like neutrophils, dendritic cells, macrophages, and lymphocytes [52]. It has been found that compared to reflux esophagitis, there is a higher concentration of Th1 cells (macrophages and CD8+ T cells) and Th2 cells (plasma cells and mast cells) in the Barrett's mucosa further suggesting the role of inflammation in metaplastic change in the esophagus [37]. Similarly, studies have established higher concentration of cytokines like TNF-α, IL-6, IL1B, IL-4, IL-8, and IL-10 in the Barrett's tissue as well as in esophageal adenocarcinoma [34, 38, 54]. Obesity-associated pro-inflammatory mediators like TNF-α and IL-1 are also known to upregulate inducible nitric oxide synthase (iNOS) leading to increased production of nitric oxide that is typically seen during inflammation and carcinogenesis in Barrett's esophagus [38]. The resulting pro-inflammatory state disrupts the balance between oxidant and antioxidant pathways, and levels of antioxidants are decreased [52]. Congruent with this, biopsy specimens from Barrett's esophagus and esophageal adenocarcinoma have shown low levels of antioxidant enzymes like glutathione S-transferase and glutathione peroxidase. This altered oxidative tissue state favors mutagenesis to promote neoplasia [39], and it is further compounded by impaired anti-oxidative defense in obese patients. There is relative deficiency of antioxidant micronutrients like vitamin C, β carotene, and lycopene in obese patients [40]. Interestingly, consumption of vegetables and fruits rich in natural antioxidant is associated with decreased risk of esophageal cancer [41].

The pro-inflammatory state, due to altered cytokine profile, along with several hormonal mediators as discussed below, together constitute a state named as metabolic syndrome [30], which provide multifaceted link between obesity and cancer progression. Therefore, being obese not only disrupts mechanical barriers that prevent reflux but also makes reflux more pro-inflammatory and oncogenic.

6.5 Obesity and Pro-neoplastic Signaling: Role of Hormonal Mediators in Growth Deregulation

Being obese is a clinical condition of endocrine derangement. Therefore, various hormonal mediators have been studied as possible links between obesity and esophageal disorders. Of particular note are insulin resistance, IGF1, leptin, and adipokines.

As discussed earlier, obesity-induced metabolic state is associated with increased circulating pro-inflammatory cytokines, many of which lead to an insulin resistance state [42]. Specifically, cytokines like TNF alpha, IL6, and IL1B released by adipocytes are thought to contribute to insulin resistance [43]. The IGF-1 axis consists of two ligands, IGF-1 and IGF-2; three cell-membrane receptors, IGF-1R, IGF-2R, and insulin receptor; and six IGF-binding proteins IGFBP-1 to IGFBP-6 [44]. Obesity and metabolic syndrome leads to deregulation of the IGF-1 axis, and it has a key role in cell proliferation, apoptosis, and tumor cell differentiation [45]. Deregulation of the IGF-1 axis leads to a decreased level of IGF-binding proteins and hence a high concentration of free IGF-1, insulin resistance, and eventually hyperinsulinemia [44, 45]. High IGF-1 concentration and overexpression of IGF-1R have been implicated in the transformation of Barrett's metaplasia to esophageal adenocarcinoma [45]. Binding of IGF-1 and IGF-2 to IGF-1R through various signaling pathways leads to cell proliferation and arrest of apoptosis [46]. High insulin level associated with insulin resistance also leads to cancer as insulin has growth-promoting effect, though it is argued that insulin probably mediates its action through IGF-1 receptors as well as insulin receptors. Binding of IGF-1 and insulin to their receptors lead to activation of extracellular-signal-regulated kinase (ERK) and phosphatidylinositol-3 kinase (PI-3K) pathways which in turn result in cell proliferation and arrest of apoptosis [46]. Doyle et al. in a study on esophageal cancer cell lines and tissue from esophageal cancer patients found relative upregulation of IGF-1 levels in esophageal adenocarcinoma compared to squamous cell carcinoma of esophagus [11]. The study also established linear relationship between visceral adiposity and IGF-1 levels; however, relationship with BMI alone was not linear [11]. Similarly in an animal study, Wu et al. found that diet-induced obese mice had increase in local tumor growth as well as metastasis whereas IGF-1-deficient mice did not have increased tumorigenicity [47]. Ounis et al. in their study on the effect of exercise training and diet restriction found that resultant weight loss was also associated with decrease in the level of IGF-1 and inflammatory markers [48].

Obesity and metabolic syndrome also lead to downregulation of adipokines like adiponectin and ghrelin and increased leptin, together these changes result in an increased inflammatory response and abnormal cell survival [46, 49, 50]. Leptin is a hormone controlled by the weight-regulating gene (ob gene) and is increased in obese patients [51]. Leptin is primarily secreted from fat cells and plays an important role in inhibiting weight gain and appetite through action on the hypothalamus [52]. Ogunwobi et al. studied the effect of leptin on Barrett's derived esophageal adenocarcinoma cell lines. They found that leptin stimulates the proliferation of esophageal adenocarcinoma cells and inhibits their apoptosis, which involves COX-2-dependent PGE2-mediated activation of EGFR and activation of c-Jun NH2-terminal kinase [53]. This was further supported by the experiment that leptin deficiency causes extreme obesity in rodents [54]. Several retrospective studies and clinical trials show that leptin levels decrease in response to weight loss achieved by either exercise and/or caloric restriction whereas they increase in response to high-caloric diet [55–59]. Insulin has a positive feedback action on leptin gene expression (whereas IGF-1 has negative control) and hence, in well-fed condition, insulin stimulates leptin release which suppress appetite [60]. The fact that visceral obesity in patients is associated with very high levels of leptin suggests resistance to its action. Leptins have been found to be mitogenic for various tissues [47]. Somasundar et al. in their study on Barrett's associated esophageal adenocarcinoma cell lines found that exogenous leptin significantly increased cell growth in cancer cell lines [51]. They also found that mitogenic effect of leptin does not affect apoptotic pathways which is in contrast to earlier study by Beales and Ogunwobi where it was found that leptin has synergistic action with acid in suppressing apoptosis and activity of caspase 3 [61].

Adiponectins and resistins are adipokines almost exclusively secreted by mature adipocytes [31]. Adiponectins through their receptors AdipoR1 and AdipoR2 regulate various metabolic processes [62]. Epidemiological studies have also found that adiponectin levels are increased after weight loss in patients either through exercise, caloric restriction, or surgical intervention, and its level is decreased in obese patients [63–66]. Certain dietary modifications without caloric restriction, like the use of fish oil, have also been found to increase the level of adiponectin which indicates these may prove to be anti-inflammatory and carcino-protective [67]. They regulate lipid and glucose metabolism by increasing tissue sensitivity to insulin [62]. Adiponectins also suppress production and action of TNF alpha and are considered anti-inflammatory [68]. By virtue of their anti-proliferative, pro-apoptotic, and anti-angiogenic action, adiponectins are considered protective against malignancies [31, 62]. Studies have suggested that obesity downregulates adiponectin levels, and low adiponectin levels are associated with esophageal cancers [49, 50]. Moreover, higher levels of adiponectins have been associated with low incidence of Barrett's esophagus. The mechanism of cancer protective role of adiponectin has been complex and unclear. However it has been proposed that adiponectins stimulate apoptosis through induction of p53, Bax expression, and suppression of Bcl-2 [69, 70]. Adiponectins also suppress Toll-like receptor-induced activation of nuclear factor κB (NF-κB) and prevent interaction of various growth factors [31]. Adiponectins also exert their tumor-inhibiting action by inactivation of MAPK kinase 1 and 3 and ERK 1 and 2 and concomitant reduced glucose uptake [70]. Anti-angiogenic action of adiponectins occurs through activation of apoptosis in vascular endothelial cells and inhibition of cell migration [31]. Resistins and RELMs (resistin-like molecules) are adipokines which are secreted from adipocytes as well as non-adipocyte sources and are believed to have a role in carcinogenesis of esophageal adenocarcinoma, but the mechanism is unclear [62]. Leptin and adiponectin receptors have also been correlated with esophageal adenocarcinoma, and it was found that upregulation of ObR (leptin receptor) was associated with advanced disease and more aggressive tumor in gastric cancer [71]. Adiponectin receptors, ADIPOR1 and ADIPOR2, were downregulated in Barrett's mucosa compared to normal esophagus [69]. However receptor levels did not correlate significantly with the serum concentrations of their ligands suggesting the complexity of molecular mechanisms involved as well as a possibility that the these adipokines might have paracrine action too [72].

Metabolism and inflammatory pathways converge on many signaling pathways which are important for homeostasis and body functions like immunity. For example, Toll-like receptors which recognize lipoproteins and lipopolysaccharides in bacterial wall also recognize nutrients like fatty acids [23]. Similarly hormones involved in metabolic pathways like leptins and adiponectins have a role in immunity [23]. Prolonged nutritional imbalances or chronic inflammation can disrupt the equilibrium between metabolism, immunity, and inflammation and lead to metabolic syndrome, obesity, and multiple cancers. The role of nuclear factor kappa B signaling pathway and JNK mitogen-activated protein kinase pathway (JNK MAPK) needs special mention as it has a central role in linking obesity to inflammation, insulin resistance, and also in oncogenesis and cell proliferation [73]. Various metabolic stressors like lipids, oxidative stress, and cytokines activate IKK/NF-κB signaling pathway and JNK MAPK pathway. IKKβ-mediated activation of NF-κB leads to macrophage recruitment, activation, and differentiation which maintains pro-inflammatory status. It also leads to insulin resistance by phosphorylation of insulin receptors. Its activation is also believed to disrupt the central leptin pathway leading to overnutrition and weight gain. It is important to note that activation of IKKβ/NF-κB creates a vicious cycle where inflammation and imbalance in nutrition lead to further inflammation and overnutrition.

Various epidemiological studies have also found that obesity is associated with pro-inflammatory state and weight loss leads to decrease in pro-inflammatory markers. In a recent randomized controlled trial on the effect of caloric restriction diet/exercise on inflammatory biomarkers in obese postmenopausal women, Imayama et al. found that caloric restriction diet and/or exercise led to decrease in hsCRP (high-sensitive C-reactive protein) as well as anthropometric parameters [74]. Cintra et al. also found that CRP level decreased in obese patients after post-bariatric abdominoplasty though the weight remained stable, indicating that diet might have a direct role in causing inflammation [75]. In a longitudinal evaluation of effect of weight loss on cancer-related biomarkers, Linkov et al. found that weight loss leads to decreased level of E-selectin and IL-6 and increased level of GH, adiponectin, and IGFBP-1 [63]. Similarly Ackermann et al. in their study on relationship between metabolic syndrome and inflammatory biomarkers found that waist circumference is positively correlated with inflammatory markers like TNF-α and IL-6 and negatively correlated with anti-inflammatory adiponectin [64]. In another study on obese/overweight breast cancer survivors, it was found that a weight loss intervention led to a fall in the levels of IL-6 and TNF-α [76].

To conclude, obesity in general, and central obesity in particular, leads to enhanced loco-regional inflammatory response in the lower esophagus by both reflux-dependent as well as reflux-independent mechanisms eventually causing Barrett's esophagus and esophageal adenocarcinoma. Adipocytes alter the circulating levels of various cytokines, growth factors, hormonal factors creating a pro-inflammatory state, and increased risk of esophageal adenocarcinoma. Study of various molecular pathways associating obesity and reflux-induced inflammatory response to the neoplastic changes in the lower esophagus is ongoing research which will likely uncover useful targets for screening and therapeutic intervention.

References

1. Brown LM, Devesa SS (2002) Epidemiologic trends in esophageal and gastric cancer in the United States. Surg Oncol Clin N Am 11(2):235–256
2. Pera M et al (2005) Epidemiology of esophageal adenocarcinoma. J Surg Oncol 92(3):151–159
3. Beddy P et al (2010) Association of visceral adiposity with oesophageal and junctional adenocarcinomas. Br J Surg 97(7):1028–1034
4. Steffen A et al (2009) Anthropometry and esophageal cancer risk in the European prospective investigation into cancer and nutrition. Cancer Epidemiol Biomarkers Prev 18(7):2079–2089
5. O'Doherty MG et al (2012) A prospective cohort study of obesity and risk of oesophageal and gastric adenocarcinoma in the NIH-AARP Diet and Health Study. Gut 61(9):1261–1268. doi:10.1136/gutjnl-2011-300551
6. Otterstatter MC et al (2012) Esophageal cancer in Canada: trends according to morphology and anatomical location. Can J Gastroenterol 26(10):723–727
7. Turati F et al (2012) A meta-analysis of body mass index and esophageal and gastric cardia adenocarcinoma. Ann Oncol 16:16
8. El-Serag HB et al (2005) Abdominal obesity and the risk of Barrett's esophagus. Am J Gastroenterol 100(10):2151–2156
9. Obesity: preventing and managing the global epidemic. Report of a WHO consultation. World Health Organ Tech Rep Ser 2000;894:i–xii, 1–253
10. Lagergren J (2011) Influence of obesity on the risk of esophageal disorders. Nat Rev Gastroenterol Hepatol 8(6):340–347
11. Kendall BJ et al (2013) The risk of Barrett's esophagus associated with abdominal obesity in males and females. Int J Cancer 132(9):2192–2199
12. Nordenstedt H, El-Serag H (2011) The influence of age, sex, and race on the incidence of esophageal cancer in the United States (1992–2006). Scand J Gastroenterol 46(5):597–602
13. Kong CY et al (2011) The impact of obesity on the rise in esophageal adenocarcinoma incidence: estimates from a disease simulation model. Cancer Epidemiol Biomarkers Prev 20(11):2450–2456
14. Nelsen EM et al (2012) Distribution of body fat and its influence on esophageal inflammation and dysplasia in patients with Barrett's esophagus. Clin Gastroenterol Hepatol 10(7):728–734; quiz e61–62. doi:10.1016/j.cgh.2012.03.007
15. Koppman JS et al (2007) Esophageal motility disorders in the morbidly obese population. Surg Endosc 21(5):761–764
16. Ortiz V et al (2006) Value of heartburn for diagnosing gastroesophageal reflux disease in severely obese patients. Obesity (Silver Spring) 14(4):696–700
17. Kuper MA et al (2009) Dysfunction of the lower esophageal sphincter and dysmotility of the tubular esophagus in morbidly obese patients. Obes Surg 19(8):1143–1149
18. Ayazi S et al (2009) Obesity and gastroesophageal reflux: quantifying the association between body mass index, esophageal acid exposure, and lower esophageal sphincter status in a large series of patients with reflux symptoms. J Gastrointest Surg 13(8):1440–1447
19. Schneider JH et al (2009) Transient lower esophageal sphincter relaxation in morbid obesity. Obes Surg 19(5):595–600
20. Pandolfino JE et al (2006) Obesity: a challenge to esophagogastric junction integrity. Gastroenterology 130(3):639–649
21. McQuaid KR et al (2011) Systematic review: the role of bile acids in the pathogenesis of gastro-oesophageal reflux disease and related neoplasia. Aliment Pharmacol Ther 34(2):146–165
22. Abdel-Latif MM et al (2009) Inflammation and esophageal carcinogenesis. Curr Opin Pharmacol 9(4):396–404
23. Tornatore L et al (2012) The nuclear factor kappa B signaling pathway: integrating metabolism with inflammation. Trends Cell Biol 22(11):557–566

24. Song S et al (2007) COX-2 induction by unconjugated bile acids involves reactive oxygen species-mediated signalling pathways in Barrett's oesophagus and oesophageal adenocarcinoma. Gut 56(11):1512–1521
25. Buttar NS et al (2002) The effect of selective cyclooxygenase-2 inhibition in Barrett's esophagus epithelium: an in vitro study. J Natl Cancer Inst 94(6):422–429
26. Buttar NS et al (2002) Chemoprevention of esophageal adenocarcinoma by COX-2 inhibitors in an animal model of Barrett's esophagus. Gastroenterology 122(4):1101–1112
27. Falk GW et al (2012) A combination of esomeprazole and aspirin reduces tissue concentrations of prostaglandin E(2) in patients with Barrett's esophagus. Gastroenterology 143(4):917–926.e1. doi:10.1053/j.gastro.2012.06.044
28. Balkwill F, Mantovani A (2001) Inflammation and cancer: back to Virchow? Lancet 357(9255):539–545
29. Groblewska M et al (2012) Interleukin 6 and C-reactive protein in esophageal cancer. Clin Chim Acta 413(19–20):1583–1590
30. Ryan AM et al (2011) Obesity, metabolic syndrome and esophageal adenocarcinoma: epidemiology, etiology and new targets. Cancer Epidemiol 35(4):309–319
31. Renehan AG, Roberts DL, Dive C (2008) Obesity and cancer: pathophysiological and biological mechanisms. Arch Physiol Biochem 114(1):71–83
32. Lysaght J et al (2011) T lymphocyte activation in visceral adipose tissue of patients with oesophageal adenocarcinoma. Br J Surg 98(7):964–974
33. Buttar N et al (2008) Infliximab-like anti-TNF alpha antibody down-regulates pro-carcinogenic signaling in Barrett's mucosa. Gastroenterology 134(4):A129
34. Tselepis C et al (2002) Tumour necrosis factor-alpha in Barrett's oesophagus: a potential novel mechanism of action. Oncogene 21(39):6071–6081
35. Hotamisligil GS, Shargill NS, Spiegelman BM (1993) Adipose expression of tumor necrosis factor-alpha: direct role in obesity-linked insulin resistance. Science 259(5091):87–91
36. Dvorakova K et al (2004) Increased expression and secretion of interleukin-6 in patients with Barrett's esophagus. Clin Cancer Res 10(6):2020–2028
37. Moons LM et al (2005) Barrett's oesophagus is characterized by a predominantly humoral inflammatory response. J Pathol 207(3):269–276
38. Fitzgerald RC et al (2002) Inflammatory gradient in Barrett's oesophagus: implications for disease complications. Gut 51(3):316–322
39. Lee OJ et al (2005) Hypermethylation and loss of expression of glutathione peroxidase-3 in Barrett's tumorigenesis. Neoplasia 7(9):854–861
40. Damms-Machado A, Weser G, Bischoff SC (2012) Micronutrient deficiency in obese subjects undergoing low calorie diet. Nutr J 11:34. doi:10.1186/1475-2891-11-34
41. Fountoulakis A et al (2004) Plasma and esophageal mucosal levels of vitamin C: role in the pathogenesis and neoplastic progression of Barrett's esophagus. Dig Dis Sci 49(6):914–919
42. Despres JP, Lemieux I (2006) Abdominal obesity and metabolic syndrome. Nature 444(7121):881–887
43. Fasshauer M, Paschke R (2003) Regulation of adipocytokines and insulin resistance. Diabetologia 46(12):1594–1603
44. Doyle SL et al (2012) IGF-1 and its receptor in esophageal cancer: association with adenocarcinoma and visceral obesity. Am J Gastroenterol 107(2):196–204
45. Iravani S et al (2003) Modification of insulin-like growth factor 1 receptor, c-Src, and Bcl-XL protein expression during the progression of Barrett's neoplasia. Hum Pathol 34(10):975–982
46. Imsumran A et al (2007) Insulin-like growth factor-I receptor as a marker for prognosis and a therapeutic target in human esophageal squamous cell carcinoma. Carcinogenesis 28(5):947–956
47. Wu Y et al (2010) Insulin-like growth factor-I regulates the liver microenvironment in obese mice and promotes liver metastasis. Cancer Res 70(1):57–67
48. Ben Ounis O et al (2010) Effect of individualized exercise training combined with diet restriction on inflammatory markers and IGF-1/IGFBP-3 in obese children. Ann Nutr Metab 56(4):260–266

49. Rubenstein JH et al (2008) A pilot study of the association of low plasma adiponectin and Barrett's esophagus. Am J Gastroenterol 103(6):1358–1364
50. Thompson OM et al (2010) Serum leptin and adiponectin levels and risk of Barrett's esophagus and intestinal metaplasia of the gastroesophageal junction. Obesity (Silver Spring) 18(11):2204–2211
51. Somasundar P et al (2003) Leptin stimulates esophageal adenocarcinoma growth by nonapoptotic mechanisms. Am J Surg 186(5):575–578
52. Halaas JL et al (1995) Weight-reducing effects of the plasma protein encoded by the obese gene. Science 269(5223):543–546
53. Ogunwobi O, Mutungi G, Beales IL (2006) Leptin stimulates proliferation and inhibits apoptosis in Barrett's esophageal adenocarcinoma cells by cyclooxygenase-2-dependent, prostaglandin-E2-mediated transactivation of the epidermal growth factor receptor and c-Jun NH2-terminal kinase activation. Endocrinology 147(9):4505–4516
54. Zhang Y et al (1994) Positional cloning of the mouse obese gene and its human homologue. Nature 372(6505):425–432
55. Rigamonti AE et al (2010) Changes in plasma levels of ghrelin, leptin, and other hormonal and metabolic parameters following standardized breakfast, lunch, and physical exercise before and after a multidisciplinary weight-reduction intervention in obese adolescents. J Endocrinol Invest 33(9):633–639
56. Spruijt-Metz D et al (2009) A high-sugar/low-fiber meal compared with a low-sugar/high-fiber meal leads to higher leptin and physical activity levels in overweight Latina females. J Am Diet Assoc 109(6):1058–1063
57. Siklova-Vitkova M et al (2012) Adipose tissue secretion and expression of adipocyte-produced and stromavascular fraction-produced adipokines vary during multiple phases of weight-reducing dietary intervention in obese women. J Clin Endocrinol Metab 97(7):E1176–E1181
58. Morel O et al (2011) Short-term very low-calorie diet in obese females improves the haemostatic balance through the reduction of leptin levels, PAI-1 concentrations and a diminished release of platelet and leukocyte-derived microparticles. Int J Obes (Lond) 35(12):1479–1486
59. Ramel A et al (2009) Effects of seafood consumption and weight loss on fasting leptin and ghrelin concentrations in overweight and obese European young adults. Eur J Nutr 48(2):107–114
60. Friedman JM, Halaas JL (1998) Leptin and the regulation of body weight in mammals. Nature 395(6704):763–770
61. Beales IL, Ogunwobi OO (2007) Leptin synergistically enhances the anti-apoptotic and growth-promoting effects of acid in OE33 oesophageal adenocarcinoma cells in culture. Mol Cell Endocrinol 274(1–2):60–68
62. Tiaka EK et al (2011) The implication of adiponectin and resistin in gastrointestinal diseases. Cytokine Growth Factor Rev 22(2):109–119
63. Linkov F et al (2012) Longitudinal evaluation of cancer-associated biomarkers before and after weight loss in RENEW study participants: implications for cancer risk reduction. Gynecol Oncol 125(1):114–119
64. Ackermann D et al (2011) Waist circumference is positively correlated with markers of inflammation and negatively with adiponectin in women with metabolic syndrome. Nutr Res 31(3):197–204
65. Gueugnon C et al (2012) Effects of an in-patient treatment program based on regular exercise and a balanced diet on high molecular weight adiponectin, resistin levels, and insulin resistance in adolescents with severe obesity. Appl Physiol Nutr Metab 37(4):672–679
66. Lang HF et al (2011) Weight loss increased serum adiponectin but decreased lipid levels in obese subjects whose body mass index was lower than 30 kg/m(2). Nutr Res 31(5):378–386
67. Gammelmark A et al (2012) Low-dose fish oil supplementation increases serum adiponectin without affecting inflammatory markers in overweight subjects. Nutr Res 32(1):15–23
68. Bruun JM et al (2003) Regulation of adiponectin by adipose tissue-derived cytokines: in vivo and in vitro investigations in humans. Am J Physiol Endocrinol Metab 285(3):E527–E533

69. Konturek PC et al (2008) Effect of adiponectin and ghrelin on apoptosis of Barrett adenocarcinoma cell line. Dig Dis Sci 53(3):597–605
70. Dieudonne MN et al (2006) Adiponectin mediates antiproliferative and apoptotic responses in human MCF7 breast cancer cells. Biochem Biophys Res Commun 345(1):271–279
71. Zhao X et al (2007) Correlation between expression of leptin and clinicopathological features and prognosis in patients with gastric cancer. J Gastroenterol Hepatol 22(8):1317–1321
72. Howard JM et al (2010) Associations between leptin and adiponectin receptor upregulation, visceral obesity and tumour stage in oesophageal and junctional adenocarcinoma. Br J Surg 97(7):1020–1027
73. Fang Y et al (2004) Bile acids induce mitochondrial ROS, which promote activation of receptor tyrosine kinases and signaling pathways in rat hepatocytes. Hepatology 40(4):961–971
74. Imayama I et al (2012) Effects of a caloric restriction weight loss diet and exercise on inflammatory biomarkers in overweight/obese postmenopausal women: a randomized controlled trial. Cancer Res 72(9):2314–2326
75. Cintra W et al (2012) C-reactive protein decrease after postbariatric abdominoplasty. Inflammation 35(1):316–320
76. Pakiz B et al (2011) Effects of a weight loss intervention on body mass, fitness, and inflammatory biomarkers in overweight or obese breast cancer survivors. Int J Behav Med 18(4):333–341

Chapter 7
Inflammation, Obesity, and Colon Cancer

Stephanie K. Doerner and Jason D. Heaney

Abstract The general focus of this chapter is to overview the contribution of obesity-induced intestinal inflammation to colorectal cancer (CRC) incidence. Using inflammatory bowel disease as a model, the mechanisms by which gastrointestinal microbes and obesity-associated, adipose-derived mediators of inflammatory increase CRC risk will be discussed. Particular emphasis will be placed on the direct impact of these factors on intestinal epithelial cell proliferation, survival, and neoplastic transformation. Additionally, the influence of inflammatory factors on immune cell function and the indirect effect of altered immunity on intestinal epithelial cell proliferation and survival, and CRC risk will be explored.

7.1 Background

Colorectal cancer (CRC) is the third most commonly diagnosed malignancy in both men and women in the United States and is the second leading cause of cancer-related deaths. In the United States alone, 141,380 colon cancer (71,850 in men, 69,360 in women) cases were diagnosed in 2011. The National Cancer Institute estimates that there will be 51,690 CRC-related deaths in 2012, which is 5 % higher than the estimated deaths in 2011 [1, 2]. The lifetime risk for

S.K. Doerner, Ph.D.
Department of Genetics, Case Western Reserve University School of Medicine,
10900 Euclid Avenue, Cleveland, OH 44106-4915, USA
e-mail: stephanie.doerner@case.edu

J.D. Heaney, Ph.D. (✉)
Department of Molecular and Human Genetics, Baylor College of Medicine,
One Baylor Plaza, MS BCM225, Houston, TX 77030, USA
e-mail: Heaney@bcm.edu

A.J. Dannenberg and N.A. Berger (eds.), *Obesity, Inflammation and Cancer*,
Energy Balance and Cancer 7, DOI 10.1007/978-1-4614-6819-6_7,

individuals to develop colon cancer is approximately 6 %, but the risk increases to 18 % among individuals who have a first-degree relative (parent, sibling, or child) with CRC [1, 3]. Most CRCs begin as foci of intestinal or colonic epithelial cells (IECs) that have accumulated mutations in oncogenes (aberrant crypt foci) [4, 5]. These cells have the capacity to transform into benign adenomas (polyps) with additional mutations or in response to environmental queues [4]. If left untreated, adenomas progress to more invasive and malignant carcinomas [6]. Many factors increase CRC risk, including genetics, diabetes, preexisting adenomas, obesity, diet, microbial profile, and inflammatory diseases [4, 7, 8]. Understanding how these factors influence tumor development will lead to new diagnostic screens, intervention paradigms, and treatment options that may reduce CRC-related deaths.

In this chapter, we will discuss the contribution of two interconnected CRC risk factors, inflammation and obesity, to cancer incidence. We will discuss the contribution of intestinal microbes to chronic inflammation in inflammatory bowel diseases (IBDs) and obesity and their contributions to increased CRC risk. Additionally, we will highlight the inflammatory mediators and mechanisms through which IBD promotes CRC risk and discuss how obesity-associated systemic inflammation similarly increases CRC susceptibility. A model will be proposed whereby obesity promotes a proinflammatory intestinal milieu that directly induces the neoplastic transformation of IECs and indirectly promotes tumor progression through activation and recruitment of immune cells.

7.2 Genetic Risk Factors for CRC

7.2.1 Hereditary Forms of Colon Cancer

It is estimated that about 15 % of colon cancer cases can be attributed to specific inherited gene mutations [9]. Genetic mutations have been identified as the cause of inherited cancer risk in some colon cancer-prone families, and these mutations are estimated to account for only 5–6 % of colon cases overall [10]. It is likely that other undiscovered genes and background genetic factors contribute to the development of familial colon cancer in conjunction with nongenetic and environmental risk factors such as diet [11]. One of several well-described hereditary forms of colon cancer is familial adenomatous polyposis (FAP), which is characterized by the presence of hundreds, sometimes thousands of benign polyps in the colon, rectum, or lower regions of the small intestine [12]. The polyps usually begin to form at puberty, and colon cancer almost always develops later in life. FAP is inherited as an autosomal-dominant trait caused by a mutation in the tumor suppressor, adenomatous polyposis coli (*APC*) [13, 14]. Individuals with FAP have a 90–100 % lifetime risk of developing colon cancer. Surgery is routinely used to remove the colon in order to prevent the development of cancer.

7.2.2 Sporadic Causes of Colon Cancer

Approximately 75–90 % of colon cancer cases are considered sporadic, with no known family history or hereditary genetic mutation [15]. It is now well established that all cancers result from an accumulation of mutations that enhance a tumor cell's ability to proliferate, evade growth suppressors, avoid immune destruction, replicate indefinitely, induce inflammation, metastasize, induce angiogenesis, accumulate additional mutations, resist cell death, and deregulate cellular energetics [16]. For example, somatic mutations can result in the inactivation of tumor suppressors (i.e., *APC*, *P53*, or *SMAD4*) and activation of oncogenes (i.e., *KRAS*, *BRAF*, *PI3KCA*, or *PTEN*) that drive proliferation and progressive transformation of normal IECs to malignant derivatives or inactivate genes necessary for the maintenance of genome stability (i.e., DNA repair factor *MLH1* or *MSH2*) [17, 18].

7.2.3 APC, Wnt Signaling, and CRC

Mutations in the Wnt signaling pathway are common in CRC patients [19, 20]. The Wnt signaling pathway is crucial for an immense number of biological processes during various stages of development [21, 22]. Activation of canonical Wnt signaling results in increased cytosolic levels of β-catenin, which then localizes to the nucleus to initiate transcription of Wnt target genes important for cellular development [21]. In the absence of Wnt activation, β-catenin is degraded to downregulate expression of genes involved in cell growth and proliferation. A complex of proteins (APC, GSK3β, AXIN, and CK1α) forms a "degradation machine" that functions to bind and tag β-catenin for destruction by the proteasome [21, 22]. Mutations in Wnt pathway proteins increase the levels, stability, and nuclear localization of β-catenin protein causing constitutive transcription of Wnt targets. Alterations in direct target genes like c-Myc (*Myc*), cyclin D1 (*CCND1*), *VEGF*, c-Jun (*JUN*), matrix metalloproteinase 7 (*MMP7*), and claudin-1 (*CLDN1*) have all been shown to induce colon cancer in humans [21, 23–25]. The influence of APC on intestinal tumor development can be modeled in mice with inactivating mutations of *Apc* [26]. In particular, the *Min* mutation of *Apc* (Apc^{Min}), which approximates the most common *APC* mutation in humans, has been useful for characterizing the pathogenesis of CRC [26]. Importantly, in both CRC patients and Apc^{Min} mice, CRC risk factors such as intestinal inflammation and obesity interact with genetic mutations in the Wnt signaling pathway to increase cancer susceptibility.

7.3 Intestinal Inflammation: A CRC Risk Factor

When pathogens invade a host and tissues undergo physical trauma, the immune system reacts with a coordinated response involving immune and nonimmune cell types (the inflammatory response). Cross talk between immune and nonimmune cells eradicates the invading pathogen and initiates a tissue repair response involving the release of cytokines (immunomodulating factors), growth factors, and extracellular matrix (ECM) remodeling proteins, which promote cell proliferation, differentiation, and migration [27]. In some instances, the immune system elicits an exaggerated and chronic response that fails to be resolved, resulting in cellular hyperplasia, disorganization of the surrounding tissue, and eventually fibrosis [27]. IBDs such as Crohn's disease (CD) and ulcerative colitis (UC) are a group of chronic inflammatory disorders of the intestine [28]. Inflammation in CD patients affects all parts of the gastrointestinal tract but usually afflicts the distal small intestine and colon. UC involves inflammation of the colon.

The prevalence for both CD and UC in the United States is greater than 200 cases per 100,000, with the total number of individuals with IBD approaching 1.5 million [29, 30]. Clinical symptoms typically involve diarrhea, abdominal pain, gastrointestinal bleeding, and weight loss [31]. In addition, development of CRC is a serious long-term complication associated with IBD. Epidemiologic and clinical studies indicate that patients affected by UC and CD have an increased risk of developing colon cancer by as much as 18 % and 3 %, respectively, with risk of CRC significantly higher among patients with long-standing disease [32, 33]. How immune responses in IBD patients alter IEC homeostasis and induce tumorigenesis is an area of intensive investigation.

7.4 Obesity: A CRC Risk Factor

In the past 50 years, the worldwide occurrence of obesity in humans has risen at an alarming rate. The World Health Organization (WHO) generally defines obesity as having a body mass index (BMI) equal to or higher than 30, as calculated by weight divided by height squared (kg/m^2). The United States has one of the highest percentages of obese individuals (BMI $\geq$30), with 33.9 % of the adult population (18 years or over) classified as obese [34, 35]. Shockingly, it has been estimated that in the United States, over one-third of children and young adults (age 6–19) and over 40 million children under the age of 5 are obese [36, 37]. Although genetic factors contribute to obesity, increased energy intake and decreased energy expenditure are the major causes of the disease and probably contribute largely to the rapid increase in obesity cases [38, 39].

Obese individuals have an increased risk for developing heart disease, hypertension, infertility, obstructive sleep apnea, dyslipidemia, nonalcoholic fatty liver disease, type 2 diabetes, and metabolic syndrome (MetS) [40, 41]. There is also a strong link between obesity and many malignancies. Several epidemiological studies have linked obesity to cancers of the esophagus, pancreas, prostate, renal, postmenopausal breast, endometrium, kidney, and colon and rectum [40, 42]. For CRC specifically, every 2.4 unit increase in BMI increases cancer risk by 7 % [43]. Additionally, a recent meta-analysis predicted that for every additional inch in waist circumference, there is a 5 % increase in CRC risk [44]. Therefore, given the increased prevalence of and lifetime exposure to obesity, it is imperative to identify the mechanisms by which obesity increases CRC risk so that effective treatment and preventative options can be delivered to a growing, at-risk population. Similar to intestinal inflammation in IBD patients, inflammation caused by increased adiposity and consumption of a high-fat diet is likely to contribute to increased CRC susceptibility in obese individuals.

7.5 The Intestinal Microbiota Contributes to Inflammation and CRC Risk

7.5.1 The Microbiota and IBD

There are approximately 10^{12} microorganisms in the human gastrointestinal track that belong to greater than 500 different species, which are collectively referred to as the microbiota [45, 46]. In both mice and humans, the majority of the microbiota consists of bacteria belonging to the Firmicutes (gram-positive bacteria) and Bacteriodetes (gram-negative bacteria) phyla [47, 48]. Probacteria (including *Helicobacter* and *Escherichia*) and Actinobacteria are also significant contributors to the gut microflora [31]. The symbiotic relationship between humans and enteric microorganisms is essential to important processes such as digestion, regeneration of the IEC barrier, and immune responses [49, 50]. While the microbiota provides beneficial protective, trophic, nutritional, and metabolic signals for the host, it may become a risk factor for disease depending on context and host susceptibility [50–52].

What were once thought to be diseases involving chronic immune response targeting self-antigens (autoimmunity), IBDs, specifically CD, are now considered to be caused by a loss of tolerance to commensal organisms and enhanced immune response to bacterial antigens [53]. In almost all rodent models of IBD, treatment with antibiotics or elimination of the intestinal microbiota (germfree rodents) significantly alleviates disease pathology [54, 55]. Similarly, in human CD patients, antibiotics and diversion of the fecal stream from the distal colon into an ileostomy facilitate disease remission [56–58]. Therefore, abnormal interactions between bowel constituents, specifically intestinal microbes, and the mucosal immune system appear to initiate CD pathology.

What underlying pathologies lead to the aberrant immune response to the commensal microbes of the gastrointestinal system? One possibility may be alterations to the makeup of the microbiota itself and alterations to intestinal physiology by specific bacterial species. In colon biopsies from CD patients, an increase in Bacteroidetes and Proteobacteria has been observed [59]. However, analysis of the microbial population in the intestines of CD, UC, and non-IBD controls demonstrated that specific flora is not enriched in the small bowel or colon of IBD patients, suggesting that there is not one specific change to the microbiota that contributes to IBD pathology in all patients [60]. Interestingly, a subset of IBD patients do have depletions in certain commensal bacteria, notably members of the phyla Firmicutes and Bacteroidetes, suggesting that, at least in some instances, particular alterations to the intestinal microflora influences disease progression [31]. Metabolic by-products that are generated from gut bacteria can have beneficial or detrimental effects on host diseases. For example, butyrate is a fermentation product of anaerobic bacteria (Clostridium strains) and has been shown to increase apoptosis, decrease proliferation, and have anti-inflammatory properties in IECs and immune cells [61, 62]. IBD patients have deficiencies in many strains of butyrate-producing Clostridium, such as *Faecalibacterium prausnitzii*, and have impaired butyrate metabolism in the colonic mucosa [63, 64]. Oral and intraperitoneal administration of *F. prausnitzii* in a mouse model of IBD reduced colitis and mortality, respectively, and increased anti-inflammatory cytokines such as IL-10 [63]. Importantly, immune responses and inflammation induced by non-commensal invasive species of *Escherichia coli* and enterotoxigenic subclasses of *Bacteroides fragilis* also enhanced inflammation and disease pathology in IBD [65, 66].

Although evidence suggesting that specific commensal microbes trigger abnormal immune responses in IBD patients is somewhat limited, there is a significant amount of experimental evidence demonstrating that altered sensitivity to the microbiota contributes to IBD. The lamina propria is a thin layer of tissue that lies beneath the intestinal epithelium [67]. Mononuclear immune cells, including macrophages and dendritic cells (antigen-presenting cells or APCs) and T cells, within the lamina propria, which are collectively referred to as LPMNCs, act as a first line of defense against pathogens that penetrate through the epithelial barrier following infection or tissue trauma. Importantly, due to their physical separation from intestinal contents by the epithelial barrier, immune cells normally do not come into contact with or react to commensal microbes. Additionally, under normal physiological conditions, IECs express low levels of toll-like receptors (TLRs), which recognize bacterial antigens to induce proinflammatory responses (e.g., gram-negative lipopolysaccharide [LPS] induction of TLR 4 signaling) [68]. However, when the intestinal epithelium is damaged, microbes can transit through the IEC barrier and come into contact with LPMNCs. Through TLR-mediated activation of APCs, phagocytosis and antigen presentation by APCs, and ultimately T cell activation, a proinflammatory response is induced that promotes wound healing [5, 49, 69]. Once the damage is repaired, the immune response normally subsides.

Importantly, in IBD patients, both genetic and environmental factors increased IEC barrier permeability, which causes chronic inflammation [70–73]. This barrier defect allows for continuous mixing of luminal contents with LPMNCs and chronic proinflammatory immune responses [74]. Interestingly, several of the proinflammatory cytokines, including tumor necrosis factor alpha (TNF-α), interferon gamma (INF-γ), interleukin (IL)-6, IL-1β, and IL-13, that are expressed in the intestines of IBD patients also induce increased barrier permeability by activating nuclear factor kappa B (NF-κB)-mediated transcription and altering the expression of tight junction proteins in IECs (Fig. 7.2) [74–77]. Thus, there is a self-amplifying cycle of barrier dysfunction, immune cell activation, cytokine release, and additional IEC barrier permeability [74]. Adding to this defect, IEC expression of TLR4 is elevated in IBD patients, allowing for LPS stimulation of NF-κB-mediated transcription and alterations in tight junction protein expression (Fig. 7.2) [68, 69, 78, 79].

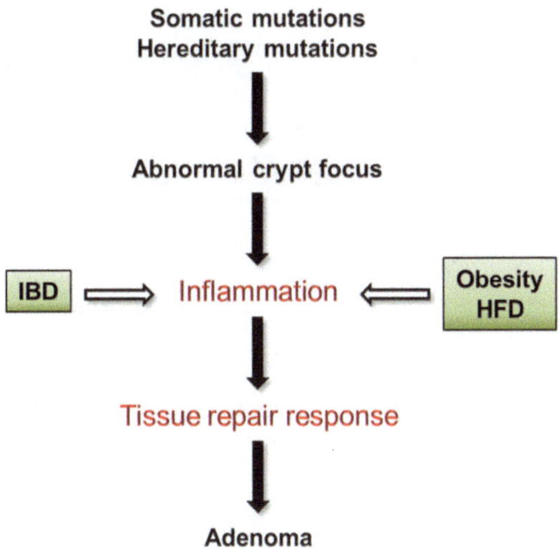

Fig. 7.1 Model for the role of intestinal inflammation in colorectal cancer progression. Colorectal cancer (CRC) susceptibility is induced by the acquisition of genetic mutations (e.g., *APC* mutations) that predispose IECs to neoplastic transformation. Preneoplastic IECs form aberrant crypt foci. Barrier defects caused by the aberrant crypt foci allow luminal contents to cross the epithelial barrier and commensal bacteria to interact with immune cells in the lamina propria to induce inflammation. Inflammation induces tissue repair responses (e.g., proliferation and migration) in IECs that are normally involved in wound healing. In pre-neoplastic IECs, tissue repair responses induced by inflammation interact with genetic predispositions to CRC to induce tumor progression to adenomas. Chronic inflammation induced by inflammatory bowel diseases (IBDs) and obesity/high-fat diets (HFD) causes additional inflammation in the intestine to further increase cancer susceptibility

7.5.2 *Inflammatory Responses to the Microbiota Increase CRC Risk*

Studies of inflammation and intestinal cancer in humans and mice have led to a model that proposes a proinflammatory intestinal environment increases CRC risk by inducing tissue repair responses (Figs. 7.1 and 7.2) [80–82]. In this model, it is hypothesized that accumulation of mutations in IECs and the formation of aberrant crypt foci damages the surrounding intestinal mucosa. The microbiota penetrates the IEC barrier and induces immune cells to release proinflammatory cytokines that elicit tissue repair responses (e.g., proliferation) in IECs [49, 83]. Normally the tissue repair response fixes damage done to the affected area of the epithelium. However, IECs found within the aberrant crypt foci have developed genetic predispositions to tumor formation, which interact with tissue repair responses to promote the progression of precancerous lesions to adenomas, adenoma growth, and additional tissue damage and inflammatory responses. Secondary conditions such as IBD exacerbate this effect by promoting a chronic, proinflammatory state in the gut that persistently induces IEC hyperplasia.

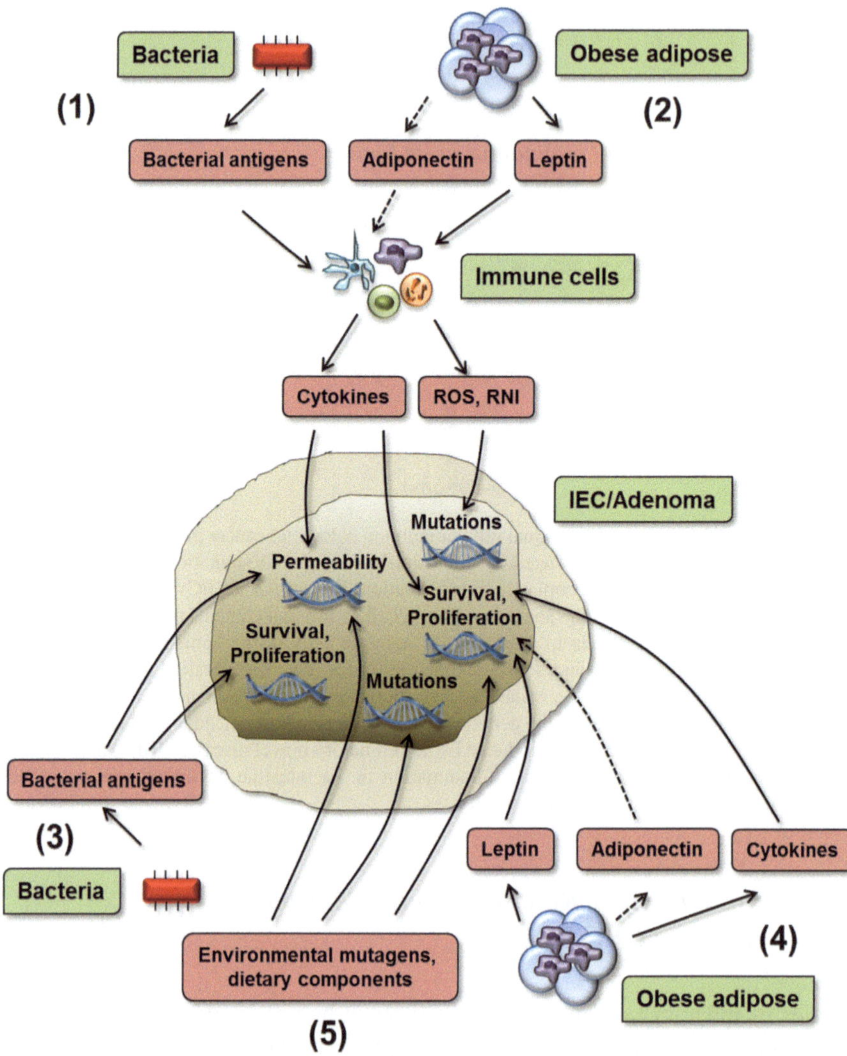

Fig. 7.2 Initiation of colorectal cancer. Susceptibility to CRC initiates from hereditary or somatic mutations that increase tumorigenic potential. Once acquired several factors interact with these genetic predispositions to increase CRC risk. (*1*) Genetic predispositions induce the formation of aberrant crypt foci, which disrupt the epithelial barrier and allow commensal bacteria in the intestinal lumen to interact with immune cells in the lamina propria to induce proinflammatory cytokine production. These cytokines induce gene expression changes in IECs and tumor cells that cause additional barrier permeability and increased survival and proliferation. Additionally, immune cells produce radical oxygen species (ROS) and radical nitrogen intermediates (RNI) that induce additional DNA damage and tumor promoting mutations. Chronic inflammation caused by IBDs increases cytokine production and CRC risk. (*2*) Obese adipose tissue releases proinflammatory leptin and decreases the production of anti-inflammatory adiponectin, resulting in immune cell activation in the intestine, proinflammatory cytokine production, and ROS and RNI release. (*3*) In addition to inducing proinflammatory responses from immune cells, bacterial antigens can interact with toll-like receptors (TLR) on IECs and tumor cells to alter the expression of tight junction

As observed with IBD, particular enterotoxic and invasive species increase CRC risk in humans and mouse models [5, 84]. However, reduced IEC barrier integrity and mixing of commensal bacteria with LMPCs are the most likely mechanism through which intestinal microbes increase CRC susceptibility. Germfree rats and mice provided the first direct evidence that the microbiota contributes to intestinal tumorigenesis. In mice treated with the carcinogen azoxymethane (AOM) to induce genetic mutations and dextran sodium sulfate (DSS) to induce CD-like inflammation, which together cause colitis-associated tumor formation, the absence of the microbiota decreases tumor burden [85]. Similar effects on tumor formation were observed in AOM-DSS-treated rats [86, 87]. Importantly, even in the absence of IBD-like inflammation microbes contribute to tumorigenesis. In the Apc^{Min} mouse model of CRC, which does not develop IBD-like pathologies, the absence of gut microbes reduces intestinal polyp burden [85]. Importantly, new evidence suggests that aberrant crypt foci and adenomas themselves decrease barrier integrity. In mice genetically engineered to be deficient for *Apc* in the colon, loss of IEC barrier integrity at the site of tumor development allowed for penetration of microbial products into lamina propria and induction of T cell-mediated proinflammatory responses that promote tumor growth [83].

7.5.3 The Microbiota, Obesity, and CRC Susceptibility

The intestinal microbiota can be manipulated by a variety of metabolic factors, changes in energy balance, or through the consumption of specific nutritional regimes. Importantly, changes in the gut microbial profile can strongly influence the development of obesity. Germfree mice are resistant to body weight gain induced by a high-fat diet, and transfer of a normal microbiota into germfree mice induced weight gain [46, 88]. Transfer of gut microbiota from an obese mouse to a lean mouse also induced weight gain in mice [45]. Together these studies suggest that the gut microbiota affects energy balance by influencing the efficiency of calorie harvest from the diet and how this harvested energy is used and stored. Human studies of lean or obese monozygotic or dizygotic twins showed that microbial profiles are shared among families on a general level (i.e., phyla), but each individual has a unique profile when examined in more detail (i.e., species) [89, 90]. However, it has been demonstrated that Bacteroidetes are found in lower numbers in obese

Fig. 7.2 (continued) proteins and further increase the permeability of the epithelial barrier, resulting in more inflammation. TLR activation also induces gene expression changes that promote tumor growth. (*4*) Leptin and proinflammatory cytokines released from obese adipose tissue can directly interact with receptors on IECs and tumor cells to promote survival and proliferation. A reduction in anti-proliferative adiponectin production also promotes cell growth. (*5*) Environmental factors, such as saturated fats, interact with IECs and tumor cells to increase barrier permeability (and inflammation) and cell growth and can induce additional genetic mutations that promote tumorigenesis

individuals and that the percentage of Firmicutes is significantly elevated with increased body fatness [91]. A similar finding was observed in obese mice [45]. In contrast, in rats fed with a high-fat diet, Bacteroidetes increased in number [92]. Thus, it is unclear if specific strains modulate obesity (i.e., increase food digestion or absorption) or if obese conditions select distinct families of microbes.

Intestinal bacteria contribute to obesity and cancer susceptibilities in humans and mice, and obesity, diet, and colon cancer can have an effect on the microbial profiles [46]. However, the relationships between host genetics, diet, obesity, susceptibility to colon cancer, and the relative numbers and kinds of intestinal bacteria have not been carefully studied. As discussed above, immune responses to particular bacterial species may be less important than increased immune cell sensitivity or exposure to the microbiota. If this hypothesis is true, a mechanism must exist by which obesity, or diets associated with obesity, increases immune cell–microbiota interactions.

Interestingly, consumption of diets high in saturated fats (HFD) induces low-grade endotoxemia (i.e., increased levels of circulating bacterial-derived LPS), which is indicative of increased IEC barrier permeability [93, 94]. Importantly, saturated fatty acids (SFAs) directly influence barrier integrity by altering the expression of claudin and occludin tight junction proteins in IECs using the same cell signaling pathways as bacterial antigens in IBD (Fig. 7.2) [92, 95]. The activity of the gram-positive and gram-negative bacterial antigens on TLR2 and TLR4 signaling, respectively, is dependent on their saturated fatty acyl moieties [96]. SFAs mimic these moieties and induce both TLR2-depedent and TLR4-depedent responses [96]. SFA stimulation of TLR4 and NF-κB activity alters the expression of tight junction proteins in IECs, resulting in reduce barrier integrity, and induces the expression of proinflammatory cytokines, such as TNF-α, IL-6, and IL-1β, that also decrease IEC barrier integrity [68, 69, 74–79]. Importantly, as discussed with inflammation induced by barrier defects in IBD patients, increased IEC barrier permeability following a meal high in saturated fat allows for the mixing of luminal contents (i.e., the microbiota) with immune cells in the lamina propria. Under normal physiological conditions and diet consumption, an HFD temporarily induces intestinal inflammation, which is resolved with reduced SFA exposure. However, in obese individuals, chronic consumption of HFDs continuously compromises the integrity of the IEC barrier, which in turn induces a chronic, proinflammatory state.

7.6 Proinflammatory Cytokines Contribute to CRC Susceptibility

7.6.1 T Cell-Mediated Inflammation Drives IBD Pathology

Studies of Crohn's disease patients and CD-like colitis in mice have established that CD is a T helper (Th)1 and Th17 cell-mediated inflammatory disease (Tables 7.1 and 7.2) [97–99]. In CD, APCs that have phagocytosed intestinal

Table 7.1 Intestinal immune cells and function

Cell type	Function
T cells (Th, NK, Tc)	Cytokine/chemokine production: Th1 (INF-γ, TNF-α, IL-2, various chemokines), Th17 (IL-17, IL-23, various chemokines), Th2 (IL-4, IL-13), NK (IL-13). Cytotoxicity and APC activation
Treg cells	Cytokine production: (IL-20, TGF-β). Anti-inflammatory, immunosuppression
APCs (macrophages and dendritic cells)	Cytokine/chemokine production: TNF-α, IL-6, IL-1β, IL-23, IL-12, IL-8, various chemokines. Phagocytosis, antigen presentation, T cell activation, granulocyte recruitment
Neutrophils	Cytokine/chemokine production: IL-6, IL-1β, IL-8. Phagocytosis, nitrogen and oxygen reactive species production, degranulation
Epithelial and tumor cells	Cytokine/chemokine production: TNF-α, IL-6, IL-1β, IL-8, PGE2, various other chemokines

Table 7.2 Intestinal immune cells and function

Cytokine/chemokine	Function and pathways in cancer and immune cells
TNF-α	Survival, growth, epithelial barrier permeability, ISEMF activation, monocyte recruitment and activation, T cell activation. MAPK, NF-κB
IL-6	Proliferation, survival, epithelial barrier permeability, ISEMF activation, Th17 cell differentiation, monocyte and granulocyte recruitment. JAK/STAT, ERK/MAPK, PI3K/AKT, NF-κB
IL-1β	Survival, growth, epithelial barrier permeability, ISEMF activation, Th17 cell activation, cytokine and chemokine production by APCs and IECs. MAPK, NF-κB
IL-23	Th17 cell differentiation, induction of IL-22 and IL-17A/F production, Treg inhibition. JAK/STAT
IL-22	Survival, tissue repair responses in IECs, ISEMF activation. JAK/STAT, MAPK, NF-κB
IL-17A/F	Survival, growth, epithelial barrier permeability, ISEMF activation, cytokine and chemokine production by APCs and IECs, monocyte and granulocyte recruitment. JAK/STAT, MAPK, NF-κB
IL-4	Proliferation, survival, metastasis, wound healing, reduced Th1 and APC production of INF-γ, reduced IL-12 production from dendritic cells, B cell class switching. JAK/STAT, MAPK
IL-13	Similar to IL-4
IL-10	Anti-inflammatory, Treg stimulation. JAK/STAT, MAPK
IL-8	Neutrophil recruitment, angiogenesis. PI3K/AKT, PLC, JAK/STAT, MAPK, NF-κB

microbes induce Th1 cell production of chemokines (immune cell recruiting molecules), such as CCL5, and cytokines, such as INF-γ and TNF-α [100]. Together Th1-derived factors recruit additional APCs and induce APCs to release additional proinflammatory cytokines (e.g., IL-6, IL-1β, TNF-α, IL-12) and chemokines (e.g., IL-8) [101]. These factors enhance APC activity (major histocompatibility class [MHC] molecule expression, release of proinflammatory cytokines, and oxygen and nitrogen reactive species production) and recruit additional immune cell types, such as granulocytes (neutrophils, basophils, and eosinophils), to the site of inflammation [74]. To control the immune reaction, immunosuppressive regulatory T (Treg) cells produce anti-inflammatory IL-10 and TGF-β to suppress APC and Th1 activation [5, 102]. Importantly, following the Th1 response, TGF-β produced by Treg cells and IL-6 and IL-23 released by APCs induce the differentiation and expansion of Th17 cells [103]. Th17 cells produce IL-17A and F, which induce the production of cytokines (e.g., IL-6, IL-1β, and TNF-α) and several neutrophil (e.g., IL-8 and CXCL1) and monocyte (e.g., CXCL2) chemokines in a variety of immune and nonimmune cell types to promote inflammation [100, 104]. Additionally, Th17 cells produce IL-22, which induces tissue repair responses in IECs [101, 103]. Importantly, transition from Th1- to Th17-mediated immunity has a significant impact on CRC risk and disease prognosis. Expression analysis of CRC specimens revealed that patients with the highest level of Th1-related genes expression had the best disease prognosis, whereas patients with the highest level of Th17-related gene expression had the worst prognosis [105].

The T cell response involved in UC was more difficult to classify because IL-5 and IL-13, two defining effector cytokines produced by Th2 cells, were produced by T cells in the lamina propria, but a third Th2-related cytokine, IL-4, was not (Tables 7.1 and 7.2) [101]. Important evidence for what type of Th cell-mediated immunity is involved in UC came from mice treated with oxazolone (OXA), which causes colitis and initially increases the expression of IL-4 [106]. Importantly, further studies using the OXA model of UC showed that, overtime, IL-13 production is stimulated in the lamina propria [107]. Interestingly, IL-13 originated from natural killer (NK) CD4+ T cells rather than conventional CD4+ T cells [107]. NK T cells have cytotoxic activity and are activated by antigens presented on CD1 molecules rather than MHC molecules [101]. In UC patients, NK T cells also appear to be the source of IL-13 production [108]. Thus, chronic inflammation in UC involves a Th2-like immune response involving IL-13, which may contribute to CRC susceptibility [105]. Importantly, unlike CD, the glycolipid (either bacterial- or self-derived) that induces NK T cell responses and inflammation has not been identified [100].

Bridging the Th1/Th17/Th2 responses in CD and UC is a group of well-known cytokines that contribute to IEC barrier defects and mediate many of the pathologies of IBD. Both APCs and IECs release TNF-α, IL-1β, and IL-6 in response to the cytokines produced by Th1, Th17, and NK T cells (Tables 7.1 and 7.2) [55, 109]. Through the activation of mitogen-activated protein kinase (MAPK) signaling pathways and the stimulation of NF-κB-mediated transcription, TNF-α, IL-1β, and IL-6 activate additional APCs. Furthermore, TNF-α, IL-1β, and IL-6 feedback on T cell function. IL-6 and possibly IL-1β induce Th17 differentiation in CD [110, 111]. Additionally, TNF-α enhances APC production of IL-12, which promotes Th1-mediated responses [101]. Importantly, these cytokines promote additional inflammation and stimulate IEC proliferation and survival, which contributes to the pathology of IBD and CRC susceptibility.

7.6.2 *Proinflammatory Cytokines Directly Influence Epithelial Cell Development and CRC Risk*

Several studies have established that the proinflammatory cytokines elevated in IBD directly interact with their respective receptors on IECs to initiate signaling cascades involved in tumor formation (Table 7.2 and Fig. 7.2). IL-6, IL-1β, and TNF-α, which are produced by APCs and IECs in the intestines of IBD and CRC patients, promote IEC proliferation through several cell signaling pathways involving Wnt, Janus kinase/signal transducer and activator (JAK/STAT), MAPK, and NF-κB [27, 112–118]. Similarly, IL-22, IL-17, and TNF-α produced by Th17 cells in CD and CRC patients induce JAK/STAT, MAPK, and NF-κB signaling in IECs to promote proliferation, survival, and tissue repair responses associated with tumorigenesis [83, 115, 117–119]. The Th2 cell-derived cytokines also modulate IEC function. IL-4 and IL-13 through MAPK-mediated cell signaling induce IEC proliferation and survival [120, 121]. Furthermore, evidence suggests that IL-4-/IL-13-mediated signaling alters colon cancer cell adhesion and contributes to cancer cell invasion and metastasis [122, 123].

Importantly, Th1-/Th17-mediated immune response cytokines found in CD (IL-22, IL-23, IL-17A/F), Th2-mediated immune response cytokines found in UC (IL-4 and IL-13), and cytokines associated with both types of T cell responses and innate immunity in IBD (TNF-α, IL-6) have all been shown to increase CRC risk in humans or tumor burden in mouse models of CRC [83, 113, 114, 120, 121, 123–130]. As discussed previously, these proinflammatory mediators likely interact with preexisting genetic abnormalities to increase cancer susceptibility (Fig. 7.1). Importantly, many of these cytokines influenced tumor burden in *Apc* mutant mice. IBD-like inflammation is not observed in *Apc* mutant mice; however, as discussed in Sect. 7.5.2, precancerous lesions can reduce intestinal barrier integrity, allow microbes to interact with LPMNCs, and induce localized inflammation, which promotes epithelial cell proliferation and survival [83]. Therefore, inflammatory cytokine production in the intestine can contribute to CRC risk even in individuals without preexisting inflammatory diseases.

7.6.3 *Epithelial TLR4 Activation Promotes IEC Proliferation and Tumorigenesis*

Similar to IBD, TLR4 receptor expression increases in CRC patients and promoted intestinal tumorigenesis in mice (Fig. 7.2) [80, 131]. Interestingly, microbial activation of TLR4 receptors on IECs promoted tumor development, in part, by inducing expression of cyclooxygenase-2 (COX-2) [131]. COX-2 is an inducible mediator of proinflammatory prostaglandin E2 (PGE2) synthesis, both of which are increased in both human and experimental models of CRC [132–137]. PGE2 through its receptors (EP family of receptors) induces proinflammatory responses in dendritic cells,

Th17 cells, Th1 cells, and IECs [138]. As with all other activators of immune cells, PGE2 promotes tumorigenesis through its proinflammatory effects. PGE2 induces IECs to express amphiregulin, a member of the epidermal growth factor (EGF) family [131]. Interestingly, amphiregulin interacts with EGFR receptors on IECs to promote proliferation and tumorigenesis [131]. Therefore, IECs can promote their own proliferation and neoplastic transformation through PGE2 synthesis.

7.6.4 Inflammation Induces Proinflammatory Cytokine Production by IECs

In addition to directly promoting IEC proliferation and tumorigenesis, cytokines can also induce proinflammatory gene expression in IECs. Through MAPK-, JAK/STAT-, and NF-κB-mediated signaling, IL-17, TNF-α, IL-6, and IL-1β induce CRC cell lines to express a variety of cytokines and chemokines that recruit and activate APCs and granulocytes [5, 139, 140]. The activation of these cell types to the intestine causes additional cytokine production and inflammation, which promotes IEC proliferation and increased tumor risk. Additionally, chronic accumulation of activated granulocytes, macrophages, and dendritic cells is accompanied by the release of oxygen and nitrogen reactive species (RONS), which promote dysplasia by inducing DNA modifications and genetic mutations in proliferating IECs (Fig. 7.2) [141, 142]. Inflammation-induced DNA damage has been linked to altered expression of genes involved in CRC such as p53, APC, KRAS, and BCL-2 [143, 144]. In the DSS mouse model of CD-like inflammation, deficiency for base-excision repair enzymes induces tumor formation, providing direct evidence that the pro-tumorigenic effects of chronic inflammation are partially mediated through DNA damage [145, 146].

7.6.5 Inflammation Induces Tumorigenesis by Altering the IEC Stem Cell Niche

Cytokine action on cells that regulate IEC homeostasis may also influence tumor development (Fig. 7.3). Intestinal subepithelial myofibroblasts (ISEMFs) are positioned subjacent to the basement membrane in the intestinal mucosa, juxtaposed against the bottom of IECs in intestinal crypts. ISEMFs have a profound influence on the intestinal stem cell niche [147]. Following infection or trauma to the intestinal mucosa and activation of inflammation, ISEMFs release factors that alter IEC homeostasis and induce a tissue repair response [147–149]. Proinflammatory cytokines such as IL-1β, TNF-α, IL-17A/F, and IL-22 induce NF-κB and MAPK signaling in ISEMFs, resulting in ISEMF proliferation and the production and release of ECM components (e.g., type I and IV collagen), basement membrane remodeling

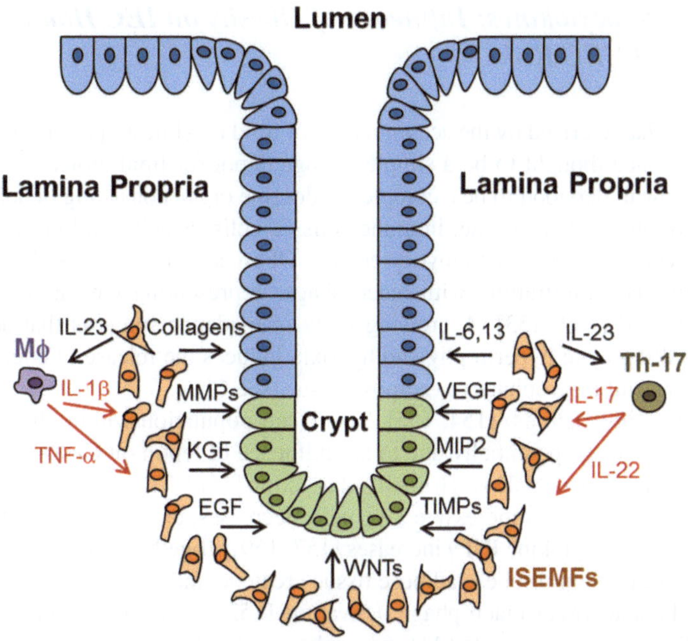

Fig. 7.3 Intestinal subepithelial myofibroblasts (ISEMFs) influence the intestinal stem cell niche during inflammatory responses. In response to proinflammatory cytokines released by macrophages and T cells, ISEMFs express and release additional proinflammatory cytokines (IL-6, IL-13, IL-23), ECM remodeling factors (matrix metalloproteinases [MMPs], tissue-specific inhibitors of MMPs [TIMPs], and collagens) and growth factors (VEGF, KGF, and EGF), and cell signaling ligands (Wnts) into the stem cell niche of the intestinal crypt. Together, the factors released by ISEMFs induce tissue repair responses in IECs that stimulate proliferation and migration. Normally this response repairs tissue damage associated with infection. In CRC-susceptible mice and humans, tissue repair responses interact with genetic predispositions to CRC to further increase cancer risk

factors (matrix metalloproteinases [MMPs] and tissue-specific inhibitors of MMPs (TIMPs)), proinflammatory cytokines (IL-6, IL-13, and IL-23), chemokines (CXCL1 and CXCL2), and growth factors (vascular endothelial growth factor [VEGF], keratinocyte growth factor [KGF], EGF, and Wnt proteins), which induce tissue remodeling and proliferation of IECs [27, 147, 150, 151]. In the chronic proinflammatory state of IBDs, cytokines cause an expansion in ISEMF numbers and induce ISEMFs to continuously release factors that result in IEC proliferation, tissue disorganization, and eventually fibrosis [27, 149]. Importantly, the uncontrolled tissue repair response induced in IECs of IBD patients is the mechanism by which inflammation is proposed to promote adenoma development. Therefore, induction of ISEMFs by inflammatory factors may be an additional mechanism through which inflammation increases CRC risk.

7.6.6 Adipocytokines: Influence of Obesity on IEC Homeostasis and CRC Risk

Obesity is characterized by the accumulation of lipid in white adipose tissue (WAT). What was once thought to be a simple storage depot for lipid molecules, adipose tissue is now understood to be a complex endocrine organ consisting of adipocytes, connective and nervous tissue, immune cells (T cells, B cells, and macrophages), chondrocytes, osteocytes, and myocytes [42]. Obese adipose tissue is characterized by immune cell infiltration, with macrophages representing the majority of cell types recruited [152, 153]. A growing body of evidence suggests that adipocyte death resulting from hypertrophy and hypoxia induces the release of chemoattractants that recruit macrophages to adipose tissue (see Chap. 6 for a detailed explanation of these interactions) [154, 155]. The T cell population within obese adipose tissue also changes, potentially due to the influence of leptin on CD8+ cytotoxic T (Tc) cell activation and CD4+ Th1 cell polarization [156]. Tc cell numbers increase, the number of immunosuppressive Treg cells decreases, and production of the Th1 proinflammatory cytokine INFγ increases [157–159]. Together changes in the T cell and cytokine content of obese adipose tissue create an inflammatory milieu conducive to classical (M1) macrophage activation [152, 153]. Once established, the obese adipose tissue-associated M1 macrophages secrete a variety of proinflammatory cytokines (IL-1β, IL-6, and TNF-α) and chemokines (monocyte chemoattractant protein-1 [MCP-1] and IL-8) and reduce expression of anti-inflammatory IL-10, which together maintain an unresolved proinflammatory state in the adipose and a subclinical, chronic inflammatory state systemically [42, 152, 153].

As discussed in the previous sections, the same cytokines released into circulation by obese adipose tissue also modulate intestinal inflammation and CRC susceptibility. Obesity and HFDs interact with DSS and AOM-DSS treatment in mice to increase the severity of intestinal inflammation and CRC incidence, respectively [160, 161]. DSS-mediated inflammation and diet-induced obesity interacted to increase LPMNC expression of proinflammatory cytokines (TNF-α, IL-6, and INF-γ), LPMNC and IEC expression of TLR4, and intestinal helper T cell number [160, 161]. However, the mechanism by which diet-induced obesity increased DSS-mediated inflammation was not clearly demonstrated. In the study of AOM-DSS mice, the increase in intestinal inflammation and CRC incidence was associated with increases in expression of leptin receptor in the colon [161]. As discussed in Sect. 7.7, leptin signaling induces inflammation and IEC proliferation, which may have increased tumor incidence. In the study of DSS-treated mice fed with a high-fat diet, intestinal inflammation was not linked to a specific mechanism. DSS and SFA likely increased the permeability of the intestinal epithelial barrier and allowed LPMNCs to interact with intestinal microbes, which may have induced proinflammatory cytokine production in the intestine. Thus, whether proinflammatory cytokines released into circulation by obese adipose contributed to intestinal inflammation in these studies is an open question. Importantly, TNF-α and IL-1β circulate at very low levels in lean and obese individuals [162]. Therefore, whether these cytokines function as endocrine factors is debatable. In contrast, circulating levels of IL-6 are substantial in obese individuals and, therefore, may have endocrine effects on distal tissues [162].

7.7 Leptin: A Proinflammatory Adipokine Associated with Increases CRC Risk

In addition to cytokines, adipocytes produce a variety of hormones (adipokines) that regulate several systemic processes, including food intake, metabolism, insulin sensitivity, reproduction, blood pressure, angiogenesis, and inflammation [163]. Leptin (LEP) is an adipokine secreted by white adipocytes in a proportion approximately linear to adipose tissue weight [42]. Leptin plays a key role as a satiety factor and regulator of energy homeostasis through binding to the lepton receptor (LEPR or Ob-R) in the hypothalamus [164]. *LEPR* encodes several splice variants called the long (Ob-Rb), short (Ob-Ra, c, and d), and soluble (Ob-Re) forms [165]. Full-length Ob-Rb retains the intracellular domains necessary to induce downstream signaling events following activation by leptin binding. The short isoform (Ob-Ra) is the most abundant in many tissues and may retain some of intracellular signaling capacities of full-length Ob-Rb. Ob-Re appears to be a carrier molecule for circulating leptin.

As adiposity increases, leptin is released from adipocytes and into circulation to reduce dietary intake and enhance energy expenditure. Interestingly, leptin signaling may have biological functions that reach beyond its normal role in energy homeostasis. Leptin receptor expression and signaling have been reported for a variety of cell and tissue types [165]. Specifically, the leptin receptor is expressed throughout the colonic epithelium, and leptin has been shown to induce proinflammatory innate and adaptive immune cell responses [156, 166, 167]. Epidemiological studies have established that obese individuals have higher circulating levels of leptin compared to normal-weight individuals [168]. Furthermore, several studies have observed an association between higher leptin levels and increased risk of colorectal adenomas [169, 170] and colon cancer [171]. Thus, given the evidence of associations between obesity, chronic inflammation, and CRC risk [172–175], leptin signaling may play an important role in mediating pathophysiological events that lead to intestinal neoplasia.

7.7.1 Leptin and Epithelial Cell Function

Previous reports have established that the leptin receptor is expressed by mouse and human colonic epithelial cells, colon cancer cells, and a variety of human epithelial colon cancer cell lines [166, 167, 176]. In vitro leptin mediates the survival, proliferation, and metastatic capacity of human colon cancer cell lines through several cell signaling pathways (Fig. 7.2). Stimulation of colon cancer cell lines with leptin induces leptin receptor tyrosine phosphorylation and the activation of a variety of cell signaling cascades [176]. Leptin increased cell number and reduced apoptosis in Caco-2 and HT-29 cultures by inducing extracellular signaling kinases (ERK1/2), PI3-kinase/protein kinase B (PKB or AKT) AKT phosphorylation, c-Jun NH(2)-terminal kinase (JNK), and JAK/STAT-mediated pathways [177, 178]. An additional study of HT-29 response to leptin demonstrated that colon cancer cell proliferation and survival are dependent on NF-κB nuclear translocation and activation [179]. Importantly, the leptin receptor is expressed by both normal and tumorigenic epithelial cells in vivo [176, 180]. However, the proliferative and anti-apoptotic effect of leptin demonstrated in vitro was contradicted by in vivo studies of the *Apc*$^{Min/+}$ and AOM mouse models of CRC. Curiously, leptin failed to increase intestinal polyp number or size in *Apc*$^{Min/+}$ mice treated with leptin [181]. Furthermore, leptin-deficient, obese *ob/ob* and leptin receptor-deficient, and obese *db/db* mice showed increased numbers of aberrant crypt foci (markers of precancerous lesions) relative to controls following AOM treatment despite the absence of leptin signaling [182].

Two studies may have resolved the differences observed in the in vitro and in vivo studies. Endo et al. examined normal and tumorigenic colonic epithelium at various stages of development in leptin-deficient, obese *ob/ob*, and wild-type mice treated with the carcinogen AOM [183]. In *ob/ob* mice normal epithelial cell proliferation and the number or precancerous aberrant crypt foci were similar to AOM-treated wild-type controls. However, when cancer adenoma polyps were examined, a significant reduction in tumor proliferation and size, but not number, was observed in the AOM-treated *ob/ob* mutants. Thus, leptin appears to promote growth subsequent to tumor initiation. This conclusion is in agreement with studies by Fenton et al. who demonstrated that stimulation of a normal mouse IEC line (YAMC), which has two functional copies of *Apc*, with leptin reduced epithelial cell proliferation and induced apoptosis [184]. In stark contrast, the pre-neoplastic mouse IEC line (IMCE), which was derived from mice harboring the *Apc*^{Min} mutation, exhibited increased proliferation and survival after leptin stimulation [184]. Thus, the influence of leptin on IECs is likely to be dependent on preexisting genetic risk factors that already initiated tumorigenesis.

7.7.2 Leptin and the Modulation of Epithelial Cell-Derived Inflammation

In addition to promoting colon cancer proliferation, survival, and migration, the role of leptin in promoting proinflammatory responses in epithelial cells may significantly contribute to CRC risk. In vitro leptin, through the activation of NF-κB-mediated transcription, induced CRC cells to express several proinflammatory cytokines and chemokines, including IL-1β, which are associated with chronic intestinal inflammation, innate immune cell infiltration, and increased CRC risk

[5, 180]. As previously described in Sects. 7.6.2 and 7.6.5, IL-1β can act directly on epithelial cells or indirectly through ISEMF to promote epithelial cell proliferation and neoplastic transformation. By contrast, leptin reduced CRC cell line expression of anti-inflammatory cytokines, such as IL-10, that downregulate immune responses to gut microbes and reduced tumorigenesis (Sect. 7.5.1) [180]. Furthermore, Fenton et al. used their normal IEC line and preneoplastic IEC line to demonstrate that leptin induces IL-6 expression from precancerous but not normal epithelial cells [184]. These results are in agreement with the postulated role of IL-6 in inducing IEC and promoting CRC risk in individuals with IBDs (Sect. 7.6.2). Therefore, induction of IEC expression of IL-1β and IL-6 by leptin may be a shared mechanism by which obesity and chronic inflammation increase CRC susceptibility.

Leptin also induces the production of chemokines in epithelial cells, which may contribute to CRC incidence. Intraperitoneal injections of leptin induced the expression of the neutrophil chemoattractant CXCL1 in the mouse colonic epithelium [185]. Neutrophils express IL-1β, IL-6, and IL-8, which are associated with IBD and CRC susceptibility, and generate nitrogen and oxygen reactive species that induce DNA damage (Sect. 7.6.4) [141, 156]. Interestingly, CXCL1 and IL-8 are both potent angiogenic compounds that may promote tumor growth through vascularization [186, 187]. Therefore, leptin may increase CRC risk by promoting neutrophil recruitment.

7.7.3 Leptin and the Activation of Immune Cells

Similar to proinflammatory effects on IECs, leptin may induce intestinal tumorigenesis by directly activating and recruiting immune cells (Fig. 7.2). In humans, it has been suggested that reduced leptin levels are associated with impaired immune function [188]. Similarly, leptin-deficient *ob/ob* mice are resistant to various experimental models of inflammation, which can be corrected by exogenous administration of leptin [189–191]. Conversely, increased leptin levels may be responsible for the altered T/B cell function, increased monocyte and granulocyte activity, and increased white blood cell counts observed in obese individuals with active leptin signaling [192]. The leptin receptor shares significant similarity to members of the class I cytokine receptor (gp130), which includes the receptor for IL-6 [193]. In fact, Ob-R stimulates many of the same downstream signaling cascades (JAK/STAT, PI3K/AKT, MAPK, and NF-κB) as IL-6, suggesting a potential function for leptin in regulating immune cell function [192, 194]. Importantly, Ob-R is expressed on the surface of several cell types involved in innate immunity [195].

Studies of leptin- and leptin receptor-deficient mice have revealed the effects of leptin on antigen-presenting cell (APC) function. Leptin signaling in macrophage induces activation, increases phagocytosis capacity, promotes proinflammatory cytokine production (e.g., IL-6, TNF-α, and granulocyte chemoattractants), and activates proliferation [192, 196, 197]. Leptin also increases the survival of dendritic cell (DC) and induce DC production of IL-8, IL-6, and TNF-α [156, 198, 199].

Leptin also downregulates DC expression of anti-inflammatory IL-10 [199]. On a functional level, leptin drives DC cells to promote cell-mediated immunity by inducing Th1 differentiation and enhancing Tc cell function [198–200].

Leptin signaling also directly modulates adaptive immune cell function. Mouse and human lymphocytes express the leptin receptor [194, 201]. However, leptin alone cannot activate lymphocytes in vitro [202, 203]. However, leptin enhances proliferation, inhibits apoptosis, and promotes the activation of primed Th cells and Tc cells [204, 205]. Leptin appears to regulate T cell development at two stages. First, immature helper T cells from *ob/ob* mice showed impaired development of both Th1 cells and Th2 cells in vitro [206]. Second, evidence suggests that leptin drives helper T cell differentiation toward Th1 cells. In mice, leptin promotes Th1 cytokine production [156, 201, 204].

Importantly, leptin modulates many immune cell function and inflammatory responses in the colon, which could have a profound influence on CRC risk. The leptin receptor is expressed by LPMNCs [207]. Intraperitoneal injections of leptin in mice induced APCs in the lamina propria to express IL-6, IL-1β, and TNF-α [185]. Together, these cytokines represent key inflammatory mediators of IBD and increased CRC risk (Sects. 7.6.2 and 7.6.5). The influence of leptin on T cell function can also promote intestinal tumorigenesis. Using the OXA-induced mouse model of Th2-like UC, Batra et al. demonstrated that leptin-deficient *ob/ob* mice are protected against colitis, which was attributed to a decrease in the production of the Th2 cytokine IL-13 in the lamina propria [206]. As discussed in Sect. 7.6.2, the two Th2-related cytokine, IL-4 and IL-13, induce CRC cell survival, proliferation, and metastasis [120–123].

7.8 Adiponectin: An Anti-inflammatory Adipokine Associated with Reduced CRC Risk

Adiponectin (APN) circulates as a low molecular weight (LMW) trimer, a middle molecular weight (MMW) hexamer, and a high molecular weight (HMW) multimer of full-length (fAb) protein [208]. An additional truncated form of APN containing only a C-terminal globular domain (gAb) is produced by proteolytic cleavage and is also found in circulation [208, 209]. All circulating forms of APN are biologically active but have different target tissues and varying biological effects that are not well delineated [210, 211]. APN induces intracellular signaling through its interactions with two receptors (ADIPOR1 and ADIPOR2). ADIPOR1 is predominately located in skeletal muscle whereas ADIPOR2 localizes to the liver [191]. APN regulates a variety of metabolic processes including glucose and fatty acid catabolism and increases glucose uptake by stimulating the translocation of the GLUT4 receptor to plasma membranes [191]. Binding of APN to its receptors activates a variety of cell signaling cascades involving ERK1/2 MAPKs, PI3-K/AKT, NF-κB, and JAK/STAT [209]. However, APN primarily exerts its effects through the activation of the AMP-activated protein kinase (AMPK) signaling cascade [209].

Unlike most other adipose-derived factors, circulating APN levels are reduced in obese individuals and negatively correlate with BMI, waist circumference, and visceral fat mass [212–215]. Although the inverse relationship between circulating APN levels and obesity makes biological sense given its influence on metabolic functions in muscle and liver, chronically low levels of APN may have more far reaching physiological consequences. Both APN receptors are expressed in a variety of tissue including IECs and immune cells [209]. Importantly, epidemiological evidence suggests that CRC risk is inversely associated with circulating APN levels; men with the highest levels of APN had a 60 % reduced CRC risk after adjusting for body size, activity, and waist circumference [216]. Thus, the reduction in APN levels associated with obesity appears to be an important factor in promoting CRC development.

7.8.1 Adiponectin and Epithelial Cell Function

IECs express ADIPOR1 and ADIPOR2 in both mice and humans [166, 217–219]. Benign adenomas and malignant carcinomas in humans and adenomas in $Apc^{Min/+}$ mice also express APN receptors [209, 218, 219]. Thus, APN signaling has the potential to regulate both normal and neoplastic epithelial cell function (Fig. 7.2). Several in vitro studies have demonstrated that APN signaling restricts the proliferation and survival of CRC cell lines. APN decreases proliferation and induces apoptosis in several CRC cell lines [220–224]. Importantly, the effect of APN on CRC cell line growth appeared to be predominately mediated by activation of AMPK signaling, which resulted in the inhibition of the tuberous sclerosis protein 2/mammalian target of rapamycin/S6 kinase (TSC2/mTOR/S6 kinase) cell signaling axis and repression of translation and cell growth [209, 220, 224, 225]. In addition, AMPK activated the G1/S cell cycle control checkpoint by promoting the expression of cell-dependent kinase inhibitors p21 and p27 and the G2/M cell cycle checkpoint by inducing p53 expression [220, 226]. Although most in vitro studies suggest an anti-proliferative and pro-apoptotic role for APN in cancer, the results of two studies imply that APN positively influences CRC cell line proliferation and survival [222, 226]. Interestingly, these contrasting responses appear to be dependent on culture conditions. The pro-proliferative response to APN occurred in the presence of fetal bovine serum (FBS), and the pro-survival effect occurred in glucose-deprived medium [222, 226]. Therefore, the influence of APN on intestinal tumorigenesis may be dependent on the nutritional and hormonal content of the tumor microenvironment.

The preponderance of evidence from mouse models of CRC also suggests that APN signaling negatively regulates intestinal cancer development. Studies of *Apn*-deficient mice harboring the Apc^{Min} mutation and *Apn*-deficient mice treated with AOM or DSS and dehydroheliotridine (DSS-DHH) demonstrated that *Apn* deficiency increases cancer burden, which was associated with reduced AMPK activation, an increase in mTOR and S6 kinase activity, increased JAK/STAT and AKT activity, and increased cell proliferation [222, 227–229]. Importantly, exogenous administration of APN to *Apc*$^{Min/+}$ mutant mice suppressed intestinal tumorigenesis, providing clear evidence of the anti-tumorigenic properties of APN [219]. Interestingly, one study reported that *Apn* deficiency reduced tumor burden in AOM-treated mice fed with a high-fat, but not a low-fat, diet [228]. As observed in vitro, the nutritional and hormonal milieu of the intestinal mucosa may determine the influence of APN on IEC growth.

7.8.2 Adiponectin and the Modulation of Colonic Epithelial Cell-Derived Inflammation

In addition to inhibiting colon cancer proliferation, survival, and migration, the regulation of inflammatory responses in IECs by APN may contribute to CRC risk. However, unlike the clear proinflammatory influence of leptin on IECs, the contribution of APN to inflammation is less clear. In one in vitro experiment, it was demonstrated that APN inhibits IL-8 production by LPS-stimulated colon cancer cells, suggesting that APN signaling has an anti-inflammatory and anti-angiogenic effect on IECs [217]. In the mouse adenoma cell line IMCE, Fenton et al. demonstrated that APN blocked the ability of leptin to induce IL-6 production [230]. Given the role of cytokines, in particular IL-6, in inducing IEC proliferation and promoting CRC risk, the anti-inflammatory influence of APN on IECs may inhibit tumor development. However, other studies suggest that APN signaling has a proinflammatory effect on IECs. Ogunwobi and Beales demonstrated that APN induces CRC cell line expression and secretion of IL-8 and MCP-1 [222]. In agreement with these findings, APN induced colonic explants from DSS-treated mice to increase IL-6 and granulocyte chemoattractant CXCL2 production [231]. It is currently unclear as to why the reported inflammatory effects of IECs are so variable. Differences in cell culture conditions, such as the isoform of APN (full length or globular) used in the medium, may have influenced results. Therefore, future experiments will need to delineate the effects of APN isoforms on IEC inflammatory responses under different culture conditions.

7.8.3 Adiponectin and the Activation of Immune Cells

In contrast to its ambiguous roles in regulating IEC inflammatory responses, the influence of APN on immune cell function is undoubtedly immunosuppressive (Fig. 7.2) [217, 232, 233]. The ADIPOR1 and ADIPOR2 receptors are expressed on the majority of monocytes, many B cells and natural killer cells, and a small percentage of T cells [234]. Mice deficient for *Apn* present with a heightened immune response in several inflammatory disease models, including microbial sepsis, transplant rejection, and DSS-induced colitis [217, 232, 233, 235]. In the DSS mouse model of CD, *Apn* deficiency causes a dramatic increase in the number of macrophages that infiltrate the affected colon, resulting in increased fibrosis, breakdown of the intestinal crypts, and destruction of the IEC barrier [217]. As discussed in Sect. 7.5, breakdown of the IEC barrier allows commensal microbes to come into contact with LPMNCs, which induces inflammation and increases CRC susceptibility. Importantly, the expression of several cytokines associated with the progression of IBD and CRC risk, including IL-1β, IL-6, and TNF-α, is higher in *Apn*-deficient mice treated with DSS [217]. Although the cell types producing these factors were not identified in this study, these results clearly demonstrate that *Apn* expression within the intestine is immunosuppressive.

APN has immunosuppressive effects on APCs. Macrophages from *Apn*-deficient mice have increased production of IL-6, TNF-α, and MCP-1 and reduce expression of anti-inflammatory IL-10 [236–239]. In agreement with these findings, several additional studies have shown that APN induces anti-inflammatory IL-10 production and inhibits LPS activation of phagocytosis and production of chemoattractants by macrophages [236–243]. A limited number of studies have identified subtle influences of APN on cell types associated with adaptive immunity. In one study, APN was shown to negatively regulate antigen-activated CD8+ T cell function by reducing INF-γ, TNF-α, and IL-2 production and to promote T cell apoptosis [244]. Supporting these findings, T cells from *Apn*-deficient mice have an increased capacity for activation and expansion [244].

Obesity is associated with lower circulating levels of APN, which may contribute to chronic inflammatory and increased CRC susceptibility in individuals genetically predisposed to tumor development. This model is supported by a study of DSS-DHH-treated, *Apn* knockout mice [227]. Both tumor number and size increased in *Apn* knockout mice exposed to DSS-DHH when compared to wild-type controls treated with DSS-DHH. Importantly, in *Apn*-deficient mice, DSS-DHH increased expression of several proinflammatory and tumorigenic cytokines, including IL-6, TNF-α, IL-1β, and COX-2, and decreased expression of anti-inflammatory and tumor suppressing IL-10 in the intestinal mucosa. Therefore, reduced APN levels are likely to increase CRC risk through the dysregulation of inflammatory responses in the gastrointestinal tract.

7.9 CRC Prevention and Therapeutics: Targeting Inflammation

Although the link between inflammation and CRC risk is well established, this knowledge is just now being translated to the clinic. Results from initial studies using anti-inflammatory agents as a component of CRC treatment are promising. Intestinal inflammation plays an important role in CRC risk and progression. Nonsteroidal anti-inflammatory drugs (NSAIDs) such as aspirin and sulindac, which inhibit COX-2 production of PGE2, are currently being used to treat CRC or prevent cancer relapse in CRC patients [245–247]. Additionally, several anti-inflammatory reagents targeting TNF-α, IL-6, IL-12/23, IL-2, and IL-17 have been shown to be effective in treating IBD patients or individuals with autoimmune disorders [5, 248, 249]. These same cytokine targets play central roles in mediating CRC progression and, therefore, may prove to be effective targets for anticancer treatment. However, the immune system possesses antitumor activity, and it will be important to select anti-inflammatory treatment options that block tumor-promoting inflammation without reducing antitumor immune responses. In addition to pharmacological interventions, modulation of the diet might prove as equally effective at reducing intestinal inflammation and CRC risk. High-fat diets increase intestinal inflammation and CRC risk. Importantly, several other dietary components, including carbohydrates, unsaturated n-3 fatty acids, vitamins, minerals, and phytochemicals (e.g., resveratrol), have been shown to reduce CRC susceptibility [250]. Although the molecular mechanisms by which these dietary factors reduce cancer risk are not well established, there is evidence to suggest that reduced intestinal inflammation plays an important role [5, 251]. Therefore, when used with existing chemotherapy and radiotherapy, agents or lifestyle changes that reduce intestinal inflammation may improve treatment outcomes and reduce tumor relapse in CRC patients.

References

1. Howlader N, Noone AM, Krapcho M, Neyman N, Aminou R, Altekruse SF, Korsay CL, Ruhl J, Tatalovich Z, Cho H, Mariotto A, Eisner MP, Lewis DR, Chen HS, Feuer EJ, Cronin KA (2012) SEER cancer statistics review, 1975–2009 (vintage 2009 populations). National Cancer Institute, Bethesda, MD.
2. Jemal A, Bray F, Center MM, Ferlay J, Ward E, Forman D (2011) Global cancer statistics. CA Cancer J Clin 61:69–90
3. Schoen RE (2000) Families at risk for colorectal cancer: risk assessment and genetic testing. J Clin Gastroenterol 31:114–120
4. Heijstek MW, Kranenburg O, Borel Rinkes IH (2005) Mouse models of colorectal cancer and liver metastases. Dig Surg 22:16–25
5. Terzic J, Grivennikov S, Karin E, Karin M (2010) Inflammation and colon cancer. Gastroenterology 138:2101–2114
6. Cho KR, Vogelstein B (1992) Genetic alterations in the adenoma—carcinoma sequence. Cancer 70:1727–1731
7. Parkin DM, Bray F, Ferlay J, Pisani P (2001) Estimating the world cancer burden: Globocan 2000. Int J Cancer 94:153–156
8. American Cancer Society (2007) Cancer facts and figures 2007. American Cancer Society, Atlanta, GA
9. Vogelstein B, Kinzler KW (1993) The multistep nature of cancer. Trends Genet 9:138–141
10. Johns LE, Houlston RS (2001) A systematic review and meta-analysis of familial colorectal cancer risk. Am J Gastroenterol 96:2992–3003
11. Lerman C, Shields AE (2004) Genetic testing for cancer susceptibility: the promise and the pitfalls. Nat Rev Cancer 4:235–241
12. Bulow S (1987) Familial polyposis coli. Dan Med Bull 34:1–15
13. Kinzler KW, Nilbert MC, Su LK et al (1991) Identification of FAP locus genes from chromosome 5q21. Science 253:661–665
14. Nishisho I, Nakamura Y, Miyoshi Y et al (1991) Mutations of chromosome 5q21 genes in FAP and colorectal cancer patients. Science 253:665–669

15. Heyer J, Yang K, Lipkin M, Edelmann W, Kucherlapati R (1999) Mouse models for colorectal cancer. Oncogene 18:5325–5333
16. Hanahan D, Weinberg RA (2011) Hallmarks of cancer: the next generation. Cell 144:646–674
17. Fearon ER (2011) Molecular genetics of colorectal cancer. Annu Rev Pathol 6:479–507
18. Thibodeau SN, French AJ, Roche PC et al (1996) Altered expression of hMSH2 and hMLH1 in tumors with microsatellite instability and genetic alterations in mismatch repair genes. Cancer Res 56:4836–4840
19. Fodde R, Smits R, Clevers H (2001) APC, signal transduction and genetic instability in colorectal cancer. Nat Rev Cancer 1:55–67
20. Suzuki H, Watkins DN, Jair KW et al (2004) Epigenetic inactivation of SFRP genes allows constitutive WNT signaling in colorectal cancer. Nat Genet 36:417–422
21. Nusse R (2005) Wnt signaling in disease and in development. Cell Res 15:28–32
22. Clevers H, Nusse R (2012) Wnt/beta-catenin signaling and disease. Cell 149:1192–1205
23. Tetsu O, McCormick F (1999) Beta-catenin regulates expression of cyclin D1 in colon carcinoma cells. Nature 398:422–426
24. Brabletz T, Jung A, Dag S, Hlubek F, Kirchner T (1999) Beta-catenin regulates the expression of the matrix metalloproteinase-7 in human colorectal cancer. Am J Pathol 155:1033–1038
25. Miwa N, Furuse M, Tsukita S, Niikawa N, Nakamura Y, Furukawa Y (2001) Involvement of claudin-1 in the beta-catenin/Tcf signaling pathway and its frequent upregulation in human colorectal cancers. Oncol Res 12:469–476
26. McCart AE, Vickaryous NK, Silver A (2008) Apc mice: models, modifiers and mutants. Pathol Res Pract 204:479–490
27. Fantini MC, Pallone F (2008) Cytokines: from gut inflammation to colorectal cancer. Curr Drug Targets 9:375–380
28. Xavier RJ, Podolsky DK (2007) Unravelling the pathogenesis of inflammatory bowel disease. Nature 448:427–434
29. Kappelman MD, Rifas-Shiman SL, Kleinman K et al (2007) The prevalence and geographic distribution of Crohn's disease and ulcerative colitis in the United States. Clin Gastroenterol Hepatol 5:1424–1429
30. Loftus EV Jr (2007) The burden of inflammatory bowel disease in the United States: a moving target? Clin Gastroenterol Hepatol 5:1383–1384
31. Rubin DC, Shaker A, Levin MS (2012) Chronic intestinal inflammation: inflammatory bowel disease and colitis-associated colon cancer. Front Immunol 3:107
32. Pohl C, Hombach A, Kruis W (2000) Chronic inflammatory bowel disease and cancer. Hepatogastroenterology 47:57–70
33. Eaden JA, Abrams KR, Mayberry JF (2001) The risk of colorectal cancer in ulcerative colitis: a meta-analysis. Gut 48:526–535
34. World Health Organization (2003) World cancer report. IARC Nonserial Publication, Geneva, Switzerland
35. World Health Organization (2006) Global database on body mass index, vol 9, No. 5. Public Health Nutrition, Geneva, Switzerland, pp 658–660
36. National Center for Health Statistics (2011) Health, United States, 2010: with special features on death and dying. Department of Health and Human Services, Hyattsville, MD
37. Ogden CL, Carroll MD, Curtin LR, Lamb MM, Flegal KM (2010) Prevalence of high body mass index in US children and adolescents, 2007–2008. JAMA 303:242–249
38. Farooqi IS, O'Rahilly S (2005) Monogenic obesity in humans. Annu Rev Med 56:443–458
39. Nkondjock A, Shatenstein B, Maisonneuve P, Ghadirian P (2003) Specific fatty acids and human colorectal cancer: an overview. Cancer Detect Prev 27:55–66
40. Wynder EL (1976) Nutrition and cancer. Fed Proc 35:1309–1315
41. Stein CJ, Colditz GA (2004) The epidemic of obesity. J Clin Endocrinol Metab 89:2522–2525
42. Yehuda-Shnaidman E, Schwartz B (2012) Mechanisms linking obesity, inflammation and altered metabolism to colon carcinogenesis. Obes Rev 13(12):1083–1095

43. Pais R, Silaghi H, Silaghi AC, Rusu ML, Dumitrascu DL (2009) Metabolic syndrome and risk of subsequent colorectal cancer. World J Gastroenterol 15:5141–5148
44. Sung MK, Yeon JY, Park SY, Park JH, Choi MS (2011) Obesity-induced metabolic stresses in breast and colon cancer. Ann N Y Acad Sci 1229:61–68
45. Turnbaugh PJ, Ley RE, Mahowald MA, Magrini V, Mardis ER, Gordon JI (2006) An obesity-associated gut microbiome with increased capacity for energy harvest. Nature 444: 1027–1031
46. Backhed F, Manchester JK, Semenkovich CF, Gordon JI (2007) Mechanisms underlying the resistance to diet-induced obesity in germ-free mice. Proc Natl Acad Sci U S A 104: 979–984
47. Backhed F, Ley RE, Sonnenburg JL, Peterson DA, Gordon JI (2005) Host-bacterial mutualism in the human intestine. Science 307:1915–1920
48. Turnbaugh PJ, Ley RE, Hamady M, Fraser-Liggett CM, Knight R, Gordon JI (2007) The human microbiome project. Nature 449:804–810
49. Pull SL, Doherty JM, Mills JC, Gordon JI, Stappenbeck TS (2005) Activated macrophages are an adaptive element of the colonic epithelial progenitor niche necessary for regenerative responses to injury. Proc Natl Acad Sci U S A 102:99–104
50. Zoetendal EG, Rajilic-Stojanovic M, de Vos WM (2008) High-throughput diversity and functionality analysis of the gastrointestinal tract microbiota. Gut 57:1605–1615
51. Ivanov II, Frutos RL, Manel N et al (2008) Specific microbiota direct the differentiation of IL-17-producing T-helper cells in the mucosa of the small intestine. Cell Host Microbe 4:337–349
52. Mazmanian SK, Round JL, Kasper DL (2008) A microbial symbiosis factor prevents intestinal inflammatory disease. Nature 453:620–625
53. Marks DJ, Segal AW (2008) Innate immunity in inflammatory bowel disease: a disease hypothesis. J Pathol 214:260–266
54. Sollid LM, Johansen FE (2008) Animal models of inflammatory bowel disease at the dawn of the new genetics era. PLoS Med 5:e198
55. Strober W, Fuss IJ, Blumberg RS (2002) The immunology of mucosal models of inflammation. Annu Rev Immunol 20:495–549
56. Sartor RB (2004) Therapeutic manipulation of the enteric microflora in inflammatory bowel diseases: antibiotics, probiotics, and prebiotics. Gastroenterology 126:1620–1633
57. Winslet MC, Allan A, Poxon V, Youngs D, Keighley MR (1994) Faecal diversion for Crohn's colitis: a model to study the role of the faecal stream in the inflammatory process. Gut 35:236–242
58. Rutgeerts P, Goboes K, Peeters M et al (1991) Effect of faecal stream diversion on recurrence of Crohn's disease in the neoterminal ileum. Lancet 338:771–774
59. Gophna U, Sommerfeld K, Gophna S, Doolittle WF, Veldhuyzen van Zanten SJ (2006) Differences between tissue-associated intestinal microfloras of patients with Crohn's disease and ulcerative colitis. J Clin Microbiol 44:4136–4141
60. Manichanh C, Rigottier-Gois L, Bonnaud E et al (2006) Reduced diversity of faecal microbiota in Crohn's disease revealed by a metagenomic approach. Gut 55:205–211
61. Hamer HM, Jonkers D, Venema K, Vanhoutvin S, Troost FJ, Brummer RJ (2008) Review article: the role of butyrate on colonic function. Aliment Pharmacol Ther 27:104–119
62. Vinolo MA, Rodrigues HG, Nachbar RT, Curi R (2011) Regulation of inflammation by short chain fatty acids. Nutrients 3:858–876
63. Sokol H, Pigneur B, Watterlot L et al (2008) Faecalibacterium prausnitzii is an anti-inflammatory commensal bacterium identified by gut microbiota analysis of Crohn disease patients. Proc Natl Acad Sci U S A 105:16731–16736
64. Thibault R, Blachier F, Darcy-Vrillon B, de Coppet P, Bourreille A, Segain JP (2010) Butyrate utilization by the colonic mucosa in inflammatory bowel diseases: a transport deficiency. Inflamm Bowel Dis 16:684–695
65. Rabizadeh S, Rhee KJ, Wu S et al (2007) Enterotoxigenic bacteroides fragilis: a potential instigator of colitis. Inflamm Bowel Dis 13:1475–1483

66. Barnich N, Darfeuille-Michaud A (2007) Adherent-invasive Escherichia coli and Crohn's disease. Curr Opin Gastroenterol 23:16–20
67. Powell DW, Pinchuk IV, Saada JI, Chen X, Mifflin RC (2011) Mesenchymal cells of the intestinal lamina propria. Annu Rev Physiol 73:213–237
68. Hanson PJ, Moran AP, Butler K (2011) Paracellular permeability is increased by basal lipopolysaccharide in a primary culture of colonic epithelial cells; an effect prevented by an activator of toll-like receptor-2. Innate Immun 17:269–282
69. Testro AG, Visvanathan K (2009) Toll-like receptors and their role in gastrointestinal disease. J Gastroenterol Hepatol 24:943–954
70. Wyatt J, Vogelsang H, Hubl W, Waldhoer T, Lochs H (1993) Intestinal permeability and the prediction of relapse in Crohn's disease. Lancet 341:1437–1439
71. D'Inca R, Di Leo L, Corrao G et al (1999) Intestinal permeability test as a predictor of clinical course in Crohn's disease. Am J Gastroenterol 94:2956–2960
72. Peeters M, Geypens B, Claus D et al (1997) Clustering of increased small intestinal permeability in families with Crohn's disease. Gastroenterology 113:802–807
73. Teahon K, Smethurst P, Levi AJ, Menzies IS, Bjarnason I (1992) Intestinal permeability in patients with Crohn's disease and their first degree relatives. Gut 33:320–323
74. Clayburgh DR, Shen L, Turner JR (2004) A porous defense: the leaky epithelial barrier in intestinal disease. Lab Invest 84:282–291
75. Bruewer M, Luegering A, Kucharzik T et al (2003) Proinflammatory cytokines disrupt epithelial barrier function by apoptosis-independent mechanisms. J Immunol 171:6164–6172
76. Yang R, Han X, Uchiyama T et al (2003) IL-6 is essential for development of gut barrier dysfunction after hemorrhagic shock and resuscitation in mice. Am J Physiol Gastrointest Liver Physiol 285:G621–G629
77. McGuckin MA, Eri R, Simms LA, Florin TH, Radford-Smith G (2009) Intestinal barrier dysfunction in inflammatory bowel diseases. Inflamm Bowel Dis 15:100–113
78. Cario E, Podolsky DK (2000) Differential alteration in intestinal epithelial cell expression of toll-like receptor 3 (TLR3) and TLR4 in inflammatory bowel disease. Infect Immun 68:7010–7017
79. Yu LC, Wang JT, Wei SC, Ni YH (2012) Host-microbial interactions and regulation of intestinal epithelial barrier function: from physiology to pathology. World J Gastrointest Pathophysiol 3:27–43
80. Rakoff-Nahoum S, Medzhitov R (2007) Regulation of spontaneous intestinal tumorigenesis through the adaptor protein MyD88. Science 317:124–127
81. Greten FR, Eckmann L, Greten TF et al (2004) IKKbeta links inflammation and tumorigenesis in a mouse model of colitis-associated cancer. Cell 118:285–296
82. Dvorak HF (1986) Tumors: wounds that do not heal. Similarities between tumor stroma generation and wound healing. N Engl J Med 315:1650–1659
83. Grivennikov SI, Wang K, Mucida D et al (2012) Adenoma-linked barrier defects and microbial products drive IL-23/IL-17-mediated tumour growth. Nature 491:254–258
84. Sussman DA, Santaolalla R, Strobel S, Dheer R, Abreu MT (2012) Cancer in inflammatory bowel disease: lessons from animal models. Curr Opin Gastroenterol 28:327–333
85. Rakoff-Nahoum S, Medzhitov R (2008) Role of toll-like receptors in tissue repair and tumorigenesis. Biochemistry (Mosc) 73:555–561
86. Vannucci L, Stepankova R, Kozakova H, Fiserova A, Rossmann P, Tlaskalova-Hogenova H (2008) Colorectal carcinogenesis in germ-free and conventionally reared rats: different intestinal environments affect the systemic immunity. Int J Oncol 32:609–617
87. Reddy BS, Narisawa T, Wright P, Vukusich D, Weisburger JH, Wynder EL (1975) Colon carcinogenesis with azoxymethane and dimethylhydrazine in germ-free rats. Cancer Res 35:287–290
88. Backhed F, Ding H, Wang T et al (2004) The gut microbiota as an environmental factor that regulates fat storage. Proc Natl Acad Sci U S A 101:15718–15723
89. Turnbaugh PJ, Hamady M, Yatsunenko T et al (2009) A core gut microbiome in obese and lean twins. Nature 457:480–484

90. Turnbaugh PJ, Gordon JI (2009) The core gut microbiome, energy balance and obesity. J Physiol 587:4153–4158
91. Ley RE, Turnbaugh PJ, Klein S, Gordon JI (2006) Microbial ecology: human gut microbes associated with obesity. Nature 444:1022–1023
92. de La Serre CB, Ellis CL, Lee J, Hartman AL, Rutledge JC, Raybould HE (2010) Propensity to high-fat diet-induced obesity in rats is associated with changes in the gut microbiota and gut inflammation. Am J Physiol Gastrointest Liver Physiol 299:G440–G448
93. Erridge C, Attina T, Spickett CM, Webb DJ (2007) A high-fat meal induces low-grade endotoxemia: evidence of a novel mechanism of postprandial inflammation. Am J Clin Nutr 86:1286–1292
94. Cani PD, Amar J, Iglesias MA et al (2007) Metabolic endotoxemia initiates obesity and insulin resistance. Diabetes 56:1761–1772
95. Suzuki T, Hara H (2010) Dietary fat and bile juice, but not obesity, are responsible for the increase in small intestinal permeability induced through the suppression of tight junction protein expression in LETO and OLETF rats. Nutr Metab (Lond) 7:19
96. Fessler MB, Rudel LL, Brown JM (2009) Toll-like receptor signaling links dietary fatty acids to the metabolic syndrome. Curr Opin Lipidol 20:379–385
97. Fuss IJ, Becker C, Yang Z et al (2006) Both IL-12p70 and IL-23 are synthesized during active Crohn's disease and are down-regulated by treatment with anti-IL-12 p40 monoclonal antibody. Inflamm Bowel Dis 12:9–15
98. Neurath MF, Fuss I, Kelsall BL, Stuber E, Strober W (1995) Antibodies to interleukin 12 abrogate established experimental colitis in mice. J Exp Med 182:1281–1290
99. Liu Z, Geboes K, Heremans H et al (2001) Role of interleukin-12 in the induction of mucosal inflammation and abrogation of regulatory T cell function in chronic experimental colitis. Eur J Immunol 31:1550–1560
100. Kikly K, Liu L, Na S, Sedgwick JD (2006) The IL-23/Th(17) axis: therapeutic targets for autoimmune inflammation. Curr Opin Immunol 18:670–675
101. Strober W, Fuss IJ (2011) Proinflammatory cytokines in the pathogenesis of inflammatory bowel diseases. Gastroenterology 140:1756–1767
102. Vignali DA, Collison LW, Workman CJ (2008) How regulatory T cells work. Nat Rev Immunol 8:523–532
103. Korn T, Bettelli E, Oukka M, Kuchroo VK (2009) IL-17 and Th17 cells. Annu Rev Immunol 27:485–517
104. Aggarwal S, Gurney AL (2002) IL-17: prototype member of an emerging cytokine family. J Leukoc Biol 71:1–8
105. Tosolini M, Kirilovsky A, Mlecnik B et al (2011) Clinical impact of different classes of infiltrating T cytotoxic and helper cells (Th1, th2, treg, th17) in patients with colorectal cancer. Cancer Res 71:1263–1271
106. Boirivant M, Fuss IJ, Chu A, Strober W (1998) Oxazolone colitis: a murine model of T helper cell type 2 colitis treatable with antibodies to interleukin 4. J Exp Med 188:1929–1939
107. Heller F, Fuss IJ, Nieuwenhuis EE, Blumberg RS, Strober W (2002) Oxazolone colitis, a Th2 colitis model resembling ulcerative colitis, is mediated by IL-13-producing NK-T cells. Immunity 17:629–638
108. Fuss IJ, Heller F, Boirivant M et al (2004) Nonclassical CD1d-restricted NK T cells that produce IL-13 characterize an atypical Th2 response in ulcerative colitis. J Clin Invest 113:1490–1497
109. Strober W, Fuss I, Mannon P (2007) The fundamental basis of inflammatory bowel disease. J Clin Invest 117:514–521
110. Veldhoen M, Hocking RJ, Atkins CJ, Locksley RM, Stockinger B (2006) TGFbeta in the context of an inflammatory cytokine milieu supports de novo differentiation of IL-17-producing T cells. Immunity 24:179–189
111. Meng G, Zhang F, Fuss I, Kitani A, Strober W (2009) A mutation in the Nlrp3 gene causing inflammasome hyperactivation potentiates Th17 cell-dominant immune responses. Immunity 30:860–874

112. Kim S, Keku TO, Martin C et al (2008) Circulating levels of inflammatory cytokines and risk of colorectal adenomas. Cancer Res 68:323–328
113. Becker C, Fantini MC, Schramm C et al (2004) TGF-beta suppresses tumor progression in colon cancer by inhibition of IL-6 trans-signaling. Immunity 21:491–501
114. Grivennikov S, Karin E, Terzic J et al (2009) IL-6 and Stat3 are required for survival of intestinal epithelial cells and development of colitis-associated cancer. Cancer Cell 15:103–113
115. Karin M (2009) NF-kappaB as a critical link between inflammation and cancer. Cold Spring Harb Perspect Biol 1:a000141
116. Kaler P, Augenlicht L, Klampfer L (2009) Macrophage-derived IL-1beta stimulates Wnt signaling and growth of colon cancer cells: a crosstalk interrupted by vitamin D3. Oncogene 28:3892–3902
117. Liu W, Reinmuth N, Stoeltzing O et al (2003) Cyclooxygenase-2 is up-regulated by interleukin-1 beta in human colorectal cancer cells via multiple signaling pathways. Cancer Res 63:3632–3636
118. Corredor J, Yan F, Shen CC et al (2003) Tumor necrosis factor regulates intestinal epithelial cell migration by receptor-dependent mechanisms. Am J Physiol Cell Physiol 284:C953–C961
119. Pickert G, Neufert C, Leppkes M et al (2009) STAT3 links IL-22 signaling in intestinal epithelial cells to mucosal wound healing. J Exp Med 206:1465–1472
120. Mandal D, Levine AD (2010) Elevated IL-13Ralpha2 in intestinal epithelial cells from ulcerative colitis or colorectal cancer initiates MAPK pathway. Inflamm Bowel Dis 16:753–764
121. Francipane MG, Alea MP, Lombardo Y, Todaro M, Medema JP, Stassi G (2008) Crucial role of interleukin-4 in the survival of colon cancer stem cells. Cancer Res 68:4022–4025
122. Kanai T, Watanabe M, Hayashi A et al (2000) Regulatory effect of interleukin-4 and interleukin-13 on colon cancer cell adhesion. Br J Cancer 82:1717–1723
123. Barderas R, Bartolome RA, Fernandez-Acenero MJ, Torres S, Casal JI (2012) High expression of IL-13 receptor alpha2 in colorectal cancer is associated with invasion, liver metastasis, and poor prognosis. Cancer Res 72:2780–2790
124. Huber S, Gagliani N, Zenewicz LA et al (2012) IL-22BP is regulated by the inflammasome and modulates tumorigenesis in the intestine. Nature 491:259–263
125. Chae WJ, Bothwell AL (2011) IL-17F deficiency inhibits small intestinal tumorigenesis in ApcMin/+ mice. Biochem Biophys Res Commun 414:31–36
126. Chae WJ, Gibson TF, Zelterman D, Hao L, Henegariu O, Bothwell AL (2010) Ablation of IL-17A abrogates progression of spontaneous intestinal tumorigenesis. Proc Natl Acad Sci U S A 107:5540–5544
127. Koller FL, Hwang DG, Dozier EA, Fingleton B (2010) Epithelial interleukin-4 receptor expression promotes colon tumor growth. Carcinogenesis 31:1010–1017
128. Old LJ (1985) Tumor necrosis factor (TNF). Science 230:630–632
129. Baltgalvis KA, Berger FG, Pena MM, Davis JM, Muga SJ, Carson JA (2008) Interleukin-6 and cachexia in ApcMin/+ mice. Am J Physiol Regul Integr Comp Physiol 294:R393–R401
130. Mudter J, Neurath MF (2007) Il-6 signaling in inflammatory bowel disease: pathophysiological role and clinical relevance. Inflamm Bowel Dis 13:1016–1023
131. Fukata M, Chen A, Vamadevan AS et al (2007) Toll-like receptor-4 promotes the development of colitis-associated colorectal tumors. Gastroenterology 133:1869–1881
132. Sheehan KM, Sheahan K, O'Donoghue DP et al (1999) The relationship between cyclooxygenase-2 expression and colorectal cancer. JAMA 282:1254–1257
133. Arber N, Eagle CJ, Spicak J et al (2006) Celecoxib for the prevention of colorectal adenomatous polyps. N Engl J Med 355:885–895
134. Taketo MM (1998) COX-2 and colon cancer. Inflamm Res 47(Suppl 2):S112–S116
135. Pugh S, Thomas GA (1994) Patients with adenomatous polyps and carcinomas have increased colonic mucosal prostaglandin E2. Gut 35:675–678
136. Kawamori T, Uchiya N, Sugimura T, Wakabayashi K (2003) Enhancement of colon carcinogenesis by prostaglandin E2 administration. Carcinogenesis 24:985–990
137. Yang VW, Shields JM, Hamilton SR et al (1998) Size-dependent increase in prostanoid levels in adenomas of patients with familial adenomatous polyposis. Cancer Res 58:1750–1753

138. Chizzolini C, Brembilla NC (2009) Prostaglandin E2: igniting the fire. Immunol Cell Biol 87:510–511
139. Awane M, Andres PG, Li DJ, Reinecker HC (1999) NF-kappa B-inducing kinase is a common mediator of IL-17-, TNF-alpha-, and IL-1 beta-induced chemokine promoter activation in intestinal epithelial cells. J Immunol 162:5337–5344
140. Lee JW, Wang P, Kattah MG et al (2008) Differential regulation of chemokines by IL-17 in colonic epithelial cells. J Immunol 181:6536–6545
141. Coussens LM, Werb Z (2002) Inflammation and cancer. Nature 420:860–867
142. Maeda H, Akaike T (1998) Nitric oxide and oxygen radicals in infection, inflammation, and cancer. Biochemistry (Mosc) 63:854–865
143. Hussain SP, Amstad P, Raja K et al (2000) Increased p53 mutation load in noncancerous colon tissue from ulcerative colitis: a cancer-prone chronic inflammatory disease. Cancer Res 60:3333–3337
144. Xie J, Itzkowitz SH (2008) Cancer in inflammatory bowel disease. World J Gastroenterol 14:378–389
145. Liao J, Seril DN, Lu GG et al (2008) Increased susceptibility of chronic ulcerative colitis-induced carcinoma development in DNA repair enzyme Ogg1 deficient mice. Mol Carcinog 47:638–646
146. Meira LB, Bugni JM, Green SL et al (2008) DNA damage induced by chronic inflammation contributes to colon carcinogenesis in mice. J Clin Invest 118:2516–2525
147. Andoh A, Bamba S, Brittan M, Fujiyama Y, Wright NA (2007) Role of intestinal subepithelial myofibroblasts in inflammation and regenerative response in the gut. Pharmacol Ther 114:94–106
148. Andoh A, Bamba S, Fujiyama Y, Brittan M, Wright NA (2005) Colonic subepithelial myofibroblasts in mucosal inflammation and repair: contribution of bone marrow-derived stem cells to the gut regenerative response. J Gastroenterol 40:1089–1099
149. Andoh A, Fujino S, Okuno T, Fujiyama Y, Bamba T (2002) Intestinal subepithelial myofibroblasts in inflammatory bowel diseases. J Gastroenterol 37(Suppl 14):33–37
150. Andoh A, Zhang Z, Inatomi O et al (2005) Interleukin-22, a member of the IL-10 subfamily, induces inflammatory responses in colonic subepithelial myofibroblasts. Gastroenterology 129:969–984
151. Sugarman BJ, Aggarwal BB, Hass PE, Figari IS, Palladino MA Jr, Shepard HM (1985) Recombinant human tumor necrosis factor-alpha: effects on proliferation of normal and transformed cells in vitro. Science 230:943–945
152. Odegaard JI, Chawla A (2011) Alternative macrophage activation and metabolism. Annu Rev Pathol 6:275–297
153. Gregor MF, Hotamisligil GS (2011) Inflammatory mechanisms in obesity. Annu Rev Immunol 29:415–445
154. Cinti S, Mitchell G, Barbatelli G et al (2005) Adipocyte death defines macrophage localization and function in adipose tissue of obese mice and humans. J Lipid Res 46:2347–2355
155. Murano I, Barbatelli G, Parisani V et al (2008) Dead adipocytes, detected as crown-like structures, are prevalent in visceral fat depots of genetically obese mice. J Lipid Res 49:1562–1568
156. Fernandez-Riejos P, Najib S, Santos-Alvarez J et al (2010) Role of leptin in the activation of immune cells. Mediators Inflamm 2010:568343
157. Nishimura S, Manabe I, Nagasaki M et al (2009) CD8+ effector T cells contribute to macrophage recruitment and adipose tissue inflammation in obesity. Nat Med 15:914–920
158. Winer S, Chan Y, Paltser G et al (2009) Normalization of obesity-associated insulin resistance through immunotherapy. Nat Med 15:921–929
159. Feuerer M, Herrero L, Cipolletta D et al (2009) Lean, but not obese, fat is enriched for a unique population of regulatory T cells that affect metabolic parameters. Nat Med 15:930–939
160. Liu Z, Brooks RS, Ciappio ED et al (2012) Diet-induced obesity elevates colonic TNF-alpha in mice and is accompanied by an activation of Wnt signaling: a mechanism for obesity-associated colorectal cancer. J Nutr Biochem 23(10):1207–1213

161. Park SY, Kim JS, Seo YR, Sung MK (2012) Effects of diet-induced obesity on colitis-associated colon tumor formation in A/J mice. Int J Obes (Lond) 36:273–280
162. Tataranni PA, Ortega E (2005) A burning question: does an adipokine-induced activation of the immune system mediate the effect of overnutrition on type 2 diabetes? Diabetes 54:917–927
163. MacDougald OA, Burant CF (2007) The rapidly expanding family of adipokines. Cell Metab 6:159–161
164. Trayhurn P, Bing C (2006) Appetite and energy balance signals from adipocytes. Philos Trans R Soc Lond B Biol Sci 361:1237–1249
165. Faggioni R, Feingold KR, Grunfeld C (2001) Leptin regulation of the immune response and the immunodeficiency of malnutrition. FASEB J 15:2565–2571
166. Drew JE, Farquharson AJ, Padidar S et al (2007) Insulin, leptin, and adiponectin receptors in colon: regulation relative to differing body adiposity independent of diet and in response to dimethylhydrazine. Am J Physiol Gastrointest Liver Physiol 293:G682–G691
167. Hardwick JC, van den Brink GR, Offerhaus GJ, Van Deventer SJ, Peppelenbosch MP (2001) Leptin is a growth factor for colonic epithelial cells. Gastroenterology 121:79–90
168. Considine RV, Sinha MK, Heiman ML et al (1996) Serum immunoreactive-leptin concentrations in normal-weight and obese humans. N Engl J Med 334:292–295
169. Stattin P, Palmqvist R, Soderberg S et al (2003) Plasma leptin and colorectal cancer risk: a prospective study in Northern Sweden. Oncol Rep 10:2015–2021
170. Stattin P, Lukanova A, Biessy C et al (2004) Obesity and colon cancer: does leptin provide a link? Int J Cancer 109:149–152
171. Chia VM, Newcomb PA, Lampe JW et al (2007) Leptin concentrations, leptin receptor polymorphisms, and colorectal adenoma risk. Cancer Epidemiol Biomarkers Prev 16:2697–2703
172. Calle EE, Kaaks R (2004) Overweight, obesity and cancer: epidemiological evidence and proposed mechanisms. Nat Rev Cancer 4:579–591
173. Harriss DJ, Atkinson G, George K et al (2009) Lifestyle factors and colorectal cancer risk (1): systematic review and meta-analysis of associations with body mass index. Colorectal Dis 11:547–563
174. Frezza EE, Wachtel MS, Chiriva-Internati M (2006) Influence of obesity on the risk of developing colon cancer. Gut 55:285–291
175. Rutter M, Saunders B, Wilkinson K et al (2004) Severity of inflammation is a risk factor for colorectal neoplasia in ulcerative colitis. Gastroenterology 126:451–459
176. Drew JE (2012) Molecular mechanisms linking adipokines to obesity-related colon cancer: focus on leptin. Proc Nutr Soc 71:175–180
177. Hoda MR, Keely SJ, Bertelsen LS, Junger WG, Dharmasena D, Barrett KE (2007) Leptin acts as a mitogenic and antiapoptotic factor for colonic cancer cells. Br J Surg 94:346–354
178. Ogunwobi OO, Beales IL (2007) The anti-apoptotic and growth stimulatory actions of leptin in human colon cancer cells involves activation of JNK mitogen activated protein kinase, JAK2 and PI3 kinase/Akt. Int J Colorectal Dis 22:401–409
179. Ogunwobi OO, Beales IL (2007) Cyclo-oxygenase-independent inhibition of apoptosis and stimulation of proliferation by leptin in human colon cancer cells. Dig Dis Sci 52:1934–1945
180. Abolhassani M, Aloulou N, Chaumette MT et al (2008) Leptin receptor-related immune response in colorectal tumors: the role of colonocytes and interleukin-8. Cancer Res 68:9423–9432
181. Aparicio T, Kotelevets L, Tsocas A et al (2005) Leptin stimulates the proliferation of human colon cancer cells in vitro but does not promote the growth of colon cancer xenografts in nude mice or intestinal tumorigenesis in Apc(Min/+) mice. Gut 54:1136–1145
182. Ealey KN, Lu S, Archer MC (2008) Development of aberrant crypt foci in the colons of ob/ob and db/db mice: evidence that leptin is not a promoter. Mol Carcinog 47:667–677
183. Endo H, Hosono K, Uchiyama T et al (2011) Leptin acts as a growth factor for colorectal tumours at stages subsequent to tumour initiation in murine colon carcinogenesis. Gut 60:1363–1371

184. Fenton JI, Hursting SD, Perkins SN, Hord NG (2006) Interleukin-6 production induced by leptin treatment promotes cell proliferation in an Apc (Min/+) colon epithelial cell line. Carcinogenesis 27:1507–1515
185. Padidar S, Farquharson AJ, Williams LM et al (2011) Leptin up-regulates pro-inflammatory cytokines in discrete cells within mouse colon. J Cell Physiol 226:2123–2130
186. Yang SK, Eckmann L, Panja A, Kagnoff MF (1997) Differential and regulated expression of C-X-C, C-C, and C-chemokines by human colon epithelial cells. Gastroenterology 113:1214–1223
187. Waugh DJ, Wilson C (2008) The interleukin-8 pathway in cancer. Clin Cancer Res 14:6735–6741
188. Palacio A, Lopez M, Perez-Bravo F, Monkeberg F, Schlesinger L (2002) Leptin levels are associated with immune response in malnourished infants. J Clin Endocrinol Metab 87:3040–3046
189. Busso N, So A, Chobaz-Peclat V et al (2002) Leptin signaling deficiency impairs humoral and cellular immune responses and attenuates experimental arthritis. J Immunol 168:875–882
190. Matarese G, Di GA, Sanna V et al (2001) Requirement for leptin in the induction and progression of autoimmune encephalomyelitis. J Immunol 166:5909–5916
191. Stofkova A (2009) Leptin and adiponectin: from energy and metabolic dysbalance to inflammation and autoimmunity. Endocr Regul 43:157–168
192. John BJ, Irukulla S, Abulafi AM, Kumar D, Mendall MA (2006) Systematic review: adipose tissue, obesity and gastrointestinal diseases. Aliment Pharmacol Ther 23:1511–1523
193. Tartaglia LA, Dembski M, Weng X et al (1995) Identification and expression cloning of a leptin receptor, OB-R. Cell 83:1263–1271
194. Sanchez-Margalet V, Martin-Romero C, Santos-Alvarez J, Goberna R, Najib S, Gonzalez-Yanes C (2003) Role of leptin as an immunomodulator of blood mononuclear cells: mechanisms of action. Clin Exp Immunol 133:11–19
195. McLean MH, Murray GI, Stewart KN et al (2011) The inflammatory microenvironment in colorectal neoplasia. PLoS One 6:e15366
196. Loffreda S, Yang SQ, Lin HZ et al (1998) Leptin regulates proinflammatory immune responses. FASEB J 12:57–65
197. Kiguchi N, Maeda T, Kobayashi Y, Fukazawa Y, Kishioka S (2009) Leptin enhances CC-chemokine ligand expression in cultured murine macrophage. Biochem Biophys Res Commun 384:311–315
198. Mattioli B, Giordani L, Quaranta MG, Viora M (2009) Leptin exerts an anti-apoptotic effect on human dendritic cells via the PI3K-Akt signaling pathway. FEBS Lett 583:1102–1106
199. Mattioli B, Straface E, Quaranta MG, Giordani L, Viora M (2005) Leptin promotes differentiation and survival of human dendritic cells and licenses them for Th1 priming. J Immunol 174:6820–6828
200. Mattioli B, Straface E, Matarrese P et al (2008) Leptin as an immunological adjuvant: enhanced migratory and CD8+ T cell stimulatory capacity of human dendritic cells exposed to leptin. FASEB J 22:2012–2022
201. Lord GM, Matarese G, Howard JK, Baker RJ, Bloom SR, Lechler RI (1998) Leptin modulates the T-cell immune response and reverses starvation-induced immunosuppression. Nature 394:897–901
202. Sanchez-Margalet V, Martin-Romero C (2001) Human leptin signaling in human peripheral blood mononuclear cells: activation of the JAK-STAT pathway. Cell Immunol 211:30–36
203. Martin-Romero C, Sanchez-Margalet V (2001) Human leptin activates PI3K and MAPK pathways in human peripheral blood mononuclear cells: possible role of Sam68. Cell Immunol 212:83–91
204. Martin-Romero C, Santos-Alvarez J, Goberna R, Sanchez-Margalet V (2000) Human leptin enhances activation and proliferation of human circulating T lymphocytes. Cell Immunol 199:15–24

205. La CA, Matarese G (2004) The weight of leptin in immunity. Nat Rev Immunol 4:371–379
206. Batra A, Okur B, Glauben R et al (2010) Leptin: a critical regulator of CD4+ T-cell polarization in vitro and in vivo. Endocrinology 151:56–62
207. Lostao MP, Urdaneta E, Martinez-Anso E, Barber A, Martinez JA (1998) Presence of leptin receptors in rat small intestine and leptin effect on sugar absorption. FEBS Lett 423:302–306
208. Tilg H, Moschen AR (2006) Adipocytokines: mediators linking adipose tissue, inflammation and immunity. Nat Rev Immunol 6:772–783
209. Dalamaga M, Diakopoulos KN, Mantzoros CS (2012) The role of adiponectin in cancer: a review of current evidence. Endocr Rev 33:547–594
210. Simpson F, Whitehead JP (2010) Adiponectin—it's all about the modifications. Int J Biochem Cell Biol 42:785–788
211. Wang Y, Lam KS, Chan L et al (2006) Post-translational modifications of the four conserved lysine residues within the collagenous domain of adiponectin are required for the formation of its high molecular weight oligomeric complex. J Biol Chem 281:16391–16400
212. Renehan AG, Tyson M, Egger M, Heller RF, Zwahlen M (2008) Body-mass index and incidence of cancer: a systematic review and meta-analysis of prospective observational studies. Lancet 371:569–578
213. Ziemke F, Mantzoros CS (2010) Adiponectin in insulin resistance: lessons from translational research. Am J Clin Nutr 91:258S–261S
214. Berg AH, Combs TP, Scherer PE (2002) ACRP30/adiponectin: an adipokine regulating glucose and lipid metabolism. Trends Endocrinol Metab 13:84–89
215. Cnop M, Havel PJ, Utzschneider KM et al (2003) Relationship of adiponectin to body fat distribution, insulin sensitivity and plasma lipoproteins: evidence for independent roles of age and sex. Diabetologia 46:459–469
216. Wei EK, Giovannucci E, Fuchs CS, Willett WC, Mantzoros CS (2005) Low plasma adiponectin levels and risk of colorectal cancer in men: a prospective study. J Natl Cancer Inst 97:1688–1694
217. Nishihara T, Matsuda M, Araki H et al (2006) Effect of adiponectin on murine colitis induced by dextran sulfate sodium. Gastroenterology 131:853–861
218. Yoneda K, Tomimoto A, Endo H et al (2008) Expression of adiponectin receptors, AdipoR1 and AdipoR2, in normal colon epithelium and colon cancer tissue. Oncol Rep 20:479–483
219. Otani K, Kitayama J, Yasuda K et al (2010) Adiponectin suppresses tumorigenesis in Apc(Min)(/+) mice. Cancer Lett 288:177–182
220. Kim AY, Lee YS, Kim KH et al (2010) Adiponectin represses colon cancer cell proliferation via AdipoR1- and -R2-mediated AMPK activation. Mol Endocrinol 24:1441–1452
221. Byeon JS, Jeong JY, Kim MJ et al (2010) Adiponectin and adiponectin receptor in relation to colorectal cancer progression. Int J Cancer 127:2758–2767
222. Ogunwobi OO, Beales IL (2006) Adiponectin stimulates proliferation and cytokine secretion in colonic epithelial cells. Regul Pept 134:105–113
223. Lang K, Ratke J (2009) Leptin and adiponectin: new players in the field of tumor cell and leukocyte migration. Cell Commun Signal 7:27
224. Sugiyama M, Takahashi H, Hosono K et al (2009) Adiponectin inhibits colorectal cancer cell growth through the AMPK/mTOR pathway. Int J Oncol 34:339–344
225. Zakikhani M, Dowling RJ, Sonenberg N, Pollak MN (2008) The effects of adiponectin and metformin on prostate and colon neoplasia involve activation of AMP-activated protein kinase. Cancer Prev Res (Phila) 1:369–375
226. Habeeb BS, Kitayama J, Nagawa H (2011) Adiponectin supports cell survival in glucose deprivation through enhancement of autophagic response in colorectal cancer cells. Cancer Sci 102:999–1006
227. Saxena A, Chumanevich A, Fletcher E et al (1822) Adiponectin deficiency: role in chronic inflammation induced colon cancer. Biochim Biophys Acta 2012:527–536
228. Fujisawa T, Endo H, Tomimoto A et al (2008) Adiponectin suppresses colorectal carcinogenesis under the high-fat diet condition. Gut 57:1531–1538

229. Nishihara T, Baba M, Matsuda M et al (2008) Adiponectin deficiency enhances colorectal carcinogenesis and liver tumor formation induced by azoxymethane in mice. World J Gastroenterol 14:6473–6480
230. Fenton JI, Birmingham JM, Hursting SD, Hord NG (2008) Adiponectin blocks multiple signaling cascades associated with leptin-induced cell proliferation in Apc Min/+ colon epithelial cells. Int J Cancer 122:2437–2445
231. Fayad R, Pini M, Sennello JA et al (2007) Adiponectin deficiency protects mice from chemically induced colonic inflammation. Gastroenterology 132:601–614
232. Uji Y, Yamamoto H, Tsuchihashi H et al (2009) Adiponectin deficiency is associated with severe polymicrobial sepsis, high inflammatory cytokine levels, and high mortality. Surgery 145:550–557
233. Kamada Y, Tamura S, Kiso S et al (2003) Enhanced carbon tetrachloride-induced liver fibrosis in mice lacking adiponectin. Gastroenterology 125:1796–1807
234. Pang TT, Narendran P (2008) The distribution of adiponectin receptors on human peripheral blood mononuclear cells. Ann N Y Acad Sci 1150:143–145
235. Okamoto Y, Christen T, Shimizu K et al (2009) Adiponectin inhibits allograft rejection in murine cardiac transplantation. Transplantation 88:879–883
236. Ohashi K, Parker JL, Ouchi N et al (2010) Adiponectin promotes macrophage polarization toward an anti-inflammatory phenotype. J Biol Chem 285:6153–6160
237. Furukawa K, Hori M, Ouchi N et al (2004) Adiponectin down-regulates acyl-coenzyme A: cholesterol acyltransferase-1 in cultured human monocyte-derived macrophages. Biochem Biophys Res Commun 317:831–836
238. Wulster-Radcliffe MC, Ajuwon KM, Wang J, Christian JA, Spurlock ME (2004) Adiponectin differentially regulates cytokines in porcine macrophages. Biochem Biophys Res Commun 316:924–929
239. Wolf AM, Wolf D, Rumpold H, Enrich B, Tilg H (2004) Adiponectin induces the anti-inflammatory cytokines IL-10 and IL-1RA in human leukocytes. Biochem Biophys Res Commun 323:630–635
240. Folco EJ, Rocha VZ, Lopez-Ilasaca M, Libby P (2009) Adiponectin inhibits pro-inflammatory signaling in human macrophages independent of interleukin-10. J Biol Chem 284:25569–25575
241. Saijo S, Nagata K, Nakano Y, Tobe T, Kobayashi Y (2005) Inhibition by adiponectin of IL-8 production by human macrophages upon coculturing with late apoptotic cells. Biochem Biophys Res Commun 334:1180–1183
242. Yokota T, Oritani K, Takahashi I et al (2000) Adiponectin, a new member of the family of soluble defense collagens, negatively regulates the growth of myelomonocytic progenitors and the functions of macrophages. Blood 96:1723–1732
243. Okamoto Y, Folco EJ, Minami M et al (2008) Adiponectin inhibits the production of CXC receptor 3 chemokine ligands in macrophages and reduces T-lymphocyte recruitment in atherogenesis. Circ Res 102:218–225
244. Wilk S, Scheibenbogen C, Bauer S et al (2011) Adiponectin is a negative regulator of antigen-activated T cells. Eur J Immunol 41:2323–2332
245. Keller JJ, Giardiello FM (2003) Chemoprevention strategies using NSAIDs and COX-2 inhibitors. Cancer Biol Ther 2:S140–S149
246. Yamamoto Y, Yin MJ, Lin KM, Gaynor RB (1999) Sulindac inhibits activation of the NF-kappaB pathway. J Biol Chem 274:27307–27314
247. Lanas A (2009) Nonsteroidal antiinflammatory drugs and cyclooxygenase inhibition in the gastrointestinal tract: a trip from peptic ulcer to colon cancer. Am J Med Sci 338:96–106
248. Waldner MJ, Neurath MF (2009) Colitis-associated cancer: the role of T cells in tumor development. Semin Immunopathol 31:249–256
249. Rutgeerts P, Vermeire S, Van AG (2009) Biological therapies for inflammatory bowel diseases. Gastroenterology 136:1182–1197
250. Kim YS, Milner JA (2007) Dietary modulation of colon cancer risk. J Nutr 137:2576S–2579S
251. Tuan D, Solomon W, Li Q, London IM (1985) The "beta-like-globin" gene domain in human erythroid cells. Proc Natl Acad Sci U S A 82:6384–6388

Chapter 8
Obesity, Inflammation, and Breast Cancer

Neil M. Iyengar, Patrick G. Morris, Clifford A. Hudis, and Andrew J. Dannenberg

Abstract Obesity, which is rising in incidence worldwide, is important with regard to the treatment of breast cancer, disease progression, and carcinogenesis. Obesity is a risk factor for the development of hormone receptor-positive breast cancer in postmenopausal women and is associated with reduced benefits from treatment. Furthermore, irrespective of breast cancer subtype, obesity is associated with worse outcomes after diagnosis. There is increasing evidence of specific biological underpinnings for these observations, including higher circulating estrogen levels, insulin resistance, altered levels of adipokines, and the consequences of chronic in-breast inflammation. Increasing adiposity also has important implications for local therapy including surgery and radiotherapy. This chapter reviews the complex interactions between obesity and breast cancer.

Abbreviations

AMPK	5′-Adenosine monophosphate-activated protein kinase
BCSS	Breast cancer-specific survival
BMI	Body mass index
CEBP-α	CCAAT/enhancer binding protein-α
CI	Confidence interval
CLS	Crown-like structures

N.M. Iyengar, M.D. • P.G. Morris, M.D. • C.A. Hudis, M.D.
Memorial Sloan-Kettering Cancer Center, New York, NY, USA
e-mail: iyengarn@mskcc.org; morrisp1@mskcc.org; hudisc@mskcc.org

A.J. Dannenberg, M.D. (✉)
Department of Medicine, Weill Cornell Medical College, 525 East 68th Street, Room F-206, New York, NY 10065, USA
e-mail: ajdannen@med.cornell.edu

A.J. Dannenberg and N.A. Berger (eds.), *Obesity, Inflammation and Cancer*, Energy Balance and Cancer 7, DOI 10.1007/978-1-4614-6819-6_8,

DFS	Disease-free survival
ER	Estrogen receptor
FFA	Free fatty acid
HER2	Human epidermal growth factor receptor-2
HR	Hazard ratio
hsCRP	High-sensitivity C-reactive protein
IGF-1	Insulin-like growth factor-1
IL-1α	Interleukin-1α
IL-1β	Interleukin-1β
IL-6	Interleukin-6
NSAID	Nonsteroidal anti-inflammatory drug
OR	Odds ratio
OS	Overall survival
PGE_2	Prostaglandin E_2
PPAR-γ	Peroxisome proliferator-activated receptor-γ
PR	Progesterone receptor
RR	Relative risk
SERM	Selective estrogen receptor modulator
SHBG	Sex hormone-binding globulin
TLR-2	Toll-like receptor-2
TLR-4	Toll-like receptor-4
TNF-α	Tumor necrosis factor-α
TRAM	Transverse rectus abdominis myocutaneous

8.1 Introduction

Worldwide, the incidence of overweight and obesity are increasing. One widely used metric of adiposity is the body mass index (BMI), calculated using the formula [(weight in kilograms)/(height in meters)^2]. This is a continuous variable but can be divided into categories. According to this scale, underweight is defined as BMI less than 18.5 kg/m^2, normal as 18.5–24.9 kg/m^2, overweight as BMI 25.0–29.9 kg/m^2, and obese as BMI 30 kg/m^2 or greater. Almost 1.5 billion adults in the world are overweight (BMI $\geq$ 25 kg/m^2), and more than 500 million of these are obese (BMI $\geq$ 30 kg/m^2) [1, 2]. In the United States, the majority (>60 %) of adults are overweight, and obesity rates are increasing [1, 2]. Recent projections estimate that as much as 65 % of the US population in several states may be obese by 2030 [3].

Obesity has a range of clinical consequences and is a risk factor for the development of several epithelial malignancies, including estrogen receptor (ER)- and progesterone receptor (PR)-positive ("hormone sensitive") breast cancer in postmenopausal women, the most common presentation of the disease [4–8]. Obesity and overweight present a number of challenges for the management of breast cancer, including potential limitations on diagnostic imaging, as well as changes in local and systemic therapeutic options. In addition, obesity is a negative prognostic variable irrespective of breast cancer subtype. In this chapter, we discuss the complex interactions between obesity and breast cancer in terms of epidemiology, biology, and treatment.

8.2 Epidemiology of Obesity and Breast Cancer

In epidemiologic studies, obesity has been linked to several subtypes of breast cancer. This is most consistently seen with regard to the most common presentation of breast cancer, hormone-sensitive (ER and/or PR positive) disease in postmenopausal women [4–10]. In a pooled analysis of eight prospective studies including over 2,000 women, the relative risk (RR) of developing breast cancer in postmenopausal females associated with each 5 kg/m^2 increase in BMI was 1.19 (95 % confidence interval [CI], 1.05–1.34) [8]. A similar effect was also observed in a larger pooled analysis of several prospective cohort studies including over 300,000 women in which the RR of developing breast cancer was 1.43 (95 % CI, 1.21–1.67) specifically in overweight postmenopausal women with BMI 27–29 kg/m^2 compared to those who were leaner (<21 kg/m^2) (Table 8.1) [6].

Normally, following menopause, circulating estradiol levels decline rapidly as the ovaries no longer produce appreciable amounts of estrogen. Instead, lower levels of estrogen are produced noncyclically from androgen precursors peripherally, primarily in the adipose tissue [11]. The critical step in this process is catalyzed by the enzyme aromatase, a cytochrome P450 enzyme encoded by the *CYP19* gene. Although ER-positive breast cancer makes up the largest proportion of the disease in all ages, the incidence of this subtype increases with age [12]. For example, in a

Table 8.1 Selected epidemiologic studies reporting significant association between overweight/obesity and the development of breast cancer in postmenopausal women

References	Subjects	Number	BMI in "normal" category (kg/m^2)	BMI in "overweight and obese" category (kg/m^2)	Magnitude of effect	95 % confidence interval	*P* value
Trentham-Dietz et al. [7]	Case Control	6,548 9,057	12.9–21.2	27.6–57.1	1.41[a]	1.25–1.60	<0.001
Van den Brandt et al. [6]	Cohort (pooled)	337,819	<21	27 to <29	1.43[b]	1.21–1.67	0.001
Key et al. [8]	Case Control (pooled)	624 1,669	<22.5	≥30	1.36[b]	1.00–1.85	0.004
La Vecchia et al. [10]	Case Control (pooled)	3,108 2,664	<21.8	>28.4	1.4[a]	1.2–1.7	<0.001

[a]Odds ratio
[b]Relative risk

population-based study in the United States, approximately 55 % of breast cancer in women in their 20s was ER positive, rising to almost 85 % in women in their 80s [12]. This leads to a paradox, whereby ER-positive breast cancer incidence increases with age despite the falling circulating levels of estrogen. Increasing adiposity with advancing age has been suggested as one of the underlying mechanisms to explain this phenomenon, but the specifics of this effect are not fully elucidated.

8.3 Obesity and Breast Carcinogenesis

Although the mechanisms remain incompletely understood, the link between obesity and ER-positive breast cancer in postmenopausal women is thought to be partially due to two related factors: increased adipose tissue and elevated aromatase expression in adipose tissue [8, 13]. Given that the incidence of ER-positive breast cancer increases with age as discussed above, it is not surprising that the majority of postmenopausal breast cancers associated with obesity are ER positive [4, 5]. In obesity, growth of hormone-sensitive tumors within the breast is dependent on locally produced estrogens and, possibly, ligand-independent activation of estrogen receptor-α or other paracrine-mediated effects that may drive carcinogenesis. If, as noted, peripheral aromatization in adipose tissue is thought to be largely responsible for estrogen production after menopause, the question remains: where do the estrogens that drive breast cancer in this age group arise? [11]. If they are produced locally, this has important implications for breast carcinogenesis as estrogens can have stimulatory effects on cell growth within the breast but also have genotoxic effects and stimulate angiogenesis (Fig. 8.1). These actions suggest a broader potential role for estrogen than merely stimulating ER in hormone-sensitive breast tumors.

Additional clinical observations suggest other linkages between obesity and breast carcinogenesis. In premenopausal women, BMI>30 kg/m^2 has also been associated with an increased risk of hormone receptor-negative breast cancers [14, 15]. This association could be related to estrogen-independent pathways such as obesity-associated altered glucose metabolism and increased levels of insulin and bioavailable insulin-like growth factor-1 (IGF-1) (Fig. 8.1). Furthermore, increased leptin and decreased adiponectin can lead to altered cellular proliferation and survival. Finally, obesity is now recognized to be a systemic inflammatory condition with known associations with elevated proinflammatory mediators that promote tumorigenesis and growth [5, 16, 17]. This complex biological interaction between obesity and breast cancer is reviewed in greater detail below.

8.3.1 Dysfunctional Adipose Tissue and Metabolism

Although perhaps not conventionally considered as such, adipose tissue is an active endocrine organ. Adipocytes secrete a variety of factors, known as adipokines, including leptin and adiponectin—which are known to have angiogenic properties—as well as several growth factors [18].

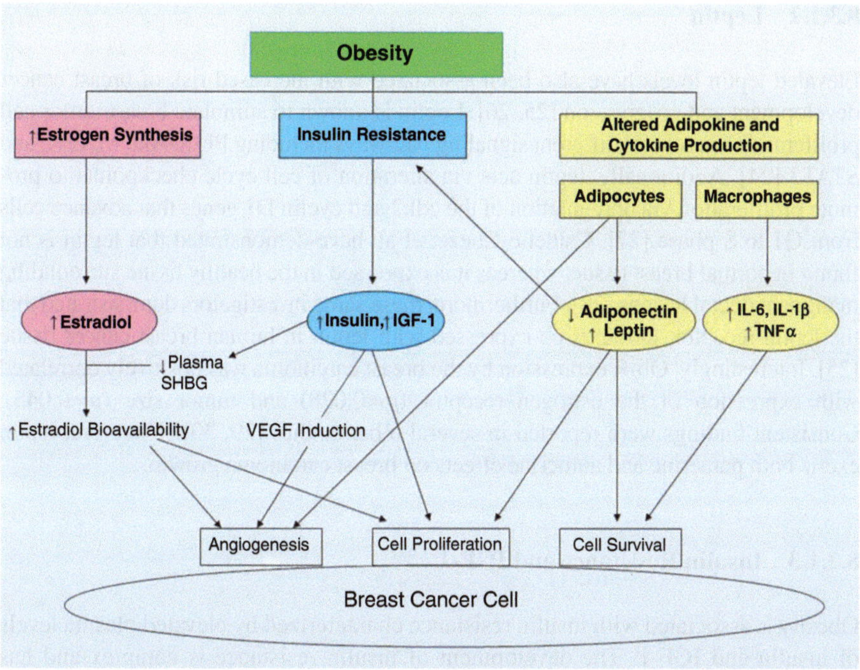

Fig. 8.1 Relationship between obesity and promotion of breast cancer. *IGF-1* insulin-like growth factor-1; *IL-6* interleukin-6; *IL-1β* interleukin-1β; *TNF-α* tumor necrosis factor-α; *SHBG* sex hormone-binding globulin; *VEGF* vascular endothelial growth factor (Sinicrope F et al.: J Clin Oncol Vol. 72(9), 2010:4–7. Reprinted with permission. © 2010 American Society of Clinical Oncology. All rights reserved)

8.3.1.1 Adiponectin

Metabolic conditions, such as obesity, that alter adipose stores modify adipokine production. Increased adiposity is associated with enlarged adipocytes [5] and the enhanced production of chemotactic factors that attract myeloid cells such as macrophages into adipose tissue [19]. This promotes an inflammatory cascade leading to decreased production of adiponectin, release of free fatty acids (FFAs), and the development of insulin resistance, characterized, clinically, by hyperinsulinemia [20]. Decreased levels of adiponectin, an adipocyte-secreted hormone, are associated with hyperinsulinemic states including obesity and type II diabetes mellitus [21]. Furthermore, decreased adiponectin levels have been associated with increasing breast cancer risk [22]. The mechanism underlying this relationship is not completely understood but adiponectin has been shown to inhibit the growth of several breast cancer cell lines in vitro [23]. Additionally, adiponectin activates the 5′-adenosine monophosphate-activated protein kinase (AMPK) pathway leading to upregulation of p53 and p21, important regulators of the cell cycle and apoptosis [5]. Other signaling pathways downstream of the adiponectin receptors, AdipoR1 and AdipoR2, have been implicated in breast cancer cell growth including regulation of aromatase expression, and the PTEN/PI3K/mTOR and MAPK pathways [24].

8.3.1.2 Leptin

Elevated leptin levels have also been associated with increased risk of breast cancer development and progression [25, 26]. Leptin is known to stimulate breast tumor cell proliferation via several different signaling pathways including PI3K/Akt, MAPK, and STAT3 [24]. Additionally, leptin acts via alteration of cell cycle checkpoints to promote proliferation via upregulation of the cdk2 and cyclin D1 genes that advance cells from G1 to S phase [27]. Caldefie-Chezet et al. have demonstrated that leptin is not found in normal breast tissue, whereas it is expressed in the healthy tissue surrounding malignant ductal lesions [28]. Furthermore, these same investigators demonstrated that the leptin receptor, Ob-R, is co-expressed with leptin in human breast cancer tissue [25]. Interestingly, Ob-R expression by the breast carcinoma was positively correlated with expression of the estrogen receptor ($p=0.028$) and tumor size ($p=0.045$). Consistent findings were reported in several other studies [29, 30]. Therefore, leptin exerts both paracrine and autocrine effects on breast carcinoma growth.

8.3.1.3 Insulin Resistance and IGF-1

Obesity is associated with insulin resistance characterized by elevated plasma levels of insulin and IGF-1. The development of insulin resistance is complex and has been implicated, in part, in the relationship between obesity and breast cancer. In a cohort of over 500 women with early-stage breast cancer, elevated fasting insulin was found to be associated with distant breast cancer recurrence (hazard ratio [HR] 2.0; 95 % CI, 1.2–3.3) and death (HR 3.1; 95 % CI, 1.7–5.7) [31]. Additionally, insulin level was correlated with BMI ($p<0.001$). Insulin has important mitogenic effects on breast cancer cells that have not yet been completely elucidated. Binding of insulin and IGF-1 to their respective receptors promotes cell proliferation and inhibits apoptosis via downstream signaling effects of the PI3K/Akt and Ras/Raf/MAPK systems [32]. Interestingly, IGF-1 has been found to be expressed at higher levels in ER-positive tumors than in cancers that do not express the estrogen receptor [33]. Additionally, IGF-1 and insulin, to a lesser degree, have been shown to stimulate aromatase activity in adipose tissue [34]. Beyond this paracrine-mediated effect on aromatase, increased serum insulin levels can stimulate androgen synthesis by the ovaries and reduce hepatic synthesis of sex hormone-binding globulin (SHBG) [5]. As suggested by the name, SHBG is a protein synthesized in the liver that tightly binds testosterone, dihydrotestosterone, and estradiol and transports these hormones in the blood in a biologically inactive state. Changes in SHBG levels affect the amount of hormone available to cells that carry the respective receptors. In a sample of over 1,000 postmenopausal women randomly selected from the Melbourne Collaborative Cohort Study, BMI was positively correlated with increased levels of plasma estradiol and associated with decreased SHBG plasma concentrations [35]. Therefore, IGF-1 and insulin, which are increased in obese patients, are involved in multiple complex systems that are associated with breast carcinogenesis and tumor progression.

8.3.2 Obesity and Inflammation

A primary function of white adipose tissue is to store energy as lipid. Excess adipose tissue is associated with elevated serum triglyceride levels. Increased adiposity is accompanied by adipocyte hyperplasia and hypertrophy, which in turn lead to adipocyte stretch and death. Indeed, increasing BMI has been correlated with adipocyte hypertrophy in the breast [36]. It is well known that obesity is associated with chronic, subclinical inflammation [5, 17, 37–39]. There is evidence that adipocyte hypertrophy leads to cell wall stretch, and subsequently, the adipocyte releases several proinflammatory cytokine mediators, including tumor necrosis factor-α (TNF-α) and interleukin-6 (IL-6). This attracts and activates immune cells including macrophages [40]. Additionally, lipolysis occurs within the stretched adipocytes resulting in release of FFAs that stimulate multiple inflammatory pathways in the activated immune cells including activation of the transcription factor NF-κB [41]. Activation of NF-κB, a contributor to inflamed adipose tissue, is mediated by Toll-like receptor-4 (TLR-4) [42].

The interactions between adipocytes and immune cells in the stromal vascular fraction of adipose tissue are complex and incompletely understood. Dead and dying adipocytes, described above, interact with macrophages and stimulate production of several proinflammatory cytokines including TNF-α, interleukin-1β (IL-1β), and IL-6 [17, 38]. Increased circulating levels of these cytokines are commonly found in obese women and have been associated with breast cancer development and progression [43–45]. In addition, FFAs are believed to activate macrophages residing in the adipose tissue via a series of receptors including Toll-like receptor 2 (TLR-2) and TLR-4 [46]. Furthermore, macrophage-adipocyte associations are histologically apparent as crown-like structures (CLS) in which macrophages encircle the dead adipocyte [47]. These CLS have been observed in the mammary gland adipose tissue of obese mice, and their presence has been associated with NF-κB-induced production of several proinflammatory mediators including TNF-α, IL-1β, and Cox-2 [48]. In both dietary and genetic mouse models of obesity, the number of CLS has been positively correlated with body weight, in addition to tissue levels of proinflammatory mediators. Importantly, the presence of CLS was associated with increased aromatase levels and activity. These findings have been translated to the human, and CLS of the human breast (CLS-B) have been recently described (Fig. 8.2) [36, 49]. The severity of breast inflammation, manifested as the CLS-B index (proportion of slides with histological evidence of CLS-B), correlated with increasing BMI. Consistent with the mouse models, the presence of CLS-B was associated with activation of NF-κB, and increased levels of proinflammatory mediators including TNF-α, IL-1β, COX-2, and COX-2 derived prostaglandin E_2 (PGE_2). Also mirroring the preclinical findings, the presence of CLS-B and elevated tissue levels of these proinflammatory cytokines paralleled increased transcription of the *CYP19* gene encoding aromatase, leading to elevated aromatase levels and activity [50]. One of the key consequences of this inflammatory pathway was increased expression of the progesterone receptor (PR). Notably,

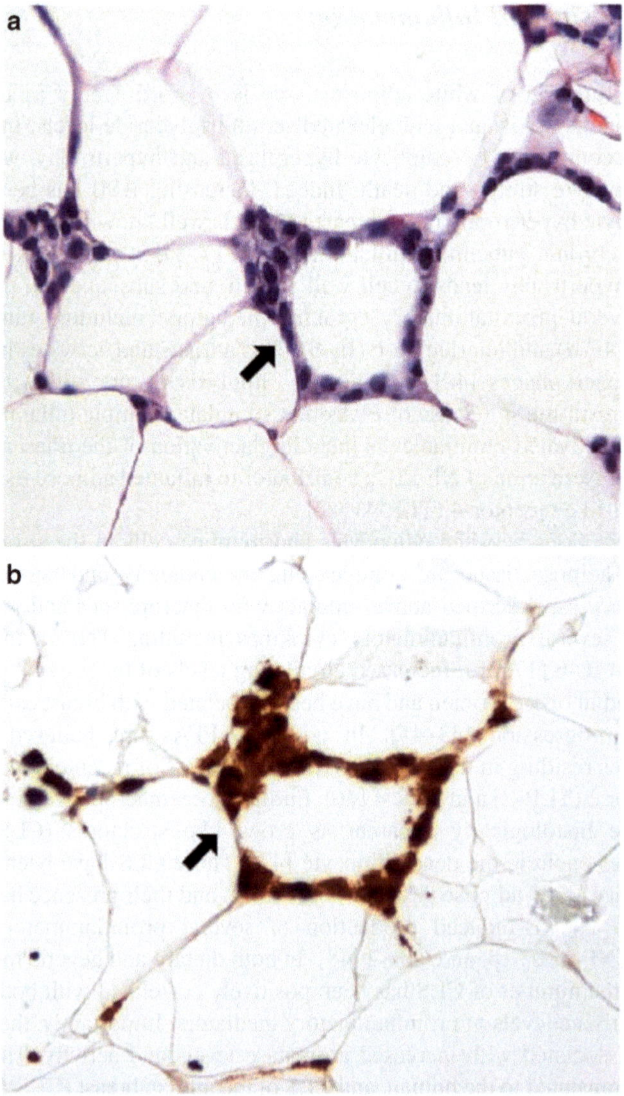

Fig. 8.2 Crown-like structure in human breast adipose tissue (CLS-B). (**a**) Hematoxylin and eosin stain. (**b**) Anti-CD68 immunostain identifies macrophages (reprinted from Morris P et al.: Cancer Prev Res Vol. 4(7), 2011:1021–9)

the extent of CLS-B had a stronger correlation than BMI with aromatase activity, implicating inflammation specifically, rather than obesity alone, as the key inducer of aromatase in the breast. These findings demonstrate an obesity-inflammation-aromatase axis that occurs in the female breast and provide a mechanistic link between obesity and the increased risk of ER-positive breast cancer in

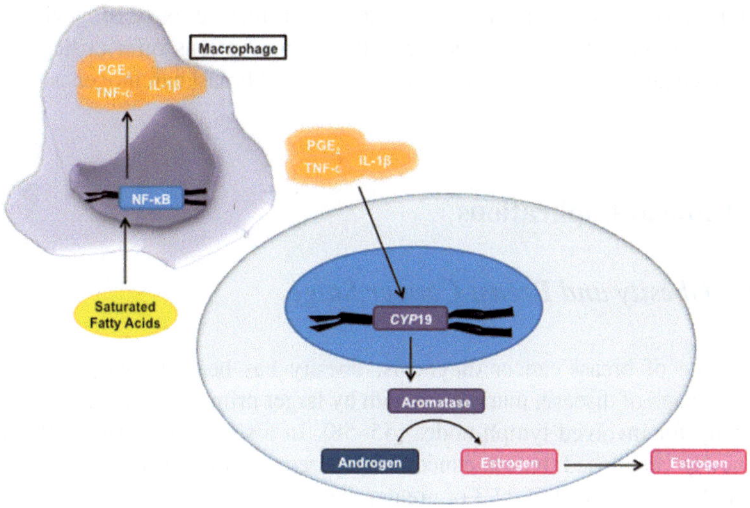

Fig. 8.3 Paracrine interactions between macrophages and other cell types establish an obesity-inflammation-aromatase axis in breast tissue of obese women and mice. *TNF-α* tumor necrosis factor-α; *IL-1β* interleukin-1β; *PGE$_2$* prostaglandin E$_2$

postmenopausal women (Fig. 8.3). In addition, since many of these same proinflammatory mediators have been linked to worse outcome after breast cancer diagnosis, these findings might offer insights into the mechanism underlying this phenomenon (discussed below).

8.3.3 Estrogen

As noted, after menopause, the ovaries are no longer very active in hormone biosynthesis, and peripheral aromatization of adrenal-produced androgens in adipose tissue is largely responsible for estrogen production [11]. In the Health, Eating, Activity and Lifestyle (HEAL) study, which included 505 postmenopausal women with early-stage breast cancer, circulating estradiol levels were found to be 130 % higher in obese women compared to women with BMI<22 ($p=0.002$) [51]. As discussed above, in addition to increased adipose mass, obesity-related inflammatory factors appear to induce aromatase in adipose tissue contributing to increased estrogen production. Transcriptional regulation of aromatase expression is complex. Tissue-specific promoters within the *CYP19* gene give rise to unique mRNA species which are translated to the same aromatase enzyme [52]. Each of these promoters is regulated by distinct signaling patterns, and increased activation of promoters I.3, II, and I.7 is observed in breast cancer in contrast to mRNA derived from promoter I.4 in normal breast tissue [53]. Several inflammatory mediators, including TNF-α, IL-6, and PGE_2, upregulate aromatase expression via specific promotor regions [50, 54]. Collectively, these observations offer critical insights into the mechanisms linking increased adiposity and breast carcinogenesis.

8.4 Clinical Implications

8.4.1 Obesity and Breast Cancer Stage

At the time of breast cancer diagnosis, obesity has been associated with more advanced stage of disease, manifested both by larger primary tumor size and greater propensity for involved lymph nodes [55–58]. In a study of nearly 1,000 women with newly diagnosed breast cancer, compared to underweight/normal weight women, those with BMI of 27.3 or greater were 1.5 times more likely to have stage II or beyond disease at diagnosis, characterized by tumors greater than 2 cm in maximum diameter, and/or those in which cancer has spread to axillary lymph nodes [58]. In a retrospective review of 88,346 consecutive screening mammograms, tumor size was found to be progressively larger with increasing adiposity. Mean tumor diameter was 11 mm in the lowest-weight cohort, compared with 19 mm in the heaviest-weight cohort ($p<0.02$) [59]. Given the fact that obesity is a risk factor for the development of breast cancer, as expected, increasing body weight was also associated with an increased cancer detection rate. Finally, obesity has also been associated greater incidence of lymph node involvement at the time of diagnosis [55–57]. Consistent with this observation, in the study by Hunt et al., the proportion of patients with stage II disease or higher at diagnosis increased with increasing weight ($p=0.046$) [59]. Collectively, these findings highlight the challenges in diagnostic imaging in obese and overweight women.

8.4.2 Challenges with Diagnostic Imaging

Increasing BMI presents a series of challenges for the diagnosis and treatment of breast cancer. In the developed world, most breast cancers are diagnosed at an early, and potentially curable, stage. Improvements in the early detection of breast cancer, leading to a decline in breast cancer mortality, have occurred partially as a result of screening mammography [60, 61]. This technique relies on 2-dimensional X-ray technology to detect calcifications or architectural distortion associated with preinvasive and small breast tumors, which may not otherwise be detectable by clinical examination. Obesity is associated with higher mammographic false-positive rates, which may result in unnecessary distress for patients and additional biopsies and other interventions. It has been suggested that greater amounts of adipose tissue within the breast of obese women interfere with the radiologist's interpretation of mammograms. The largest study to date to examine the diagnostic performance of mammography reported more than a 20 % increased risk of false-positive findings in obese women compared with normal and underweight women [62]. In the study by Hunt et al., discussed above, increasing body weight was associated with increased biopsy and recall rates (the proportion of women undergoing screening who have to return for further assessment due to suspicious findings) [59].

8.4.3 Obesity and Breast Cancer Prognosis

In addition to being a risk factor for breast carcinogenesis and being associated with higher stage at diagnosis, obesity is a poor prognostic factor after breast cancer diagnosis [9, 13, 63–67]. This effect appears to be independent of both menopausal status and of tumor subtype [9, 63, 65–67]. Furthermore, this association cannot simply be explained by the observation that obese women are more likely to present with higher-stage disease at diagnosis as obesity has been shown to be independently associated with an increased risk of developing distant metastases and breast cancer-specific mortality [68].

Seminal observations from key studies are discussed below (Table 8.2). In a study of 3,924 younger women, age 20–54, Whiteman et al. report a significantly higher likelihood of breast cancer-related death in obese women [63]. At a median follow-up of 14.6 years, women with BMI $\geq$ 30 kg/m^2 were more likely to die from breast cancer than lean women (HR 1.34; 95 % CI, 1.09–1.65) [63]. The inclusion of younger women in this study suggests that obesity has broad implications for breast cancer prognosis, irrespective of other factors. Similar results linking obesity to adverse outcomes were seen in the largest cohort study published to date [9]. In this study of nearly half a million women, elevated BMI was associated with increased death rates for all cancers combined as well as for individual tumors including breast cancer [9]. Importantly, Calle et al. report an RR of death from breast cancer in overweight (BMI 25.0–29.9 kg/m^2) and extremely obese women (BMI 35–39.9 kg/m^2) of 1.34 (95 % CI, 1.23–1.46) and 1.70 (95 % CI, 1.33–2.17), respectively, compared to lean individuals ($p < 0.001$). This rigorous analysis provides the evidence in detail that increasing BMI is associated with breast cancer-specific mortality, as opposed to increased death from other obesity-related illnesses such as cardiovascular disease. Second, this important study shows clear evidence of a dose–response effect for increasing BMI on mortality.

In a large retrospective study from the Danish Breast Cancer Cooperative Group database of 53,816 patients, BMI data were available for 18,967 patients. As expected, obese patients (BMI $\geq$ 30 kg/m^2) were more likely to have high-risk disease at the time of breast cancer diagnosis (including larger primary tumors of higher grade and more positive lymph nodes) as compared to women with BMI < 25 kg/m^2. After 10 years, the risk of developing distant metastases was significantly increased by 46 % for patients with BMI > 25 kg/m^2 when compared to women with BMI < 25 kg/m^2 (HR 1.46; 95 % CI, 1.11–1.92; $p = 0.007$) [68]. In a

Table 8.2 Selected studies reporting BMI and outcomes after breast cancer diagnosis

References	Number	Median follow-up (years)	BMI in "normal" group (kg/m^2)	BMI in "overweight and obese" group (kg/m^2)	Outcome	Magnitude of effect	95 % confidence interval	*P* value	Adjustment variables
Whiteman et al. [63]	3,924	14.6	≤22.9	≥30	Breast cancer death	1.34[a]	1.09–1.65	<0.0001	Age, race, radiation therapy, history of benign breast disease, education level, menopausal status, stage
Majed et al. [65]	14,709	8	<25	≥30	Distant metastases	1.17[a]	1.04–1.31	<0.01	Age, tumor size, clinical node involvement, menopausal status, diagnosis year, ER/PR status, clinical tumor extension, nodal status, tumor grade
					OS	1.15[a]	1.02–1.29	<0.05	
Ewertz et al. [68]	18,967	7.1 (estimated potential)	<25	>30	Distant metastases (5–10 years)	1.46[a]	1.11–1.92	0.007	Age, menopausal status, tumor size, nodal status, deep fascia invasion, histologic type and grade, ER/PR status, protocol year, systemic therapy
					Breast cancer death (10+ years)	1.38[a]	1.11–1.71	0.003	

Berclaz et al. [69]	6,370	14	≤24.9	≥30	10-yr DFS	1.10[a]	1.10–1.20	0.04	Menopausal status, nodal status, tumor size, vessel invasion, ER/PR status, tumor grade, treatment
					10-yr OS	1.14[a]	1.03–1.27	<0.01	
Chen et al. [70]	4,989	3.8	18.5–25	≥30	Relapse and disease-specific mortality	1.44[a]	1.02–2.03	<0.05	Age, education, income, marital status, comorbidity, exercise, meat intake, vegetable intake, soy intake, time from diagnosis to enrollment, menopausal status, menopausal symptoms, treatment, stage, ER/PR status
Kroenke et al. [72]	5,204	9	21–22	25–29	Breast cancer death (nonsmokers)	1.83[b]	1.09–3.05	<0.001	Age, OCP use, birth index, menopausal status, age at menopause, HRT use, protein intake, tumor size, nodal status, chemotherapy, tamoxifen
			21–22	>30		1.96[b]	1.30–2.97	<0.001	

(continued)

Table 8.2 (continued)

References	Number	Median follow-up (years)	BMI in "normal" group (kg/m²)	BMI in "overweight and obese" group (kg/m²)	Outcome	Magnitude of effect	95 % confidence interval	*P* value	Adjustment variables
Calle et al. [9]	147,583	16	18.5–24.9	>40	Breast cancer death	2.12[b]	1.41–3.19	<0.001	Age, education, smoking status, physical activity, alcohol use, marital status, race, aspirin use, HRT, fat consumption, vegetable consumption
Daling et al. [14]	1,177	Up to 17 years	15.8–20.6	25.8–52.6	5-yr survival	1.7[a]	1.0–2.9	<0.05	Age, diagnosis year, tumor size, nodal status, ER/PR status, other tumor characteristics

OCP oral contraceptive pills; *HRT* hormone replacement therapy; *BMI* body mass index; *HR* hazard ratio; *RR* relative risk; *OS* overall survival; *DFS* disease-free survival; *ER* estrogen receptor; *PR* progesterone receptor; *yr* year

[a]Hazard ratio

[b]Relative risk

multivariate analysis adjusted for stage at presentation, obesity emerged as an independent predictor of distant recurrence. At 3 years of follow-up, risk of distant metastases began to show a clear trend of increasing risk with increasing BMI. At 10 years, the cumulative incidence of distant metastases for patients with a BMI of 30 kg/m^2 or more was 24.3 % (95 % CI, 22.1–26.5 %) compared to 20.1 % in patients with a BMI less than 25 kg/m^2 (95 % CI, 19.2–20.9 %). Strikingly, at 30 years of follow-up, the cumulative risks of breast cancer-related death were 57.2 % for patients with BMI of 30 kg/m^2 or more (95 % CI, 51.8–62.2 %) compared with 46.4 % for patients with BMI less than 25 kg/m^2 (95 % CI, 44.8–48.0 %). For overweight women with a BMI of 25–29 kg/m^2, mortality was also higher compared to women with BMI less than 25 kg/m^2 with a 53.4 % (95 % CI, 50.5–56.2 %) cumulative incidence of cancer-related death at 30 years follow-up. Overall the risk of dying as a result of breast cancer after 10 years was significantly increased for obese women compared with women with a BMI of less than 25 kg/m^2 (HR 1.38; 95 % CI, 1.11–1.71). Notably, the risk of dying as a result of other causes not related to breast cancer was also increased after 10 years in obese patients (HR 1.31; 95 % CI, 1.05–1.63).

These findings are consistent with several other population studies. In a French population of over 14,000 patients who were prospectively followed after early-stage breast cancer diagnosis, more advanced tumor at the time of diagnosis was associated with BMI over 30 kg/m^2 [65]. Additionally, obesity was associated with higher rates of distant relapse (HR 1.32; 95 % CI, 1.19–1.48), shorter disease-free interval (HR 1.20; 95 % CI, 1.08–1.32), and shorter OS (HR 1.43; 95 % CI, 1.28–1.60). The International Breast Cancer Study Group reported similar findings in over 6,000 patients with a median follow-up time of 14 years [69]. Women with BMI of 30 kg/m^2 or greater had a 10 % higher risk of distant relapse ($p=0.04$) and 14 % higher risk of death ($p<0.01$) when compared to women with normal BMI. Obesity was associated with worse 10-year disease-free survival (DFS) (HR 1.17; 95 % CI, 1.07–1.28; $p<0.01$) and OS rates (HR 1.25; 95 % CI, 1.13–1.38; $p<0.01$) when compared to women of normal weight. Additionally, elevated BMI was associated with more advanced tumor at the time of breast cancer diagnosis (specifically tumor size of 2 cm or greater), as well as greater incidence of lymph node involvement. Furthermore, Daling et al. report an association between larger tumors in obese women and higher tumor proliferation rate, characterized by the Ki-67 proliferative index [14]. In this population-based study of just over 1,000 women with breast cancer identified from the Cancer Surveillance System of the Surveillance, Epidemiology and End Results (SEER) cancer registry, the heaviest quartile of women (BMI of 25.8 kg/m^2 or greater) were 2.5 times more likely to succumb to breast cancer-related death than women in the lowest quartile of BMI (20.6 kg/m^2 or less) (95 % CI, 1.6–3.9). Women in the heaviest quartile presented with larger tumor size (2 to less than 5 cm odds ratio (OR) 2.2; 95 % CI, 1.5–3.1, and 5 cm or greater OR 2.7; 95 % CI, 1.5–4.8). The authors suggest that this may be related to faster growing tumors in heavier women. Compared to the lowest quartile of body weight, tumors 2 cm or larger in the obese women were more likely to have a Ki-67 proliferative index of 25 % or greater (OR 1.9; 95 % CI, 1.1–3.3) and a high mitotic rate (OR 1.6; 95 % CI, 0.9–3.1). Notably, lymph node status did not vary by BMI in this study.

8.4.4 Impact of Weight Change After Breast Cancer Diagnosis

In addition to the negative effect of pre-morbid obesity on cancer-specific mortality, weight change after breast cancer diagnosis may also have prognostic significance [70–72]. Consistent with other reports, in a study of 5,042 patients identified through the population-based Shanghai Cancer Registry, women who were obese 1 year prior to cancer diagnosis in this study had 1.6 times higher total mortality than women with a normal BMI. As expected, shorter OS and DFS associated with obesity was independent of hormone receptor status. Critically, women who gained 5 kg (approximately 10 lb) or more within 18 months of breast cancer diagnosis had higher breast cancer-specific (HR 1.90; 95 % CI, 1.23–2.93) and total mortality (HR 1.71; 95 % CI, 1.12–2.60) rates compared with women who maintained their weight within 1 kg (approximately 2 lb). Similar results were seen in the Nurses' Health Study, which included 5,204 patients diagnosed with non-metastatic breast cancer at the time of accrual. In this study, nonsmoking women whose BMI increased more than 2.0 kg/m^2 (approximately 11 lb for a woman of an average height of 64 in.) during a median follow-up time of 9 years after diagnosis had an elevated risk of death as a result of breast cancer compared with women whose weight remained stable (RR 1.64; 95 % CI, 1.07–2.51) [72]. Critically, this association was particularly accentuated in premenopausal vs. postmenopausal women. Weight gain after diagnosis was also associated with worse outcomes in premenopausal women undergoing treatment for early breast cancer in a study by Camoriano et al. [71]. In this study of 646 women with node-positive disease, patients who received adjuvant chemotherapy gained more weight compared to those who did not [71]. Among women treated with chemotherapy, weight change at 60 weeks was greater for women who were premenopausal (median gain 5.9 kg) than postmenopausal (median gain 3.6 kg) ($p<0.001$). Among premenopausal women treated with chemotherapy, those who surpassed the median weight gain at 60 weeks had a 1.5 times higher risk of distant relapse (covariate $p=0.17$) and 1.6 times greater risk of death (covariate $p=0.04$) at median follow-up of 6.6 years. In addition, similar trends in increased relapse and shorter OS were seen in postmenopausal women, although these results did not reach statistical significance. Conversely, a stronger association between weight gain and poorer outcomes by menopausal status was not observed in the Shanghai study [70]. Although definitive conclusions from these studies are limited by their design, these findings collectively suggest that weight gain in the first year after breast cancer diagnosis may have a deleterious effect on breast cancer-specific outcomes. In some studies, these effects appear to be independent of the negative effects of preexisting obesity [70, 72]. Although the effect of long-term weight gain cannot be delineated from these studies, it is likely to be similarly detrimental, given the known biological consequences of weight gain.

There are studies that do not confirm any negative effect on breast cancer-specific outcomes associated with weight gain after diagnosis [73, 74]. In a study investigating the psychologic and behavioral functioning related to exercise and eating habits in women undergoing adjuvant chemotherapy for breast cancer, an average weight gain of approximately 6 kg at 2 years follow-up was reported. However, this weight gain did not translate into a greater risk of breast cancer recurrence, albeit in a very limited sample ($n=32$) [73]. Similarly, weight gain up to 4 years after breast cancer diagnosis was not associated with increased risk of relapse or death in a larger prospective cohort study of 1,692 breast cancer survivors [74]. It is important to note that consistent with other studies, obesity 1 year before diagnosis was associated with increased risk of death from any cause (HR 1.6; 95 % CI, 1.1–2.3) and an increased risk of breast cancer-related death (HR 1.6; 95 % CI, 0.9–2.7), although the latter result was not statistically significant. Therefore, while it is clear that obesity is associated with worse breast cancer-related outcomes, not all studies demonstrate a significant effect for weight gain after breast cancer diagnosis. Nonetheless, available evidence shows consistent trends towards a negative effect of weight gain on breast cancer outcomes.

8.5 Obesity and Breast Cancer Treatment

8.5.1 Breast Surgery

Obesity is associated with greater rates of complications after breast cancer surgery and breast reconstruction [75–85]. In a retrospective analysis of potential factors that affected postoperative hospital stay for 73 patients who underwent modified radical mastectomy, patients with stays under 5 days had a significantly lower BMI than that of patients who required a longer hospitalization, although obesity was not found to be an independent predictor of longer hospital stay in multivariate analysis [75]. In a study conducted by the National Surgical Quality Improvement Program Patient Safety in Surgery, El-Tamer et al. reported morbidity and mortality data for 1,660 women undergoing mastectomy and 1,447 women undergoing lumpectomy with an axillary procedure [76]. Wound infection was the most frequent morbid complication and occurred at a higher rate in patients undergoing mastectomy (4.34 %) vs. lumpectomy (1.97 %). Importantly, BMI greater than 30 kg/m^2 was an independent predictor of wound infection. With specific regard to axillary lymph node procedures, lymphedema (swelling of the arm) is a well-known complication after axillary node dissection and may occur years after the procedure. In a report of 137 patients with breast cancer who underwent sentinel lymph node dissection, of whom 85 underwent immediate completion axillary lymph node dissection, BMI greater than 30 kg/m^2 was associated with a greater risk of developing lymphedema within 2 years of surgery compared to patients with BMI less than 25 (OR 2.93; 95 % CI, 1.03–8.31; $p=0.003$) [77]. Recently, neo-adipogenesis and adipocyte hypertrophy have been implicated in the development of lymphedema in animal models [86, 87]. Using a mouse tail model, Zampell et al. demonstrated a twofold increase in fat thickness ($p<0.01$) in response to lymphatic fluid stasis [86]. Additionally, a marked inflammatory response consisting of a fivefold increase in

CD45+ mononuclear cells ($p < 0.001$) and a greater than threefold increase in F4/80-stained monocytes and macrophages ($p < 0.001$) in the subcutaneous mouse-tail fat was associated with adipogenesis in response to lymphatic fluid stasis. In part two of this study, Aschen et al. went on to demonstrate concurrent upregulation of several adipogenic factors including adiponectin, CCAAT/enhancer binding protein-α (CEBP-α), and peroxisome proliferator-activated receptor-γ (PPAR-γ) in response to lymphatic fluid stasis [87]. Both CEBP-α and PPAR-γ are key transcription factors involved in adipogenesis and promote adipocyte differentiation. Therefore, adipogenesis, adipocyte hypertrophy, and upregulation of adipogenic factors appear to be integrally involved in the development of lymphedema and in turn promote an inflammatory response within subcutaneous fat.

Breast reconstruction after breast cancer surgery may also be challenging in obese and overweight patients. In a retrospective analysis of 1,195 abdominal flap reconstruction procedures performed in 952 patients, Mehrara et al. reported that obesity, defined as greater than 25 % of ideal body weight, was the major independent predictor of postoperative complications with up to a threefold increased incidence of adverse events [78]. Specifically, in patients who underwent free transverse rectus abdominis myocutaneous (TRAM) flap reconstruction, BMI ≥ 30 kg/m^2 was significantly associated with partial flap loss (OR 2.6, $p < 0.03$), donor-site complications (OR 3.0, $p < 0.03$), and hernia and/or laxity (OR 2.5, $p = 0.05$). In univariate analysis, obesity was associated with fat necrosis, a benign complication of surgery or trauma, but this association was not significant on multivariate analysis. Strikingly, obesity was associated with significantly increased risk of arterial thrombosis (OR 1.7, $p = 0.06$) although this did not translate to a higher rate of flap loss. It has been hypothesized that increased rates of arterial thrombosis in obese patients may be related to the increased difficulty of deep microsurgical anastomosis within the axilla. Similar findings were reported by Chang et al. in 718 patients who underwent TRAM flap breast reconstruction [85]. Obese and overweight patients had higher rates of total flap loss, flap hematoma, flap seroma, mastectomy skin flap necrosis, donor-site infection, donor-site seroma, and hernia compared with patients of normal weight. These results suggest that morbidly obese patients are at very high risk of flap failure and other complications and that procedures such as TRAM flap breast reconstruction should be avoided in such patients. In an attempt to minimize these complications, an alternative approach to breast reconstruction in overweight and obese patients might be the use of a deep inferior epigastric perforator flap [88].

Elevated BMI can also complicate other reconstruction surgical techniques. Obesity is an independent risk factor for the development of perioperative complications in patients undergoing expander/implant reconstruction (OR 1.8; 95 % CI, 1.1–3.0; $p = 0.02$) [80]. These complications include mastectomy flap necrosis, seroma/hematoma, infection, failed expansion, and expander/implant exposure.

It is important to note, however, that the overall rate of complications associated with reconstructive surgery is low and that the majority of overweight and even obese patients complete reconstruction successfully. Most investigators agree that overweight and obese women should be counseled and informed about risks and benefits of reconstruction, including higher complication rates associated with elevated BMI despite its overall safety, to aide in their choice of approach [78, 85]. It is useful to note that in a study of patient satisfaction in US patients undergoing breast reconstruction, obese patients with expander/implants were significantly less satisfied with aesthetic outcome than normal weight patients (OR 0.14, $p=0.02$) [83]. There was no significant difference in aesthetic satisfaction between obese and normal weight women undergoing TRAM flap reconstruction. Taken together, these data underscore the importance of careful patient counseling, including a detailed discussion of the risks and benefits involved.

8.5.2 Breast Radiation

In addition to greater rates of postoperative complications, increasing BMI and large breast size have been associated with greater toxicity following adjuvant radiotherapy [89–94]. Pragmatically speaking, very morbid obesity may preclude the use of radiation therapy given the upper weight limit that currently manufactured treatment tables are capable of supporting. For those who do undergo radiation, several studies have demonstrated higher complication rates in overweight and obese patients. Allen et al. report a case control study of women who underwent adjuvant radiotherapy after lumpectomy or mastectomy followed by chemotherapy [89]. In total, 200 patients received radiotherapy to the breast and/or chest wall, with or without inclusion of regional lymph nodes. This analysis identified BMI as the only significant factor associated with the development of radiation pneumonitis (multivariate $p<0.01$). The mean BMI for 14 (7 %) patients who developed radiation pneumonitis was 31.2 vs. 26.3 for patients whose course was not complicated by radiation pneumonitis. These findings could be explained by heterogeneity in dose delivery as a result of increased chest wall separation in obese patients, resulting in higher maximum radiation dose delivered. Indeed, in this study, maximum delivered dose correlated with BMI. Additionally, the known link between obesity, asthma, and other pulmonary comorbidities may predispose this population to development of radiation pneumonitis. It is also possible that circulating proinflammatory mediators associated with asthma and other atopic conditions may contribute to the increased radiation pneumonitis risk. For example, elevated circulating IL-6 and interleukin-1α (IL-1α) levels prior to radiotherapy have been identified as markers of increased radiation pneumonitis risk [95, 96]. As discussed above, elevated circulating IL-6 levels are found in overweight and obese individuals. Therefore, it is possible that obesity-related inflammation may be a key mechanism underlying the increased risk of radiation pneumonitis in obese patients who undergo breast irradiation.

As with breast surgery, elevated BMI increases the risk of ipsilateral arm edema following radiotherapy [90]. In a series of 282 patients with stage I or II breast cancer who underwent breast surgery followed by radiation at Memorial Sloan-Kettering Cancer Center, Werner et al. reported BMI as the most important factor associated with posttreatment arm edema [90]. Larger breast size was also a significant independent risk factor in the development of this complication. Greater breast volume has also been associated with more fibrosis and poorer cosmetic outcome [91] as well as increased skin toxicity [92] following radiotherapy. Finally, increased radiation dose heterogeneity within larger-volume breasts may result in increased maximum dose, or "hot spots" within the radiated breast, although the incidence of this problem with more modern radiotherapy techniques is less clear [93, 94].

8.5.3 Systemic Therapies

Irrespective of the possible interaction between BMI and cancer biology, overweight and obese patients are more likely to have less favorable outcomes with adjuvant chemotherapy than women of normal weight. A number of contributory factors have been identified and largely fall into two categories—underdosing and decreased efficacy. Most cytotoxic chemotherapy is dosed based on body surface area, which, like BMI, is calculated based on height and weight. Many clinicians limit chemotherapy dosing for some drugs in obese patients, by using a maximum weight, due to concerns regarding an increased risk of toxicity in obese patients if treated with doses based on actual body weight. In a US survey of 1,243 community-based practitioners that extracted data from 20,799 early-stage breast cancer patients, Lyman et al. observed that due to dose reductions and/or treatment delays, 55.5 % of all patients received a relative dose intensity of less than 85 %, a threshold that has been linked to inferior efficacy [97]. In this study, increasing BMI was associated with greater dose reductions ($p<0.001$) although there were no differences in the need for unplanned dose modifications due to toxicity. Importantly, body surface area greater than 2 m^2 emerged as an independent predictor of dose reduction (OR 1.71; 95 % CI, 1.59–1.85; $p<0.001$). Consistent with these observations, another community-based study of 9,672 women treated with anthracycline-based chemotherapy found increasing BMI to be independently associated with first-cycle dose reduction, i.e., the administration of doses below that from the calculated body surface area [98]. First-cycle dose reduction of 10 % or more was seen in 11 %, 20 %, and 37 % of overweight, obese, and severely obese women, respectively, compared with 9 % of normal weight women. After initial dose reduction, it was uncommon (3 % patients) for dose escalation (to a level based on actual weight) to occur in subsequent cycles for obese and overweight patients even if chemotherapy was well tolerated. Contrary to concerns regarding excessive toxicity, an increased hospital admission rate for febrile neutropenia was associated with higher administered dose rather than being overweight or obese. In fact, severe obesity (BMI $\geq$ 35) was associated with lower likelihood of hospitalization for febrile neutropenia (OR 0.61; 95 % CI, 0.38–0.97), although this could have been because the majority of these patients received attenuated doses.

Other studies have also demonstrated that overweight and obese patients who receive doses based on actual body weight do not experience excessive toxicity

Table 8.3 Cytotoxic chemotherapy dosing for obese adult patients with cancer

American Society of Clinical Oncology clinical practice guideline
Use full weight-based chemotherapy dosing for obese patients, particularly when treatment intent is curative
Respond to treatment-related toxicity similarly for both obese and nonobese patients
Dose reductions, if needed, should be done similarly for obese and nonobese patients; consider resuming full weight-based dose on subsequent cycles if feasible
Fixed dosing should only be used for select agents (e.g., carboplatin, bleomycin)

compared with women of normal weight [99–104]. For example, in the Cancer and Leukemia Group B (CALGB) 8541 randomized trial, which examined the schedule and dose of adjuvant anthracycline-based chemotherapy in 1,435 women with node-positive breast cancer, no evidence of increased toxicity was experienced by obese women who received chemotherapy based on actual body weight [100]. In this retrospective analysis, obese women who received cycle 1 doses below 95 % of the calculated weight-based dose had shorter DFS than women who received within 5 % of actual body weight-based doses. The risk of treatment failure for women in the 5 % group was two-thirds the risk of obese women whose dose was reduced below 95 %, although this effect did not reach statistical significance in multivariate analysis (HR 0.67: 95 % CI, 0.38–1.20; $p=0.25$). Taken together, these findings suggest that dosing of most commonly used drugs based on actual body weight does not increase toxicity and may result in improved outcomes in obese patients. The importance of these observations led to a new guideline from an expert panel convened by the American Society of Clinical Oncology (ASCO) [105]. The main recommendations from this panel include the dosing of chemotherapy by actual body weight in the treatment of obese cancer patients, particularly when prescribed with curative intent (Table 8.3) [105].

8.5.4 Preoperative Chemotherapy in Obese Patients

Apart from the underdosing of chemotherapy in overweight and obese patients, it appears that current preoperative systemic therapies may be less effective in these patients compared with women of normal weight [68, 99, 106–108]. In a study of over 1,000 breast cancer patients who received neoadjuvant (preoperative) chemotherapy at the MD Anderson Cancer Center, overweight and obese patients were less likely to achieve a pathologic complete response (pCR) at the time of breast surgery than normal-weight patients (OR 0.67; 95 % CI, 0.45–0.99) [106]. However, this observation should be interpreted cautiously as chemotherapy dosing for these patients could not be verified in this study. Given the prevalence of underdosing for overweight and obese patients, the lower rates of pCR may be a function of chemotherapy exposure. In a different study, 307 Chinese women with breast cancer treated at the Shanghai Cancer Hospital all underwent neoadjuvant chemotherapy dosed by actual weight [107]. Women with a BMI of 25 kg/m^2 or greater were less likely to achieve a pCR at the time of breast surgery after neoadjuvant chemotherapy compared to women with BMI less than 25 kg/m^2 (OR 0.45; 95 % CI, 0.22–0.94; $p=0.03$). Notably, BMI was identified as an independent predictor of pCR after neoadjuvant chemotherapy in postmenopausal and hormone receptor-negative subgroups ($p=0.004$ and $p=0.038$, respectively).

8.5.5 Postoperative Chemotherapy in Obese Patients

Less favorable responses to adjuvant (postoperative) breast cancer treatment have also been reported [68, 99, 108]. In the Danish Breast Cancer Cooperative Group study, Ewertz et al. report shorter OS after adjuvant therapy that became evident after 10 years of follow-up in patients with BMI of 30 kg/m^2 [68]. Again, in this study, chemotherapy dosing was not addressed limiting our ability to interpret the results. However, several other studies have reported decreased OS for obese and overweight patients after completing adjuvant therapy [99, 108]. Bastarrachea et al. report a greater risk of disease recurrence after adjuvant chemotherapy dosed by actual body weight in obese vs. nonobese patients (HR 1.33; 95 % CI, 1.05–1.68) [99]. In a post hoc analysis of the Breast International Group (BIG) 02-98 study, which was a randomized phase III trial in which docetaxel was added to adjuvant anthracycline-based chemotherapy, BMI of 30 kg/m^2 or greater was associated with shorter 5-year OS and DFS (HR 1.34, $p=0.013$, and HR 1.20, $p=0.041$) [108]. Taken together, these studies suggest a decreased efficacy of chemotherapy in obese patients; however, underdosing remains a potential confounder. As the ASCO clinical practice guidelines for the dosing of chemotherapy in obese patients become increasingly incorporated into clinical practice, the efficacy of chemotherapy in overweight and obese patients will likely become more apparent.

8.5.6 Obesity and Endocrine-Targeted Therapy

For patients with ER-positive breast cancer, endocrine therapy, which aims to disrupt the tumor supply of estrogen or change the receptor-ligand interactions, is an important aspect of treatment. For obese patients, concerns about the dosing and resultant efficacy have been raised with regard to the use of endocrine therapy [109]. Importantly, in the randomized phase III Arimidex, Tamoxifen, Alone or in Combination (ATAC) trial, proportionately worse outcomes were seen in obese vs. lean women who were treated with the aromatase inhibitor, anastrozole [109]. Treatment with anastrozole was less efficacious in women with BMI$\geq$35 kg/m^2 when compared to women with BMI<23 kg/m^2 in multivariate analysis (HR 1.53; 95 % CI, 1.01–2.32; $p=0.001$). In contrast, tamoxifen appeared to be equally effective across all BMI groups. While recurrence rates were overall lower in the postmenopausal women treated with anastrozole compared to tamoxifen, the benefit of anastrozole was greater in leaner women. The adjusted HR comparing anastrozole with tamoxifen for distant plus local recurrences in the lowest BMI group was 0.64 (95 % CI, 0.45–0.91) in contrast to 0.84 (95 % CI, 0.61–1.14) for the heaviest women.

Reassuringly, use of the selective estrogen receptor modulator (SERM), tamoxifen, was also found to be equally efficacious in obese and nonobese women with node-negative ER-positive breast cancer in a cohort of 3,385 women enrolled in the National Surgical Adjuvant Breast and Bowel Project (NSABP) B-14 protocol [110]. However, obesity was found to be associated with increased risk of contralateral breast cancer, second primary cancers, and increased overall mortality. Therefore, these findings are consistent with the hypothesis that there is a worsened prognosis for obese patients, but suggest that tamoxifen may be relatively more effective in obese women. To be precise, the efficacy advantages of aromatase inhibition over tamoxifen appear to diminish with increasing weight. However, anastrozole was not inferior to tamoxifen, and, in particular among overweight patients, its toxicity profile may make it preferable despite the lack of superior efficacy in the obese patients.

In terms of overall outcomes, Sparano et al. reported inferior DFS, defined as the length of time after completing cancer treatment that a patient lives without any signs or symptoms of that cancer, and OS, defined as the proportion of patients still alive after starting cancer treatment, in obese vs. lean women with operable breast cancer who received chemotherapy followed by endocrine therapy with tamoxifen [111]. In this pooled outcomes analysis of three adjuvant trials conducted by the Eastern Cooperative Oncology Group (ECOG), all patients received anthracycline-based chemotherapy dosed by actual body weight. There were no significant differences in dose delivered to obese vs. nonobese patients, except in the group that received weekly or every 3-week paclitaxel. In one of the three included ECOG trials (E1199), patients received either tamoxifen or tamoxifen followed by an aromatase inhibitor after adjuvant chemotherapy. In multivariate modeling, BMI > 30 kg/m^2 vs. BMI < 30 kg/m^2 was associated with inferior DFS (HR 1.24; 95 % CI, 1.06–1.46; $p=0.0079$), inferior OS (HR 1.37; 95 % CI, 1.13–1.67; $p=0.0015$), and inferior breast cancer-specific survival (BCSS) (HR 1.40; 95 % CI, 1.11–1.76; $p=0.0042$) in women with hormone receptor-positive, human epidermal growth factor receptor-2 (HER2)-negative (or unknown) disease. Similar findings were observed in the E5188 trial in which premenopausal women with ER-positive breast cancer received no endocrine therapy, ovarian suppression, or ovarian suppression plus tamoxifen after chemotherapy. Survival rates were significantly worse for obese patients (DFS HR 1.41, 95 % CI 1.19–1.67, $p<0.0001$; OS HR 1.51, 95 % CI 1.24–1.83, $p<0.0001$; BCSS HR 1.54, 95 % CI 1.26–1.88, $p<0.0001$). In sum, these findings demonstrate worse survival rates in obese women treated with chemoendocrine therapy, including tamoxifen, for hormone receptor-positive breast cancer, but they do not indicate a superior treatment approach for such patients.

A likely explanation for the reduced efficacy of hormone therapies in obese women is that elevated estrogen levels, which are commonly seen in obese patients, may not be adequately suppressed by aromatase inhibitors such as anastrozole. In support of this, using sensitive estradiol assays, a recent study has demonstrated that higher levels of circulating estrogens are commonly seen in obese women [112]. In this study, Folkerd et al. reported a positive correlation between circulating estradiol level and increasing BMI ($\rho=0.57$, $p<0.001$), noting that values were almost 3 times higher in women with BMI>35 kg/m^2 compared to women with BMI<25 kg/m^2. Suppression of estrogens was greater with the newer aromatase inhibitor, letrozole, compared with anastrozole. This raises the question of whether dose escalation or other strategies, such as using more potent aromatase inhibitors, would be useful in these patients. Prospective clinical trials are needed to adequately address these issues and develop a risk-stratified approach for obese patients. Another possible explanation for the reduced efficacy of aromatase inhibitors in the obese is ligand-independent activation of ER-α. This could occur for a variety of reasons including elevated levels of circulating IGF-1 or the effects of proinflammatory mediators and warrants investigation.

Additionally, dosing adequacy remains a concern for obese patients treated with molecular and other targeted therapies (e.g., the monoclonal antibody trastuzumab and the tyrosine kinase inhibitor lapatinib). Several of these agents are delivered as predetermined doses that are not weight based (e.g., lapatinib). Appropriate dosing of these agents is a subject of ongoing investigation and has not yet been addressed by current guidelines [105].

8.6 Interventional Strategies and Future Directions

Having reviewed the pathogenesis, poor outcomes, and treatment implications conferred by increased adiposity on women with breast cancer, we now address potential strategies to reduce these risks. These approaches range from behavioral interventions, including lifestyle modifications such as diet and exercise, to pharmaceutical interventions that target the pathophysiology of dysregulated metabolism and obesity-related inflammation. In terms of breast cancer prevention, there is extensive interest in research to better characterize specific points along the route of carcinogenesis where interventions may be beneficial. Similarly, much work is being done to understand the reasons for poorer prognosis after breast cancer diagnosis, many of which have been discussed above, in order to determine specific interventions that may be used as an adjunct to cancer-directed therapy to improve outcomes.

8.6.1 Weight Loss

Weight loss or maintenance of a normal BMI, achieved by dietary modification and/or physical activity, has been associated with improved outcomes after breast cancer diagnosis [113–121]. Much of these data, however, are observational, which will hopefully set the stage for the development of adequately powered randomized clinical trials. In a prospective observational study reported by Holmes et al., exercise was associated with improved outcome after breast cancer diagnosis [117].

Nearly 3,000 nurses with early breast cancer (stage I to III) who enrolled in the Nurses' Health Study were surveyed about their levels of physical activity and were followed for breast cancer mortality [117]. Women whose physical activity was the equivalent of walking 3–5 h per week at an average pace had the greatest benefit in terms of breast cancer-related death (RR 0.50; 95 % CI, 0.31–0.82). A consistent association between physical activity and reduced breast cancer mortality has been reported in similar observational studies [118–120].

With regard to the effects of dietary interventions on breast cancer-specific outcomes, the Women's Intervention Nutrition Study (WINS) randomly assigned women to a low-fat diet, in which fat accounted for 20 % of total calories, vs. normal diet [121]. Women in the low-fat diet group lost an average of 6 lb and, after 5 years of follow-up, had a 24 % reduction in the risk of relapse (HR 0.76; 95 % CI, 0.60–0.98; $p=0.034$). Although studies have failed to confirm a consistent benefit of diet on breast cancer-specific outcomes, the results of this randomized trial do suggest that diet is one important strategy to reverse the negative effects of obesity on breast cancer mortality. Importantly, reduction in breast cancer recurrence was greater in women with hormone receptor-negative breast cancer than in women with hormone receptor-positive disease. This again suggests that estrogen-independent mechanisms are involved in regulating the growth of breast cancer, such as insulin signaling as discussed above.

Importantly, several studies demonstrate significant associations between weight loss and/or dietary modification with improvements in circulating biomarkers that have been implicated in breast cancer development and progression. A weight loss of 7.1 % after consumption of a hypocaloric diet, defined by a 250–350 kcal/day deficit, for 6 months has been associated with significant decreases in circulating levels of leptin and insulin [122]. In a recent study, weight loss of 5 % or more within 12 months via caloric restriction and/or exercise was associated with significant reductions in circulating markers of inflammation including high-sensitivity C-reactive protein (hsCRP), IL-6, and serum amyloid A [123]. Furthermore, reductions in circulating estradiol levels and elevation of SHBG levels have been associated with diet and exercise [115]. In the Nutrition and Exercise for Women (NEW) trial, circulating estradiol decreased by 20.3 % ($p<0.001$) and SHBG increased by 25.8 % ($p<0.001$) in postmenopausal women with BMI$>$25 kg/m^2 who were randomized to reduced-calorie diet and moderate- to vigorous-intensity aerobic exercise [115]. Reduced serum levels of insulin, CRP, and leptin and increased adiponectin were also observed. Additionally, TNF-α levels and insulin release after glucose challenge significantly fell after diet-induced weight loss [44]. The magnitude of this decline is associated with baseline body weight, again suggesting that obese women in particular benefit most from weight loss. Reductions in serum IL-6 and CRP and a rise in adiponectin levels have also been associated with diet-induced weight loss [124, 125]. Additionally, reduced levels of inflammatory markers within subcutaneous white adipose tissue have been demonstrated after weight loss [37]. Collectively, these findings suggest that diet-induced weight loss can have desirable effects on inflammatory mediators, known to be involved in breast cancer development and progression.

With our current understanding of obesity-related inflammation and breast cancer risk and prognosis, these changes induced by weight loss help elucidate specific pathogenic targets for possible intervention. Importantly, the feasibility of long-term weight loss interventions in cancer survivors was recently demonstrated in the Reach Out to Enhance Wellness (RENEW) trial [114]. In this study, 641 overweight and obese patients with a history of locoregional colorectal, breast, and prostate cancer underwent diet and exercise counseling via mailed printed material and telephone. Improvements in diet quality, physical activity, and BMI were demonstrated at 2 years of follow-up. Therefore, the observational data associating weight loss with improved survival and biomarkers of breast cancer prognosis, coupled with the durability of weight loss intervention, set the stage for a prospective randomized trial investigating the effect of dietary and exercise interventions on breast cancer outcomes in selected subsets of patients, using biomarkers known to be dysregulated in overweight and obese women with breast cancer.

8.6.2 Medications

The use of several medications has been associated with attenuation of processes involved in breast cancer development and progression. One attractive investigational agent is metformin, which has beneficial effects on insulin levels. Use of metformin has been suggested to both reduce the risk of breast cancer and improve prognosis. In the preclinical setting, when human breast adipose stromal cells are treated with metformin, aromatase expression is inhibited via activation of AMPK [126]. Furthermore, metformin has been demonstrated to downregulate breast-specific aromatase expression by inhibiting *CYP19* gene promoters I.3 and II [127]. Goodwin et al. reported a small clinical trial of 22 women with early-stage breast cancer in which treatment with 6 months of metformin resulted in significant reduction in fasting insulin level [128]. In addition, diabetic patients treated with metformin were found to achieve pCR at the time of breast surgery more often than diabetic patients who were not treated with metformin after undergoing neoadjuvant chemotherapy for early-stage breast cancer [129]. In a retrospective analysis, 2,529 patients who received neoadjuvant chemotherapy were identified as nondiabetic patients ($n=2{,}374$), diabetic patients taking metformin ($n=68$), and diabetic patients not taking metformin ($n=87$). Patients taking metformin had a significantly higher proportion of pCR at the time of surgery (24 %; 95 % CI, 13–34 %) compared to diabetic patients not taking metformin (8.0 %; 95 % CI, 2.3–14 %) and nondiabetic patients (16 %; 95 % CI, 15–18 %; $p=0.02$). In multivariate analysis adjusting for diabetes status, neoadjuvant metformin use was an independent predictor of pCR (OR 2.95; 95 % CI, 1.07–8.17; $p=0.04$). This clinical evidence in conjunction with the known preclinical antiproliferative effects of metformin has generated significant interest in investigating metformin in ongoing randomized, prospective clinical trials. A significant challenge in determining the optimum use of metformin in breast cancer will be developing biomarkers to select the patients most likely to benefit from this agent.

Use of nonsteroidal anti-inflammatory drugs (NSAIDs) for the prevention of breast cancer is an attractive approach given the role of inflammation in breast carcinogenesis and the reduced risk of other tumors such as colon cancer associated with NSAID use [130–133]. Epidemiologic data, however, are conflicting with regard to the benefits of aspirin and COX-2 inhibitors in breast cancer prevention. In a large cohort analysis of the Nurses' Health Study, Zhang et al. prospectively observed 84,602 postmenopausal women for the development of breast cancer with a follow-up period of 28 years [134]. A total of 4,734 cases of invasive breast cancer were identified, and no statistically significant difference was observed between regular aspirin users and nonusers. The lack of benefit with aspirin consumption was independent of tumor phenotype, dose, or either past or current aspirin use. The authors also examined use of acetaminophen and other NSAIDs and similarly failed to demonstrate any risk reduction. Notably, however, a moderately lower risk of hormone receptor-positive breast cancer was observed in women who consumed six or more tablets of aspirin per week for at least 10 years. While this prospective study provides important information, the question remains as to why a preventative effect associated with NSAIDs is observed in several other epidemiologic studies [135, 136]. Aside from study design limitations, one possible explanation for this inconsistency is the heterogeneity of patients studied. It is possible that a specific population, such as those with breast inflammation and elevated tissue levels of COX-2, is most likely to derive particular benefit from the use of NSAIDs and specific COX-2 inhibitors [137].

A number of other potential therapeutic strategies and targets are at different stages of evaluation. Statins possess anti-inflammatory properties and are the subject of ongoing investigation as data regarding their use and impact on cancer development and evolution are currently conflicting [138–146]. In preclinical ER+ breast cancer cell lines, use of small molecule MEK inhibitors has been shown to counter the anti-apoptotic effects of IGF-1 [147]. Other approaches to IGF-1 inhibition include the use of antibodies or small molecules directed against the IGF-1 receptor [148–150].

8.7 Conclusion

The increasing worldwide problem of overweight and obesity has critical implications for breast cancer. First, increasing adiposity is a risk factor for the development of hormone receptor-positive breast cancer in postmenopausal women. Second, irrespective of breast cancer subtype, obese women have a worse prognosis after breast cancer diagnosis. As obesity rates continue to rise, we are likely to see a concomitant rise in both breast cancer incidence and mortality rates, unless this trend can be reversed. Overweight and obesity have important implications for the local treatment of breast cancer, including surgery and radiotherapy, as well as for the use of systemic therapies such as chemotherapy, endocrine therapy, and novel targeted agents. Critical to the development of effective strategies to prevent increasing mortality with greater obesity prevalence is an understanding of the biological mechanisms by which rising BMI promotes breast cancer development and progression. As such, a wealth of recent studies have identified several mechanisms, which mediate this link, including dysregulated metabolism, insulin resistance, altered adipokine levels, as well as the emerging role of chronic inflammation of the breast. Tailoring currently available interventions to circumvent the pathways by which inflammation and dysfunctional metabolism cause treatment resistance may be one strategy to improve outcomes in obese and overweight patients.

The development of effective lifestyle modifications may also be important in risk reduction and improving breast cancer-specific outcomes. Finally, identifying specific patient populations that may benefit from pharmaceutical intervention is a subject of ongoing research. It is likely that a comprehensive, multimodality approach that incorporates all of these strategies will prove to provide the greatest benefit and redefine breast cancer treatment in the coming era. Hopefully, in the future, a greater understanding of obesity-breast cancer biology will translate into improvements in patient outcomes. Furthermore, a better understanding of obesity-mediated inflammation may offer broad implications for scientific progress in the many other diseases associated with inflammation and obesity.

Acknowledgments This work was supported by NCI 1R01CA154481, the Breast Cancer Research Foundation, and the Botwinick-Wolfensohn Foundation (in memory of Mr. and Mrs. Benjamin Botwinick).

References

1. Wang YC, McPherson K, Marsh T, Gortmaker SL, Brown M (2011) Health and economic burden of the projected obesity trends in the USA and the UK. Lancet 378(9793):815–825. doi:S0140-6736(11)60814-3 [pii] 10.1016/S0140-6736(11)60814-3 [doi]
2. Centers for disease control and prevention: overweight and obesity. http://wwwcdcgov/obesity/indexhtml
3. Levi J, Segal LM, St. Laurent R, Lang A, Rayburn J (2012) F as in fat: how obesity threatens America's future 2012. Trust for America's Health. http://healthyamericans.org/assets/files/TFAH2012FasInFatFnlRv.pdf. Accessed 28 Sept 2012.
4. Cleary MP, Grossmann ME (2009) Minireview: obesity and breast cancer: the estrogen connection. Endocrinology 150(6):2537–2542. doi:10.1210/en.2009-0070
5. van Kruijsdijk RC, van der Wall E, Visseren FL (2009) Obesity and cancer: the role of dysfunctional adipose tissue. Cancer Epidemiol Biomarkers Prev 18(10):2569–2578. doi:10.1158/1055-9965.EPI-09-0372
6. van den Brandt PA, Spiegelman D, Yaun SS, Adami HO, Beeson L, Folsom AR, Fraser G, Goldbohm RA, Graham S, Kushi L, Marshall JR, Miller AB, Rohan T, Smith-Warner SA, Speizer FE, Willett WC, Wolk A, Hunter DJ (2000) Pooled analysis of prospective cohort studies on height, weight, and breast cancer risk. Am J Epidemiol 152(6):514–527
7. Trentham-Dietz A, Newcomb PA, Storer BE, Longnecker MP, Baron J, Greenberg ER, Willett WC (1997) Body size and risk of breast cancer. Am J Epidemiol 145(11):1011–1019
8. Key TJ, Appleby PN, Reeves GK, Roddam A, Dorgan JF, Longcope C, Stanczyk FZ, Stephenson HE Jr, Falk RT, Miller R, Schatzkin A, Allen DS, Fentiman IS, Wang DY,

Dowsett M, Thomas HV, Hankinson SE, Toniolo P, Akhmedkhanov A, Koenig K, Shore RE, Zeleniuch-Jacquotte A, Berrino F, Muti P, Micheli A, Krogh V, Sieri S, Pala V, Venturelli E, Secreto G, Barrett-Connor E, Laughlin GA, Kabuto M, Akiba S, Stevens RG, Neriishi K, Land CE, Cauley JA, Kuller LH, Cummings SR, Helzlsouer KJ, Alberg AJ, Bush TL, Comstock GW, Gordon GB, Miller SR (2003) Body mass index, serum sex hormones, and breast cancer risk in postmenopausal women. J Natl Cancer Inst 95(16):1218–1226

9. Calle EE, Rodriguez C, Walker-Thurmond K, Thun MJ (2003) Overweight, obesity, and mortality from cancer in a prospectively studied cohort of U.S. adults. N Engl J Med 348(17):1625–1638. doi:10.1056/NEJMoa021423
10. La Vecchia C, Negri E, Franceschi S, Talamini R, Bruzzi P, Palli D, Decarli A (1997) Body mass index and post-menopausal breast cancer: an age-specific analysis. Br J Cancer 75(3):441–444
11. Lorincz AM, Sukumar S (2006) Molecular links between obesity and breast cancer. Endocr Relat Cancer 13(2):279–292. doi:10.1677/erc.1.00729
12. Li CI, Daling JR, Malone KE (2003) Incidence of invasive breast cancer by hormone receptor status from 1992 to 1998. J Clin Oncol 21(1):28–34
13. Wake DJ, Strand M, Rask E, Westerbacka J, Livingstone DE, Soderberg S, Andrew R, Yki-Jarvinen H, Olsson T, Walker BR (2007) Intra-adipose sex steroid metabolism and body fat distribution in idiopathic human obesity. Clin Endocrinol 66(3):440–446. doi:10.1111/j.1365-2265.2007.02755.x
14. Daling JR, Malone KE, Doody DR, Johnson LG, Gralow JR, Porter PL (2001) Relation of body mass index to tumor markers and survival among young women with invasive ductal breast carcinoma. Cancer 92(4):720–729
15. Fagherazzi G, Chabbert-Buffet N, Fabre A, Guillas G, Boutron-Ruault MC, Mesrine S, Clavel-Chapelon F (2012) Hip circumference is associated with the risk of premenopausal ER-/PR- breast cancer. Int J Obes (Lond) 36(3):431–439. doi:10.1038/ijo.2011.66
16. Pierce BL, Ballard-Barbash R, Bernstein L, Baumgartner RN, Neuhouser ML, Wener MH, Baumgartner KB, Gilliland FD, Sorensen BE, McTiernan A, Ulrich CM (2009) Elevated biomarkers of inflammation are associated with reduced survival among breast cancer patients. J Clin Oncol 27(21):3437–3444. doi:10.1200/JCO.2008.18.9068
17. Olefsky JM, Glass CK (2010) Macrophages, inflammation, and insulin resistance. Annu Rev Physiol 72:219–246. doi:10.1146/annurev-physiol-021909-135846
18. Tilg H, Moschen AR (2006) Adipocytokines: mediators linking adipose tissue, inflammation and immunity. Nat Rev Immunol 6(10):772–783. doi:10.1038/nri1937
19. Xu H, Barnes GT, Yang Q, Tan G, Yang D, Chou CJ, Sole J, Nichols A, Ross JS, Tartaglia LA, Chen H (2003) Chronic inflammation in fat plays a crucial role in the development of obesity-related insulin resistance. J Clin Invest 112(12):1821–1830. doi:10.1172/JCI19451
20. Skurk T, Alberti-Huber C, Herder C, Hauner H (2007) Relationship between adipocyte size and adipokine expression and secretion. J Clin Endocrinol Metab 92(3):1023–1033. doi:10.1210/jc.2006-1055
21. Weyer C, Funahashi T, Tanaka S, Hotta K, Matsuzawa Y, Pratley RE, Tataranni PA (2001) Hypoadiponectinemia in obesity and type 2 diabetes: close association with insulin resistance and hyperinsulinemia. J Clin Endocrinol Metab 86(5):1930–1935
22. Mantzoros C, Petridou E, Dessypris N, Chavelas C, Dalamaga M, Alexe DM, Papadiamantis Y, Markopoulos C, Spanos E, Chrousos G, Trichopoulos D (2004) Adiponectin and breast cancer risk. J Clin Endocrinol Metab 89(3):1102–1107
23. Grossmann ME, Nkhata KJ, Mizuno NK, Ray A, Cleary MP (2008) Effects of adiponectin on breast cancer cell growth and signaling. Br J Cancer 98(2):370–379. doi:10.1038/sj.bjc.6604166
24. Jarde T, Perrier S, Vasson MP, Caldefie-Chezet F (2011) Molecular mechanisms of leptin and adiponectin in breast cancer. Eur J Cancer 47(1):33–43. doi:10.1016/j.ejca.2010.09.005
25. Jarde T, Caldefie-Chezet F, Damez M, Mishellany F, Penault-Llorca F, Guillot J, Vasson MP (2008) Leptin and leptin receptor involvement in cancer development: a study on human primary breast carcinoma. Oncol Rep 19(4):905–911

26. Revillion F, Charlier M, Lhotellier V, Hornez L, Giard S, Baranzelli MC, Djiane J, Peyrat JP (2006) Messenger RNA expression of leptin and leptin receptors and their prognostic value in 322 human primary breast cancers. Clin Cancer Res 12(7 pt 1):2088–2094. doi:10.1158/1078-0432.CCR-05-1904
27. Okumura M, Yamamoto M, Sakuma H, Kojima T, Maruyama T, Jamali M, Cooper DR, Yasuda K (2002) Leptin and high glucose stimulate cell proliferation in MCF-7 human breast cancer cells: reciprocal involvement of PKC-alpha and PPAR expression. Biochim Biophys Acta 1592(2):107–116
28. Caldefie-Chezet F, Damez M, de Latour M, Konska G, Mishellani F, Fusillier C, Guerry M, Penault-Llorca F, Guillot J, Vasson MP (2005) Leptin: a proliferative factor for breast cancer? Study on human ductal carcinoma. Biochem Biophys Res Commun 334(3):737–741. doi:10.1016/j.bbrc.2005.06.077
29. Garofalo C, Koda M, Cascio S, Sulkowska M, Kanczuga-Koda L, Golaszewska J, Russo A, Sulkowski S, Surmacz E (2006) Increased expression of leptin and the leptin receptor as a marker of breast cancer progression: possible role of obesity-related stimuli. Clin Cancer Res 12(5):1447–1453. doi:10.1158/1078-0432.CCR-05-1913
30. Ishikawa M, Kitayama J, Nagawa H (2004) Enhanced expression of leptin and leptin receptor (OB-R) in human breast cancer. Clin Cancer Res 10(13):4325–4331. doi:10.1158/1078-0432.CCR-03-0749
31. Goodwin PJ, Ennis M, Pritchard KI, Trudeau ME, Koo J, Madarnas Y, Hartwick W, Hoffman B, Hood N (2002) Fasting insulin and outcome in early-stage breast cancer: results of a prospective cohort study. J Clin Oncol 20(1):42–51
32. Pollak MN, Schernhammer ES, Hankinson SE (2004) Insulin-like growth factors and neoplasia. Nat Rev Cancer 4(7):505–518. doi:10.1038/nrc1387
33. Chong YM, Colston K, Jiang WG, Sharma AK, Mokbel K (2006) The relationship between the insulin-like growth factor-1 system and the oestrogen metabolising enzymes in breast cancer tissue and its adjacent non-cancerous tissue. Breast Cancer Res Treat 99(3):275–288. doi:10.1007/s10549-006-9215-y
34. Lueprasitsakul P, Latour D, Longcope C (1990) Aromatase activity in human adipose tissue stromal cells: effect of growth factors. Steroids 55(12):540–544
35. Baglietto L, English DR, Hopper JL, MacInnis RJ, Morris HA, Tilley WD, Krishnan K, Giles GG (2009) Circulating steroid hormone concentrations in postmenopausal women in relation to body size and composition. Breast Cancer Res Treat 115(1):171–179. doi:10.1007/s10549-008-0069-3
36. Morris PG, Hudis CA, Giri D, Morrow M, Falcone DJ, Zhou XK, Du B, Brogi E, Crawford CB, Kopelovich L, Subbaramaiah K, Dannenberg AJ (2011) Inflammation and increased aromatase expression occur in the breast tissue of obese women with breast cancer. Cancer Prev Res (Phila) 4(7):1021–1029. doi:10.1158/1940-6207.CAPR-11-0110
37. Cancello R, Henegar C, Viguerie N, Taleb S, Poitou C, Rouault C, Coupaye M, Pelloux V, Hugol D, Bouillot JL, Bouloumie A, Barbatelli G, Cinti S, Svensson PA, Barsh GS, Zucker JD, Basdevant A, Langin D, Clement K (2005) Reduction of macrophage infiltration and chemoattractant gene expression changes in white adipose tissue of morbidly obese subjects after surgery-induced weight loss. Diabetes 54(8):2277–2286
38. Cinti S, Mitchell G, Barbatelli G, Murano I, Ceresi E, Faloia E, Wang S, Fortier M, Greenberg AS, Obin MS (2005) Adipocyte death defines macrophage localization and function in adipose tissue of obese mice and humans. J Lipid Res 46(11):2347–2355. doi:10.1194/jlr.M500294-JLR200
39. Mathis D, Shoelson SE (2011) Immunometabolism: an emerging frontier. Nat Rev Immunol 11(2):81. doi:10.1038/nri2922
40. Monteiro R, Azevedo I (2010) Chronic inflammation in obesity and the metabolic syndrome. Mediators Inflamm 2010. doi:10.1155/2010/289645
41. Hotamisligil GS (2010) Endoplasmic reticulum stress and the inflammatory basis of metabolic disease. Cell 140(6):900–917. doi:10.1016/j.cell.2010.02.034

42. Shi H, Kokoeva MV, Inouye K, Tzameli I, Yin H, Flier JS (2006) TLR4 links innate immunity and fatty acid-induced insulin resistance. J Clin Invest 116(11):3015–3025. doi:10.1172/JCI28898
43. Bachelot T, Ray-Coquard I, Menetrier-Caux C, Rastkha M, Duc A, Blay JY (2003) Prognostic value of serum levels of interleukin 6 and of serum and plasma levels of vascular endothelial growth factor in hormone-refractory metastatic breast cancer patients. Br J Cancer 88(11):1721–1726. doi:10.1038/sj.bjc.6600956
44. Dandona P, Weinstock R, Thusu K, Abdel-Rahman E, Aljada A, Wadden T (1998) Tumor necrosis factor-alpha in sera of obese patients: fall with weight loss. J Clin Endocrinol Metab 83(8):2907–2910
45. Vozarova B, Weyer C, Hanson K, Tataranni PA, Bogardus C, Pratley RE (2001) Circulating interleukin-6 in relation to adiposity, insulin action, and insulin secretion. Obes Res 9(7):414–417. doi:10.1038/oby.2001.54
46. Nguyen MT, Favelyukis S, Nguyen AK, Reichart D, Scott PA, Jenn A, Liu-Bryan R, Glass CK, Neels JG, Olefsky JM (2007) A subpopulation of macrophages infiltrates hypertrophic adipose tissue and is activated by free fatty acids via Toll-like receptors 2 and 4 and JNK-dependent pathways. J Biol Chem 282(48):35279–35292. doi:10.1074/jbc.M706762200
47. Murano I, Barbatelli G, Parisani V, Latini C, Muzzonigro G, Castellucci M, Cinti S (2008) Dead adipocytes, detected as crown-like structures, are prevalent in visceral fat depots of genetically obese mice. J Lipid Res 49(7):1562–1568. doi:10.1194/jlr.M800019-JLR200
48. Subbaramaiah K, Howe LR, Bhardwaj P, Du B, Gravaghi C, Yantiss RK, Zhou XK, Blaho VA, Hla T, Yang P, Kopelovich L, Hudis CA, Dannenberg AJ (2011) Obesity is associated with inflammation and elevated aromatase expression in the mouse mammary gland. Cancer Prev Res (Phila) 4(3):329–346. doi:10.1158/1940-6207.CAPR-10-0381
49. Sun X, Casbas-Hernandez P, Bigelow C, Makowski L, Joseph Jerry D, Smith Schneider S, Troester MA (2012) Normal breast tissue of obese women is enriched for macrophage markers and macrophage-associated gene expression. Breast Cancer Res Treat 131(3):1003–1012. doi:10.1007/s10549-011-1789-3
50. Subbaramaiah K, Morris PG, Zhou XK, Morrow M, Du B, Giri D, Kopelovich L, Hudis CA, Dannenberg AJ (2012) Increased levels of COX-2 and prostaglandin E2 contribute to elevated aromatase expression in inflamed breast tissue of obese women. Cancer Discov 2(4):356–365. doi:10.1158/2159-8290.CD-11-0241
51. McTiernan A, Rajan KB, Tworoger SS, Irwin M, Bernstein L, Baumgartner R, Gilliland F, Stanczyk FZ, Yasui Y, Ballard-Barbash R (2003) Adiposity and sex hormones in postmenopausal breast cancer survivors. J Clin Oncol 21(10):1961–1966. doi:10.1200/JCO.2003.07.057
52. Bulun SE, Lin Z, Imir G, Amin S, Demura M, Yilmaz B, Martin R, Utsunomiya H, Thung S, Gurates B, Tamura M, Langoi D, Deb S (2005) Regulation of aromatase expression in estrogen-responsive breast and uterine disease: from bench to treatment. Pharmacol Rev 57(3):359–383. doi:10.1124/pr.57.3.6
53. Irahara N, Miyoshi Y, Taguchi T, Tamaki Y, Noguchi S (2006) Quantitative analysis of aromatase mRNA expression derived from various promoters (I.4, I.3, PII and I.7) and its association with expression of TNF-alpha, IL-6 and COX-2 mRNAs in human breast cancer. Int J Cancer 118(8):1915–1921. doi:10.1002/ijc.21562
54. Zhao Y, Agarwal VR, Mendelson CR, Simpson ER (1996) Estrogen biosynthesis proximal to a breast tumor is stimulated by PGE2 via cyclic AMP, leading to activation of promoter II of the CYP19 (aromatase) gene. Endocrinology 137(12):5739–5742
55. Daniell HW (1988) Increased lymph node metastases at mastectomy for breast cancer associated with host obesity, cigarette smoking, age, and large tumor size. Cancer 62(2):429–435
56. Sohrabi A, Sandoz J, Spratt JS, Polk HC Jr (1980) Recurrence of breast cancer. Obesity, tumor size, and axillary lymph node metastases. JAMA 244(3):264–265
57. Boyd NF, Campbell JE, Germanson T, Thomson DB, Sutherland DJ, Meakin JW (1981) Body weight and prognosis in breast cancer. J Natl Cancer Inst 67(4):785–789
58. Cui Y, Whiteman MK, Flaws JA, Langenberg P, Tkaczuk KH, Bush TL (2002) Body mass and stage of breast cancer at diagnosis. Int J Cancer 98(2):279–283

59. Hunt KA, Sickles EA (2000) Effect of obesity on screening mammography: outcomes analysis of 88,346 consecutive examinations. AJR Am J Roentgenol 174(5):1251–1255
60. Nelson HD, Tyne K, Naik A, Bougatsos C, Chan BK, Humphrey L (2009) Screening for breast cancer: an update for the U.S. Preventive Services Task Force. Ann Inter Med 151(10):727–737, W237–W742. doi:10.1059/0003-4819-151-10-200911170-00009
61. Nystrom L, Andersson I, Bjurstam N, Frisell J, Nordenskjold B, Rutqvist LE (2002) Long-term effects of mammography screening: updated overview of the Swedish randomised trials. Lancet 359(9310):909–919. doi:10.1016/S0140-6736(02)08020-0
62. Elmore JG, Carney PA, Abraham LA, Barlow WE, Egger JR, Fosse JS, Cutter GR, Hendrick RE, D'Orsi CJ, Paliwal P, Taplin SH (2004) The association between obesity and screening mammography accuracy. Arch Intern Med 164(10):1140–1147. doi:10.1001/archinte.164.10.1140
63. Whiteman MK, Hillis SD, Curtis KM, McDonald JA, Wingo PA, Marchbanks PA (2005) Body mass and mortality after breast cancer diagnosis. Cancer Epidemiol Biomarkers Prev 14(8):2009–2014. doi:10.1158/1055-9965.EPI-05-0106
64. Sinicrope FA, Dannenberg AJ (2011) Obesity and breast cancer prognosis: weight of the evidence. J Clin Oncol 29(1):4–7. doi:10.1200/JCO.2010.32.1752
65. Majed B, Moreau T, Senouci K, Salmon RJ, Fourquet A, Asselain B (2008) Is obesity an independent prognosis factor in woman breast cancer? Breast Cancer Res Treat 111(2):329–342. doi:10.1007/s10549-007-9785-3
66. Rose DP, Vona-Davis L (2009) Influence of obesity on breast cancer receptor status and prognosis. Expert Rev Anticancer Ther 9(8):1091–1101. doi:10.1586/era.09.71
67. Protani M, Coory M, Martin JH (2010) Effect of obesity on survival of women with breast cancer: systematic review and meta-analysis. Breast Cancer Res Treat 123(3):627–635
68. Ewertz M, Jensen MB, Gunnarsdottir KA, Hojris I, Jakobsen EH, Nielsen D, Stenbygaard LE, Tange UB, Cold S (2011) Effect of obesity on prognosis after early-stage breast cancer. J Clin Oncol 29(1):25–31. doi:10.1200/JCO.2010.29.7614
69. Berclaz G, Li S, Price KN, Coates AS, Castiglione-Gertsch M, Rudenstam CM, Holmberg SB, Lindtner J, Erien D, Collins J, Snyder R, Thurlimann B, Fey MF, Mendiola C, Werner ID, Simoncini E, Crivellari D, Gelber RD, Goldhirsch A (2004) Body mass index as a prognostic feature in operable breast cancer: the International Breast Cancer Study Group experience. Ann Oncol 15(6):875–884
70. Chen X, Lu W, Zheng W, Gu K, Chen Z, Zheng Y, Shu XO (2010) Obesity and weight change in relation to breast cancer survival. Breast Cancer Res Treat 122(3):823–833. doi:10.1007/s10549-009-0708-3
71. Camoriano JK, Loprinzi CL, Ingle JN, Therneau TM, Krook JE, Veeder MH (1990) Weight change in women treated with adjuvant therapy or observed following mastectomy for node-positive breast cancer. J Clin Oncol 8(8):1327–1334
72. Kroenke CH, Chen WY, Rosner B, Holmes MD (2005) Weight, weight gain, and survival after breast cancer diagnosis. J Clin Oncol 23(7):1370–1378. doi:10.1200/JCO.2005.01.079
73. Levine EG, Raczynski JM, Carpenter JT (1991) Weight gain with breast cancer adjuvant treatment. Cancer 67(7):1954–1959
74. Caan BJ, Kwan ML, Hartzell G, Castillo A, Slattery ML, Sternfeld B, Weltzien E (2008) Pre-diagnosis body mass index, post-diagnosis weight change, and prognosis among women with early stage breast cancer. Cancer Causes Control 19(10):1319–1328. doi:10.1007/s10552-008-9203-0
75. Tartter PI, Beck G, Fuchs K (1994) Determinants of hospital stay after modified radical mastectomy. Am J Surg 168(4):320–324
76. El-Tamer MB, Ward BM, Schifftner T, Neumayer L, Khuri S, Henderson W (2007) Morbidity and mortality following breast cancer surgery in women: national benchmarks for standards of care. Ann Surg 245(5):665–671. doi:10.1097/01.sla.0000245833.48399.9a
77. Helyer LK, Varnic M, Le LW, Leong W, McCready D (2010) Obesity is a risk factor for developing postoperative lymphedema in breast cancer patients. Breast J 16(1):48–54. doi:10.1111/j.1524-4741.2009.00855.x

78. Mehrara BJ, Santoro TD, Arcilla E, Watson JP, Shaw WW, Da Lio AL (2006) Complications after microvascular breast reconstruction: experience with 1195 flaps. Plast Reconstr Surg 118(5):1100–1109; discussion 1110–1101. doi:10.1097/01.prs.0000236898.87398.d6
79. Spear SL, Ducic I, Cuoco F, Taylor N (2007) Effect of obesity on flap and donor-site complications in pedicled TRAM flap breast reconstruction. Plast Reconstr Surg 119(3):788–795. doi:10.1097/01.prs.0000252003.14537.d2
80. McCarthy CM, Mehrara BJ, Riedel E, Davidge K, Hinson A, Disa JJ, Cordeiro PG, Pusic AL (2008) Predicting complications following expander/implant breast reconstruction: an outcomes analysis based on preoperative clinical risk. Plast Reconstr Surg 121(6):1886–1892. doi:10.1097/PRS.0b013e31817151c4
81. Kroll SS, Netscher DT (1989) Complications of TRAM flap breast reconstruction in obese patients. Plast Reconstr Surg 84(6):886–892
82. Greco JA III, Castaldo ET, Nanney LB, Wu YC, Donahue R, Wendel JJ, Hagan KF, Shack RB (2008) Autologous breast reconstruction: the Vanderbilt experience (1998 to 2005) of independent predictors of displeasing outcomes. J Am Coll Surg 207(1):49–56. doi:10.1016/j.jamcollsurg.2007.12.052
83. Atisha DM, Alderman AK, Kuhn LE, Wilkins EG (2008) The impact of obesity on patient satisfaction with breast reconstruction. Plast Reconstr Surg 121(6):1893–1899. doi:10.1097/PRS.0b013e3181715198
84. Alderman AK, Wilkins EG, Kim HM, Lowery JC (2002) Complications in postmastectomy breast reconstruction: two-year results of the Michigan Breast Reconstruction Outcome Study. Plast Reconstr Surg 109(7):2265–2274
85. Chang DW, Wang B, Robb GL, Reece GP, Miller MJ, Evans GR, Langstein HN, Kroll SS (2000) Effect of obesity on flap and donor-site complications in free transverse rectus abdominis myocutaneous flap breast reconstruction. Plast Reconstr Surg 105(5):1640–1648
86. Zampell JC, Aschen S, Weitman ES, Yan A, Elhadad S, De Brot M, Mehrara BJ (2012) Regulation of adipogenesis by lymphatic fluid stasis: part I. Adipogenesis, fibrosis, and inflammation. Plast Reconstr Surg 129(4):825–834. doi:10.1097/PRS.0b013e3182450b2d
87. Aschen S, Zampell JC, Elhadad S, Weitman E, De Brot M, Mehrara BJ (2012) Regulation of adipogenesis by lymphatic fluid stasis: part II. Expression of adipose differentiation genes. Plast Reconstr Surg 129(4):838–847. doi:10.1097/PRS.0b013e3182450b47
88. Garvey PB, Buchel EW, Pockaj BA, Gray RJ, Samson TD (2005) The deep inferior epigastric perforator flap for breast reconstruction in overweight and obese patients. Plast Reconstr Surg 115(2):447–457
89. Allen AM, Prosnitz RG, Ten Haken RK, Normolle DP, Yu X, Zhou SM, Marsh R, Marks LB, Pierce LJ (2005) Body mass index predicts the incidence of radiation pneumonitis in breast cancer patients. Cancer J 11(5):390–398
90. Werner RS, McCormick B, Petrek J, Cox L, Cirrincione C, Gray JR, Yahalom J (1991) Arm edema in conservatively managed breast cancer: obesity is a major predictive factor. Radiology 180(1):177–184
91. Murphy C, Anderson PR, Li T, Bleicher RJ, Sigurdson ER, Goldstein LJ, Swaby R, Denlinger C, Dushkin H, Nicolaou N, Freedman GM (2011) Impact of the radiation boost on outcomes after breast-conserving surgery and radiation. Int J Radiat Oncol Biol Phys 81(1):69–76. doi:10.1016/j.ijrobp.2010.04.067
92. Dorn PL, Corbin KS, Al-Hallaq H, Hasan Y, Chmura SJ (2012) Feasibility and acute toxicity of hypofractionated radiation in large-breasted patients. Int J Radiat Oncol Biol Phys 83(1):79–83. doi:10.1016/j.ijrobp.2011.05.074
93. Neal AJ, Torr M, Helyer S, Yarnold JR (1995) Correlation of breast dose heterogeneity with breast size using 3D CT planning and dose-volume histograms. Radiother Oncol 34(3):210–218
94. Das IJ, Cheng CW, Fein DA, Fowble B (1997) Patterns of dose variability in radiation prescription of breast cancer. Radiother Oncol 44(1):83–89
95. Chen Y, Rubin P, Williams J, Hernady E, Smudzin T, Okunieff P (2001) Circulating IL-6 as a predictor of radiation pneumonitis. Int J Radiat Oncol Biol Phys 49(3):641–648

96. Chen Y, Williams J, Ding I, Hernady E, Liu W, Smudzin T, Finkelstein JN, Rubin P, Okunieff P (2002) Radiation pneumonitis and early circulatory cytokine markers. Semin Radiat Oncol 12(1 suppl 1):26–33
97. Lyman GH, Dale DC, Crawford J (2003) Incidence and predictors of low dose-intensity in adjuvant breast cancer chemotherapy: a nationwide study of community practices. J Clin Oncol 21(24):4524–4531. doi:10.1200/JCO.2003.05.002
98. Griggs JJ, Sorbero ME, Lyman GH (2005) Undertreatment of obese women receiving breast cancer chemotherapy. Arch Intern Med 165(11):1267–1273. doi:10.1001/archinte.165.11.1267
99. Bastarrachea J, Hortobagyi GN, Smith TL, Kau SW, Buzdar AU (1994) Obesity as an adverse prognostic factor for patients receiving adjuvant chemotherapy for breast cancer. Ann Intern Med 120(1):18–25
100. Rosner GL, Hargis JB, Hollis DR, Budman DR, Weiss RB, Henderson IC, Schilsky RL (1996) Relationship between toxicity and obesity in women receiving adjuvant chemotherapy for breast cancer: results from cancer and leukemia group B study 8541. J Clin Oncol 14(11):3000–3008
101. Georgiadis MS, Steinberg SM, Hankins LA, Ihde DC, Johnson BE (1995) Obesity and therapy-related toxicity in patients treated for small-cell lung cancer. J Natl Cancer Inst 87(5):361–366
102. Gelman RS, Tormey DC, Betensky R, Mansour EG, Falkson HC, Falkson G, Creech RH, Haller DG (1987) Actual versus ideal weight in the calculation of surface area: effects on dose of 11 chemotherapy agents. Cancer Treat Rep 71(10):907–911
103. Smith TJ, Desch CE (1991) Neutropenia-wise and pound-foolish: safe and effective chemotherapy in massively obese patients. South Med J 84(7):883–885
104. Poikonen P, Blomqvist C, Joensuu H (2001) Effect of obesity on the leukocyte nadir in women treated with adjuvant cyclophosphamide, methotrexate, and fluorouracil dosed according to body surface area. Acta Oncol 40(1):67–71
105. Griggs JJ, Mangu PB, Anderson H, Balaban EP, Dignam JJ, Hryniuk WM, Morrison VA, Pini TM, Runowicz CD, Rosner GL, Shayne M, Sparreboom A, Sucheston LE, Lyman GH (2012) Appropriate chemotherapy dosing for obese adult patients with cancer: American Society of Clinical Oncology clinical practice guideline. J Clin Oncol 30(13):1553–1561. doi:10.1200/JCO.2011.39.9436
106. Litton JK, Gonzalez-Angulo AM, Warneke CL, Buzdar AU, Kau SW, Bondy M, Mahabir S, Hortobagyi GN, Brewster AM (2008) Relationship between obesity and pathologic response to neoadjuvant chemotherapy among women with operable breast cancer. J Clin Oncol 26(25):4072–4077. doi:10.1200/JCO.2007.14.4527
107. Chen S, Chen CM, Zhou Y, Zhou RJ, Yu KD, Shao ZM (2012) Obesity or overweight is associated with worse pathological response to neoadjuvant chemotherapy among chinese women with breast cancer. PLoS One 7(7):e41380. doi:10.1371/journal.pone.0041380
108. de Azambuja E, McCaskill-Stevens W, Francis P, Quinaux E, Crown JP, Vicente M, Giuliani R, Nordenskjold B, Gutierez J, Andersson M, Vila MM, Jakesz R, Demol J, Dewar J, Santoro A, Lluch A, Olsen S, Gelber RD, Di Leo A, Piccart-Gebhart M (2010) The effect of body mass index on overall and disease-free survival in node-positive breast cancer patients treated with docetaxel and doxorubicin-containing adjuvant chemotherapy: the experience of the BIG 02-98 trial. Breast Cancer Res Treat 119(1):145–153. doi:10.1007/s10549-009-0512-0
109. Sestak I, Distler W, Forbes JF, Dowsett M, Howell A, Cuzick J (2010) Effect of body mass index on recurrences in tamoxifen and anastrozole treated women: an exploratory analysis from the ATAC trial. J Clin Oncol 28(21):3411–3415. doi:10.1200/JCO.2009.27.2021
110. Dignam JJ, Wieand K, Johnson KA, Fisher B, Xu L, Mamounas EP (2003) Obesity, tamoxifen use, and outcomes in women with estrogen receptor-positive early-stage breast cancer. J Natl Cancer Inst 95(19):1467–1476
111. Sparano JA, Wang M, Zhao F, Stearns V, Martino S, Ligibel JA, Perez EA, Saphner T, Wolff AC, Sledge GW Jr, Wood WC, Fetting J, Davidson NE (2012) Obesity at diagnosis is associated with inferior outcomes in hormone receptor-positive operable breast cancer. Cancer 118(23):5937–5946. doi:10.1002/cncr.27527

112. Folkerd EJ, Dixon JM, Renshaw L, A'Hern RP, Dowsett M (2012) Suppression of plasma estrogen levels by letrozole and anastrozole is related to body mass index in patients with breast cancer. J Clin Oncol 30(24):2977–2980. doi:10.1200/JCO.2012.42.0273
113. Ligibel JA, Goodwin PJ (2012) NEW and RENEW: building the case for weight loss in breast cancer. J Clin Oncol 30(19):2294–2296. doi:10.1200/JCO.2012.42.5496
114. Demark-Wahnefried W, Morey MC, Sloane R, Snyder DC, Miller PE, Hartman TJ, Cohen HJ (2012) Reach out to enhance wellness home-based diet-exercise intervention promotes reproducible and sustainable long-term improvements in health behaviors, body weight, and physical functioning in older, overweight/obese cancer survivors. J Clin Oncol 30(19):2354–2361. doi:10.1200/JCO.2011.40.0895
115. Campbell KL, Foster-Schubert KE, Alfano CM, Wang CC, Wang CY, Duggan CR, Mason C, Imayama I, Kong A, Xiao L, Bain CE, Blackburn GL, Stanczyk FZ, McTiernan A (2012) Reduced-calorie dietary weight loss, exercise, and sex hormones in postmenopausal women: randomized controlled trial. J Clin Oncol 30(19):2314–2326. doi:10.1200/JCO.2011.37.9792
116. Chlebowski RT, Aiello E, McTiernan A (2002) Weight loss in breast cancer patient management. J Clin Oncol 20(4):1128–1143
117. Holmes MD, Chen WY, Feskanich D, Kroenke CH, Colditz GA (2005) Physical activity and survival after breast cancer diagnosis. JAMA 293(20):2479–2486. doi:10.1001/jama.293.20.2479
118. Holick CN, Newcomb PA, Trentham-Dietz A, Titus-Ernstoff L, Bersch AJ, Stampfer MJ, Baron JA, Egan KM, Willett WC (2008) Physical activity and survival after diagnosis of invasive breast cancer. Cancer Epidemiol Biomarkers Prev 17(2):379–386. doi:10.1158/1055-9965.EPI-07-0771
119. Irwin ML, Smith AW, McTiernan A, Ballard-Barbash R, Cronin K, Gilliland FD, Baumgartner RN, Baumgartner KB, Bernstein L (2008) Influence of pre- and postdiagnosis physical activity on mortality in breast cancer survivors: the health, eating, activity, and lifestyle study. J Clin Oncol 26(24):3958–3964. doi:10.1200/JCO.2007.15.9822
120. Chen X, Lu W, Zheng W, Gu K, Matthews CE, Chen Z, Zheng Y, Shu XO (2011) Exercise after diagnosis of breast cancer in association with survival. Cancer Prev Res (Phila) 4(9):1409–1418. doi:10.1158/1940-6207.CAPR-10-0355
121. Chlebowski RT, Blackburn GL, Thomson CA, Nixon DW, Shapiro A, Hoy MK, Goodman MT, Giuliano AE, Karanja N, McAndrew P, Hudis C, Butler J, Merkel D, Kristal A, Caan B, Michaelson R, Vinciguerra V, Del Prete S, Winkler M, Hall R, Simon M, Winters BL, Elashoff RM (2006) Dietary fat reduction and breast cancer outcome: interim efficacy results from the Women's Intervention Nutrition Study. J Natl Cancer Inst 98(24):1767–1776. doi:10.1093/jnci/djj494
122. van Rossum EF, Nicklas BJ, Dennis KE, Berman DM, Goldberg AP (2000) Leptin responses to weight loss in postmenopausal women: relationship to sex-hormone binding globulin and visceral obesity. Obes Res 8(1):29–35. doi:10.1038/oby.2000.5
123. Imayama I, Ulrich CM, Alfano CM, Wang C, Xiao L, Wener MH, Campbell KL, Duggan C, Foster-Schubert KE, Kong A, Mason CE, Wang CY, Blackburn GL, Bain CE, Thompson HJ, McTiernan A (2012) Effects of a caloric restriction weight loss diet and exercise on inflammatory biomarkers in overweight/obese postmenopausal women: a randomized controlled trial. Cancer Res 72(9):2314–2326. doi:10.1158/0008-5472.CAN-11-3092
124. Esposito K, Pontillo A, Di Palo C, Giugliano G, Masella M, Marfella R, Giugliano D (2003) Effect of weight loss and lifestyle changes on vascular inflammatory markers in obese women: a randomized trial. JAMA 289(14):1799–1804. doi:10.1001/jama.289.14.1799
125. Bastard JP, Jardel C, Bruckert E, Blondy P, Capeau J, Laville M, Vidal H, Hainque B (2000) Elevated levels of interleukin 6 are reduced in serum and subcutaneous adipose tissue of obese women after weight loss. J Clin Endocrinol Metab 85(9):3338–3342
126. Brown KA, Hunger NI, Docanto M, Simpson ER (2010) Metformin inhibits aromatase expression in human breast adipose stromal cells via stimulation of AMP-activated protein kinase. Breast Cancer Res Treat 123(2):591–596. doi:10.1007/s10549-010-0834-y

127. Samarajeewa NU, Ham S, Yang F, Simpson ER, Brown KA (2011) Promoter-specific effects of metformin on aromatase transcript expression. Steroids 76(8):768–771. doi:10.1016/j.steroids.2011.02.041
128. Goodwin PJ, Pritchard KI, Ennis M, Clemons M, Graham M, Fantus IG (2008) Insulin-lowering effects of metformin in women with early breast cancer. Clin Breast Cancer 8(6):501–505. doi:10.3816/CBC.2008.n.060
129. Jiralerspong S, Palla SL, Giordano SH, Meric-Bernstam F, Liedtke C, Barnett CM, Hsu L, Hung MC, Hortobagyi GN, Gonzalez-Angulo AM (2009) Metformin and pathologic complete responses to neoadjuvant chemotherapy in diabetic patients with breast cancer. J Clin Oncol 27(20):3297–3302. doi:10.1200/JCO.2009.19.6410
130. Dannenberg AJ, Altorki NK, Boyle JO, Dang C, Howe LR, Weksler BB, Subbaramaiah K (2001) Cyclo-oxygenase 2: a pharmacological target for the prevention of cancer. Lancet Oncol 2(9):544–551. doi:10.1016/S1470-2045(01)00488-0
131. Howe LR, Subbaramaiah K, Brown AM, Dannenberg AJ (2001) Cyclooxygenase-2: a target for the prevention and treatment of breast cancer. Endocr Relat Cancer 8(2):97–114
132. Flossmann E, Rothwell PM (2007) Effect of aspirin on long-term risk of colorectal cancer: consistent evidence from randomised and observational studies. Lancet 369(9573):1603–1613. doi:10.1016/S0140-6736(07)60747-8
133. Howe LR (2007) Inflammation and breast cancer. Cyclooxygenase/prostaglandin signaling and breast cancer. Breast Cancer Res 9(4):210
134. Zhang X, Smith-Warner SA, Collins LC, Rosner B, Willett WC, Hankinson SE (2012) Use of aspirin, other nonsteroidal anti-inflammatory drugs, and acetaminophen and postmenopausal breast cancer incidence. J Clin Oncol 30(28):3468–3477. doi:10.1200/JCO.2012.42.2006
135. Terry MB, Gammon MD, Zhang FF, Tawfik H, Teitelbaum SL, Britton JA, Subbaramaiah K, Dannenberg AJ, Neugut AI (2004) Association of frequency and duration of aspirin use and hormone receptor status with breast cancer risk. JAMA 291(20):2433–2440. doi:10.1001/jama.291.20.2433
136. Takkouche B, Regueira-Mendez C, Etminan M (2008) Breast cancer and use of nonsteroidal anti-inflammatory drugs: a meta-analysis. J Natl Cancer Inst 100(20):1439–1447. doi:10.1093/jnci/djn324
137. Hudis CA, Subbaramaiah K, Morris PG, Dannenberg AJ (2012) Breast cancer risk reduction: no pain, no gain? J Clin Oncol 30(28):3436–3438. doi:10.1200/JCO.2012.44.8597
138. Shellman YG, Ribble D, Miller L, Gendall J, Vanbuskirk K, Kelly D, Norris DA, Dellavalle RP (2005) Lovastatin-induced apoptosis in human melanoma cell lines. Melanoma Res 15(2):83–89
139. Koyuturk M, Ersoz M, Altiok N (2004) Simvastatin induces proliferation inhibition and apoptosis in C6 glioma cells via c-jun N-terminal kinase. Neurosci Lett 370(2–3):212–217. doi:10.1016/j.neulet.2004.08.020
140. Girgert R, Vogt Y, Becke D, Bruchelt G, Schweizer P (1999) Growth inhibition of neuroblastoma cells by lovastatin and L-ascorbic acid is based on different mechanisms. Cancer Lett 137(2):167–172
141. Dimitroulakos J, Thai S, Wasfy GH, Hedley DW, Minden MD, Penn LZ (2000) Lovastatin induces a pronounced differentiation response in acute myeloid leukemias. Leuk Lymphoma 40(1–2):167–178. doi:10.3109/10428190009054894
142. Kawata S, Yamasaki E, Nagase T, Inui Y, Ito N, Matsuda Y, Inada M, Tamura S, Noda S, Imai Y, Matsuzawa Y (2001) Effect of pravastatin on survival in patients with advanced hepatocellular carcinoma. A randomized controlled trial. Br J Cancer 84(7):886–891. doi:10.1054/bjoc.2000.1716
143. Kim WS, Kim MM, Choi HJ, Yoon SS, Lee MH, Park K, Park CH, Kang WK (2001) Phase II study of high-dose lovastatin in patients with advanced gastric adenocarcinoma. Invest New Drugs 19(1):81–83

144. Knox JJ, Siu LL, Chen E, Dimitroulakos J, Kamel-Reid S, Moore MJ, Chin S, Irish J, LaFramboise S, Oza AM (2005) A Phase I trial of prolonged administration of lovastatin in patients with recurrent or metastatic squamous cell carcinoma of the head and neck or of the cervix. Eur J Cancer 41(4):523–530. doi:10.1016/j.ejca.2004.12.013
145. Larner J, Jane J, Laws E, Packer R, Myers C, Shaffrey M (1998) A phase I-II trial of lovastatin for anaplastic astrocytoma and glioblastoma multiforme. Am J Clin Oncol 21(6):579–583
146. Thibault A, Samid D, Tompkins AC, Fig. WD, Cooper MR, Hohl RJ, Trepel J, Liang B, Patronas N, Venzon DJ, Reed E, Myers CE (1996) Phase I study of lovastatin, an inhibitor of the mevalonate pathway, in patients with cancer. Clin Cancer Res 2(3):483–491
147. Periyasamy-Thandavan S, Takhar S, Singer A, Dohn MR, Jackson WH, Welborn AE, Leroith D, Marrero M, Thangaraju M, Huang S, Schoenlein PV (2012) Insulin-like growth factor 1 attenuates antiestrogen- and antiprogestin-induced apoptosis in ER+ breast cancer cells by MEK1 regulation of the BH3-only pro-apoptotic protein Bim. Breast Cancer Res 14(2):R52. doi:10.1186/bcr3153
148. Sachdev D, Li SL, Hartell JS, Fujita-Yamaguchi Y, Miller JS, Yee D (2003) A chimeric humanized single-chain antibody against the type I insulin-like growth factor (IGF) receptor renders breast cancer cells refractory to the mitogenic effects of IGF-I. Cancer Res 63(3):627–635
149. Ye JJ, Liang SJ, Guo N, Li SL, Wu AM, Giannini S, Sachdev D, Yee D, Brunner N, Ikle D, Fujita-Yamaguchi Y (2003) Combined effects of tamoxifen and a chimeric humanized single chain antibody against the type I IGF receptor on breast tumor growth in vivo. Horm Metab Res 35(11–12):836–842. doi:10.1055/s-2004-814145
150. Sachdev D, Singh R, Fujita-Yamaguchi Y, Yee D (2006) Down-regulation of insulin receptor by antibodies against the type I insulin-like growth factor receptor: implications for anti-insulin-like growth factor therapy in breast cancer. Cancer Res 66(4):2391–2402. doi:10.1158/0008-5472.CAN-05-3126

Chapter 9
Obesity, Inflammation, Nonalcoholic Fatty Liver Disease, and Hepatocellular Carcinoma

Naim Alkhouri and Arthur McCullough

Abstract

- Hepatocellular carcinoma (HCC) is a common malignancy worldwide that is increasing in incidence in the United States.
- Viral and alcohol-related liver diseases account for most cases; however, a significant number of patients with HCC do not have a known underlying chronic liver disease.
- New evidence suggests that obesity and type 2 diabetes may play a significant role in the development of HCC.
- Nonalcoholic fatty liver disease (NAFLD) is very common in patients with obesity and diabetes and may progress to cirrhosis and HCC.
- Insulin resistance, adipose tissue inflammation, adipokines, and inflammatory cytokines comprise mechanistic pathways that link obesity with NAFLD and HCC.

9.1 Introduction: The Changing Epidemiologic Pattern of HCC

Primary liver cancer is the sixth most common liver cancer worldwide and the third most common cause of cancer-related death [1]. Hepatocellular carcinoma (HCC) accounts for approximately 90 % of all primary liver cancers, and these two terms are frequently used interchangeably [2]. Each year, 20,000 new cases of HCC are diagnosed in the United States with an incidence that has tripled over the past 2 decades making HCC the fastest growing cause of cancer death in the male population [3].

N. Alkhouri, M.D. (✉) • A. McCullough, M.D.
Digestive Disease Institute, Cleveland Clinic Foundation,
9500 Euclid Avenue, Cleveland, OH 44105, USA
e-mail: alkhoun@ccf.org

A.J. Dannenberg and N.A. Berger (eds.), *Obesity, Inflammation and Cancer*,
Energy Balance and Cancer 7, DOI 10.1007/978-1-4614-6819-6_9,

Liver cirrhosis is present in approximately 70–90 % of HCC cases. Although chronic hepatitis B virus (HBV) infection is the major risk factor for HCC worldwide, infection with hepatitis C virus (HCV) is the causal factor in the majority of cases in the United States and Japan [4]. Other etiologies of chronic liver disease/cirrhosis associated with HCC include alcoholic liver disease, exposure to aflatoxin, hereditary hemochromatosis, α1-antitrypsin deficiency, primary biliary cirrhosis, and autoimmune hepatitis. However, a significant number of patients with HCC (15–50 %) do not have a known underlying chronic liver disease (cryptogenic) [5].

New evidence suggests that obesity and obesity-related complications, such as type 2 diabetes (DM2), metabolic syndrome (MetS), and nonalcoholic fatty liver disease (NAFLD), play a significant role in the development of HCC in susceptible patients.

In fact, epidemiological data have demonstrated a parallel increase in prevalence of HCC and obesity. In multiple population-based cohort studies from the USA and Europe [6–8], HCC was approximately twice as likely to develop in obese individuals as in those who were not obese, and similar results were found in patients with DM2 compared to nondiabetic subjects [9, 10]. NAFLD is the hepatic manifestation of MetS, and its prevalence (33 % of the adult US population) has been increasing with the growing epidemics of obesity and DM2. In addition, NAFLD appears to be a significant factor in the tumorigenesis of HCC in the setting of obesity [11].

9.2 Mechanistic Pathways Liking Obesity and Inflammation to Chronic Liver Disease and HCC

It has become clear that a state of low-grade chronic inflammation is typically associated with obesity and plays a crucial role in the development of insulin resistance (IR) [12, 13]. An important initiator of this inflammatory response is the adipose tissue, which actively secretes a variety of products such as cytokines, adipokines, and fatty acids into the circulation [14]. Abnormal adipose tissue metabolism has also been identified as a critical mechanistic link between obesity and NAFLD [15, 16]. As shown in Fig. 9.1, the increased release of free fatty acids (FFA) from adipose tissue due to a state of insulin resistance, which is characteristic of obesity, represents the main source of FFA during development of hepatic steatosis and progressive liver injury in NAFLD [17]. A surplus of FFA in non-adipose cells (ectopic fat deposition) may enter deleterious pathways leading to hepatocyte dysfunction (lipotoxicity) and apoptotic cell death (lipoapoptosis) [18, 19] which can occur through death receptors, the mitochondrial-lysosomal pathway, and endoplasmic reticulum (ER) stress [17]. The ensuing responses of cell repair, inflammation, regeneration, and fibrosis may all be triggered by apoptosis of adjacent cells [20]. Of these processes in the liver, hepatic fibrosis has the potential to be the most deleterious, as progressive fibrosis may result in cirrhosis and end-stage liver disease [21, 22].

Liver fibrosis is strongly associated with the development of HCC; in fact, up to 90 % of HCC cases arise in cirrhotic livers [23]. As shown in Fig. 9.2, excess

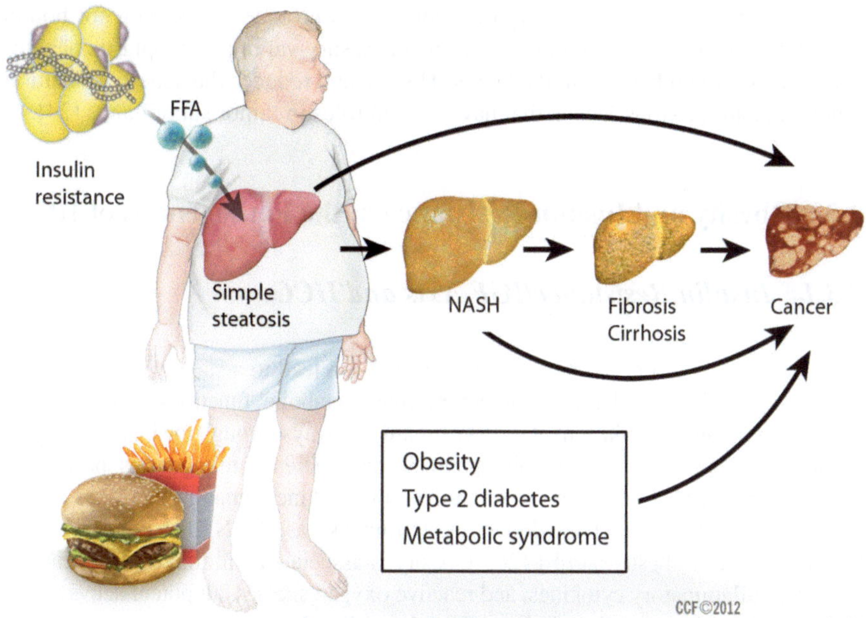

Fig. 9.1 The role of insulin resistance, free fatty acids, obesity, and diabetes in the development and progression of nonalcoholic fatty liver disease (NAFLD)

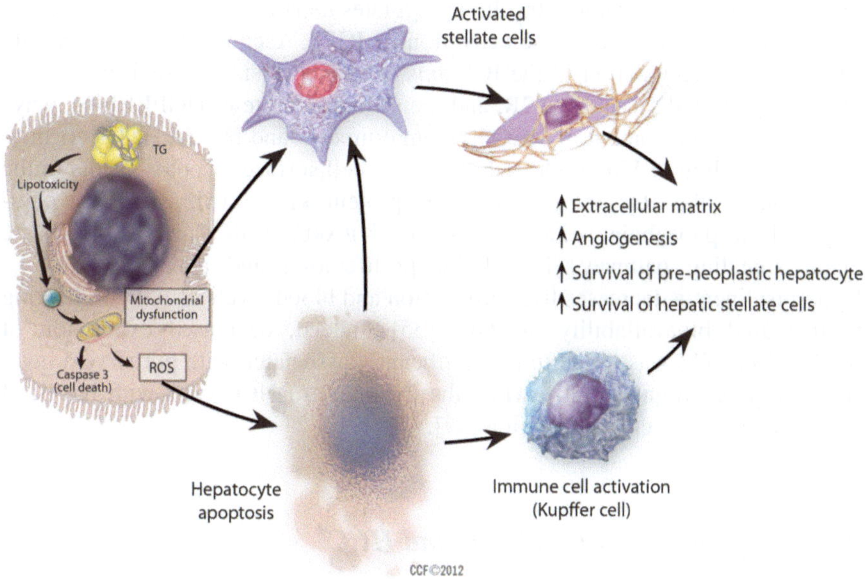

Fig. 9.2 The importance of fibrosis and the immune system for creating an environment in the liver conducive for the development of hepatocellular carcinoma (HCC)

extracellular matrix and stiffness in cirrhotic livers provides a reservoir for bound growth factors, promotes angiogenesis, enhances survival of pre-neoplastic hepatocytes and activated hepatic stellate cells (HSCs), and reduces the activity of natural killer and natural killer T cells that have critical roles in tumor surveillance [24].

9.3 Obesity and Insulin Resistance in the Pathogenesis of HCC

9.3.1 Insulin Resistance/IGF Axis and HCC

IR is frequently present in obese individuals and can lead to a state of compensatory hyperinsulinemia in order to maintain normal metabolic functions. As shown in Fig. 9.3, hyperinsulinemia activates downstream pathways through the insulin receptor substrate-1 (IRS-1), which inhibit apoptosis and increase mitogenesis promoting tumorigenesis. IRS-1 plays an important role in cytokine signaling pathways and is upregulated in HCC [25]. The c-Jun amino-terminal kinase 1 (JNK1) has emerged as a key link between obesity and IR [26]. Obesity is associated with increased release of FFA, pro-inflammatory cytokines, and reactive oxygen species, all potent activators of JNK, which lead to phosphorylation of IRS-1 and IR. JNK1 plays a significant role in the development of nonalcoholic steatohepatitis (NASH) [27] and was found recently to be associated with the development of HCC as more than 50 % of human HCC samples were observed to have increased activation of JNK1 [11, 28, 29].

The liver synthesizes and secretes insulin-like growth factors (IGFs), in response to growth hormone. Insulin upregulates hepatic growth hormone receptors, which in turn increase the activation of IGFs. Accumulating nascent data suggest that dysregulation of the IGF axis, which consists of IGF-1 and IGF-2, their receptors IGF1R and IGF2R, and their binding proteins (IGFBP1–6), plays a role in the development of HCC in animal models and human HCC cell lines [30, 31]. Binding of IGF-1 and IGF-2 to their cell surface receptor results in the activation of the Ras-mitogen-activated protein kinase (MAPK)-ERK pathway and the phosphoinositide 3-kinase (PI3K)/Akt/mTOR pathway, which are major signaling pathways in cellular proliferation and apoptosis [32–34]. Hyperinsulinemia decreases liver production and blood levels of IGFBPs leading to increased bioavailability of IGFs [35]. Indeed, decreased expression of IGFBP-3 in HCC was found to be significantly associated with portal vein invasion and poor prognosis [36], while the treatment of HCC cells with IGFBP-3 in vitro controlled cell proliferation [37].

9.3.2 Inflammatory Cytokines and HCC

Obesity is associated with the upregulation of pro-inflammatory cytokines such as tumor necrosis factor-α (TNF-α) and interleukin 6 (IL-6), both of which can stimulate

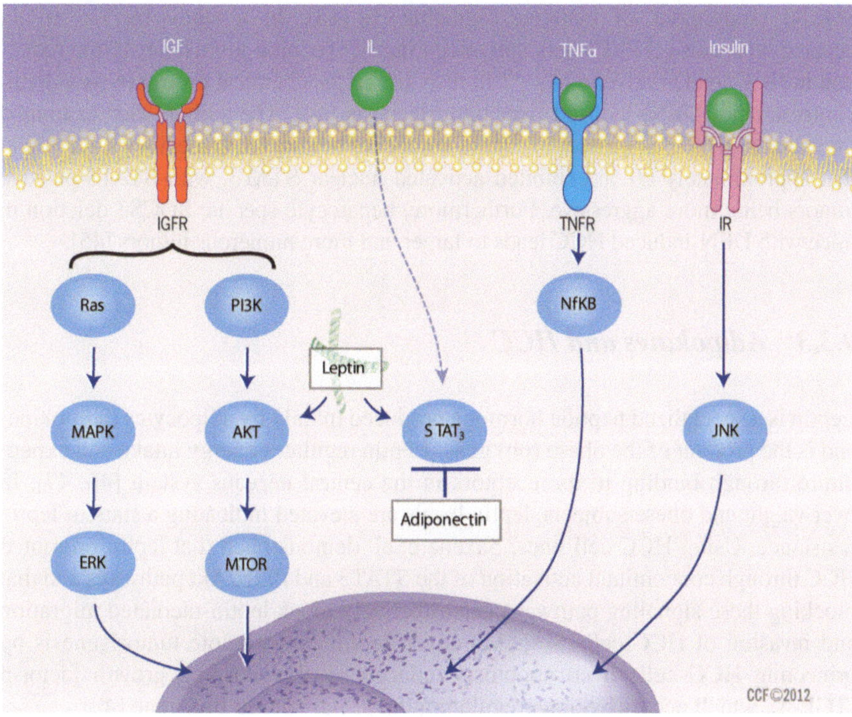

Fig. 9.3 The transcriptional factors and adipokines currently considered the most important carcinogenic in NAFLD

normal and malignant hepatocyte proliferation [38, 39]. The binding of TNF-α to its receptor activates the nuclear factor-κB pathway (NF-κB), which is considered a major factor in determining the ability of pre-neoplastic and malignant cells to avoid apoptosis and therefore providing a link between inflammation and cancer [40]. After activation, NF-κB translocate dimers to the nucleus where they affect transcriptional activation of hundreds of target genes. By using the Mdr-2 knockout mouse model, which spontaneously develops cholestatic hepatitis leading to dysplasia and eventually HCC, Pikarsky et al. demonstrated that TNF-α expression was upregulated in adjacent inflammatory and endothelial cells, triggering NF-κB signaling in hepatocytes, thereby promoting survival and the ultimate formation of HCC [41]. Furthermore, anti-TNF treatment suppressed NF-κB activation resulting in apoptosis of transformed hepatocytes and failure to progress to HCC. In contrast with the tumor-promoting role of NF-κB pathway in the animal model described above, the inactivation of the pathway at an early stage in other models can also promote carcinogenesis because of increased hepatocyte apoptosis which triggers compensatory hyperplasia, regeneration, and fibrosis leading to malignant transformation [42].

Another important inflammation-associated pathway is the signal transducer and activator of transcription 3 (STAT3) which is triggered by IL-6 and inhibited by

SOCS3 (suppressor of cytokine signaling 3) [43]. In a study by He et al., hepatocyte-specific STAT3-deficient mice (*Stat3*$^{\Delta hep}$) exhibit more than sixfold reduction in HCC load relative to *Stat3*$^{F/F}$ in response to the chemical procarcinogen diethylnitrosamine (DEN)-induced liver tumorigenesis [44]. The authors also examined the status of STAT3 activation in a large number of human HCC specimens and found that approximately 60 % exhibited activated nuclear STAT3, with STAT3-positive tumors being more aggressive. Furthermore, hepatocyte-specific SOCS3 deletion in mice with DEN-induced HCC leads to larger and more numerous tumors [45].

9.3.3 Adipokines and HCC

Leptin is a specialized peptide hormone produced mainly by adipocytes (adipokine) and is the product of the obese (ob) gene. Leptin regulates energy intake and expenditure through binding to its receptors in the central nervous system [46, 47]. In overweight and obese subjects, leptin levels are elevated indicating a state of leptin resistance. Using HCC cell lines, Saxena et al. demonstrated that leptin promoted HCC through concomitant activation of the STAT3 and PI3K/Akt pathways and that blocking these signaling pathways can effectively block leptin-mediated migration and invasion of HCC cells [48]. Moreover, leptin can promote tumorigenesis by protecting HCC cells from apoptosis induced by transforming growth factor-β (TGF-β), which occurs by downregulating the pro-apoptotic Bax gene [49].

Adiponectin is another adipokine mainly secreted by visceral adipose tissue. Adiponectin has three receptors, AdipoR1, AdipoR2, and T-cadherin, with AdipoR2 being the one predominantly expressed in the liver [50]. Unlike leptin, adiponectin levels are markedly decreased in obesity, DM2, NAFLD, and atherosclerosis indicating a protective effect of adiponectin against obesity-related disorders [51–53]. Similarly, reduced plasma or hepatic adiponectin correlates closely with the development of HCC with an inverse relationship between adiponectin expression and tumor size [54, 55]. Interestingly, adiponectin may inhibit tumor growth induced by leptin by reducing STAT3 and PI3K/Akt activity and increasing SOCS3 which is a negative regulator of leptin signaling [56].

9.4 HCC and Obesity

9.4.1 Obesity and HCC in the General Population

Obesity has been established as a risk factor for HCC by multiple large cohort studies [57]. This increased risk may be related to the increased prevalence of NAFLD and NASH in obese individuals, and the carcinogenic potential exerted by obesity alone, as discussed above. In a large Danish study of nearly 44,000 obese individuals, Moller et al. found an overall 16 % increased incidence of cancers and a relative risk of 1.9 for liver cancer [7]. Wolk et al. found that obesity was associated with a threefold increase in HCC risk in a large Swedish study that included more than 28,000 obese patients [8]. A Korean study that followed 781,283 males without a prior diagnosis of cancer for over a 10-year period found a relative risk for HCC of 1.53 in obese patients compared to normal controls after controlling for HBV infection (the most common cause of HCC in Korea) [58]. A mortality study conducted by the American Cancer Society that followed more than 900,000 US adults initially free from cancer for 16 years demonstrated that the relative risk of dying from liver cancer was 1.68 times higher in women and 4.52 times higher in men who had a BMI $\geq$ 35 kg/m^2 compared to lean controls with normal BMI (18.5–24.9 kg/m^2) [6]. Of note, the relative risk of liver cancer mortality among males was the highest of all the cancers studied. A meta-analysis that included 11 cohort studies found that compared to normal-weight individuals, the relative risk of liver cancer was 1.17 (95 % CI, 1.02–1.34) for overweight patients and 1.89 (95 % CI, 1.51–2.36) for obese patients [57].

9.4.2 Obesity and HCC in Patients with Cirrhosis

In addition to the association between obesity and HCC in the general population, obesity is an independent risk factor for HCC in patients with cirrhosis of different etiologies. A review of the United Network of Organ Sharing (UNOS) database that included 19,271 cirrhotic patients who had liver transplantation between 1991 and 2000 revealed that the overall incidence of HCC was higher in obese patients compared to non-obese patients (4.0 % vs. 3.0 %, $p = 0.013$) [59]. In the multivariate analysis of this study, obesity was an independent predictor of HCC in cryptogenic cirrhosis (OR = 11.1, 95 % CI, 1.5–87.4) and alcoholic cirrhosis (OR = 3.2, 95 % CI, 1.5–6.6) but not in cirrhosis due to viral hepatitis or autoimmune liver disease. A prospective 7-year study in France that followed 771 patients with alcohol- or HCV-related cirrhosis demonstrated that a BMI $\geq$ 30 kg/m^2 was associated with a hazard ratio for HCC of 2.8 [60].

Beyond the effect of obesity (assessed by BMI) on the risk of HCC, it appears that visceral fat accumulation per se may play a role in tumor initiation and progression. Visceral adipose tissue is a highly active endocrine organ that produces multiple pro-inflammatory cytokines (such as TNF-α and IL-6), promotes insulin resistance, and causes hepatocyte fat accumulation with ensuing hepatocyte injury and possibly carcinogenesis through FFA toxicity [31]. Ohki et al. studied the effect of visceral fat area, as measured by contrast-enhanced dynamic CT, on the recurrence of HCC after curative ablation [61]. They found that patients with high visceral fat area (defined as >130 cm^2 in male and >90 cm^2 in female) had a cumulative recurrence rates at 1 and 3 years of 15.9 % and 75.1 %, respectively, compared to 9.7 and 43.1 % in the control group ($p = 0.018$).

9.5 HCC and Type 2 Diabetes Mellitus (DM2)

Establishing the causal relationship between diabetes and HCC has proven difficult for multiple reasons. First, cirrhosis and end-stage liver disease can cause impaired glucose tolerance and overt diabetes [62] introducing potential bias in case–control studies. Second, some etiologies of chronic liver disease such as HCV and hemochromatosis are associated with increased diabetes risk [63]. Finally, diabetes is a strong risk factor for NAFLD and NASH which can progress to cirrhosis and HCC [64]. However, several large cohort studies have established the presence of DM2 as an independent risk factor for the development of HCC. Earlier population-based studies from Sweden and Denmark [65, 66] have shown a significant increase in HCC risk among patients with diabetes alone and in the presence of viral hepatitis, alcoholic liver disease, and cirrhosis. A large longitudinal study from the USA followed 173,643 diabetic patients and 650,620 patients without diabetes for 10–15 years and found that diabetes was associated with a hazard ratio for HCC of 2.16 (95 % confidence interval [CI], 1.86–1.09; $p<0.0001$) even after excluding patients with HCV, HBV, and alcoholic liver disease [67]. Importantly, this study provided evidence for a causal association between DM2 and HCC by demonstrating that diabetes preceded the development of chronic liver disease, as well as an increased risk for HCC in patients with longer duration of diabetes. Similar findings were reported from a hospital-based case–control study with diabetes being present in 87 % of cases before the diagnosis of HCC and patients with diabetes duration >10 years having an adjusted hazard ratio for HCC of 2.2 (95 % CI, 1.2–4.8) compared to those with a diabetes duration of 2–5 years. More robust support for diabetes as a risk factor for HCC came from a meta-analysis of 13 cohort studies that aimed to assess the association between diabetes and HCC. The analysis revealed a statistically significant 2.5-fold increase in HCC incidence among diabetic patients (95 % CI, 1.9–3.2; $p<0.01$) [9]. A more recent meta-analysis that included a total of 17 case–control studies and 32 cohort studies confirmed the association between diabetes and risk of HCC with a pooled relative risk of 2.31 for HCC among diabetic patients (95 % CI, 1.87–2.84) [68].

9.5.1 Association of Diabetes Treatment with HCC Risk

Several lines of evidence suggest that the type of antidiabetic therapy may increase or decrease HCC risk contingent upon mechanism of action [68].

On one hand, insulin sensitizers such as metformin can improve insulin sensitivity and suppress cell growth by activating the AMP-activated protein kinase yielding a protective effect on cancer risk [69]. A retrospective study from Italy that included 465 patients with HCC, 618 cirrhotics, and 490 controls found a statistically significant reduction in HCC risk for diabetics receiving metformin (OR = 0.33; 95 % CI, 0.1–0.61) [70]. Another study from the University of Texas reported a similar decrease in the risk for HCC with metformin (OR = 0.3; 95 % CI, 0.2–0.6) [71]. Moreover, a recent study demonstrated that metformin decreased the risk of HCC in diabetic patients in a dose-dependent manner with a 7 % risk reduction of HCC for each incremental year increase in metformin use [72]. Interestingly, metformin inhibited hepatoma cell line proliferation by inducing cell cycle arrest at the G0/G1 phase via AMP-activated protein kinase. Furthermore, metformin had chemosensitizing effect in combination with doxorubicin to accelerate HCC regression in a mouse xenograft model.

On the other hand, exogenous insulin and sulfonylureas lead to an increase in circulating insulin levels promoting carcinogenesis and increasing the risk for HCC. Donadon et al. found a statistically significant increase in HCC risk for diabetics receiving sulfonylureas or insulin (OR = 2.99; 95 % CI, 1.34–6.65), and similar findings were also reported by Hassan et al. (insulin use was associated with HCC OR of 1.9; 95 % CI, 0.8–4.6, and sulfonylurea use was associated with HCC OR of 7.1; 95 % CI, 2.9–16.9) [70, 71].

Overall, in the meta-analysis by Wang et al., the pooled risk estimates for developing HCC were 0.31 (95 % CI, 0.19–0.49) for diabetics treated with metformin and 4.0 (95 % CI, 1.94–8.24) for those treated with sulfonylurea or exogenous insulin [68].

In addition to the effects of antidiabetic therapy on the risk for liver cancer, statin use in diabetics appears to have a protective effect against HCC. Proposed mechanisms include the inhibition of downstream products of the mevalonate pathway which are important for the growth of malignant cells [73] and the inhibition of HSC proliferation and their production of collagen [74]. In a nested, matched, case–control study in diabetics, El-Serag et al. examined 1,303 cases and 5,212 controls and found a risk reduction for the development of HCC that ranged between 25 and 40 % providing evidence of the cancer-preventive effect of statins specific to HCC [75].

9.6 HCC and Nonalcoholic Fatty Liver Disease

9.6.1 The Epidemic of NAFLD

NAFLD is considered the hepatic manifestation of MetS and as such has become the most common form of chronic liver disease in the world [76–78]. NAFLD encompasses a wide histological spectrum of disease ranging from simple steatosis characterized by lipid accumulation in the liver in the form of triglyceride (TG) to NASH characterized by the association of lipid accumulation with evidence of hepatocyte injury, inflammation, and various degrees of fibrosis [79]. NASH is a serious condition that can progress to cirrhosis and its feared complications of portal hypertension and end-stage liver disease requiring liver transplantation. It is estimated that in the United States, NAFLD affects 30–45 % of the general population and as high as 90 % of the morbidly obese [80, 81]. Furthermore, the aggressive form of NASH affects between 5 and 13 % of Americans, making this entity a rising cause of liver-related morbidity and mortality and potentially the main indication for liver transplantation in the next decade [82]. A recent study found that among diabetic patients, NAFLD was present in 74 % and NASH in 22.2 % [81].

9.6.2 HCC and Cryptogenic Cirrhosis

When NASH progresses to cirrhosis, most of the classic histological features of the disease disappear, making the diagnosis of the underlying etiology difficult [83]. It has been proposed that NASH accounts for the majority of cases of cryptogenic cirrhosis (CC) because these patients have a significantly higher prevalence of conditions associated with NASH including obesity and diabetes compared to patients with cirrhosis of well-defined etiology [84, 85]. Three landmark studies published in 2002 expanded the spectrum of NAFLD from CC to HCC. In the first study, Bugianesi et al. retrospectively identified 646 Italian patients with cirrhosis-associated HCC and found that the prevalence of CC was 6.9 % compared to 55 % for HCV-cirrhosis, 16 % for HBV-cirrhosis, and 13 % for alcoholic cirrhosis [86]. Interestingly, the prevalence of pre-cirrhotic obesity, DM2, and dyslipidemia were more than twice as prevalent in patients who had CC as in the control group (p value <0.05 for all metabolic factors). The second study by Marrero et al. demonstrated that CC was the second most common etiology of underlying liver disease (29 %) after HCV (51 %) in a group of 105 patients with HCC from Michigan likely related to the high prevalence of obesity in the United States [87]. The third study by Ratziu et al. in France found CC in 27 % of HCC patients and corroborated the association between DM2, IR, and dyslipidemia with both CC and HCC [88].

9.6.3 HCC and NASH-Cirrhosis

Several studies have directly examined the incidence of HCC in patients with NAFLD or NASH-cirrhosis (Table 9.1). In a large cohort study that included 7,326 patients discharged with a diagnosis of fatty liver from Danish hospitals over a 16-year period, Sorensen et al. found that the risk for primary liver cancer was significantly elevated in NAFLD patients compared to the Danish general population with a standardized incidence ratio of 4.4 (95 % CI, 1.2–11.4) [89]. A large prospective US study compared 152 patients with NASH-cirrhosis (median age of 55 years) with 150 matched patients with HCV-cirrhosis. Those with NASH-cirrhosis had significantly lower risk of developing HCC over a 10-year follow-up period (6.7 % vs. 17 %, respectively, $p<0.01$) [90]. Hashimoto et al. conducted a case-controlled study that included 34 NASH patients with HCC and 348 patients NASH patients without HCC [91]. Risk factors for HCC in this study included older age, low AST level, low NAFLD activity score, and advanced fibrosis stage. A prospective cohort

Table 9.1 Studies that evaluated the incidence of hepatocellular carcinoma (HCC) in patients with nonalcoholic fatty liver disease (NAFLD)

References	NASH patients (*n*)	Follow-up (years)	HCC incidence (%)	Risk factors
Sanyal et al. [90]	152	10	6.7	None stated
Hashimoto et al. [91]	34 (case-controlled study) 137 (prospective study)	5	7.6	Older age Low AST Low NAFLD activity score Advanced fibrosis
Ascha et al. [92]	195	3.2	12.8	Older age Any alcohol consumption

study of 137 NASH patients with advanced fibrosis was included in that same study; this prospective cohort demonstrated that the 5-year cumulative incidence of HCC was 7.6 % [91]. Finally, a study from the Cleveland Clinic compared 195 patients with NASH-cirrhosis to 315 patients with HCV-cirrhosis that were evaluated for liver transplantation between 2003 and 2007 [92]. The yearly cumulative incidence of HCC in cirrhotic patients with NASH was 2.6 % compared to 4.0 % for those with hepatitis C infection. Interestingly, older age at the time of cirrhosis diagnosis and any alcohol consumption were independently associated with the development of HCC in the multivariate analysis (hazard ratio of 1.08 and 3.8, respectively; $p<0.005$), supporting the notion that alcohol intake even in small quantities may increase the risk of HCC development in the setting of cirrhosis.

9.6.4 HCC and Non-cirrhotic NAFLD

New data indicate that HCC can arise in steatotic livers in the absence of cirrhosis. In fact, several case series have described the occurrence of HCC in non-cirrhotic NAFLD patients. In a Japanese study that included 9 patients with HCC in the setting of NAFLD, one third of patients only had mild hepatic fibrosis and no evidence of cirrhosis [93]. Hashimoto et al. confirmed these findings by demonstrating that 12 % of their patients with NASH-HCC had mild fibrosis (fibrosis stage 1–2) [91].

In a large nationwide survey of 14,530 HCC patients in Japan, NAFLD-HCC and unknown HCC accounted for 7.1 % of all cases, but only 62 % of NAFLD-HCC had cirrhosis [94]. A European study by Ertle et al. that enrolled 150 patients with HCC (including 36 with NAFLD) intriguingly found that only 58 % of patients with NAFLD-HCC had evidence of cirrhosis [95]. These studies underscore the importance of HCC screening in high-risk NAFLD patients such as those with family history of HCC.

9.7 Summary

NAFLD has emerged as the most prevalent chronic liver disease. Population-based studies in the United States estimate that the prevalence of NAFLD is at least 30 % in adults and as high as 80 % in obese patients and in patients with type 2 diabetes. Of additional concern is the fact that the incidence of NAFLD has increased from 4.5/100,000 to 38/100,000 over the epoch of 1980–1999. Consistent with these data is the recognition that obesity and diabetes are risk factors for both NAFLD and HCC. Although viral and alcohol-related liver disease account for the majority of HCC, the increased incidence of HCC has occurred in parallel with incidences of NAFLD, obesity, and diabetes, with the latter two being risk factors for both the development and progression of NAFLD. This increased incidence of HCC in NAFLD and the fact that there is a predilection for HCC to develop in NAFLD even in the absence of cirrhosis has important clinical implications that may change the current paradigms for cancer surveillance in these patients. Finally, our increased knowledge of the mechanistic pathways that link obesity and diabetes to the pathogenesis of HCC in NAFLD (discussed above) provide strategies for both the clinical management of NAFLD and the development of targeted treatments for HCC associated with NAFLD.

References

1. Ferlay J, Shin HR, Bray F et al (2010) Estimates of worldwide burden of cancer in 2008: GLOBOCAN 2008. Int J Cancer 127:2893–2917
2. Nordenstedt H, White DL, El-Serag HB (2010) The changing pattern of epidemiology in hepatocellular carcinoma. Dig Liver Dis 42(suppl 3):S206–S214
3. El-Serag HB (2011) Hepatocellular carcinoma. N Engl J Med 365:1118–1127
4. McGlynn KA, London WT (2011) The global epidemiology of hepatocellular carcinoma: present and future. Clin Liver Dis 15:223–243, vii–x
5. Page JM, Harrison SA (2009) NASH and HCC. Clin Liver Dis 13:631–647
6. Calle EE, Rodriguez C, Walker-Thurmond K et al (2003) Overweight, obesity, and mortality from cancer in a prospectively studied cohort of U.S. adults. N Engl J Med 348:1625–1638
7. Moller H, Mellemgaard A, Lindvig K et al (1994) Obesity and cancer risk: a Danish record-linkage study. Eur J Cancer 30A:344–350
8. Wolk A, Gridley G, Svensson M et al (2001) A prospective study of obesity and cancer risk (Sweden). Cancer Causes Control 12:13–21
9. El-Serag HB, Hampel H, Javadi F (2006) The association between diabetes and hepatocellular carcinoma: a systematic review of epidemiologic evidence. Clin Gastroenterol Hepatol 4:369–380
10. El-Serag HB, Tran T, Everhart JE (2004) Diabetes increases the risk of chronic liver disease and hepatocellular carcinoma. Gastroenterology 126:460–468
11. Starley BQ, Calcagno CJ, Harrison SA (2010) Nonalcoholic fatty liver disease and hepatocellular carcinoma: a weighty connection. Hepatology 51:1820–1832
12. Wellen KE, Hotamisligil GS (2005) Inflammation, stress, and diabetes. J Clin Invest 115:1111–1119
13. Hotamisligil GS (2006) Inflammation and metabolic disorders. Nature 444:860–867

14. Weisberg SP, McCann D, Desai M et al (2003) Obesity is associated with macrophage accumulation in adipose tissue. J Clin Invest 112:1796–1808
15. Jou J, Choi SS, Diehl AM (2008) Mechanisms of disease progression in nonalcoholic fatty liver disease. Semin Liver Dis 28:370–379
16. Wree A, Kahraman A, Gerken G et al (2010) Obesity affects the liver—the link between adipocytes and hepatocytes. Digestion 83:124–133
17. Alkhouri N, Dixon LJ, Feldstein AE (2009) Lipotoxicity in nonalcoholic fatty liver disease: not all lipids are created equal. Expert Rev Gastroenterol Hepatol 3:445–451
18. Listenberger LL, Han X, Lewis SE et al (2003) Triglyceride accumulation protects against fatty acid-induced lipotoxicity. Proc Natl Acad Sci U S A 100:3077–3082
19. Unger RH (2002) Lipotoxic diseases. Annu Rev Med 53:319–336
20. Guicciardi ME, Gores GJ (2005) Apoptosis: a mechanism of acute and chronic liver injury. Gut 54:1024–1033
21. Bataller R, Brenner DA (2005) Liver fibrosis. J Clin Invest 115:209–218
22. Friedman SL (2003) Liver fibrosis—from bench to bedside. J Hepatol 38(suppl 1):S38–S53
23. Seitz HK, Stickel F (2006) Risk factors and mechanisms of hepatocarcinogenesis with special emphasis on alcohol and oxidative stress. Biol Chem 387:349–360
24. Zhang DY, Friedman SL (2012) Fibrosis-dependent mechanisms of hepatocarcinogenesis. Hepatology 56:769–775
25. Tanaka S, Mohr L, Schmidt EV et al (1997) Biological effects of human insulin receptor substrate-1 overexpression in hepatocytes. Hepatology 26:598–604
26. Hirosumi J, Tuncman G, Chang L et al (2002) A central role for JNK in obesity and insulin resistance. Nature 420:333–336
27. Puri P, Mirshahi F, Cheung O et al (2008) Activation and dysregulation of the unfolded protein response in nonalcoholic fatty liver disease. Gastroenterology 134:568–576
28. Chang Q, Zhang Y, Beezhold KJ et al (2009) Sustained JNK1 activation is associated with altered histone H3 methylations in human liver cancer. J Hepatol 50:323–333
29. Hui L, Zatloukal K, Scheuch H et al (2008) Proliferation of human HCC cells and chemically induced mouse liver cancers requires JNK1-dependent p21 downregulation. J Clin Invest 118:3943–3953
30. Shen C, Zhao CY, Zhang R et al (2012) Obesity-related hepatocellular carcinoma: roles of risk factors altered in obesity. Front Biosci 17:2356–2370
31. Siddique A, Kowdley KV (2011) Insulin resistance and other metabolic risk factors in the pathogenesis of hepatocellular carcinoma. Clin Liver Dis 15:281–296, vii–x
32. Kurmasheva RT, Houghton PJ (2006) IGF-I mediated survival pathways in normal and malignant cells. Biochim Biophys Acta 1766:1–22
33. Chen JS, Wang Q, Fu XH et al (2009) Involvement of PI3K/PTEN/AKT/mTOR pathway in invasion and metastasis in hepatocellular carcinoma: association with MMP-9. Hepatol Res 39:177–186
34. Lawlor MA, Alessi DR (2001) PKB/Akt: a key mediator of cell proliferation, survival and insulin responses? J Cell Sci 114:2903–2910
35. Nam SY, Lee EJ, Kim KR et al (1997) Effect of obesity on total and free insulin-like growth factor (IGF)-1, and their relationship to IGF-binding protein (BP)-1, IGFBP-2, IGFBP-3, insulin, and growth hormone. Int J Obes Relat Metab Disord 21:355–359
36. Aishima S, Basaki Y, Oda Y et al (2006) High expression of insulin-like growth factor binding protein-3 is correlated with lower portal invasion and better prognosis in human hepatocellular carcinoma. Cancer Sci 97:1182–1190
37. Huynh H, Chow PK, Ooi LL et al (2002) A possible role for insulin-like growth factor-binding protein-3 autocrine/paracrine loops in controlling hepatocellular carcinoma cell proliferation. Cell Growth Differ 13:115–122
38. Alison MR, Nicholson LJ, Lin WR (2011) Chronic inflammation and hepatocellular carcinoma. Recent results in cancer research. Fortschr Krebsforsch 185:135–148
39. Park EJ, Lee JH, Yu GY et al (2010) Dietary and genetic obesity promote liver inflammation and tumorigenesis by enhancing IL-6 and TNF expression. Cell 140:197–208

40. Ben-Neriah Y, Karin M (2011) Inflammation meets cancer, with NF-kappaB as the matchmaker. Nat Immunol 12:715–723
41. Pikarsky E, Porat RM, Stein I et al (2004) NF-kappaB functions as a tumour promoter in inflammation-associated cancer. Nature 431:461–466
42. Luedde T, Beraza N, Kotsikoris V et al (2007) Deletion of NEMO/IKKgamma in liver parenchymal cells causes steatohepatitis and hepatocellular carcinoma. Cancer Cell 11:119–132
43. He G, Karin M (2011) NF-kappaB and STAT3—key players in liver inflammation and cancer. Cell Res 21:159–168
44. He G, Yu GY, Temkin V et al (2010) Hepatocyte IKKbeta/NF-kappaB inhibits tumor promotion and progression by preventing oxidative stress-driven STAT3 activation. Cancer Cell 17:286–297
45. Ogata H, Kobayashi T, Chinen T et al (2006) Deletion of the SOCS3 gene in liver parenchymal cells promotes hepatitis-induced hepatocarcinogenesis. Gastroenterology 131:179–193
46. Jequier E (2002) Leptin signaling, adiposity, and energy balance. Ann N Y Acad Sci 967:379–388
47. Sun B, Karin M (2012) Obesity, inflammation, and liver cancer. J Hepatol 56:704–713
48. Saxena NK, Sharma D, Ding X et al (2007) Concomitant activation of the JAK/STAT, PI3K/AKT, and ERK signaling is involved in leptin-mediated promotion of invasion and migration of hepatocellular carcinoma cells. Cancer Res 67:2497–2507
49. Chen C, Chang YC, Liu CL et al (2007) Leptin induces proliferation and anti-apoptosis in human hepatocarcinoma cells by up-regulating cyclin D1 and down-regulating Bax via a Janus kinase 2-linked pathway. Endocr Relat Cancer 14:513–529
50. Kadowaki T, Yamauchi T (2005) Adiponectin and adiponectin receptors. Endocr Rev 26:439–451
51. Hotta K, Funahashi T, Arita Y et al (2000) Plasma concentrations of a novel, adipose-specific protein, adiponectin, in type 2 diabetic patients. Arterioscler Thromb Vasc Biol 20:1595–1599
52. Jiang LL, Li L, Hong XF et al (2009) Patients with nonalcoholic fatty liver disease display increased serum resistin levels and decreased adiponectin levels. Eur J Gastroenterol Hepatol 21:662–666
53. Skrabal CA, Czaja J, Honz K et al (2011) Adiponectin—its potential to predict and prevent coronary artery disease. Thorac Cardiovasc Surg 59:201–206
54. Man K, Ng KT, Xu A et al (2010) Suppression of liver tumor growth and metastasis by adiponectin in nude mice through inhibition of tumor angiogenesis and downregulation of Rho kinase/IFN-inducible protein 10/matrix metalloproteinase 9 signaling. Clin Cancer Res 16:967–977
55. Saxena NK, Fu PP, Nagalingam A et al (2010) Adiponectin modulates C-jun N-terminal kinase and mammalian target of rapamycin and inhibits hepatocellular carcinoma. Gastroenterology 139:1762–1773, 1773e1–5
56. Sharma D, Wang J, Fu PP et al (2010) Adiponectin antagonizes the oncogenic actions of leptin in hepatocellular carcinogenesis. Hepatology 52:1713–1722
57. Larsson SC, Wolk A (2007) Overweight, obesity and risk of liver cancer: a meta-analysis of cohort studies. Br J Cancer 97:1005–1008
58. Oh SW, Yoon YS, Shin SA (2005) Effects of excess weight on cancer incidences depending on cancer sites and histologic findings among men: Korea National Health Insurance Corporation Study. J Clin Oncol 23:4742–4754
59. Nair S, Mason A, Eason J et al (2002) Is obesity an independent risk factor for hepatocellular carcinoma in cirrhosis? Hepatology 36:150–155
60. N'Kontchou G, Paries J, Htar MT et al (2006) Risk factors for hepatocellular carcinoma in patients with alcoholic or viral C cirrhosis. Clin Gastroenterol Hepatol 4:1062–1068
61. Ohki T, Tateishi R, Shiina S et al (2009) Visceral fat accumulation is an independent risk factor for hepatocellular carcinoma recurrence after curative treatment in patients with suspected NASH. Gut 58:839–844
62. Petrides AS, Vogt C, Schulze-Berge D et al (1994) Pathogenesis of glucose intolerance and diabetes mellitus in cirrhosis. Hepatology 19:616–627

63. Zein CO, Levy C, Basu A et al (2005) Chronic hepatitis C and type II diabetes mellitus: a prospective cross-sectional study. Am J Gastroenterol 100:48–55
64. Angulo P, Keach JC, Batts KP et al (1999) Independent predictors of liver fibrosis in patients with nonalcoholic steatohepatitis. Hepatology 30:1356–1362
65. Adami HO, Chow WH, Nyren O et al (1996) Excess risk of primary liver cancer in patients with diabetes mellitus. J Natl Cancer Inst 88:1472–1477
66. Wideroff L, Gridley G, Mellemkjaer L et al (1997) Cancer incidence in a population-based cohort of patients hospitalized with diabetes mellitus in Denmark. J Natl Cancer Inst 89:1360–1365
67. El-Serag HB (2004) Hepatocellular carcinoma: recent trends in the United States. Gastroenterology 127:S27–S34
68. Wang P, Kang D, Cao W et al (2012) Diabetes mellitus and risk of hepatocellular carcinoma: a systematic review and meta-analysis. Diabetes Metab Res Rev 28:109–122
69. Hardie DG (2007) AMP-activated protein kinase as a drug target. Annu Rev Pharmacol Toxicol 47:185–210
70. Donadon V, Balbi M, Ghersetti M et al (2009) Antidiabetic therapy and increased risk of hepatocellular carcinoma in chronic liver disease. World J Gastroenterol 15:2506–2511
71. Hassan MM, Curley SA, Li D et al (2010) Association of diabetes duration and diabetes treatment with the risk of hepatocellular carcinoma. Cancer 116:1938–1946
72. Chen HP, Shieh JJ, Chang CC et al (2013) Metformin decreases hepatocellular carcinoma risk in a dose-dependent manner: population-based and in vitro studies. Gut 62(4):606–615
73. Blanco-Colio LM, Villa A, Ortego M et al (2002) 3-Hydroxy-3-methyl-glutaryl coenzyme A reductase inhibitors, atorvastatin and simvastatin, induce apoptosis of vascular smooth muscle cells by downregulation of Bcl-2 expression and Rho A prenylation. Atherosclerosis 161:17–26
74. Rombouts K, Kisanga E, Hellemans K et al (2003) Effect of HMG-CoA reductase inhibitors on proliferation and protein synthesis by rat hepatic stellate cells. J Hepatol 38:564–572
75. El-Serag HB, Johnson ML, Hachem C et al (2009) Statins are associated with a reduced risk of hepatocellular carcinoma in a large cohort of patients with diabetes. Gastroenterology 136:1601–1608
76. Angulo P (2002) Nonalcoholic fatty liver disease. N Engl J Med 346:1221–1231
77. Browning JD, Szczepaniak LS, Dobbins R et al (2004) Prevalence of hepatic steatosis in an urban population in the United States: impact of ethnicity. Hepatology 40:1387–1395
78. Fan JG, Zhu J, Li XJ et al (2005) Prevalence of and risk factors for fatty liver in a general population of Shanghai, China. J Hepatol 43:508–514
79. Brunt EM (2005) Pathology of nonalcoholic steatohepatitis. Hepatol Res 33:68–71
80. Torres DM, Harrison SA (2008) Diagnosis and therapy of nonalcoholic steatohepatitis. Gastroenterology 134:1682–1698
81. Williams CD, Stengel J, Asike MI et al (2011) Prevalence of nonalcoholic fatty liver disease and nonalcoholic steatohepatitis among a largely middle-aged population utilizing ultrasound and liver biopsy: a prospective study. Gastroenterology 140:124–131
82. Charlton MR, Burns JM, Pedersen RA, Watt KD, Heimbach JK, Dierkhising RA (2011) Frequency and outcomes of Liver transplantation for nonalcoholic streatohepatitis in the United States. Gastroenterology 141:1249–1253
83. Falck-Ytter Y, Younossi ZM, Marchesini G et al (2001) Clinical features and natural history of nonalcoholic steatosis syndromes. Semin Liver Dis 21:17–26
84. Caldwell SH, Oelsner DH, Iezzoni JC et al (1999) Cryptogenic cirrhosis: clinical characterization and risk factors for underlying disease. Hepatology 29:664–669
85. Poonawala A, Nair SP, Thuluvath PJ (2000) Prevalence of obesity and diabetes in patients with cryptogenic cirrhosis: a case–control study. Hepatology 32:689–692
86. Bugianesi E, Leone N, Vanni E et al (2002) Expanding the natural history of nonalcoholic steatohepatitis: from cryptogenic cirrhosis to hepatocellular carcinoma. Gastroenterology 123:134–140
87. Marrero JA, Fontana RJ, Su GL et al (2002) NAFLD may be a common underlying liver disease in patients with hepatocellular carcinoma in the United States. Hepatology 36:1349–1354

88. Ratziu V, Bonyhay L, Di Martino V et al (2002) Survival, liver failure, and hepatocellular carcinoma in obesity-related cryptogenic cirrhosis. Hepatology 35:1485–1493
89. Sorensen HT, Mellemkjaer L, Jepsen P et al (2003) Risk of cancer in patients hospitalized with fatty liver: a Danish cohort study. J Clin Gastroenterol 36:356–359
90. Sanyal AJ, Banas C, Sargeant C et al (2006) Similarities and differences in outcomes of cirrhosis due to nonalcoholic steatohepatitis and hepatitis C. Hepatology 43:682–689
91. Hashimoto E, Yatsuji S, Tobari M et al (2009) Hepatocellular carcinoma in patients with nonalcoholic steatohepatitis. J Gastroenterol 44(suppl 19):89–95
92. Ascha MS, Hanouneh IA, Lopez R et al (2010) The incidence and risk factors of hepatocellular carcinoma in patients with nonalcoholic steatohepatitis. Hepatology 51:1972–1978
93. Hashizume H, Sato K, Takagi H et al (2007) Primary liver cancers with nonalcoholic steatohepatitis. Eur J Gastroenterol Hepatol 19:827–834
94. Tokushige K, Hashimoto E, Horie Y et al (2011) Hepatocellular carcinoma in Japanese patients with nonalcoholic fatty liver disease, alcoholic liver disease, and chronic liver disease of unknown etiology: report of the nationwide survey. J Gastroenterol 46:1230–1237
95. Ertle J, Dechene A, Sowa JP et al (2011) Non-alcoholic fatty liver disease progresses to hepatocellular carcinoma in the absence of apparent cirrhosis. Int J Cancer 128:2436–2443

Chapter 10
Obesity, Inflammation, and Prostate Cancer

Jorge Blando, Achinto Saha, Kaoru Kiguchi, and John DiGiovanni

Abstract Obesity and the associated metabolic syndrome produce a complex set of alterations both systemically and locally in tissues that support cancer development and progression. In prostate cancer (PCa), the weight of evidence suggests that obesity is primarily associated with more aggressive disease and increased risk of biochemical failure following prostatectomy or radiation treatment. Inflammation processes and inflammation-associated signaling pathways are upregulated in the obese state, and both human and mouse studies support an important role for inflammation in obesity-driven PCa progression. Inflammation signaling pathways along with other signaling pathways (e.g., growth factor signaling pathways) altered in the obese state represent promising targets for both lifestyle and pharmacologic interventions to prevent or control PCa progression.

J. Blando • A. Saha • K. Kiguchi
Division of Pharmacology and Toxicology, College of Pharmacy,
The University of Texas at Austin, Austin, TX 78723, USA

J. DiGiovanni, Ph.D. (✉)
Division of Pharmacology and Toxicology, College of Pharmacy,
The University of Texas at Austin, Austin, TX 78723, USA

Department of Nutritional Sciences, College of Natural Sciences,
The University of Texas at Austin, Austin, TX 78723, USA

Dell Pediatric Research Institute, The University of Texas at Austin,
1400 Barbara Jordan Boulevard, Austin, TX 78723, USA
e-mail: john.digiovanni@austin.utexas.edu

A.J. Dannenberg and N.A. Berger (eds.), *Obesity, Inflammation and Cancer*,
Energy Balance and Cancer 7, DOI 10.1007/978-1-4614-6819-6_10,

10.1 Introduction

10.1.1 Obesity

The prevalence of obesity, defined as a body mass index (BMI) ≥30 kg/m², has increased dramatically in recent decades in the United States, and nearly 35 % of adults and 20 % of children are now obese [1]. Worldwide, an estimated 1.1 billion adults are overweight and 500 million adults are obese (http://www.iaso.org/policy/aboutobesity). The obese state is characterized by an excessive expansion of adipose tissue mass, which manifests as adipocyte hypertrophy (increased size), hyperplasia (increased number), and increased intracellular lipids. Excessive adiposity per se can exert untoward structural and biomechanical effects on organs (such as the lungs, liver, and pancreas), blood vessels, musculoskeletal system, and other tissues [2]. In addition, the resulting adipocyte hyperplasia and hypertrophy are associated with adipocyte dysfunction that can trigger local and systemic changes characteristic of the metabolic syndrome that increase risk and worsen prognosis of several cancers and other chronic diseases [3].

Among obese adults, approximately 60 % meet the criteria for the metabolic syndrome, a state of metabolic dysregulation characterized by insulin resistance, hyperglycemia, dyslipidemias (particularly hypertriglyceridemia), and hypertension [4]. In obesity and/or metabolic syndrome, alterations also occur in circulating levels of insulin, bioavailable insulin-like growth factor (IGF)-1, adipokines (e.g., leptin, adiponectin, monocyte chemotactic factor), inflammatory factors (e.g., cytokines such as interleukin 6 [IL-6]), and angiogenesis factors (e.g., vascular endothelial growth factor [VEGF] and plasminogen activator inhibitor [PAI]-1) [5, 6]. Through these mediators, obesity and metabolic syndrome are linked to various chronic diseases [5, 7] such as cardiovascular disease, type II diabetes, and cancer, including prostate cancer (PCa) that is the focus of this chapter.

10.1.2 Obesity and Cancer

Overall, an estimated 15–20 % of all cancer deaths in the USA are attributable to overweight and obese body types [8]. Obesity is associated with increased mortality from cancer of the prostate and stomach in men; breast (postmenopausal), endometrium, cervix, uterus, and ovaries in women; and kidney (renal cell), colon, esophagus (adenocarcinoma), pancreas, gallbladder, and liver in both genders [8]. While the relationships between metabolic syndrome and specific cancers are less well established, first reports from the Metabolic Syndrome and Cancer Project, a European cohort study of approximately 580,000 adults, confirm associations between obesity (or BMI) in metabolic syndrome and risks of colorectal, thyroid, and cervical cancer [9].

In this chapter, we discuss possible mechanisms underlying the links between obesity, metabolic syndrome, and PCa, with an emphasis on obesity-associated adipocyte dysfunction, inflammation, and growth factor signaling. Specifically, we describe the dysregulation of growth signals (including insulin, IGF-1, and downstream signaling pathways), adipokines (including leptin and adiponectin), inflammatory cytokines (including IL-6 and tumor necrosis factor-α [TNFα]), and angiogenesis factors (including VEGF and PAI-1) in the obese state that may contribute to more aggressive disease and higher mortality in PCa patients.

10.2 Relationship Between Obesity and PCa

10.2.1 Summary of Data from Epidemiologic Studies

Worldwide, PCa is the second most commonly diagnosed non-cutaneous cancer in men and the sixth most common cause of death [10]. In the United States, it is the most frequently diagnosed non-cutaneous cancer and the second leading cause of cancer-related mortality in this population [11, 12]. An estimated 241,740 new cases will be diagnosed in 2012.

Data associating obesity and PCa risk have been inconclusive [13, 14]; however, recent studies have shown increased risk of biochemical failure and metastasis, as well as poorer survival among obese PCa patients with androgen-dependent tumors, especially those who experienced rapid weight gain [15]. Men with low-volume PCa have a lower BMI, less body fat, and a smaller waist-to-hip ratio than men with high-volume PCa, which agrees with other reported findings [16–18]. Fat is the most energy dense component and has been the focus of most PCa dietary epidemiologic investigations. High consumption of energy and fat, especially saturated fat [19–21], is associated with advanced stage PCa and mortality [19, 22]. Several biologic mechanisms have been postulated to explain the role of obesity in PCa progression and these will be discussed in more detail in the sections that follow.

10.3 Mechanisms Associated with Obesity and PCa

As noted above, several biological mechanisms have been postulated to explain the association between obesity and aggressive disease in PCa patients, including increases in circulating levels of growth factors (i.e., IGF-1 and leptin), hyperinsulinemia, and inflammation as well as diet-induced alterations in adiponectin, steroid hormones, and possibly other factors (e.g., angiogenesis-related factors). These mechanisms will be discussed in more detail below.

10.3.1 Growth Factor Signaling

10.3.1.1 Insulin and Insulin-Like Growth Factor 1

Insulin is a peptide hormone, produced by beta cells of the pancreas, and is critical for the regulation of glucose and fat metabolism in the body. Metabolic dysfunction, mainly hyperinsulinemia and insulin resistance, is commonly associated with Western lifestyle and obesity and is considered as one of the risk factors for more aggressive PCa [23–26]. Insulin promotes cell division and proliferation in PCa and facilitates the expression of other growth factors and their regulators and acts as a cell survival factor inhibiting apoptosis [27–32]. Evidence suggests that serum level of C-peptide, a marker of insulin secretion, is positively correlated with obesity, and higher levels of this marker have been associated with high-grade PCa and PCa-specific mortality [33]. In obesity, adipose tissue releases increased amounts of free fatty acids, TNFα, resistin, and reduced amounts of adiponectin, which leads to the development of insulin resistance and chronic hyperinsulinemia. The higher levels of insulin in turn decrease the synthesis of insulin-like growth factor-binding protein 1 (IGFBP1) from the liver and possibly from other tissues and also decrease its blood levels. Hyperinsulinemia is also associated with reduced blood levels of IGFBP2. The IGFBPs normally bind and inhibit the action of IGF-1. So the resultant reduction in IGFBPs leads to increased levels of bioavailable, free IGF-1 in the circulation. Both the insulin and IGF-1 receptors (IGF-1Rs) are expressed in normal and neoplastic prostatic tissue [34–36], and physiological doses of insulin and IGF-1 act as mitogens for a variety of cancer cells, including PCa cells. Insulin and IGF-1 signal through the insulin receptors (IRs) and IGF-1R, respectively, to promote cell proliferation and inhibition of apoptosis, contributing to cancer progression. Binding of IGF-I or insulin to their cognate receptors leads to activation of a number of downstream signaling pathways following phosphorylation of insulin receptor substrates (e.g., IRS-1, IRS-2) and Src homology 2 domain-containing adapter protein (Shc). Phosphorylated Shc activates the Ras/Raf/mitogen-activated protein kinase (MAPK) pathway and ultimately stimulates cell growth and proliferation. Phosphorylated IRS-1 activates the phosphatidylinositol 3′ kinase (PI3K)/Akt pathway, leading to inhibition of apoptosis and stimulation of cell proliferation through downstream mediators including BAD, mTOR, P70S6K, and nuclear factor-κB (NF-κB). Akt is normally inhibited by the tumor suppressor gene phosphatase and tensin homolog (PTEN), which is often lost in advanced PCa [32, 34, 37]. Thus, changes in this signaling pathway associated with obesity may be exacerbated in the presence of PTEN deficiency.

10.3.1.2 Sex Steroids

Androgens are essential for normal development, differentiation, and proliferation of prostatic tissue [38], and it is generally believed that higher concentrations of androgens are associated with increased PCa risk. Testosterone and its metabolite dihydrotestosterone (DHT) bind to the androgen receptor (AR) which then binds to androgen response elements in the DNA of prostate cells to initiate changes in gene expression [39]. Adipose tissue produces the enzyme aromatase, which catalyzes the conversion of androgens to estrogens (note that DHT is not converted to estrogens by aromatase). Thus, obesity is generally associated with modestly lower serum concentrations of testosterone, lower concentrations of sex hormone-binding globulin, and higher concentrations of estrogens [40–42]. Interestingly, recent data suggest that higher serum testosterone levels are associated with a reduced risk of high-grade PCa but an increased risk of low-grade tumors [43]. It has been hypothesized that lower androgen concentrations may provide a microenvironment, which favors more aggressive and/or androgen-independent tumor growth and ultimately disease progression [32]. Another explanation is that individuals with low serum testosterone levels are at an increased risk of developing metabolic syndrome. However, it is not clear whether low serum testosterone levels associated with obesity and insulin resistance or insulin alone, in the absence of high serum testosterone, is sufficient to trigger progression of PCa to higher-grade tumors. Alternatively, lower testosterone concentrations are associated with a reduced risk of low-grade tumors because low-grade tumors are slow growing and need androgen stimulation for progression [44]. Note that intraprostatic conversion of testosterone to DHT, which is not strongly related to circulating androgen levels, may be more influential in PCa progression than androgens in the circulation.

10.3.2 Adipokines

Two kinds of adipose tissues are present in mammals: white adipose tissue (WAT) and brown adipose tissue. WAT is the most abundant and is considered as the major site of energy storage [45]. In addition to adipocytes, adipose tissue also contains pre-adipocytes, endothelial cells, fibroblasts, leukocytes, and, most importantly, macrophages. Macrophages infiltrate adipose tissue and they remain in increased numbers in association with obesity [46]. Adipose tissue is no longer considered to be an inert tissue functioning only as energy storage, but plays an important role in the regulation of many pathological processes. Polypeptide hormones derived from adipocytes are known as adipokines. Currently more than 50 different types of adipokines are recognized such as leptin, adiponectin, resistin, serpin, lipocalin 2, retinol-binding protein 4 (RBP4), zinc-α2, glycoprotein, vaspin, visfatin, omentin, apelin, TNFα, IL-6, IL-1, CC chemokine ligand 2, and mediators of the clotting process, such as PAI-1 and chemerin [45, 47]. Adipokines may exert their biologic effects either at a local level via autocrine/paracrine pathways or in an endocrine manner by entering the circulation and activating receptors on more distant target cells. Paracrine effects of adipokines are important in cases of PCa progression where extracapsular extension and invasion of the retropubic fat pad occurs [48]. This could result in the exposure of malignant cells to high concentrations of potentially proangiogenic, survival, and proliferative factors, which may enhance their capacity for further growth and metastasis. Adiponectin and leptin are the most abundant and most studied adipokines produced by adipocytes.

10.3.2.1 Leptin

Leptin, first described by Zhang et al. in 1994 [49], is a 16-kD adipokine produced predominantly by adipocytes in WAT. Circulating leptin concentrations exhibit a positive correlation with total body fat, so that serum leptin is elevated in obese individuals compared to lean individuals [50]. Although numerous epidemiologic studies have investigated the relationship between obesity, circulating leptin levels, and PCa [51–57], results from these studies are still inconclusive. The risk for PCa in individuals with higher levels of leptin has not been demonstrated [51–57]. However, it was found that higher leptin levels were linked to tumor progression and advanced disease [51, 52, 54–57]. Collectively, the current data suggest that leptin may be predominantly involved in PCa progression.

Several in vitro studies have shown that leptin induces proliferation, cell migration, and invasion and/or prevents apoptosis in androgen-independent human PCa cell lines [58]. Leptin also induces expression of several angiogenic growth factors including VEGF, transforming growth factor-beta1 (TGF-β1), and basic fibroblast growth factor (bFGF) that may contribute to its action in PCa progression [59, 60]. These responses to activation of the leptin receptor are mediated by several signaling pathways including the Janus kinase-signal transducer and activator of transcription (JAK/STAT) (especially Stat3), phosphatidylinositol 3-kinase (PI3-K)/Akt, and c-Jun NH2-terminal kinase (JNK) pathways [61]. Alterations in these signaling pathways are not only critical in prostate carcinogenesis and malignant transformation, but also important in obesity, diabetes, and insulin resistance [59, 62].

10.3.2.2 Adiponectin

Adiponectin, also known as 30-kD adipocyte complement-related protein (Acrp30), adipoQ, APM-1, or gelatin-binding protein 28 (GBP28), is the most abundant circulating adipokine synthesized mainly by adipocytes but also expressed by skeletal muscle cells, cardiac myocytes, and endothelial cells [45, 48]. In contrast to other adipokines, circulating levels of adiponectin are negatively correlated with central obesity, BMI, visceral fat accumulation, and insulin resistance. Epidemiological studies show more consistent inverse associations between circulating adiponectin levels and cancer [47]. Reductions in the levels of plasma adiponectin have also been observed in obesity-related conditions such as type 2 diabetes, cardiovascular disease, hypertension, and metabolic syndrome [63]. Adiponectin has antiproliferative, anti-angiogenesis, and proapoptotic activity, and the plasma levels are found to be lower in patients with PCa compared to patients with benign prostate hyperplasia (BPH) [47].

Although the exact mechanism of antitumor activity of adiponectin is not clearly understood, it is thought to act by modulation of signaling pathways including AMP-activated protein kinase (AMPK), MAPK, NF-κB, Stat3, and p53 [47, 48, 62]. Two receptor isoforms are known to exist for adiponectin: adiponectin receptor 1 (Adipo-R1) and adiponectin receptor 2 (Adipo-R2). Activation of the receptor(s) by adiponectin stimulates the activation of AMPK, peroxisome proliferator-activated receptor-α (PPARα), and p38 MAPK. Activation of AMPK is considered to be one of the most important signaling events for metabolic effects of adiponectin and has also been implicated in prostate carcinogenesis [48]. Adiponectin regulates the expression of several pro- and anti-inflammatory cytokines. Its main anti-inflammatory function might be related to the suppression of the synthesis of TNFα and interferon-γ (IFNγ) and to induce the production of anti-inflammatory cytokines such as interleukin-10 (IL-10) and IL-1 receptor antagonist (IL-1RA). Activation of PPARs also exerts anti-inflammatory effects through inhibition of the transcriptional activation of proinflammatory response genes [45]. Recent evidence suggests that the ratio of adiponectin to leptin levels is critical for the overall effects of these two important adipokines [64].

10.3.3 Angiogenesis Factors

10.3.3.1 VEGF

VEGF, a heparin-binding glycoprotein produced by adipocytes and tumor cells, has angiogenic, mitogenic, and vascular permeability-enhancing activities specific for endothelial cells [65]. Circulating levels of VEGF are increased in obese, relative to lean, humans and animals, and increased tumoral expression of VEGF is associated with poor prognosis in several obesity-related cancers [66]. The need for nutrients and oxygen triggers tumor cells to produce VEGF, which leads to the formation of new blood vessels to nourish the rapidly growing tumor and facilitate the metastatic spread of tumor cells [65]. Adipocytes communicate with endothelial cells by producing a variety of proangiogenic and vascular permeability-enhancing factors. These include VEGF, IGF-1, PAI-1, leptin, hepatocyte growth factor, and fibroblast growth factor-2 [67]. In the obese, nontumor setting, these factors stimulate neovascularization in support of the expanding fat mass. These adipose-derived factors may also contribute to obesity-associated enhancement of tumor angiogenesis.

In PCa, VEGF plays a major role in endothelial cell differentiation, proliferation, migration, and vessel formation [68–71]. In addition to adipocytes, VEGF is produced by PCa cells and tumor-infiltrating lymphocytes [72, 73]. Systemic VEGF was found to be significantly elevated in PCa patients [74] and patients with metastatic disease [70]. In a cohort of 50 radical prostatectomy specimens, El-Gohary et al. [75] showed that VEGF expression correlated significantly with angiolymphatic invasion and Gleason score. Clinical studies comparing PCa with BPH revealed that VEGF expression was correlated with increased levels of angiogenesis [76]. The levels of VEGF in serum, plasma, or urine are correlated with patient outcome in both localized and disseminated PCa [70, 77, 78]. In addition, the levels of the VEGFR were correlated with a poorer grade of tumor differentiation and prognosis in PCa [79]. Normal prostate tissue expresses minimal to no VEGF, unlike PCa tissue that stains positively for VEGF in areas of increased microvessel density (MVD) [73]. PCa cells express VEGF in vitro and in vivo [80]. VEGF expression by PCa specimens [81] and human PCa cell lines (LNCaP, PC-3, and DU145) is far greater than that by stromal cells of the normal prostate [73, 82]. VEGF expression may also contribute to PCa-induced osteoblastic activity in vivo. Using a TRAMP model, Isayeva et al. demonstrated that inhibitors of the VEGFR-2 delayed tumor progression only when administered in the early stages of PCa, before a significant rise in VEGF levels was observed [83]. In a preclinical study using a xenograft model, injecting human DU145 cells in nude mice, inhibition of VEGF resulted in decreased tumor proliferation [84].

10.3.3.2 Plasminogen Activator Inhibitor-1

PAI-1 is a serine protease inhibitor produced by endothelial cells, stromal cells, and adipocytes in visceral WAT [85]. Increased circulating PAI-1 levels, frequently found in obese subjects, are associated with increased risk of atherogenesis and cardiovascular disease, diabetes, and several cancers [6, 85]. PAI-1, through its inhibition of urokinase-type and tissue-type plasminogen activators, regulates fibrinolysis and integrity of the extracellular matrix. PAI-1 is also involved in angiogenesis and thus may contribute to obesity-driven tumor cell growth, invasion, and metastasis [6]. Although PAI-1 levels in obese individuals may be reduced via weight loss or TNFα blockade [86, 87], the role of PAI-1 in tumorigenesis remains controversial [85]. Plasminogen activator activities are elevated in PCa compared with BPH [88]. Swiercz et al. demonstrated that exogenously applied recombinant PAI-1 is a powerful inhibitor of angiogenesis in three different in vitro models and is a powerful anticancer agent in a SCID mice inoculated with LNCaP cells [89]. Therefore, further study is warranted to determine the exact role of PAI-1 in PCa and other cancers.

10.3.4 Inflammation and Inflammatory Mediators

Chronic inflammation is now considered one of the major risk factors linking obesity and the development of PCa [90–92]. The inflammatory response is characterized by increased synthesis of various cytokines, acute-phase reactants such as C-reactive protein (CRP), and the activation of proinflammatory signaling pathways [45]. Adipose tissues in obese individuals and in animal models of obesity are infiltrated by a large number of macrophages, and this is associated with systemic inflammation and insulin resistance as noted above [93]. Macrophages are also the main source of soluble mediators such as TNFα. In addition to macrophages, adipocytes are also responsible for the production of IL-6, CCL2, and other inflammatory mediators that further contribute to macrophage infiltration and proinflammatory cytokine production from the adipose tissue.

IL-6 is a proinflammatory cytokine and is involved in the regulation of various cellular functions including proliferation, apoptosis, angiogenesis, differentiation, and regulation of the immune response [94]. Clinically, plasma IL-6 levels positively correlate with adiposity and insulin resistance in human populations [93]. Although not exclusively secreted by adipocytes, as much as one-third of circulating IL-6 originates from adipose tissue, and expression and secretion of IL-6 is 2–3 times greater in visceral relative to subcutaneous adipose tissue [48].

In PCa patients, serum IL-6 levels are higher in those patients with metastatic disease compared to those with localized disease [46, 95]. It has also been reported that serum IL-6 levels >7 pg/mL are associated with a poor prognosis in men with PCa [96].

Various in vitro studies have shown that IL-6 is secreted by androgen-independent PCa cells suggesting an autocrine effect on tumor cells and paracrine effects on normal cells in the tumor microenvironment [94]. High expression of IL-6 in various organs particularly lung, liver, brain, and bone attracts the circulating tumor cells and therefore promotes metastasis [97]. In one study it was reported that IL-6 levels were significantly higher in PCa patients with bone metastasis compared to patients without metastasis [98]. In another study, serum IL-6 levels were found to be higher in castration-resistant PCa compared to normal controls as well as patients with BPH, prostatitis, and localized or recurrent disease [99]. Higher IL-6 levels in obese individuals also contribute to PCa progression. The periprostatic adipose tissue (adipose tissue surrounding the prostate) is frequently invaded by prostate tumor cells. Studies have suggested that periprostatic adipose tissue produces significantly higher local IL-6 and that may contribute to PCa progression in obese individuals [100, 101].

The biological activity of IL-6 is mediated by binding with the IL-6R/gp130 receptor complex. Dimerization of IL-6/IL-6R/gp130 leads to the initiation of intracellular signaling, through the JAK/STAT, mitogen-activated protein kinase (MAPK), and phosphatidylinositol 3-kinase/Akt kinase (PI3-K/AKT) pathways [102]. This leads to the expression of various genes involved in inflammation and cancer development. Binding of IL-6 to its receptor also leads to increased receptor activator of nuclear factor kappa B ligand (RANKL) expression. IL-6 may also activate AKT via increased JAK-dependent PI3K activity resulting in cell survival and anti-apoptosis signaling. Concomitantly, increased MAPK activity downstream of JAK activation can lead to upregulated cell growth, proliferation, and mitosis. In some cases, IL-6 also acts through soluble IL-6R which binds to gp130 and initiates signal transduction [94]. Recent studies have also revealed a role of IL-6 in regulation of the AR [103–105]. In the absence of androgen, IL-6 causes activation of the AR. In low-concentration conditions, androgen action is potentiated by IL-6, leading to synergistic activation of the AR [106]. These observations demonstrate a cross talk between the IL-6 pathway and AR that may play a significant role during PCa progression leading to androgen independence.

TNFα is a cytokine involved in systemic and acute inflammation. TNFα is synthesized as a 26-kD membrane-bound protein and cleaved into a 17-kD soluble protein by TNF-converting enzyme (TACE) [107]. TNFα is predominantly produced by macrophages, CD4+ T cells, and natural killer (NK) cells, although nonimmune cells such as fibroblasts, smooth muscle, and tumor cells have also been reported to secrete low amounts of this cytokine [108]. TNFα is regulated by the proteolytic activity of stromal metalloproteinases, secreted from the membranes of somatic cells [109], and is directly produced by tumor-associated macrophages (TAMs) [110]. Significantly elevated levels of TNFα are found in patients with metastatic disease compared to those with localized disease [95]. According to Nakashima et al. [111], serum TNFα activity was positive in 76 % of the patients with relapsed disease who had a significantly higher mortality rate than those with undetectable serum TNFα levels. Obesity-related TNFα levels could thus potentially enhance PCa progression.

NF-κB has been shown to be activated by reactive oxygen species and many carcinogens, in addition to proinflammatory chemokines and cytokines such as TNFα [112–114]. Recently, Huerta-Yepez et al. showed that TNFα has a role in conveying resistance to Fas-induced apoptosis through a similar pathway involving NF-κB in human PCa cells [112]. Cyclooxygenase-2 (COX-2), upregulated during the inflammatory response, is thought to be inducible by a number of proinflammatory cytokines, including TNFα [115, 116]. Subbarayan et al. found that TNFα could induce COX-2 expression in both normal prostate and androgen-unresponsive PCa cells where upregulation of COX-2 was related to increases in cellular levels of prostaglandins [117]. COX-2 expression has been suggested to be dependent on inflammatory cytokine/NF-κB pathways in PCa [118], supporting a possible positive feedback between inflammatory mediators, prostaglandin production, and PCa progression. Wang et al. [119] found that foci of chronic inflammation within human samples of BPH were associated with accumulation of inflammatory cells. Moderate to high accumulation of these cells, particularly T lymphocytes and macrophages, was highly correlated with positive staining for COX-2.

10.4 Lessons Learned from Animal Models of PCa

10.4.1 Animal Model Studies on Obesity and PCa

PCa research has been greatly aided by the development and use of relevant animal models, particularly transgenic and knockout mouse models [120–123]. Only a few reports have appeared dealing with the effects of obesity, per se, on PCa development and progression using animal models. In earlier studies using the TRAMP model, mice with higher body weight and body fat content had larger and more aggressive tumors compared to those with lower body weight and body fat content [124]. Although the TRAMP mice in this study were not "obese," the mice with the highest body fat content had higher levels of leptin and lower levels of adiponectin. Additional studies using TRAMP mice fed a "Western-type" diet enriched in both fat and cholesterol showed accelerated prostate tumor growth and increased histological grade of tumors [125]. This diet also increased lung metastases. It is not clear whether there were differences in body weight between the "Western diet" group and control diet group in this study as body weight and fat mass data were not reported. However, these data would appear to support epidemiological studies showing that fat, and, in particular, saturated fat, is an independent predictor of disease progression [19, 21, 22]. These data are also supported by another study using the HiMyc mouse model of PCa [126]. In this regard, Kobayashi et al. [127] showed that if HiMyc mice were maintained on a low-fat (LF, 12 kcal %) diet vs. a high-fat isocaloric (HF, 42 kcal %) diet, the transition from prostate intraepithelial neoplasia (PIN) to PCa was delayed, and there was a decrease in prostatic Akt activity as well as a decrease in phosphorylation of the downstream targets p-70S6K and GSK3β. Again, there were no differences in body weight between the HF and LF diet groups in this study.

We recently reported a study to evaluate the impact of dietary energy balance on PCa progression using HiMyc mice [126]. In this study, we compared the effect of different body phenotypes from lean to obese using diets of different caloric density. Diet-induced obesity (DIO) enhanced PCa progression in the ventral prostate (VP) of HiMyc mice [126]. Notably, DIO (mice fed a diet containing 60 kcal % fat) increased the incidence of invasive adenocarcinomas as well as the severity of these tumors compared to HiMyc mice on control diet (i.e., modified AIN76 diet containing 10 kcal % fat). In contrast, calorie restriction (30 % CR) completely prevented the formation of invasive adenocarcinomas in these mice. These dietary energy balance-induced alterations in PCa progression in HiMyc mice were associated with differences in circulating levels of insulin, IGF-1, leptin, adiponectin, and resistin. In particular, obese mice had higher levels of insulin, IGF-1, leptin, and resistin and lower levels of adiponectin compared to lean mice on a 30 % CR diet. Further analysis revealed that both growth factor (i.e., Akt/mTOR) and especially inflammatory signaling (i.e., Stat3, NF-κB, IL-1, IL-6, IL-7, IL-23, IL-27, NF-κB1, and TNFα) pathways were upregulated in VP of mice on the obesity-induced diet. Dramatic increases in inflammatory cell infiltrates (macrophages and T cells) were also observed in the VP of HiMyc mice on the obesity-induced diet. In addition to these changes, there were dramatic increases in angiogenesis (increased number and size of blood vessels in the VP) accompanied by dramatic increased in expression of all VEGF family members. These data provide strong support for a role of local inflammation in the effects of obesity on PCa progression using this model system.

Ribeiro et al. [128] evaluated the influence of genetically induced obesity on PCa cell growth and angiogenesis in vivo. In these studies, the androgen-insensitive murine PCa cell line, RM1, when inoculated in obese, hyperleptinemic *db/db* mice produced small tumors with a low proliferation index and angiogenesis. In contrast, when these cells were injected into obese, leptin-deficient *ob/ob* mice, larger tumors were produced with a higher proliferation index when compared to control mice. These data suggested that high leptin levels are not favorable for the androgen-insensitive PCa cell line and it may limit the growth of these tumor cells. While these studies do not shed light on the role of inflammation in PCa progression, they demonstrate the complexity of genetic models of obesity for such studies. In addition, evaluating serum levels of individual factors such as leptin may not adequately predict local changes in prostate tissues. In a recent study, Zhang et al. [129] reported that WAT can directly mediate cancer progression by serving as a source of cells (adipose stromal cells, ASCs) that migrate to tumors and promote neovascularization. Recruitment of ASCs by tumors was sufficient to promote tumor growth. In this study, ASCs injected subcutaneously at a distant site migrated to the site of DU145 xenografts promoting tumor growth and angiogenesis.

In other studies, serum obtained from obese Zucker rats that harbor a leptin receptor mutation was shown to stimulate proliferation of human PCa cells (i.e., LNCaP cells) but not rat PCa cells (AT3B1) [130]. This was attributed to higher concentrations of VEGF in the serum of the obese rats. In further studies using C57BL/6 mice on an obesity-induced diet, serum from these mice was found to contain higher levels of leptin, VEGF, PAI-1, and IL-6 and lower levels of testosterone compared to mice fed a control diet [131]. Serum from these obese mice increased proliferation of LNCaP and PacMetUT1 cells but had no effect on PrEc and DU145 cells. Furthermore, serum from obese mice induced increased invasion, migration, and MMP-9 activity in LNCaP, PacMetUT1, and DU145 cells and showed increased expression of EMT-related markers. These later two studies emphasize the potential local effects of circulating factors, including inflammatory cytokines, altered in the obese state on PCa cells localized within the prostate or at other sites.

10.5 Strategies for Reversing the Effects of Obesity on PCa

Obesity is clearly associated with increased morbidity and mortality. Weight loss in overweight and obese individuals reduces risk factors for diabetes and cardiovascular disease. Weight loss and calorie restriction may also be an important strategy for preventing or reducing progression of PCa. A variety of effective options exist for reducing weight including dietary therapies such as low-calorie diets and lower-fat diets, exercise, behavior therapy techniques, medications, surgery, and combinations of these techniques [132]. Losing weight can reduce the risk of many cancers, even a weight reduction of only 5–10 % may be beneficial and improve health [133].

Although the relationship between physical activity and obesity is undisputed, the putative association between physical activity and PCa is inconclusive [134–136]. Postulated mechanisms include the ability of physical activity to modulate testosterone levels and immune function, insulin sensitivity, and reduce obesity [137, 138]. Physical activity increases insulin sensitivity and decreases actual insulin secretion. Increased serum levels of IGFBP-3 are also associated with greater physical activity. Some studies suggested an association between physical activity and leptin levels [139–141]. Exercise appears to lower circulating levels of IGF-1 and insulin and increase levels of sex hormone-binding globulin [142]. In vitro studies have shown that serum from men participating in exercise intervention programs can induce apoptosis and decrease the overall growth of prostate carcinoma (LNCaP) cells [143]. In a study of PCa patients on "watchful waiting," Ornish et al. reported that those in the experimental group (vegan diet, moderate aerobic exercise, vitamin supplements, and a stress management program) had lower levels of PSA than did those in the control group, suggesting that intensive lifestyle changes may affect the progression of early, low-grade PCa [144].

Weight loss strategies and increased physical activity could provide a decrease in obesity and parameters associated with metabolic syndrome with potentially beneficial effects on PCa progression in certain individuals. However, it may not be possible for most individuals to undergo extensive dietary manipulation and/or perform rigorous physical exercise. Thus, another important approach to offsetting the effects of obesity on PCa and other cancers is to identify novel agents that can produce some of the same beneficial effects of CR. Such agents or combination of agents could be combined with lifestyle changes as an optimal prevention strategy. An ideal agent for this purpose should produce metabolic, hormonal, and physiological effects similar to CR, without significant reduction in long-term food intake, and activate stress response pathways similar to CR and finally extend life span and reduce or delay the onset of age-related diseases [145]. Several compounds, especially naturally occurring phytochemicals, have been examined as promising agents in this regard. These include resveratrol, curcumin, epigallocatechin gallate (EGCG), quercetin, ursolic acid, rapamycin, and metformin.

Resveratrol, a plant-derived polyphenol in the skin of red grapes, is one of the most studied CR mimetics. Resveratrol has been shown to activate SIRT1 in cells and thus mimic the actions of CR without restriction of food intake. Treatment with resveratrol produces a transcriptional response in cells similar to CR and in obese mice; it has been shown to increase insulin sensitivity [146, 147]. A recent report suggested that dietary resveratrol prevents high-grade PIN development in the PTEN knockout mouse model of PCa by modulating mTORC1 and SIRT1 [148]. Curcumin is a major chemical component of turmeric (*Curcuma longa*) and widely used as a food additive in the Indian subcontinent. Various in vitro and in vivo data have suggested a potential chemopreventive role of curcumin in PCa. For example, curcumin was found to inhibit tumor development in TRAMP model either alone or in combination with phenethyl isothiocyanate (PEITC) [149]. The green tea polyphenol EGCG, another potential CR mimetic agent, decreased PCa progression in the TRAMP mouse model by inhibiting IGF-1 signaling [150]. All these observations indicate that phytochemicals can be considered as potential CR mimetics that might be beneficial in reversing the development and/or progression of obesity-related PCa.

Recently, it has been shown that diabetic patients taking metformin have reduced cancer incidence for a number of cancers [151–153]. Metformin is a biguanide class of antidiabetic agent widely used for the treatment of type II diabetes for its insulin-sensitizing effects. Recent studies also suggest that metformin has beneficial effects on PCa [153]. For example, in one study use of metformin produced 44 % reduction in PCa cases in Caucasian men compared to the control population [152]. Various in vitro and animal studies have also demonstrated the anticancer activity of metformin in PCa [154, 155]. Metformin activates AMPK which is thought to be the primary mechanism through which it exerts its anticancer effects. Thus, metformin may be a promising CR mimetic compound for reversing the effects of obesity on PCa as well as other cancers.

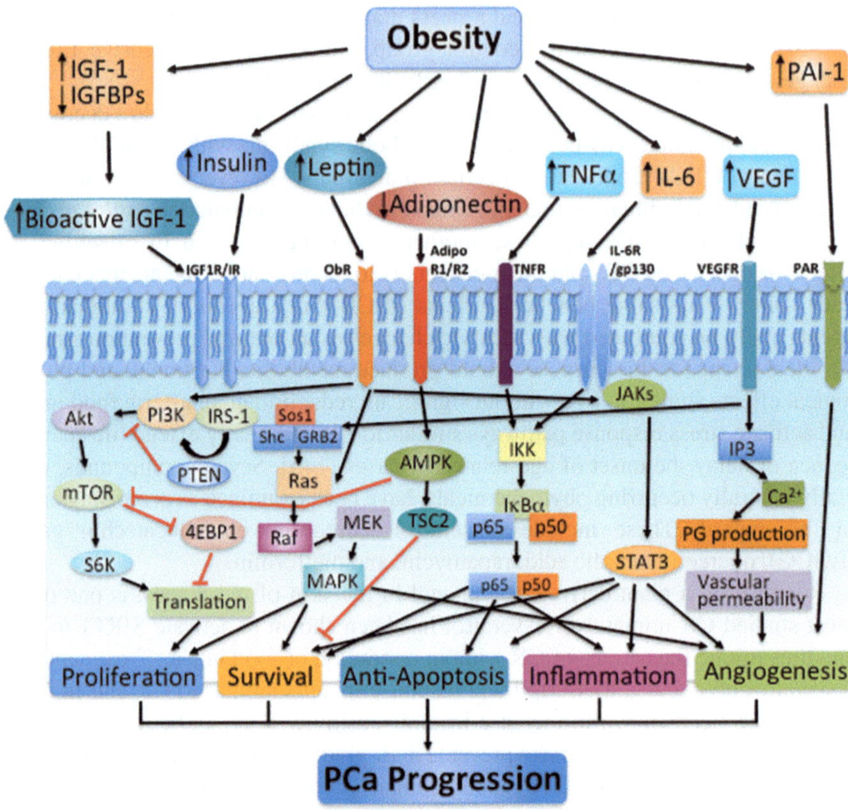

Fig. 10.1 Summary of growth factor and inflammation-associated signaling pathways that may represent targets for reversing the effects of obesity on PCa progression

10.6 Concluding Remarks

Obesity and the associated metabolic syndrome produce a complex set of alterations both systemically and locally in tissues that support cancer development and progression. Adipose tissue dysfunction, along with multiple hormones, growth factors, inflammatory cytokines, and other mediators associated with the obese state, enables cross talk between macrophages, adipocytes, endothelial cells, and epithelial cells that contribute to cancer-related processes (including growth and proliferation signaling, inflammation, and vascular alterations). Inflammation processes and inflammation-associated signaling pathways are upregulated in the obese state, and both human and mouse studies support an important role for inflammation in obesity-driven PCa progression. Inflammation signaling pathways along with other signaling pathways altered in the obese state represent promising targets for both lifestyle and pharmacologic interventions to prevent or control PCa progression. Figure 10.1 summarizes some of these pathways that may represent targets for offsetting the effects of obesity on cancer and in particular PCa progression.

A particular promising approach is the search for, and development of, CR mimetic compounds. Combinatorial approaches (e.g., lifestyle and pharmacologic) may ultimately lead to rational approaches to prevent obesity-driven PCa progression.

Acknowledgments Achinto Saha is supported by a Cancer Prevention Research Institute of Texas postdoctoral fellowship award RP101501 from the State of Texas.

References

1. Flegal KM, Carroll MD, Kit BK, Ogden CL (2012) Prevalence of obesity and trends in the distribution of body mass index among US adults, 1999–2010. JAMA 307:491–497
2. Grassi G, Seravalle G, Scopelliti F, Dell'Oro R, Fattori L, Quarti-Trevano F et al (2010) Structural and functional alterations of subcutaneous small resistance arteries in severe human obesity. Obesity (Silver Spring) 18:92–98
3. Gottschling-Zeller H, Birgel M, Scriba D, Blum WF, Hauner H (1999) Depot-specific release of leptin from subcutaneous and omental adipocytes in suspension culture: effect of tumor necrosis factor-alpha and transforming growth factor-beta1. Eur J Endocrinol 141:436–442
4. Ford ES, Li C, Zhao G (2010) Prevalence and correlates of metabolic syndrome based on a harmonious definition among adults in the US. J Diabetes 2:180–193
5. Hursting SD, Berger NA (2010) Energy balance, host-related factors, and cancer progression. J Clin Oncol 28:4058–4065
6. Carter JC, Church FC (2009) Obesity and breast cancer: the roles of peroxisome proliferator-activated receptor-gamma and plasminogen activator inhibitor-1. PPAR Res 2009:345320
7. Poirier P, Giles TD, Bray GA, Hong Y, Stern JS, Pi-Sunyer FX et al (2006) Obesity and cardiovascular disease: pathophysiology, evaluation, and effect of weight loss. Arterioscler Thromb Vasc Biol 26:968–976
8. Calle EE, Rodriguez C, Walker-Thurmond K, Thun MJ (2003) Overweight, obesity, and mortality from cancer in a prospectively studied cohort of U.S. adults. N Engl J Med 348:1625–1638
9. Stocks T, Borena W, Strohmaier S, Bjorge T, Manjer J, Engeland A et al (2010) Cohort profile: the metabolic syndrome and cancer project (Me-Can). Int J Epidemiol 39:660–667
10. Aus G, Abbou CC, Bolla M, Heidenreich A, Schmid HP, van Poppel H et al (2005) EAU guidelines on prostate cancer. Eur Urol 48:546–551
11. Stephenson RA, Stanford JL (1997) Population-based prostate cancer trends in the United States: patterns of change in the era of prostate-specific antigen. World J Urol 15:331–335
12. Jemal A, Bray F, Center MM, Ferlay J, Ward E, Forman D (2011) Global cancer statistics. CA Cancer J Clin 61:69–90
13. Nomura AM (2001) Body size and prostate cancer. Epidemiol Rev 23:126–131
14. Porter MP, Stanford JL (2005) Obesity and the risk of prostate cancer. Prostate 62:316–321
15. Flavin R, Zadra G, Loda M (2011) Metabolic alterations and targeted therapies in prostate cancer. J Pathol 223:283–294
16. Allott EH, Masko EM, Freedland SJ (2012) Obesity and prostate cancer: weighing the evidence. Eur Urol. 2012. pii:S0302–2838(12)01344-9. doi:10.1016/j.eururo.2012.11.013
17. Spitz MR, Strom SS, Yamamura Y, Troncoso P, Babaian RJ, Scardino PT et al (2000) Epidemiologic determinants of clinically relevant prostate cancer. Int J Cancer 89:259–264
18. Giovannucci E, Rimm EB, Stampfer M, Colditz GA, Willett W (1997) Height, body weight and risk of prostate cancer. Cancer Epidemiol Biomarkers Prev 6(8):557–563
19. Graham S, Haughey B, Marshall J, Priore R, Byers T, Rzepka T et al (1983) Diet in the epidemiology of carcinoma of the prostate gland. J Natl Cancer Inst 70:687–692
20. Mettlin C, Selenskas S, Natarajan N, Huben R (1989) Beta-carotene and animal fats and their relationship to prostate cancer risk. A case–control study. Cancer 64:605–612

21. Strom SS, Yamamura Y, Forman MR, Pettaway CA, Barrera SL, DiGiovanni J (2008) Saturated fat intake predicts biochemical failure after prostatectomy. Int J Cancer 122:2581–2585
22. Kondo Y, Homma Y, Aso Y, Kakizoe T (1994) Promotional effect of two-generation exposure to a high-fat diet on prostate carcinogenesis in ACI/Seg rats. Cancer Res 54:6129–6132
23. Neuhouser ML, Till C, Kristal A, Goodman P, Hoque A, Platz EA et al (2010) Finasteride modifies the relation between serum C-peptide and prostate cancer risk: results from the Prostate Cancer Prevention Trial. Cancer Prev Res (Phila) 3:279–289
24. Hsu IR, Kim SP, Kabir M, Bergman RN (2007) Metabolic syndrome, hyperinsulinemia, and cancer. Am J Clin Nutr 86:s867–s871
25. Zhou JR, Blackburn GL, Walker WA (2007) Symposium introduction: metabolic syndrome and the onset of cancer. Am J Clin Nutr 86:s817–s819
26. Laukkanen JA, Laaksonen DE, Niskanen L, Pukkala E, Hakkarainen A, Salonen JT (2004) Metabolic syndrome and the risk of prostate cancer in Finnish men: a population-based study. Cancer Epidemiol Biomarkers Prev 13:1646–1650
27. White MF, Kahn CR (1994) The insulin signaling system. J Biol Chem 269:1–4
28. Venkateswaran V, Haddad AQ, Fleshner NE, Fan R, Sugar LM, Nam R et al (2007) Association of diet-induced hyperinsulinemia with accelerated growth of prostate cancer (LNCaP) xenografts. J Natl Cancer Inst 99:1793–1800
29. Kahn BB, Flier JS (2000) Obesity and insulin resistance. J Clin Invest 106:473–481
30. Gottlob K, Majewski N, Kennedy S, Kandel E, Robey RB, Hay N (2001) Inhibition of early apoptotic events by Akt/PKB is dependent on the first committed step of glycolysis and mitochondrial hexokinase. Genes Dev 15:1406–1418
31. Chan JM, Stampfer MJ, Ma J, Gann P, Gaziano JM, Pollak M et al (2002) Insulin-like growth factor-I (IGF-I) and IGF binding protein-3 as predictors of advanced-stage prostate cancer. J Natl Cancer Inst 94:1099–1106
32. Burton AJ, Tilling KM, Holly JM, Hamdy FC, Rowlands MA, Donovan JL et al (2010) Metabolic imbalance and prostate cancer progression. Int J Mol Epidemiol Genet 1:248–271
33. Ma J, Li H, Giovannucci E, Mucci L, Qiu W, Nguyen PL et al (2008) Prediagnostic body-mass index, plasma C-peptide concentration, and prostate cancer-specific mortality in men with prostate cancer: a long-term survival analysis. Lancet Oncol 9:1039–1047
34. Pollak M (2008) Insulin and insulin-like growth factor signalling in neoplasia. Nat Rev Cancer 8:915–928
35. Hellawell GO, Turner GD, Davies DR, Poulsom R, Brewster SF, Macaulay VM (2002) Expression of the type 1 insulin-like growth factor receptor is up-regulated in primary prostate cancer and commonly persists in metastatic disease. Cancer Res 62:2942–2950
36. Cox ME, Gleave ME, Zakikhani M, Bell RH, Piura E, Vickers E et al (2009) Insulin receptor expression by human prostate cancers. Prostate 69:33–40
37. Renehan AG, Roberts DL, Dive C (2008) Obesity and cancer: pathophysiological and biological mechanisms. Arch Physiol Biochem 114:71–83
38. Platz EA, Giovannucci E (2004) The epidemiology of sex steroid hormones and their signaling and metabolic pathways in the etiology of prostate cancer. J Steroid Biochem Mol Biol 92:237–253
39. Hsing AW, Reichardt JK, Stanczyk FZ (2002) Hormones and prostate cancer: current perspectives and future directions. Prostate 52:213–235
40. Gapstur SM, Kopp P, Gann PH, Chiu BC, Colangelo LA, Liu K (2007) Changes in BMI modulate age-associated changes in sex hormone binding globulin and total testosterone, but not bioavailable testosterone in young adult men: the CARDIA Male Hormone Study. Int J Obes (Lond) 31:685–691
41. Kupelian V, Page ST, Araujo AB, Travison TG, Bremner WJ, McKinlay JB (2006) Low sex hormone-binding globulin, total testosterone, and symptomatic androgen deficiency are associated with development of the metabolic syndrome in nonobese men. J Clin Endocrinol Metab 91:843–850

42. Goldstein BJ (2003) Insulin resistance: from benign to type 2 diabetes mellitus. Rev Cardiovasc Med 4(suppl 6):S3–S10
43. Littman AJ, White E, Kristal AR (2007) Anthropometrics and prostate cancer risk. Am J Epidemiol 165:1271–1279
44. Hsing AW, Sakoda LC, Chua S Jr (2007) Obesity, metabolic syndrome, and prostate cancer. Am J Clin Nutr 86:s843–s857
45. Tilg H, Moschen AR (2006) Adipocytokines: mediators linking adipose tissue, inflammation and immunity. Nat Rev Immunol 6:772–783
46. Baillargeon J, Rose DP (2006) Obesity, adipokines, and prostate cancer (review). Int J Oncol 28:737–745
47. Roberts DL, Dive C, Renehan AG (2010) Biological mechanisms linking obesity and cancer risk: new perspectives. Annu Rev Med 61:301–316
48. Mistry T, Digby JE, Desai KM, Randeva HS (2007) Obesity and prostate cancer: a role for adipokines. Eur Urol 52:46–53
49. Zhang Y, Proenca R, Maffei M, Barone M, Leopold L, Friedman JM (1994) Positional cloning of the mouse obese gene and its human homologue. Nature 372:425–432
50. Considine RV, Sinha MK, Heiman ML, Kriauciunas A, Stephens TW, Nyce MR et al (1996) Serum immunoreactive-leptin concentrations in normal-weight and obese humans. N Engl J Med 334:292–295
51. Chang S, Hursting SD, Contois JH, Strom SS, Yamamura Y, Babaian RJ et al (2001) Leptin and prostate cancer. Prostate 46:62–67
52. Freedland SJ, Sokoll LJ, Platz EA, Mangold LA, Bruzek DJ, Mohr P et al (2005) Association between serum adiponectin, and pathological stage and grade in men undergoing radical prostatectomy. J Urol 174:1266–1270
53. Hsing AW, Chua S Jr, Gao YT, Gentzschein E, Chang L, Deng J et al (2001) Prostate cancer risk and serum levels of insulin and leptin: a population-based study. J Natl Cancer Inst 93:783–789
54. Kaaks R, Lukanova A, Rinaldi S, Biessy C, Soderberg S, Olsson T et al (2003) Interrelationships between plasma testosterone, SHBG, IGF-I, insulin and leptin in prostate cancer cases and controls. Eur J Cancer Prev 12:309–315
55. Lagiou P, Signorello LB, Trichopoulos D, Tzonou A, Trichopoulou A, Mantzoros CS (1998) Leptin in relation to prostate cancer and benign prostatic hyperplasia. Int J Cancer 76:25–28
56. Saglam K, Aydur E, Yilmaz M, Goktas S (2003) Leptin influences cellular differentiation and progression in prostate cancer. J Urol 169:1308–1311
57. Stattin P, Kaaks R, Johansson R, Gislefoss R, Soderberg S, Alfthan H et al (2003) Plasma leptin is not associated with prostate cancer risk. Cancer Epidemiol Biomarkers Prev 12:474–475
58. Ribeiro R, Lopes C, Medeiros R (2004) Leptin and prostate: implications for cancer prevention—overview of genetics and molecular interactions. Eur J Cancer Prev 13:359–368
59. Frankenberry KA, Somasundar P, McFadden DW, Vona-Davis LC (2004) Leptin induces cell migration and the expression of growth factors in human prostate cancer cells. Am J Surg 188:560–565
60. Sierra-Honigmann MR, Nath AK, Murakami C, Garcia-Cardena G, Papapetropoulos A, Sessa WC et al (1998) Biological action of leptin as an angiogenic factor. Science 281:1683–1686
61. Cirillo D, Rachiglio AM, la Montagna R, Giordano A, Normanno N (2008) Leptin signaling in breast cancer: an overview. J Cell Biochem 105:956–964
62. Hoda MR, Popken G (2008) Mitogenic and anti-apoptotic actions of adipocyte-derived hormone leptin in prostate cancer cells. BJU Int 102:383–388
63. Kadowaki T, Yamauchi T (2005) Adiponectin and adiponectin receptors. Endocr Rev 26:439–451
64. Hursting SD, Digiovanni J, Dannenberg AJ, Azrad M, Leroith D, Demark-Wahnefried W et al (2012) Obesity, energy balance, and cancer: new opportunities for prevention. Cancer Prev Res (Phila) 5:1260–1272

65. Byrne AM, Bouchier-Hayes DJ, Harmey JH (2005) Angiogenic and cell survival functions of vascular endothelial growth factor (VEGF). J Cell Mol Med 9:777–794
66. Liu Y, Tamimi RM, Collins LC, Schnitt SJ, Gilmore HL, Connolly JL et al (2011) The association between vascular endothelial growth factor expression in invasive breast cancer and survival varies with intrinsic subtypes and use of adjuvant systemic therapy: results from the Nurses' Health Study. Breast Cancer Res Treat 129:175–184
67. Cao Y (2007) Angiogenesis modulates adipogenesis and obesity. J Clin Invest 117:2362–2368
68. Fong GH, Rossant J, Gertsenstein M, Breitman ML (1995) Role of the Flt-1 receptor tyrosine kinase in regulating the assembly of vascular endothelium. Nature 376:66–70
69. Weidner N, Carroll PR, Flax J, Blumenfeld W, Folkman J (1993) Tumor angiogenesis correlates with metastasis in invasive prostate carcinoma. Am J Pathol 143:401–409
70. Duque JL, Loughlin KR, Adam RM, Kantoff PW, Zurakowski D, Freeman MR (1999) Plasma levels of vascular endothelial growth factor are increased in patients with metastatic prostate cancer. Urology 54:523–527
71. Woollard DJ, Opeskin K, Coso S, Wu D, Baldwin ME, Williams ED (2012) Differential expression of VEGF ligands and receptors in prostate cancer. Prostate. doi:10.1002/pros.22596
72. Freeman MR, Schneck FX, Gagnon ML, Corless C, Soker S, Niknejad K et al (1995) Peripheral blood T lymphocytes and lymphocytes infiltrating human cancers express vascular endothelial growth factor: a potential role for T cells in angiogenesis. Cancer Res 55:4140–4145
73. Ferrer FA, Miller LJ, Andrawis RI, Kurtzman SH, Albertsen PC, Laudone VP et al (1997) Vascular endothelial growth factor (VEGF) expression in human prostate cancer: in situ and in vitro expression of VEGF by human prostate cancer cells. J Urol 157:2329–2333
74. George DJ, Regan MM, Oh WK, Tay MH, Manola J, Decalo N et al (2004) Radical prostatectomy lowers plasma vascular endothelial growth factor levels in patients with prostate cancer. Urology 63:327–332
75. El-Gohary YM, Silverman JF, Olson PR, Liu YL, Cohen JK, Miller R et al (2007) Endoglin (CD105) and vascular endothelial growth factor as prognostic markers in prostatic adenocarcinoma. Am J Clin Pathol 127:572–579
76. Stefanou D, Batistatou A, Kamina S, Arkoumani E, Papachristou DJ, Agnantis NJ (2004) Expression of vascular endothelial growth factor (VEGF) and association with microvessel density in benign prostatic hyperplasia and prostate cancer. In Vivo 18:155–160
77. Bok RA, Halabi S, Fei DT, Rodriquez CR, Hayes DF, Vogelzang NJ et al (2001) Vascular endothelial growth factor and basic fibroblast growth factor urine levels as predictors of outcome in hormone-refractory prostate cancer patients: a cancer and leukemia group B study. Cancer Res 61:2533–2536
78. George DJ, Halabi S, Shepard TF, Vogelzang NJ, Hayes DF, Small EJ et al (2001) Prognostic significance of plasma vascular endothelial growth factor levels in patients with hormone-refractory prostate cancer treated on Cancer and Leukemia Group B 9480. Clin Cancer Res 7:1932–1936
79. Huss WJ, Hanrahan CF, Barrios RJ, Simons JW, Greenberg NM (2001) Angiogenesis and prostate cancer: identification of a molecular progression switch. Cancer Res 61:2736–2743
80. Turner K, Jones A (2000) Vascular endothelial growth factor in prostate cancer. Urology 56:183
81. Jackson MW, Bentel JM, Tilley WD (1997) Vascular endothelial growth factor (VEGF) expression in prostate cancer and benign prostatic hyperplasia. J Urol 157:2323–2328
82. Harper ME, Glynne-Jones E, Goddard L, Thurston VJ, Griffiths K (1996) Vascular endothelial growth factor (VEGF) expression in prostatic tumours and its relationship to neuroendocrine cells. Br J Cancer 74:910–916
83. Kitagawa Y, Dai J, Zhang J, Keller JM, Nor J, Yao Z et al (2005) Vascular endothelial growth factor contributes to prostate cancer-mediated osteoblastic activity. Cancer Res 65:10921–10929

84. Borgstrom P, Bourdon MA, Hillan KJ, Sriramarao P, Ferrara N (1998) Neutralizing anti-vascular endothelial growth factor antibody completely inhibits angiogenesis and growth of human prostate carcinoma micro tumors in vivo. Prostate 35:1–10
85. Iwaki T, Urano T, Umemura K (2012) PAI-1, progress in understanding the clinical problem and its aetiology. Br J Haematol 157:291–298
86. Skurk T, Hauner H (2004) Obesity and impaired fibrinolysis: role of adipose production of plasminogen activator inhibitor-1. Int J Obes Relat Metab Disord 28:1357–1364
87. Muldowney JA III, Chen Q, Blakemore DL, Vaughan DE (2012) Pentoxifylline lowers plasminogen activator inhibitor 1 levels in obese individuals: a pilot study. Angiology 63:429–434
88. Festuccia C, Vincentini C, di Pasquale AB, Aceto G, Zazzeroni F, Miano L et al (1995) Plasminogen activator activities in short-term tissue cultures of benign prostatic hyperplasia and prostatic carcinoma. Oncol Res 7:131–138
89. Swiercz R, Keck RW, Skrzypczak-Jankun E, Selman SH, Jankun J (2001) Recombinant PAI-1 inhibits angiogenesis and reduces size of LNCaP prostate cancer xenografts in SCID mice. Oncol Rep 8:463–470
90. Sfanos KS, De Marzo AM (2012) Prostate cancer and inflammation: the evidence. Histopathology 60:199–215
91. Gueron G, De Siervi A, Vazquez E (2012) Advanced prostate cancer: reinforcing the strings between inflammation and the metastatic behavior. Prostate Cancer Prostatic Dis 15:213–221
92. Koul HK, Kumar B, Koul S, Deb AA, Hwa JS, Maroni P et al (2010) The role of inflammation and infection in prostate cancer: Importance in prevention, diagnosis and treatment. Drugs Today (Barc) 46:929–943
93. Ouchi N, Parker JL, Lugus JJ, Walsh K (2011) Adipokines in inflammation and metabolic disease. Nat Rev Immunol 11:85–97
94. Azevedo A, Cunha V, Teixeira AL, Medeiros R (2011) IL-6/IL-6R as a potential key signaling pathway in prostate cancer development. World J Clin Oncol 2:384–396
95. Michalaki V, Syrigos K, Charles P, Waxman J (2004) Serum levels of IL-6 and TNF-alpha correlate with clinicopathological features and patient survival in patients with prostate cancer. Br J Cancer 90:2312–2316
96. Nakashima J, Tachibana M, Horiguchi Y, Oya M, Ohigashi T, Asakura H et al (2000) Serum interleukin 6 as a prognostic factor in patients with prostate cancer. Clin Cancer Res 6:2702–2706
97. Knupfer H, Preiss R (2008) sIL-6R: more than an agonist? Immunol Cell Biol 86:87–91
98. Akimoto S, Okumura A, Fuse H (1998) Relationship between serum levels of interleukin-6, tumor necrosis factor-alpha and bone turnover markers in prostate cancer patients. Endocr J 45:183–189
99. Drachenberg DE, Elgamal AA, Rowbotham R, Peterson M, Murphy GP (1999) Circulating levels of interleukin-6 in patients with hormone refractory prostate cancer. Prostate 41:127–133
100. Finley DS, Calvert VS, Inokuchi J, Lau A, Narula N, Petricoin EF et al (2009) Periprostatic adipose tissue as a modulator of prostate cancer aggressiveness. J Urol 182:1621–1627
101. Akira S (1997) IL-6-regulated transcription factors. Int J Biochem Cell Biol 29:1401–1418
102. Rose-John S (2001) Coordination of interleukin-6 biology by membrane bound and soluble receptors. Adv Exp Med Biol 495:145–151
103. Lin DL, Whitney MC, Yao Z, Keller ET (2001) Interleukin-6 induces androgen responsiveness in prostate cancer cells through up-regulation of androgen receptor expression. Clin Cancer Res 7:1773–1781
104. Chen T, Wang LH, Farrar WL (2000) Interleukin 6 activates androgen receptor-mediated gene expression through a signal transducer and activator of transcription 3-dependent pathway in LNCaP prostate cancer cells. Cancer Res 60:2132–2135
105. Hobisch A, Eder IE, Putz T, Horninger W, Bartsch G, Klocker H et al (1998) Interleukin-6 regulates prostate-specific protein expression in prostate carcinoma cells by activation of the androgen receptor. Cancer Res 58:4640–4645

106. Culig Z (2003) Role of the androgen receptor axis in prostate cancer. Urology 62:21–26
107. Szlosarek PW, Balkwill FR (2003) Tumour necrosis factor alpha: a potential target for the therapy of solid tumours. Lancet Oncol 4:565–573
108. van Horssen R, Ten Hagen TL, Eggermont AM (2006) TNF-alpha in cancer treatment: molecular insights, antitumor effects, and clinical utility. Oncologist 11:397–408
109. Robinson SC, Coussens LM (2005) Soluble mediators of inflammation during tumor development. Adv Cancer Res 93:159–187
110. Pollard JW (2004) Tumour-educated macrophages promote tumour progression and metastasis. Nat Rev Cancer 4:71–78
111. Nakashima J, Tachibana M, Ueno M, Miyajima A, Baba S, Murai M (1998) Association between tumor necrosis factor in serum and cachexia in patients with prostate cancer. Clin Cancer Res 4:1743–1748
112. Huerta-Yepez S, Vega M, Garban H, Bonavida B (2006) Involvement of the TNF-alpha autocrine-paracrine loop, via NF-kappaB and YY1, in the regulation of tumor cell resistance to Fas-induced apoptosis. Clin Immunol 120:297–309
113. Ohshima H, Tazawa H, Sylla BS, Sawa T (2005) Prevention of human cancer by modulation of chronic inflammatory processes. Mutat Res 591:110–122
114. Mantovani A, Schioppa T, Porta C, Allavena P, Sica A (2006) Role of tumor-associated macrophages in tumor progression and invasion. Cancer Metastasis Rev 25:315–322
115. Tsatsanis C, Androulidaki A, Venihaki M, Margioris AN (2006) Signalling networks regulating cyclooxygenase-2. Int J Biochem Cell Biol 38:1654–1661
116. Zha S, Yegnasubramanian V, Nelson WG, Isaacs WB, De Marzo AM (2004) Cyclooxygenases in cancer: progress and perspective. Cancer Lett 215:1–20
117. Subbarayan V, Sabichi AL, Llansa N, Lippman SM, Menter DG (2001) Differential expression of cyclooxygenase-2 and its regulation by tumor necrosis factor-alpha in normal and malignant prostate cells. Cancer Res 61:2720–2726
118. Munoz-Espada AC, Watkins BA (2006) Cyanidin attenuates PGE2 production and cyclooxygenase-2 expression in LNCaP human prostate cancer cells. J Nutr Biochem 17:589–596
119. Wang W, Bergh A, Damber JE (2005) Cyclooxygenase-2 expression correlates with local chronic inflammation and tumor neovascularization in human prostate cancer. Clin Cancer Res 11:3250–3256
120. Hensley PJ, Kyprianou N (2012) Modeling prostate cancer in mice: limitations and opportunities. J Androl 33:133–144
121. Wang F (2011) Modeling human prostate cancer in genetically engineered mice. Prog Mol Biol Transl Sci 100:1–49
122. Ahmad I, Sansom OJ, Leung HY (2008) Advances in mouse models of prostate cancer. Expert Rev Mol Med 10:e16
123. Szymanska H (2007) [Genetically engineered mice: mouse models for cancer research]. Postepy Hig Med Dosw (Online) 61:639–645
124. Huffman DM, Johnson MS, Watts A, Elgavish A, Eltoum IA, Nagy TR (2007) Cancer progression in the transgenic adenocarcinoma of mouse prostate mouse is related to energy balance, body mass, and body composition, but not food intake. Cancer Res 67:417–424
125. Llaverias G, Danilo C, Wang Y, Witkiewicz AK, Daumer K, Lisanti MP et al (2010) A Western-type diet accelerates tumor progression in an autochthonous mouse model of prostate cancer. Am J Pathol 177:3180–3191
126. Blando J, Moore T, Hursting S, Jiang G, Saha A, Beltran L et al (2011) Dietary energy balance modulates prostate cancer progression in Hi-Myc mice. Cancer Prev Res (Phila) 4:2002–2014
127. Kobayashi N, Barnard RJ, Said J, Hong-Gonzalez J, Corman DM, Ku M et al (2008) Effect of low-fat diet on development of prostate cancer and Akt phosphorylation in the Hi-Myc transgenic mouse model. Cancer Res 68:3066–3073
128. Ribeiro AM, Andrade S, Pinho F, Monteiro JD, Costa M, Lopes C et al (2010) Prostate cancer cell proliferation and angiogenesis in different obese mice models. Int J Exp Pathol 91:374–386

129. Zhang Y, Daquinag A, Traktuev DO, Amaya-Manzanares F, Simmons PJ, March KL et al (2009) White adipose tissue cells are recruited by experimental tumors and promote cancer progression in mouse models. Cancer Res 69:5259–5266
130. Lamarre NS, Ruggieri MR Sr, Braverman AS, Gerstein MI, Mydlo JH (2007) Effect of obese and lean Zucker rat sera on human and rat prostate cancer cells: implications in obesity-related prostate tumor biology. Urology 69:191–195
131. Price RS, Cavazos DA, De Angel RE, Hursting SD, de Graffenried LA (2012) Obesity-related systemic factors promote an invasive phenotype in prostate cancer cells. Prostate Cancer Prostatic Dis 15:135–143
132. National Institutes of Health (1998) Clinical guidelines on the identification, evaluation, and treatment of overweight and obesity in adults—the evidence report. Obes Res 6(suppl 2):51S–209S
133. Vainio H, Kaaks R, Bianchini F (2002) Weight control and physical activity in cancer prevention: international evaluation of the evidence. Eur J Cancer Prev 11(suppl 2):S94–S100
134. Cerhan JR, Torner JC, Lynch CF, Rubenstein LM, Lemke JH, Cohen MB et al (1997) Association of smoking, body mass, and physical activity with risk of prostate cancer in the Iowa 65+ Rural Health Study (United States). Cancer Causes Control 8:229–238
135. Hartman TJ, Albanes D, Rautalahti M, Tangrea JA, Virtamo J, Stolzenberg R et al (1998) Physical activity and prostate cancer in the Alpha-Tocopherol, Beta-Carotene (ATBC) Cancer Prevention Study (Finland). Cancer Causes Control 9:11–18
136. Oliveria SA, Lee IM (1997) Is exercise beneficial in the prevention of prostate cancer? Sports Med 23:271–278
137. Kelley DE, Goodpaster BH (1999) Effects of physical activity on insulin action and glucose tolerance in obesity. Med Sci Sports Exerc 31:S619–S623
138. Chang S, Wu X, Yu H, Spitz MR (2002) Plasma concentrations of insulin-like growth factors among healthy adult men and postmenopausal women: associations with body composition, lifestyle, and reproductive factors. Cancer Epidemiol Biomarkers Prev 11:758–766
139. Kraemer RR, Chu H, Castracane VD (2002) Leptin and exercise. Exp Biol Med (Maywood) 227:701–708
140. Hulver MW, Berggren JR, Cortright RN, Dudek RW, Thompson RP, Pories WJ et al (2003) Skeletal muscle lipid metabolism with obesity. Am J Physiol Endocrinol Metab 284:E741–E747
141. Nindl BC, Kraemer WJ, Arciero PJ, Samatallee N, Leone CD, Mayo MF et al (2002) Leptin concentrations experience a delayed reduction after resistance exercise in men. Med Sci Sports Exerc 34:608–613
142. Barnard RJ, Ngo TH, Leung PS, Aronson WJ, Golding LA (2003) A low-fat diet and/or strenuous exercise alters the IGF axis in vivo and reduces prostate tumor cell growth in vitro. Prostate 56:201–206
143. Leung PS, Aronson WJ, Ngo TH, Golding LA, Barnard RJ (2004) Exercise alters the IGF axis in vivo and increases p53 protein in prostate tumor cells in vitro. J Appl Physiol 96:450–454
144. Ornish D, Weidner G, Fair WR, Marlin R, Pettengill EB, Raisin CJ et al (2005) Intensive lifestyle changes may affect the progression of prostate cancer. J Urol 174:1065–1069; discussion 1069–1070
145. Ribaric S (2012) Diet and aging. Oxid Med Cell Longev 2012:741468
146. Barger JL, Kayo T, Vann JM, Arias EB, Wang J, Hacker TA et al (2008) A low dose of dietary resveratrol partially mimics caloric restriction and retards aging parameters in mice. PLoS One 3:e2264
147. Pearson KJ, Baur JA, Lewis KN, Peshkin L, Price NL, Labinskyy N et al (2008) Resveratrol delays age-related deterioration and mimics transcriptional aspects of dietary restriction without extending life span. Cell Metab 8:157–168
148. Li G, Rivas P, Bedolla R, Thapa D, Reddick RL, Ghosh R et al (2012) Dietary resveratrol prevents development of high-grade prostatic intraepithelial neoplastic lesions: involvement of SIRT1/S6K axis. Cancer Prev Res (Phila) 6(1):27–39

149. Kwon KH, Barve A, Yu S, Huang MT, Kong AN (2007) Cancer chemoprevention by phytochemicals: potential molecular targets, biomarkers and animal models. Acta Pharmacol Sin 28:1409–1421
150. Adhami VM, Siddiqui IA, Ahmad N, Gupta S, Mukhtar H (2004) Oral consumption of green tea polyphenols inhibits insulin-like growth factor-I-induced signaling in an autochthonous mouse model of prostate cancer. Cancer Res 64:8715–8722
151. Bowker SL, Majumdar SR, Veugelers P, Johnson JA (2006) Increased cancer-related mortality for patients with type 2 diabetes who use sulfonylureas or insulin: response to Farooki and Schneider. Diabetes Care 29:1990–1991
152. Wright JL, Stanford JL (2009) Metformin use and prostate cancer in Caucasian men: results from a population-based case–control study. Cancer Causes Control 20:1617–1622
153. Zadra G, Priolo C, Patnaik A, Loda M (2010) New strategies in prostate cancer: targeting lipogenic pathways and the energy sensor AMPK. Clin Cancer Res 16:3322–3328
154. Zakikhani M, Dowling RJ, Sonenberg N, Pollak MN (2008) The effects of adiponectin and metformin on prostate and colon neoplasia involve activation of AMP-activated protein kinase. Cancer Prev Res (Phila) 1:369–375
155. Ben Sahra I, Laurent K, Loubat A, Giorgetti-Peraldi S, Colosetti P, Auberger P et al (2008) The antidiabetic drug metformin exerts an antitumoral effect in vitro and in vivo through a decrease of cyclin D1 level. Oncogene 27:3576–3586

Chapter 11
Pharmacologic Interventions with NSAIDs

Louise R. Howe

Abstract Obesity as a determinant of increased cancer risk and poorer cancer outcome is well established for cancers of several organ sites, including colorectal and postmenopausal breast cancer. Obesity-associated adipose inflammation leads to local and systemic accumulation of inflammatory mediators and hormones, which have multiple proneoplastic effects. Key among these from a pharmacological perspective are cyclooxygenase (COX)-derived prostaglandins (PGs), since COX enzymes are the primary target for nonsteroidal anti-inflammatory drugs (NSAIDs). Overexpression of the inducible PG synthase COX-2 occurs in the majority of colorectal neoplasias and ~40 % of breast cancers and is also evident in inflamed adipose tissue from obese mice and humans. COX/PG signaling has multiple protumorigenic consequences, which provide at least a partial explanation for epidemiologic and experimental observations of reduced cancer risk associated with NSAID use. Notably, COX/PG-mediated upregulation of estrogen biosynthesis and signaling offers a plausible target for NSAID-mediated risk reduction with respect to breast and other hormone-sensitive cancers. Additionally, "off-target" NSAID effects including modulation of NFκB and AMP kinase activity may be of particular significance in the context of obesity. NSAID-mediated amelioration of obesity-related metabolic dysfunction has been reported, and it seems likely that NSAIDs will be similarly protective for obesity-associated carcinogenesis.

L.R. Howe, Ph.D. (✉)
Department of Cell and Developmental Biology, Weill Cornell Medical College,
1300 York Avenue, New York, NY 10065, USA
e-mail: lrhowe@med.cornell.edu

A.J. Dannenberg and N.A. Berger (eds.), *Obesity, Inflammation and Cancer*, Energy Balance and Cancer 7, DOI 10.1007/978-1-4614-6819-6_11,

11.1 Introduction

It is becoming increasingly apparent that the molecular aberrations characterizing chronic inflammation may impact many, if not all, of the hallmark capacities intrinsic to the neoplastic process. Thus, logically, anti-inflammatory approaches may be of benefit in abrogating the increased cancer risk associated with inflammation. This may be of particular relevance in the context of obesity, an established risk factor for cancers of many organ sites [1–3], and now understood to be an inflammatory condition [4].

The classic signs of acute inflammation (pain, heat, redness, and swelling) reflect the ingress and activation of myeloid cells, with consequent release of inflammatory mediators that elicit pain, vasodilatation, and exudation of blood vessel contents into the surrounding tissue. Initial response to injury or pathogenic insult is orchestrated by resident macrophages and recruited antimicrobial granulocytes, whereas persistent inflammation is characterized by monocytic and lymphocytic accumulation. On a molecular level, inflamed tissues exhibit elevated levels of cytokines, chemokines, and eicosanoids. Key among these from a pharmacological perspective are cyclooxygenase (COX)-derived prostaglandins, which can elicit vasodilation, pain, and fever. Inhibition of COX enzymes is the primary mechanism of analgesics of the nonsteroidal anti-inflammatory drug class (NSAIDs).

Significantly, obesity is now recognized as an inflammatory condition, with leukocyte infiltration of white adipose tissue depots resulting in increased production of inflammatory mediators, including adipokines, cytokines, and prostanoids, as well as increased release of fatty acids [4, 5]. Adverse systemic consequences of adipose inflammation include both insulin resistance, leading to type II diabetes, and increased cancer risk. Notably, NSAIDs have been reported to both ameliorate insulin resistance and attenuate cancer risk (discussed below). While intuitively this might appear predictable given the relationship between these diseases and inflammation, the mechanistic basis by which NSAIDs afford protection is likely complex and multifactorial and not predicated solely on COX inhibition. This chapter reviews the evidence for NSAIDs as anticancer agents, considers likely mechanisms, and discusses the relevance of these in the context of obesity-driven neoplasia.

11.2 Cyclooxygenases and Cancer

The established anticancer activity of NSAIDs, with respect to both human disease and experimental models, is best understood in terms of their activity as COX inhibitors. COX enzymes are prostaglandin (PG) synthases, responsible for conversion of arachidonic acid (AA) to the intermediary prostanoid PGH_2, which then serves as substrate for a range of synthases, ultimately resulting in production of numerous eicosanoids, including PGD_2, PGE_2, $PGF_{2\alpha}$, PGI_2 (prostacyclin), and thromboxanes (TXs) [6]. The complexion of prostanoid production by a given cell type is predicated by the unique expression pattern of terminal synthases. Thus platelets are a rich source of TXA_2, whereas endothelium is an important site of prostacyclin synthesis. PGE_2, the isoform most strongly implicated in pain, inflammation, and neoplasia, is produced both by inflammatory and epithelial cells. COX expression levels are also important determinants of prostanoid production. COX-1 (encoded by the *PTGS1* gene) is constitutively expressed and believed to serve an important function in protecting the gastrointestinal (GI) mucosa, NSAID-mediated impairment of which is associated with GI bleeds. In contrast, the *COX-2* (or *PTGS2*) gene exhibits a highly restricted pattern of constitutive expression, notably in specific regions of kidney and brain, but induction is elicited by a wide variety of stimuli including cytokines, growth factors, and oncogenes [7–9]. COX-2-derived eicosanoids play key roles in the inflammatory response and thus provide a major focus for pharmacological intervention.

A driving role for COX-2 in neoplasia was initially suggested by parallel findings of COX-2 overexpression in cancers and epidemiological data linking analgesic use with reduced cancer incidence. Notably, the most consistent associations between NSAID use and diminished cancer risk have been identified for colorectal cancer both in epidemiological and interventional studies. Thus this chapter focuses predominantly on two main cancer sites, colorectal, based on the epidemiological evidence, and breast, based on plausible mechanisms for NSAID-mediated protection in the context of obesity for this latter disease.

11.2.1 Cyclooxygenase Expression in Intestinal and Breast Tissues

A general theme with respect to COX expression and epithelial cancers has emerged: *COX-1* expression tends to be relatively constant in normal and cancerous tissues, while *COX-2* is frequently upregulated in precancers and cancers relative to benign epithelium. For example, evaluation of *COX* expression in human colorectal neoplasms revealed unaltered expression levels of *COX-1* in colorectal carcinomas relative to adjacent normal mucosa [10–12], whereas *COX-2* is virtually undetectable in normal colonic mucosa, but highly expressed in adenocarcinomas, with detectable expression in epithelial, endothelial, and stromal components [10–13]. Consistent with human data, *COX-2* upregulation is also evident in intestinal tumors from both genetic (*Apc* mutant) and carcinogen-driven rodent models [14–18]. Given the present focus on inflammation as a driver of carcinogenesis, it is noteworthy that, while in adenocarcinomas *COX-2* expression is predominantly epithelial, in the precursor polyps COX-2 is primarily localized in inflammatory cells in the lamina propria [15, 19, 20]. These data suggest potential paracrine interactions between *COX-2*-expressing macrophages and enterocytes as early drivers of intestinal neoplasia. Intriguingly, a single study has identified an association between rectal mucosal PGE_2 levels and body mass index (BMI) in patients with a history of polyps, although the cell type from which PGE_2 derived was not identified [21]. Nevertheless, these data provide provocative evidence for a link between BMI and upregulated prostaglandin production in the intestine, potentially mediated via increased inflammation.

In breast carcinomas, *COX-2* overexpression is less prevalent than in CRC, although elevated PG levels were first reported in human breast tumors over 30 years ago [22–24]. In aggregate, immunohistochemical (IHC) studies have established that approximately 40 % of breast carcinomas exhibit epithelial *COX-2* overexpression, with negligible COX-2 in benign epithelium [25–38]. Increased frequencies of *COX-2* overexpression (63–85 %) are evident in precursor ductal carcinoma in situ (DCIS) lesions relative to invasive disease [25, 29, 33, 34, 36, 39, 40], and COX-2 has also been observed in normal-looking epithelium adjacent to neoplastic tissue [28, 29, 34, 36, 39–41]. Additionally, COX-2 is expressed in morphologically normal breast epithelium in association with $p16^{INK4a}$ hypermethylation, perhaps identifying very early foci of precancerous cells [42], and is associated with high mammographic density [43], an index of increased breast cancer risk.

Several studies identify COX-2 overexpression in breast cancer epithelium as a poor prognostic indicator, and COX-2 is associated with large tumor size, high grade, and the presence of nodal metastases [26, 28, 32, 35, 44]. COX-2 overexpression in DCIS precancers also has prognostic significance in association with $p16^{INK4a}$ and Ki67 status [45]. Additional correlations have been observed with overexpression of HER2/neu (human epidermal growth factor receptor 2) and with hormone receptor-negative status of carcinomas, both of which are independently associated with poor prognosis [25, 28, 32, 37, 46], although one recent study identified selective expression of COX-2 in estrogen receptor (ER)-positive breast cancers [44]. In contrast, COX-1 is ubiquitously expressed in mammary tissue and is not of prognostic significance [28, 34, 38]. Rodent models provide parallel observations with respect to COX expression in breast epithelium, including ubiquitous COX-1 expression [47], and upregulated COX-2 expression in mammary tumors from both transgenic and carcinogen-induced models [48–52].

Most recently, attention has focused on COX-2 expression in normal breast tissue as a function of obesity. Specifically, both in human and murine breast adipose tissue, PGE_2 and COX-2 levels increase with increasing BMI [53, 54]. Striking correlations are evident between COX-2/PGE_2 levels and the presence of inflammatory foci, comprising adipocytes encircled by macrophages [54]. These crown-like structures (CLS), first identified in white adipose tissues of non-mammary origin from obese mice and humans, have emerged as hallmark lesions of obesity-associated inflammation [55–57]. In both human and rodent breast tissue, CLS abundance increases as a function of body mass [53, 58, 59]. Simplistically, macrophages can be divided into two classes. Classically activated, or M1-polarized, macrophages secrete proinflammatory mediators including PGE_2, interleukin-1β (IL-1β), IL-6, and tumor necrosis factor-α (TNFα), whereas alternatively activated (M2) macrophages triggered by Th2 cytokines are associated with tissue remodeling and immunosuppression. While resident macrophages in lean adipose tend to be M2-polarized, inflammatory M1 macrophages predominate in CLS. Importantly, these morphologic entities likely correspond to the functional unit responsible for a significant component of obesity-related pathologies. Paracrine interaction between necrotic adipocytes and associated M1-polarized macrophages results in increased synthesis of PGE_2, IL-1β, IL-6, and TNFα in the macrophage compartment [53]. These proinflammatory mediators can have both local and systemic actions, which likely contribute both to insulin resistance and to the increased cancer incidence associated with overweight/obesity [1, 2].

11.2.2 Animal Models: Genetic Evidence for COX Contributions to Cancer

The fact of COX-2 overexpression in colorectal, breast, and other cancers, and the observed correlations between epithelial COX-2 overexpression and prognosis, argue strongly for a protumorigenic role of COX-2 in neoplasia. This thesis is supported by a wealth of data from animal models including several definitive genetic studies, as well as pharmacological studies (described below; Sect. 11.3.1). The Taketo lab pioneered the genetic approach, demonstrating substantial reduction (86 %) in intestinal tumor incidence in an *Apc* mutant mouse strain ($Apc^{\Delta 716}$) consequent on homozygous inactivation of the *COX-2* gene [15]. Of note, *Apc* mutant strains are widely used to model human colorectal neoplasia because mutational inactivation of the *APC* tumor suppressor gene is the initiating event in the vast majority of human sporadic colorectal neoplasms, as well as in some familial polyposis syndromes.

We subsequently employed a genetic strategy modeled on the Taketo approach to explore the contribution of COX-2 to breast cancer, using a mouse strain with mammary-targeted *HER2/neu* transgene expression that develops multiple DCIS lesions in each mammary gland to model the HER2-COX-2 relationship observed in the human disease. Decreased mammary tumor multiplicity was observed in the context of *COX-2* deficiency in our study [60]. *COX-2* nullizyogosity was associated with reduced DCIS tumor formation and growth, and with substantial diminution in mammary gland vascularization, unequivocally demonstrating a role for COX-2 in mammary neoplasia.

Further knockout studies have implicated both COX isoforms in experimental neoplasia. Similar magnitudes of intestinal and epidermal tumor abrogation can be achieved by disruption of either *COX-1* or *COX-2*, suggesting that perhaps the sum total of COX activity is the key determinant of tumorigenesis, rather than unique contributions of either enzyme [61, 62].

Hla and colleagues adopted a genetic overexpression approach to provide elegant proof of the in vivo oncogenicity of COX-2 through generation of a *COX-2* transgenic mouse strain. Mammary-targeted expression of a *COX-2* transgene drives tumor formation in multiparous mice [63]. Increased microvessel density and angiogenic gene expression are evident prior to discernible tumor formation, and

COX-2 overexpression retards mammary involution in post-weaning dams via suppression of apoptosis, providing mechanistic insights into COX-driven neoplasia [63, 64]. Furthermore, transgenic *COX-2* overexpression in skin increases sensitivity to carcinogen-induced neoplasia, again providing evidence for a proneoplastic role of COX-2 [65]. In aggregate, genetic knockout and overexpression approaches, together with pharmacological data discussed below, firmly substantiate the role of COXs and COX-derived prostanoids in cancer.

Follow-up studies have focused on evaluating alternative molecular targets for anti-inflammatory intervention that retain the efficacy of NSAIDs while minimizing corollary toxicity. A central role for PGE_2 as an effector of COX-mediated tumorigenesis has been established. PGE_2 administration reverses aspirin-induced adenoma regression and enhances carcinogen-induced tumor incidence, whereas genetic deletion of PGE_2 receptors (EPs) confers resistance to formation of aberrant crypt foci, polyps, and cancers in the intestinal tract [66]. Importantly, genetic deletion of the terminal enzyme responsible for PGE_2 synthesis (microsomal PGE_2 synthase-1, mPGES-1) suppresses the growth of both intestinal [67] and mammary tumors (Howe et al., manuscript in preparation), identifying mPGES-1 as a candidate target for antineoplastic intervention.

11.2.3 Cyclooxygenases and Cancer: Mechanisms

COX/PG signaling impacts many of the capabilities highlighted by Hanahan and Weinberg as essential for tumor growth and metastatic dissemination [68], including increased proliferative potential, resistance to apoptosis, enhanced invasiveness, immune suppression, and angiogenesis (Fig. 11.1). Self-evidently, as a key component of the inflammatory response, the "enabling characteristic" of tumor-promoting inflammation is also encompassed by COX-2 overexpression.

11.2.3.1 Prostanoid Effects on Proliferation and Immune Suppression

Prostaglandin stimulation of proliferation is well established for multiple cell types, such as mammary epithelial cells and fibroblasts [69–71]. PGE_2 can also stimulate growth of human colorectal adenoma cells, although it has been suggested that response may vary as a function of EP4 receptor expression level [72]. Notably, reported PG-mediated stimulation of stem cells [73] may have important ramifications for cancer, and indeed recent studies suggest that PGE_2 released from stromal fibroblasts plays a key role in paracrine regulation of cancer stem cells [74, 75]. Numerous pro-proliferative signaling pathways are implicated in mediating PG effects including activation of phosphatidylinositol 3-kinase (PI3K)/AKT and Wnt/β-catenin signaling pathways [76] and activation of epidermal growth factor receptor (EGFR) signaling [77, 78].

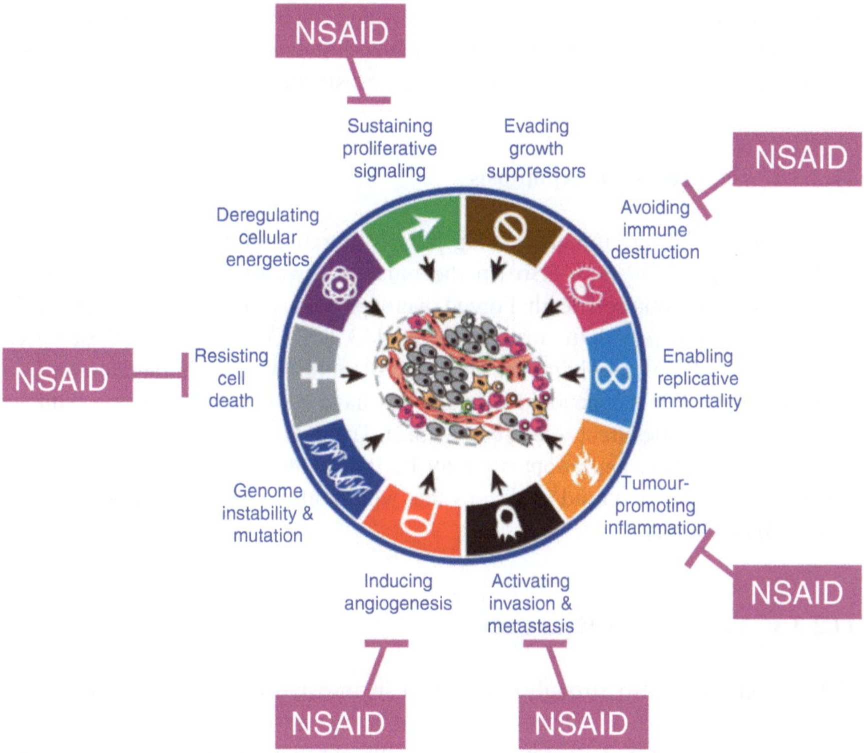

Fig. 11.1 Nonsteroidal anti-inflammatory drug (NSAID) effects on cancer hallmarks. Cyclooxygenase/prostaglandin signaling impacts many of the capabilities suggested by Hanahan and Weinberg to be essential for tumor growth and metastatic dissemination [68]. In turn, NSAID-mediated COX inhibition affords the possibility of reversing the majority of these protumorigenic traits. Persuasive evidence supports positive effects of COX/PG signaling and hence reciprocal NSAID-mediated suppression on the indicated parameters. Additional data suggest the possibility that NSAIDs may suppress genome instability via reducing microsatellite instability [316] and may also antagonize replicative immortality through downregulating telomerase expression [317]

In contrast to these pro-proliferative PG effects, certain immune cells respond to PGs with attenuated proliferative responses, which may be parlayed into an immune evasion strategy in the neoplastic setting [79–81]. Both B and T lymphocytes are subject to PGE_2-mediated growth suppression. PGE_2 also induces biased cytokine production from T cells towards Th2 cytokine release, downregulates Th1-associated cytokine synthesis, and promotes IL-10 expression, all of which may favor immune suppression. Furthermore, COX-2 is implicated in skewing macrophages towards a tumor-promoting M2 phenotype and averting the potential protection afforded by M1 polarization [82, 83]. Of note, while macrophages comprising CLS in inflamed adipose tend towards M1 polarization, tumor-associated macrophages are M2-like and are thought to favor cancer progression by promoting tissue remodeling and immune suppression [84]. Prostanoids also impact the function and development of dendritic cells [85, 86] and enhance accumulation and activity of myeloid-derived suppressor cells (MDSCs), which can inhibit both innate and adaptive immunity [80]. Thus, undermining the host immune response is likely to be a key protumorigenic property of COX-derived prostanoids.

11.2.3.2 Suppression of Apoptosis

In addition to eliciting tumor-favoring pro-proliferative effects in epithelial cells, COX overexpression may also tip the balance towards cell survival through suppressing apoptotic cell death. Forced overexpression of *COX-2* in epithelial cells suppresses apoptosis both in vitro and in vivo [63, 87]. Consistent with these observations, NSAIDs commonly induce apoptosis of cultured cells, although both COX-mediated and COX-independent mechanisms have been proposed to account for NSAID proapoptotic effects (see Sect. 11.5.2). The dual effects of COX overexpression on proliferation and apoptosis may be important for promoting survival of damaged cells with acquired mutations that would otherwise trigger cytoprotective cell death machinery.

11.2.3.3 Invasion and Metastasis

COX signaling is also strongly implicated in invasiveness and hence metastatic potential. In vitro studies have established increased cellular invasiveness as a consequence of *COX-2* overexpression, with corollary upregulation of matrix metalloproteinases which can promote extracellular matrix remodeling and hence tumor invasion [88, 89]. Consistent with a prometastatic function, correlations are evident between *COX-2* expression and both lymph node metastases and prognosis in human breast cancer [26, 28, 32, 35], and pulmonary metastasis of breast cancer xenografts and allografts can be suppressed by COX inhibitors [90–92]. Furthermore, COX-2 is also implicated in osteoclast differentiation, a key step in bone metastasis, which requires osteoclast-mediated bone resorption to provide a niche for tumor cells. PGE_2 stimulates both osteoclast differentiation and bone resorption, and osteoclastogenesis is impaired in bone marrow from *COX-2*-deficient mice [93–95]. Additionally, in an experimental model of sarcoma bone invasion, COX-2 inhibition reduces osteoclast proliferation and hypertrophy, tumor-induced bone resorption, and tumor burden [96]. These data are consistent with early observations from Powles et al. that aspirin treatment reduced the in vitro osteolysis-inducing activity of human breast carcinomas [97]. In aggregate, these data suggest that COX-derived PGs may regulate cancer cell invasion and metastasis.

11.2.3.4 Angiogenesis

Angiogenesis, the process of tumor neovascularization that ensures adequate blood supply, is another intrinsic feature of neoplasia in which COX/PG signaling plays a role. *COX-2*-deficient mice exhibit substantial reductions in mammary gland vascularization and expression of proangiogenic genes, with reduced vessel density evident not only in mammary tumors but also in morphologically normal mammary gland, implying a broad role for COX-2 in vascularization of the mammary gland [60]. In contrast, knockout studies using intestinal cancer models suggest that in the intestine *COX-2* contributes primarily to growth and vascularization of tumors beyond 1 mm in diameter [98, 99]. Contribution of stromal *COX-2* is well illustrated by the impaired growth and diminished vascular density of Lewis lung carcinoma xenografts when implanted into *COX-2*-null hosts [100]. Nevertheless, epithelial COX-2 can also promote angiogenesis: transgenic *COX-2* overexpression in mammary epithelium induces substantial remodeling of mammary gland vascular architecture [64]. Together these genetic approaches demonstrate a clear role for COX-2 in angiogenesis. Pharmacological approaches yield similar data: in vivo angiogenesis is reduced by administration of non-selective NSAIDs or selective COX-2 inhibitors (COXibs) in multiple models [64, 90, 91, 101–107], while in vitro COXibs decrease endothelial tube formation [108, 109]. COX-1 is also implicated, since NSAIDs can have similar effects in experimental systems lacking COX-2 expression [106, 108, 109]. COX-regulated production of proangiogenic factors such as vascular endothelial growth factor (VEGF) provides a clear mechanistic basis for the COX-angiogenesis relationship [110, 111] and likely explains observed correlations between COX-2 and microvessel density/VEGF expression in human breast cancers [26, 27, 31, 41].

The COX-2/VEGF interrelationship might also provide insight into the increased cancer incidence and poorer outcome associated with obesity. Human studies have identified positive correlations between circulating levels of VEGF and BMI [112–114]. This could reflect increased adiposity, given that adipose tissue is a rich source of VEGF. Additionally, based on the increased COX-2 levels observed in visceral and mammary adipose depots from obese subjects [53, 54], COX-dependent upregulation of VEGF could induce elevated adipose VEGF expression per wet tissue weight in obese vs. lean adipose. Indeed both adipose and circulating VEGF levels are elevated in a mouse model of postmenopausal obesity [115]. Furthermore, Hung and colleagues recently reported that murine mammary tumor sensitivity to the VEGF antagonist bevacizumab is increased in the context of obesity [116]. Thus, obesity-associated disproportionate increases in adipose VEGF synthesis may contribute to tumor angiogenesis.

Evidence is also available to support a role for COX-2 in regulation of lymphangiogenesis, the formation of new lymphatic vessels. COX-2 expression correlates with that of VEGF-C, the primary driver of lymphangiogenesis, in breast cancer cell lines and tissues, and COXib treatment suppresses tumor expression of a lymphangiogenic marker in vivo [117, 118].

11.2.3.5 Regulation of Peripheral Estrogen Biosynthesis

COX-2 overexpression may have consequences of particular significance for breast neoplasia, based on the ability of prostaglandins to regulate transcription of the *CYP19* gene, which encodes the estrogen synthase cytochrome P-450 aromatase.

Approximately two-thirds of human breast carcinomas express estrogen receptor-α (ERα), and the importance of estrogen in breast neoplasia is well illustrated by the demonstrated clinical efficacy of antiestrogen approaches including selective estrogen receptor modulators (SERMs) and aromatase inhibitors (AIs) as breast cancer therapeutics.

Prior to menopause, the ovary is the primary site of estrogen synthesis from androgens. However, post climacteric, peripheral sources assume increased importance, particularly adipose tissue as well as breast cancer epithelium [119–123]. Notably, a switch in *CYP19* promoter usage occurs in breast cancers and cancer-proximal stromal tissue, involving selective utilization of cAMP-sensitive promoters [124–127]. Consistent with the ability of PGE_2 to elicit cAMP accumulation [128], PGE_2 signaling stimulates aromatase transcription [129–133]. Furthermore, positive correlations between *COX* and *CYP19* expression have been identified in human breast cancers [134–136]. Definitive in vivo evidence for a causal basis for these findings is provided by transgenic and knockdown approaches: aromatase activity is increased in the mammary glands of *COX-2* transgenic mice and, conversely, decreased in *HER2/neu*-overexpressing mammary tissues that are *Cox-2* nullizygous [137]. Together these findings establish COX-2 as an important determinant of aromatase expression in mammary tissues and hence likely to regulate neoplastic growth through increasing tumor-proximal estrogen tissue levels. A further prediction from these data is that COX-inhibiting NSAIDs may regulate circulating hormone levels in postmenopausal women, and indeed some (though not all) studies have identified correlations between NSAID use and serum estradiol in postmenopausal women [138, 139].

Importantly, the COX-aromatase-estrogen pathway may not only be relevant for driving breast cancer progression but may also be a component of the increased breast cancer risk associated with obesity. Based on the importance of adipose as an estrogen source in postmenopausal women, it has long been assumed that increased adiposity should be associated with elevated estrogen synthesis, and indeed BMI is an established determinant of serum estradiol in postmenopausal women [140–142]. However, it is becomingly increasingly apparent that not only the quantity but also the quality of fat is altered in obesity, and this has important ramifications for adipose estrogen production. As described above, white adipose tissue from obese mice and humans is distinguished by the presence of inflammatory foci, or CLS, consisting of macrophage-encircled adipocytes [55–57]. These are characteristic not only in visceral fat but also in breast adipose tissue [53, 54, 58, 59]. Activated M1-polarized macrophages in these foci provide a rich source of proinflammatory factors, including PGE_2 as well as TNFα and IL-1β, which can induce increased aromatase transcription in the surrounding adipose tissue (Fig. 11.2). Thus, obesity drives adipose inflammation leading to local induction of aromatase expression in breast tissue. Delineation of this pathway provides a clear explanation for the earlier observation that BMI is a determinant of aromatase transcript levels in visceral fat [143] and offers mechanistic insight into the positive correlation between postmenopausal obesity and breast cancer risk.

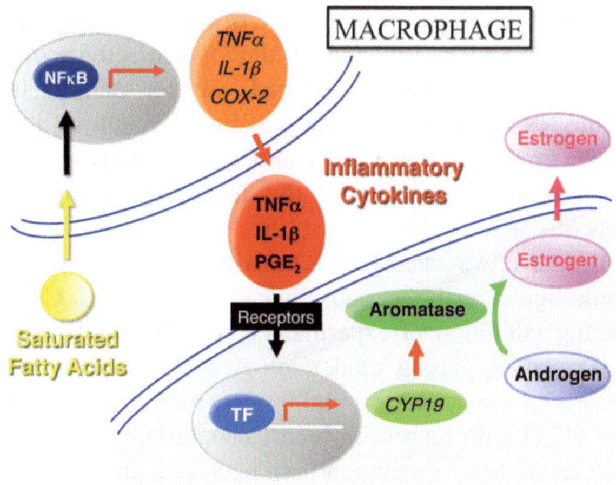

Fig. 11.2 Paracrine interactions between macrophages and other cell types establish an inflammatory milieu in obese breast adipose tissue, resulting in elevated estrogen biosynthesis. Saturated fatty acids, liberated from adipocytes as a result of obesity-associated lipolysis, stimulate NFκB-driven induction of inflammatory mediators in macrophages including TNFα, IL-1β, and COX-2. Additionally, bacterial toxins entering systemic circulation as a consequence of obesity-induced impairment of gut mucosal integrity may elicit NFκB activation through TLR4 ligation. Cytokines and COX-2-derived PGE_2 activate transcription of the *CYP19* gene encoding aromatase in neighboring cells, including preadipocytes, leading to elevated expression and activity of aromatase. Consequently, estrogen biosynthesis is enhanced, leading to increased expression of ER target genes, including PR

11.2.3.6 Mutagen Synthesis

A final procarcinogenic consequence of COX-2 overexpression, independent of prostanoid production, hinges on the ability of COX-2 to generate mutagens. For example, malondialdehyde produced from PGH_2 can induce frameshifts and base-pair substitutions through adduct formation with deoxynucleosides and thus drive mutagenesis [144]. COX-2-mediated oxidation of aromatic amines, heterocyclic amines, and dihydrodiol derivatives of polycyclic hydrocarbons also generates carcinogens. Thus, COX-2-dependent DNA damage may play a role at the earliest stages of neoplasia.

11.3 NSAID-Mediated Suppression of Neoplasia

NSAIDs can be divided into broad functional classes based on their relative selectivity for COX-1 vs. COX-2. Conventional NSAIDs (e.g., indomethacin, ibuprofen) inhibit both COX isoforms, although relative potencies for each enzyme vary between individual drugs. COX-2-selective inhibitors (celecoxib, rofecoxib, nimesulide, etc.) were developed with the goal of circumventing GI toxicity associated with prototypic NSAIDs. By contrast, the marked activity of aspirin towards COX-1 provides the basis for widespread use for antiplatelet prophylaxis. Intriguingly, some NSAID metabolites such as sulindac sulfone, the hepatic oxidation product of the prodrug sulindac, have anticancer activity in the absence of discernible COX inhibition, highlighting the functional significance of NSAID off-target effects (discussed below, Sect. 11.5).

Myriad studies have evaluated NSAID-mediated tumor suppression in animal models, epidemiological analyses, and clinical interventions. Consensus findings include convincing inhibition of experimental intestinal and breast neoplasia by multiple NSAID classes, strong epidemiological support for NSAID-mediated reduction in colorectal cancer risk, similar but less robust epidemiological evidence for a protective effect with respect to breast neoplasia, and compelling evidence from interventional studies focusing on both neoplasia and cardiovascular disease as primary endpoints. These data are reviewed in the following sections.

11.3.1 NSAID-Mediated Tumor Suppression in Animal Models

COX inhibitor efficacy for suppression of experimental neoplasia has been evaluated in a variety of animal models (for detailed reviews, see [9, 145–147]). For assessing intestinal cancer prophylaxis, the range of model systems encompasses carcinogen-treated rodents (AOM, azoxymethane; DMH, dimethylhydrazine), *Apc* mutant mouse strains (*Apc*^{Min}, *Apc*$^{\Delta 716}$, *Apc*$^{\Delta 474}$, and *Apc1638*), and mismatch repair-deficient mice (*Msh2*-deficient). Experimental endpoints include aberrant crypt foci, adenomas, and colon carcinomas in carcinogen-treated rodents, and predominantly small intestinal polyps in *Apc* mutant mice. Early studies of aspirin and nonselective NSAIDs (e.g., sulindac, piroxicam) were succeeded by analyses of selective COX-2 inhibitors (nimesulide, MF tricyclic, celecoxib, NS-398, JTE-522, rofecoxib) as these became available and most recently NSAID phospho- and NO derivatives. Substantial chemopreventive efficacy with respect to experimental intestinal neoplasia has been reported for all NSAID classes, consistent with human epidemiologic observations (see Sect. 11.3.2.1), although dosing issues in some cases preclude direct interspecies comparisons [8].

Increasing interest has also focused on combination drug approaches, with dual goals of maximizing anticancer efficacy and minimizing NSAID cardiovascular and GI toxicity. COXibs have been tested in combination with EGFR inhibitors [148, 149], the HER2/neu-neutralizing antibody trastuzumab [150], fish oil [151], atorvastatin [152, 153], as well as the ornithine decarboxylase (ODC) inhibitor difluoromethylornithine (DFMO) [154], with generally promising combinatorial efficacy.

Comparable analyses have been performed in rodent breast cancer models, again with a prominent emphasis on chemoprevention, although therapeutic efficacy has also been tested [9, 147]. Protective effects of conventional NSAIDs (e.g., indomethacin, flurbiprofen) with respect to carcinogen-induced breast cancer were first established 30 years ago, with subsequent validation of selective COX-2 inhibitors in carcinogen models. Additionally, using *HER2/neu* transgenic rodents to model the correlation between HER2/neu and COX-2 overexpression observed in the human disease, celecoxib has been shown to effect modest retardation of HER2/neu-driven tumor formation [50, 155, 156]. As in intestinal cancer models, combination approaches have been explored. For example, coadministration of celecoxib with the retinoid X receptor ligand LGD1069 achieved a greater degree of tumor retardation than either agent singly [157]. The demonstrated chemopreventive efficacy of COX inhibitors in breast cancer models is consistent with general findings from human epidemiologic analyses (see Sect. 11.3.3.1).

Increasing appreciation of the relationship between obesity and elevated cancer risk is likely to prompt additional evaluation of NSAIDs with respect to obesity-associated cancer, both experimentally and epidemiologically. This is particularly germane given that COX-2/PG levels are elevated in breast adipose tissue of obese mice and humans in association with inflammation [53, 54]. To date, no reports of NSAID efficacy in obesity-driven intestinal cancer models are available. However, emerging data are starting to address this issue in breast cancer. The accelerated growth rate of E0771 mammary tumors implanted into diet-induced obese mice, relative to that in lean mice, is reversed by aspirin treatment [116], implying a selective anti-inflammatory action of aspirin in the obese cohort. Future studies will no doubt substantiate this finding in other models and provide mechanistic insight.

11.3.2 NSAIDs and Human Colorectal Neoplasia

Epidemiological analyses of NSAID use and colorectal cancer incidence yield findings consonant with those from rodent studies. Numerous analyses of both large prospective cohorts and smaller case–control studies have been performed, which variously consider adenoma or carcinoma as an endpoint. Inverse associations between NSAID use and risk of CRC have been fairly consistently identified, with the most abundant datasets for aspirin, based in part on widespread aspirin use for cardiovascular prophylaxis [158]. Randomized clinical trials also provide compelling evidence of suppression of colorectal neoplasia by NSAIDs.

11.3.2.1 Epidemiologic Findings

Bosetti et al.'s recent meta-analysis of aspirin and cancer risk provides an updated overview of observational studies [159]. Thirty studies which considered aspirin use and CRC were evaluated: 15 case–control studies (21,414 total cases) and 15 cohort studies (16,105 cases). Overall, regular aspirin use was associated with a 27 % reduced risk of CRC (relative risk [RR], 0.73; 95 % confidence interval [CI],

0.67–0.79; $P < 0.001$). In general, stronger protective effects were observed in case–control studies (RR, 0.63; 95 % CI, 0.56–0.70) than in cohort studies (RR, 0.82; 95 % CI, 0.75–0.89). Similar protection was afforded by aspirin with respect to colon and rectal cancer. Bosetti et al. concluded from their analysis that use of regular/high-strength aspirin for at least 5 years is necessary to confer protection against colorectal cancer [159]. Similar or greater risk reductions have also been identified in aspirin users considering adenomas as an endpoint. Of note, although many observational studies to date have focused solely on aspirin, others have considered combined aspirin and NSAID use. Comparative analyses suggest that non-aspirin NSAIDs afford similar protection [160–163]. Furthermore, use of selective COX-2 inhibitors (rofecoxib, celecoxib) is also associated with reduced risk of colorectal neoplasia [162, 164, 165].

11.3.2.2 Findings from Randomized Clinical Trials

Complementing the epidemiological findings, interventional approaches have also identified reduced colorectal neoplasia in groups assigned to NSAID. Aspirin has been evaluated on colorectal adenoma and carcinoma endpoints in a variety of clinical settings, including patients with prior adenoma, prior carcinoma, and Lynch syndrome patients (i.e., individuals with hereditary predisposition to colorectal cancer due to mutations in DNA repair genes) [158]. Additionally, cancer outcome data have recently become available from several large-scale clinical trials of aspirin primarily intended to evaluate cardiovascular (CV) disease prevention [166–168]. Effects on colorectal adenoma/carcinoma endpoints of several non-aspirin NSAIDs have also been evaluated, most notably sulindac and selective COX-2 inhibitors (see Sect. 11.3.2.5).

Early proof-of-principle data for NSAID-mediated suppression of colorectal neoplasia were provided by demonstration of sulindac-mediated reduction in polyp burden in familial adenomatous polyposis (FAP) patients over a 9 month treatment period [169]. Importantly in mechanistic terms, sulindac treatment caused significant reductions in rectal mucosal prostanoid levels (PGE_2, $PGF_{2\alpha}$, TXB_2) [170]. More recently, substantial suppression of adenoma recurrence has been achieved using sulindac in combination with the ODC inhibitor DFMO for 3 years [171].

Aspirin efficacy for suppression of colorectal adenomas has been evaluated in four placebo-controlled, randomized trials in patients with a history of colorectal adenoma or carcinoma [172–175]. Meta-analysis of these trials showed that aspirin at daily doses ranging from 81 to 325 mg reduced the risk of colorectal adenoma by 17 % (RR, 0.83; 95 % CI, 0.72–0.96) over a median post-randomization follow-up of 33 months [176], although one of the four trials subsequently reported negative findings with 4 years follow-up [177] contrasting with the positive 1-year findings included in the meta-analysis [176]. Notably, aspirin is capable of suppressing mucosal PGE_2 levels over this dose range, when administered for 2 weeks to 3 months [178–182].

In contrast to studies of colorectal adenoma recurrence, few trials have evaluated aspirin with carcinoma as the primary endpoint. One notable exception is the CAPP2 trial, in which 600 mg/day aspirin has been shown to reduce CRC incidence in Lynch syndrome patients who completed 2 years of intervention (hazard ratio (HR), 0.41; 95 % CI, 0.19–0.86; $P=0.02$) [183]. Importantly, recent meta-analyses of several randomized trials of aspirin for CV disease reduction have identified reduced risk in aspirin cohorts not only of colon cancer incidence but also of metastasis and mortality [166–168]. Initial analysis of four trials (Thrombosis Prevention Trial, British Doctors Aspirin Trial, Swedish Aspirin Low-Dose Trial, and UK-TIA Aspirin Trial) identified a reduced 20-year risk of colon cancer incidence and mortality (incidence: HR, 0.76; 95 % CI, 0.60–0.96; $P=0.02$; mortality: HR, 0.65; 95 % CI, 0.48–0.88; $P=0.005$), with no evidence for increased benefit at doses greater than 75 mg/day [167]. Increasing benefit was associated with longer durations of scheduled treatment. Consistent data for death due to several common cancers were obtained through analysis of eight aspirin trials [166]. Further analysis of five UK-based trials of aspirin for CV disease prevention revealed aspirin-associated reduction in adenocarcinoma metastasis, both at initial diagnosis and on subsequent follow-up, particularly in patients with CRC [168].

At odds with data from the CV prevention trials, two large US-based intervention studies of aspirin failed to see reduced CRC risk [184, 185]. The basis for the discrepant observations in the Physicians' Health Study and Women's Health Study relative to other studies remains unclear, but may be a function of insufficient exposure. Both studies used an alternate day dosing schedule. Additionally, insufficient duration of treatment or follow-up may have limited the capacity to detect protective effects. This possibility is supported by the clear signal of increased efficacy as a function of increased duration of aspirin use in both observational and intervention studies [159, 166, 167, 186–189].

11.3.2.3 NSAIDs and CRC Mortality: Determinants of Sensitivity

As mentioned above, it is increasingly apparent that aspirin/NSAID use not only decreases colorectal adenoma and carcinoma occurrence but may also be associated with reduced CRC mortality. In the Nurses' Health Study, regular aspirin use was associated with reduced risk of death from CRC (RR, 0.72; 95 % CI, 0.56–0.92), although greater protection was afforded for death from CV disease (RR, 0.62; 95 % CI, 0.55–0.71) [190]. Several analyses have identified an association specifically between prediagnosis NSAID use and CRC-specific survival [191, 192], although findings are not consistent in all studies [191, 193]. Associations between postdiagnosis aspirin use and reduced CRC-specific mortality have also been observed in several cohorts [194–196]. Consistent with the protective effect of aspirin/NSAIDs observed in epidemiologic analyses, the recent Rothwell meta-analyses of multiple randomized clinical trials with a primary CV endpoint have also revealed reduced colon cancer mortality in patients allocated to aspirin [166, 167].

Tumor COX-2 expression is emerging as a molecular determinant of CRC sensitivity to aspirin. In the Health Professionals Follow-Up and Nurses' Health Study cohorts, aspirin-mediated risk reduction was apparent for COX-2-overexpressing cancers, but not for those with weak or absent COX-2 expression [197]. Subsequent analysis of these cohorts was designed to determine the relationship between aspirin use post diagnosis, CRC-specific mortality, and tumor COX-2 expression. Strikingly, regular aspirin use after diagnosis was associated with a lower risk of CRC-specific mortality among patients whose primary tumors overexpressed COX-2 (HR, 0.39; 95 % CI, 0.20–0.76), but not in patients with primary tumors with weak or absent expression (HR, 1.22; 95 % CI, 0.36–4.18) [194]. Stromal COX-2 expression has also been identified as predictive of adenoma recurrence, although in this study adenoma COX-2 expression did not correlate with aspirin responsiveness [198].

Contrasting with the CRC data, COX-2 expression in breast carcinoma has not been found to be a determinant of aspirin sensitivity. In the Nurses' Health Study, tumor COX-2 expression was associated with higher diagnostic stage [44], consistent with earlier studies in which COX-2 expression in breast cancers was identified as an indicator of poor prognosis [26, 28, 32, 35]. However, the relative risk of breast cancer death for current aspirin use was similar for patients with COX-2-positive and COX-2-negative disease [44].

A recently reported analysis of the Health Professionals Follow-Up and Nurses' Health Study cohorts suggests that mutational activation of PI3K signaling may also confer aspirin sensitivity [199]. Specifically, mutation analysis of the gene encoding the catalytic α subunit (*PIK3CA*) was performed on colorectal cancers from 964 patients, and survival was compared with aspirin use as a function of mutational status. Use of postdiagnosis aspirin was found to selectively confer a survival advantage for patients with *PIK3CA*-mutant tumors. Thus, in aggregate, molecular analyses from the Health Professionals Follow-Up and Nurses' Health Study suggest both COX-2 expression and *PI3KCA* mutation in colorectal cancers as determinants of sensitivity to aspirin [194, 197, 199].

11.3.2.4 Obesity and NSAID-Mediated Suppression of Human Colorectal Neoplasia

To date, little epidemiologic data exists addressing obesity as a modifier of, or sensitizer to, NSAID action in human colorectal neoplasia, although, as discussed below (Sect. 11.5.4), there are abundant potential functional intersections between NSAID actions and obesity-related procarcinogenic mechanisms. Furthermore, an association between BMI and rectal mucosal PGE_2 levels has been identified in patients with a history of polyps, implying increased activation of COX/PG signaling, the primary target of NSAIDs, in association with obesity [21]. Nevertheless, the relationship between BMI and NSAID efficacy remains largely unaddressed in observational studies. Specifically, stratification by BMI with comparative analysis of relative risk in each BMI category has rarely been reported. One puzzling report from the Prostate, Lung, Colorectal, and Ovarian Cancer Screening Trial identified protective NSAID effects with respect to polyp formation in women limited to those with BMI < 25 [200]. Furthermore, reductions in colon cancer risk associated with aspirin use were not found to be modified by BMI in the Health Professionals Follow-Up Study or the Nurses' Health Study [201]. In contrast, the Aspirin/Folate Polyp Prevention Study suggested that aspirin may be more effective in preventing colorectal adenomas in patients with higher BMI [202]. Among obese subjects (i.e., BMI ≥ 30), the risk for advanced adenomas in aspirin users (325 mg/day) compared with placebo was 0.44 (95 % CI, 0.17–1.10) vs. RR = 1.23 (95 % CI 0.55–2.77) among those with normal weight (BMI < 25). These data suggest a selective protective effect of aspirin in obese individuals and offer a possible path forward for amelioration of elevated cancer risk in association with obesity.

11.3.2.5 Clinical Evaluation of COXibs for Suppression of Colorectal Neoplasia

The associated GI toxicity of aspirin and conventional NSAIDs represents a clear limitation and provided rationale for development of selective COX-2 inhibitors for anti-inflammatory applications. Evaluation of these latter agents for cancer prophylaxis was logical, based on the substantial epidemiologic evidence for anti-CRC efficacy of NSAIDs. Initial evaluation of celecoxib in FAP patients demonstrated a substantial reduction in polyp multiplicity with 400 mg twice daily for 6 months (28 % vs. 4.5 % reduction in placebo cohort) [203]. Three subsequent large-scale trials, the Adenoma Prevention with Celecoxib (APC), PRevention of Colorectal Sporadic Adenomatous Polyps (PreSAP), and Adenomatous Polyp Prevention On Vioxx (APPROVe) trials, all demonstrated COXib efficacy (both celecoxib and rofecoxib) for suppression of recurrent adenomas [204–206]. Relative risks achieved with 3 years treatment were 0.55 (95 % CI, 0.48–0.64; $P < 0.001$) with 400 mg celecoxib twice daily in the APC trial [206], 0.64 (95 % CI, 0.56–0.75; $P < 0.001$) with 400 mg celecoxib once daily in the PreSAP trial [204], and 0.76 (95 % CI, 0.69–0.83; $P < 0.0001$) with 25 mg rofecoxib daily in the APPROVe trial [205]. Suppression of advanced adenomas was also observed.

Analysis of patient tissue samples from the APC trial revealed a novel determinant of sensitivity to celecoxib—expression of the 15-hydroxyprostaglandin dehydrogenase (15-PGDH) enzyme [207]. 15-PGDH catalyzes conversion of PGE_2 to a relatively inactive keto derivative, 15-keto-PGE_2, and hence terminates PGE_2 signaling. In the APC trial, in all cases tested, those individuals who developed new adenomas while taking celecoxib uniformly exhibited low colonic mucosal *15-PGDH* transcript levels. These data suggest that both adequate *15-PGDH* expression and pharmacological COX inhibition are required to achieve sufficient PGE_2 suppression to effectively prevent polyp formation. Intriguingly, several NSAIDs reportedly elevate *15-PGDH* expression [208], which may be an integral component of their anticancer action.

Excitement generated by positive findings in the COXib polyp prevention trials was attenuated by the revelations of CV toxicity associated with the use of selective COX-2 inhibitors [209–213]. In aggregate these data led the US Preventive Services Task Force to conclude that the risk-benefit ratio did not favor chemoprevention in average-risk individuals [214]. Nevertheless, it may be possible to identify patients with a low CV risk profile for whom COXib-based prophylaxis of colorectal neoplasia is associated with a favorable risk-benefit ratio. Post hoc analysis of the APC trial identified the circulating inflammatory marker high-sensitivity C-reactive protein (hsCRP) as a potential predictor of celecoxib-associated CV toxicity [215]. Among patients with high hsCRP, the RR of CV events compared with placebo was 2.27 (95 % CI, 0.72–7.14) for those in the 200 mg celecoxib cohort and 3.28 (95 % CI, 1.09–9.91) in the 400 mg cohort, whereas no increase in CV risk was associated with assignment to celecoxib in patients with low hsCRP. Thus, it may be possible to identify patients for whom COXib use for chemoprevention is associated with minimal increase in CV risk.

11.3.3 NSAIDs and Human Breast Cancer

Broadly similar epidemiologic findings have been made with respect to NSAIDs and breast cancer as those discussed above for colorectal neoplasia. One important caveat is that the magnitude of risk reduction tends to be less, which could reflect the reduced prevalence of COX-2 overexpression in breast cancer vs. CRC. Furthermore, few data are available from interventional studies. Nevertheless, several recent meta-analyses of observational studies have identified a consistent signal of protection associated with NSAID use [159, 216, 217].

11.3.3.1 Epidemiologic Findings

Bosetti et al.'s recent meta-analysis of aspirin and cancer risk considered 10 case–control studies and 22 cohort studies of breast cancer, including a total of 25,835 and 27,091 breast cancer cases, respectively, and observed a modest but highly significant RR of 0.90 (95 % CI, 0.85–0.95; $P<0.001$) associated with aspirin use [159]. A similar magnitude of aspirin protection was identified in a separate meta-analysis (odds ratio (OR), 0.86; 95 % CI, 0.81–0.92) which included 13 case–control studies, 19 cohort studies, and 1 randomized clinical trial [216]. The findings of both these studies were consistent with an earlier report from Takkouche et al., in which not only aspirin but other NSAIDs were considered [217]. Use of any NSAID was associated with reduced breast cancer risk (RR, 0.88; 95 % CI, 0.84–0.93), with protective effects for both aspirin and ibuprofen when considered individually (RR, 0.87, and RR, 0.79, respectively). In contrast, the vast majority of studies have reported null associations for breast cancer risk and acetaminophen, consistent with this drug having a distinct analgesic mechanism from aspirin and commonly used

NSAIDs [218–222], although a couple of studies have identified acetaminophen-associated protection [223, 224].

Several studies have identified reduced breast cancer risk associated with selective COX-2 inhibitor use [218, 220, 225], although null associations have also been reported, and a single study reported increased risk in COXib users [226]. Consistent with preclinical studies, Valsecchi et al. observed a reduced frequency of bone metastasis in users of COXibs [227]. In high-risk patients, the calculated OR for bone metastases was 0.10 (95 % CI, 0.01–0.78).

Of note, although meta-analyses reveal a consistent protective NSAID effect, several individual studies have reported negative findings, or even increased breast cancer risk associated with NSAID use [228–230]. A population-based analysis of Danish women failed to observe associations between breast cancer incidence and NSAID use [231]. Furthermore, no NSAID-mediated protection was observed in the Nurses' Health Study [232–234]. Interestingly, the Nurses' Health Study cohorts included predominantly premenopausal women, suggesting that NSAIDs could mediate protection selectively in the postmenopausal setting. Mechanistically, this could relate to the ability of COX-derived prostaglandins to regulate estrogen biosynthesis. As discussed above (Sect. 11.2.3.5), PG-mediated upregulation of aromatase expression likely assumes greater significance in postmenopausal women after cessation of ovarian estrogen synthesis. Thus, if suppression of COX-dependent estrogen synthesis is a key mechanism for NSAID-mediated protection with respect to breast cancer, superior NSAID efficacy in postmenopausal women would be anticipated.

The hypothesis that NSAID-mediated suppression of postmenopausal estrogen synthesis is central to NSAID efficacy in the context of breast cancer provides the basis for several further predictions. Most notably, one might expect that NSAID protection would be abrogated by postmenopausal hormone replacement therapy (HRT). Consistent with this notion, neither aspirin use nor total NSAID use was associated with breast cancer incidence in the Cancer Prevention Study II Nutrition Cohort, of which over half were users of hormone therapy [235]. Similarly, in the Danish Diet, Cancer and Health Cohort, NSAID protection was also not detected, but almost 50 % of women were HRT users, with higher hormone use in women using NSAIDs [229]. These data support a model in which NSAIDs are selectively effective in postmenopausal women who are not users of hormonal therapy, although clearly further data are required to validate this concept.

If NSAID-mediated decreases in estrogen biosynthesis are genuinely mechanistically important, one might also anticipate selective suppression of those tumors that express hormone receptors (HRs). This possibility has been addressed in several epidemiologic studies to date, which have yielded mixed results. In the Long Island Breast Cancer Study, aspirin-associated breast cancer risk reduction was selectively observed in women with HR-positive tumors (OR, 0.74; 95 % CI, 0.60–0.93), whereas no benefit was detected for HR-negative tumors (OR, 0.97; 95 % CI, 0.67–1.40) [222]. A similar selective benefit of daily aspirin use for risk reduction of ER-positive breast cancer was identified in the NIH-AARP Diet and Health study (RR, 0.84; 95 % CI, 0.71–0.98) [236]. Additionally, although overall protective efficacy of aspirin was not observed in the Nurses' Health Study, there was some signal of response of HR-positive breast cancers to high-dose, long-term aspirin exposure [234]. Other studies have also provided suggestive evidence for HR status as a determinant of aspirin sensitivity [237, 238], although aspirin/NSAIDs have not been uniformly observed to confer selective protection against hormone receptor-positive tumors [228, 239–242]. Nevertheless, in aggregate the data suggest that NSAIDs may afford increased protection with respect to HR-expressing tumors, which could reflect NSAID-mediated decreases in estrogen biosynthesis.

Finally, given the observation that breast adipose tissue exhibits inflammation in the context of obesity leading to increased estrogenic signaling [53, 54, 58, 59], it seems likely that NSAIDs might afford increased breast cancer risk reduction in obese vs. lean postmenopausal women. BMI is an established determinant of serum estradiol in postmenopausal women [140–142]. Thus, obesity-associated increases in aromatase expression and hence estrogen signaling in breast tissue might provide a target for NSAIDs. Additional inflammation-associated pathways may also be subject to NSAID-mediated amelioration (discussed below in Sect. 11.5.4). Stratification of NSAID effects according to BMI in cohort and case–control studies may furnish insights into the relative importance of obesity as a determinant of NSAID sensitivity in breast cancer.

11.3.3.2 Interventional Studies

A consensus finding from the epidemiologic studies discussed above is that NSAID use is associated with reduced breast cancer risk, albeit with lesser magnitude of protection than observed for colorectal neoplasia. However, in contrast to CRC, substantial randomized clinical trial data are not available to support these findings. As discussed above (Sect. 11.3.2.2), recent meta-analyses of trials with the primary goal of evaluating aspirin for CV disease prevention have yielded exciting data concerning aspirin protective effects on CRC incidence, mortality, and metastasis [166–168]. However, too few women were included in these trials with long-term follow-up to determine the effect of aspirin on breast or gynecological cancers in the Rothwell analyses. Thus, the major prospective trial of aspirin with breast cancer as a measured endpoint remains the Women's Health Study, a prospective trial of 100 mg aspirin every other day vs. placebo for an average of 10 years. In this trial, no effect of aspirin was observed on total cancer (RR, 1.01; 95 % CI, 0.94–1.08; $P=0.87$), colorectal cancer (RR, 0.97; 95 % CI, 0.77–1.24; $P=0.83$), or breast cancer (RR, 0.98; 95 % CI, 0.87–1.09; $P=0.68$) [185]. Null results in this study have been widely interpreted as indicating that 100 mg aspirin on alternate days is an insufficient dose to achieve meaningful risk reduction. Consistent with this interpretation, several epidemiologic analyses have failed to observe protective effects with respect to breast cancer of low-dose (<100 mg/day) aspirin [220, 221, 225]. Also worth noting with respect to the Women's Health Study was the high proportion of HRT use, approximately 55 %, which could also have obscured protective effects contingent on modulation of estrogenic signals. Nevertheless, at present there is a dearth of clinical trial data to substantiate epidemiologic findings of NSAID-mediated breast cancer risk reduction.

11.3.3.3 NSAIDs and Breast Cancer Recurrence/Mortality

Associations have been identified between NSAID use and not only decreased breast cancer risk but also improved outcome following breast cancer diagnosis. Several studies have identified decreased breast cancer recurrence and/or mortality in breast cancer patients who used NSAIDs following diagnosis [243–245]. In the Life After Cancer Epidemiology study, use of ibuprofen and non-aspirin NSAIDs was associated with decreased breast cancer recurrence, although aspirin did not afford protection [245]. In contrast, in the Nurses' Health Study, aspirin use was associated with a decreased risk of breast cancer mortality [244]. Interestingly, the association was not modified by menopausal status, BMI, or tumor ER status, and as discussed above, tumor COX-2 expression was also not a determinant of aspirin sensitivity [44, 244], providing few clues as to the responsible mechanisms underlying aspirin effects. Of note, conflicting data were provided by the Collaborative Women's Longevity Study and the Western New York Exposures and Breast Cancer study, neither of which identified a relationship between NSAID use and either all-cause or breast cancer-specific mortality [246, 247]. Further studies will undoubtedly provide greater clarification concerning NSAID use and breast cancer outcome.

11.3.3.4 Therapeutic Interventions with COX-2 Inhibitors

A handful of trials have tested COXibs as therapeutics or performed biomarker studies. Celecoxib has been evaluated in early phase tolerability/efficacy trials in combination with the HER2/neu-directed monoclonal antibody trastuzumab, the aromatase inhibitor exemestane, or the antimetabolite capecitabine in patients with advanced breast cancer [248–252]. Where comparator arms were included, no increase in time to progression (TTP) resulted from addition of celecoxib [250]; however, one study identified a longer TTP in celecoxib/capecitabine-treated patients whose tumors overexpressed COX-2 [251]. Window trials comparing molecular endpoints pre and post intervention have been relatively uninformative to date. Celecoxib-mediated decreases in nipple aspirate fluid PGE_2 levels have been reported in postmenopausal women and in women with newly diagnosed breast cancer [253]. However, neither proliferation nor apoptosis in ER-positive DCIS was altered by 14-day celecoxib treatment, whereas the AI exemestane significantly suppressed proliferation [254]. Similarly, 14-day celecoxib treatment did not effect a significant alteration in proliferation (Ki67 staining) in breast carcinoma tissue relative to that observed in the no-treatment group and also did not significantly affect apoptosis, COX-2, ER, or progesterone receptor (PR) expression [255]. However, statistically significant decreases in carcinoma ER, PR, Ki-67, and COX-2 expression have been achieved by coadministration of celecoxib with exemestane, although aromatase protein levels were unaltered by exemestane or combination therapy [256]. Given the link between obesity, inflammation, and neoplasia, it would obviously be of interest to determine if BMI is a determinant of clinical response of breast cancers to anti-inflammatory drugs.

11.4 Dysregulated Signaling in Obesity: Implications for Neoplasia

Many of the signaling pathways dysregulated in obesity likely contribute to the associated increase in cancer risk and moreover provide targets for NSAID-mediated intervention. Thus, this section provides a simplified overview of signaling abnormalities associated with obesity and inflammation to generate a framework for considering obesity-related mechanisms of NSAID anticancer action. For increased detail, the reader is directed to accompanying chapters in this book. Since mechanistic contributions of COX/PG signaling to neoplasia are extensively discussed above (Sect. 11.2.3), these are largely not reiterated in this section.

Obesity is defined by increased BMI ($\geq$30) and characterized by disproportionate adipose accumulation. Importantly, as discussed above, adipose tissue from obese individuals exhibits marked inflammation, defined by infiltration of leucocytes, including macrophages, as well as CD8-positive T lymphocytes and mast cells [4]. Inflammatory foci consisting of M1-like macrophages surrounding necrotic adipocytes, the so-called CLS, are observed in breast and other adipose depots from obese mice and humans and increase as a function of body mass [53, 55–59]. Adipose tissue macrophages can comprise up to 40 % of cells in obese adipose tissue and represent a rich source of cytokines, chemokines, and fatty acids, which are key mediators of the increased risk of insulin resistance and neoplasia associated with obesity [4]. Proinflammatory molecules released from activated macrophages include COX-2-derived PGE_2, TNFα, and IL-1β (Fig. 11.2), synthesis of which is driven by increased NFκB (nuclear factor of kappa light polypeptide gene enhancer in B cells) signaling induced by ligation of cell-surface toll-like receptor 4 (TLR4). These cytokines may have both local and systemic effects. Key consequences include attenuation of insulin signaling through kinase-mediated covalent modification of insulin receptor substrate-1 (IRS-1) and decreased expression of the GLUT4 glucose transporter. Thus, obesity-associated cytokine overproduction can drive insulin resistance, hyperinsulinemia, and ultimately type II diabetes. Importantly, tissue inflammation is evident not only in adipose but in all classical insulin target tissues, including liver and muscle.

Obesity-induced cytokine overproduction has important proneoplastic sequelae [257]. Cytokines and chemokines can induce DNA damage via reactive oxygen species and inhibit DNA repair. Furthermore, they may act as paracrine growth and survival factors for malignant cells, as well as inducing angiogenesis via microenvironment remodeling and upregulation of proangiogenic factors. Tissue remodeling may also contribute to cancer cell migration and invasion. Additionally, obesity-induced cytokines may be important in subverting host immune responses and hence optimizing tumor cell survival. Thus, excessive levels of proinflammatory mediators may impact both tumor initiation and progression.

Key signal transduction pathways activated by PGE_2, TNFα, and IL-1β include protein kinase A, mitogen-activated protein kinase (MAPK), and NFκB signaling, all of which are implicated in carcinogenesis. Additionally, as previously discussed, an important local effect in obese adipose tissue is paracrine induction of aromatase expression in neighboring cells (Fig. 11.2), likely increasing breast cancer risk via upregulated estrogenic signaling. Increased circulating estradiol levels in obesity may also promote neoplasia in other hormone-responsive tissues including uterine endometrium.

In parallel with hyperinsulinemia in obesity, increased production of other growth factors and endocrine agents is observed, including the satiety hormone leptin and insulin-like growth factor 1 (IGF-1) [258]. Increased IGF-1 synthesis and decreased expression of IGF-binding proteins lead to a net increase in bioavailable IGF-1. Both insulin and IGF-1 can induce cell growth and proliferation via PI3K/AKT signaling, a pathway that is of central importance in neoplasia [259, 260]. Additionally, both ligands can induce β-catenin/TCF signaling [261], another highly important pathway in cancer [262].

The peptide hormone leptin emanates from adipocytes and thus is positively correlated with adipose stores. In lean animals, leptin functions as a satiety factor, but leptin overproduction in obesity is accompanied by diminished hypothalamic responsiveness to the leptin signal. In contrast, leptin-regulated JAK/STAT, MAP kinase, and PI3K signaling in other tissues can promote cancer progression. Animal studies have identified leptin signaling as important for mammary tumorigenesis, and it appears that the ratio of leptin to adiponectin (which decreases with increasing BMI) may be a key determinant of breast cancer risk [258, 263].

Several signaling pathways are thus emerging as key mediators of obesity-associated inflammation, both for insulin resistance and cancer risk. NFκB is a central regulator of adipose cytokine synthesis. Saturated fatty acids, and potentially bacterial toxins ingressing via leaky gut mucosa, induce NFκB activation via TLR4, resulting in increased production of cytokines, many of which in turn signal through NFκB. Central importance of NFκB in systemic insulin resistance has been demonstrated via genetic manipulation in rodents, which has further established myeloid NFκB as the key determinant [264–267]. Hyperactivation of PI3K/AKT signaling to mTOR (mammalian target of rapamycin) is apparent in obesity, consequent on overproduction of inflammatory cytokines and hormones (e.g., insulin, IGF-1, leptin). mTOR is a central regulator of cell growth, responsible for integrating growth factor signaling, nutrient, energy, and oxygen availability and translating the net signal to provide the appropriate level of translational activity within the cell [268]. Balancing the progrowth signals initiated by hormones and growth factors, AMP kinase (AMPK) serves as an energy sensor activated by increased AMP:ATP ratio that can signal the need for decreased metabolic activity by providing critical input to mTOR. AMPK activation initiates a switch from glucose to fat utilization and shuts down energy-intense processes. Of note, antidiabetic agents including metformin function in part through activation of AMPK via decreasing oxidative phosphorylation and hence ATP production. Altered balance between PI3K and AMPK signaling in obesity may prove to be a key regulator of the neoplastic process via control of tumor cell growth.

Thus, obesity is associated with activation of multiple key proneoplastic pathways including NFκB, PI3K/AKT/mTOR, Wnt/β-catenin, JAK/STAT, and MAP kinase signaling and may also effect protumorigenic changes by skewing the cellular energy balance in favor of continued cell growth. Notably these processes are deemed most likely to contribute to cancer progression after acquisition of precursor mutations that induce cell transformation, although inflammatory cytokines may also regulate tumor initiation via increasing DNA damage.

11.5 NSAID Anticancer Mechanisms

11.5.1 NSAIDs Suppress Hallmark Characteristics of Cancer

Consideration of the key signaling abnormalities associated with obesity provides a foundation for reviewing NSAID anticancer mechanisms, although clearly many of these mechanisms are generalizable to the carcinogenic process independent of BMI. As discussed above, COX/PG signaling regulates a number of the intrinsic, or "hallmark", capabilities suggested as essential for tumor growth and metastatic dissemination; thus, NSAID-mediated PG suppression can impact neoplasia via multiple effector pathways (Fig. 11.1). Intriguingly, NSAIDs can decrease PG levels not only via modulation of their synthesis (i.e., COX inhibition) but also through increased catabolism achieved via *15-PGDH* upregulation [208]. Nevertheless, considerable controversy has focused on the mechanistic basis for the protective efficacy of NSAIDs with respect to cancer, based in part on the existence of NSAID derivatives such as sulindac sulfone that lack COX inhibitory activity but are still active in animal tumor models [269], as well as demonstrated activity of NSAIDs towards COX-2-null cells [270–272]. This has been resolved by genetic approaches which provide incontrovertible evidence for the importance of COX/PG signaling in rodent neoplasia (Sect. 11.2.2). Facets of tumorigenesis that can be regulated by prostaglandins and are thus susceptible to NSAID-mediated attenuation include cell proliferation and apoptosis avoidance, invasiveness, angiogenesis, immune suppression, and regulation of estrogen biosynthesis (Sect. 11.2.3; Fig. 11.1). Additionally, recent data identify a key role for stromal fibroblast-derived PGE_2 in regulating cancer stem cells [74, 75]. Nevertheless, it is evident that NSAIDs suppress tumorigenesis through a combination of both COX-dependent and COX-independent effects. NSAID regulation of programmed cell death provides a key illustration of utilization of parallel COX-dependent and COX-independent pathways to effect a response.

11.5.2 NSAID-Mediated Suppression of Apoptosis

COX-2 overexpression in epithelial cells suppresses apoptosis both in vitro and in vivo [63, 87], providing a clear mechanism for NSAID-mediated apoptosis via COX inhibition. Antiapoptotic consequences of elevated COX-2/PG include increased expression of the antiapoptotic gene *Bcl-2* [273] and decreased arachidonate-driven apoptosis consequent on increased conversion of arachidonic acid to PGs [274, 275]. Furthermore, COX-derived prostacyclin may act to suppress apoptosis via direct interaction with the nuclear receptor peroxisome proliferator-activated receptor (PPAR) δ [276, 277]. However, a multiplicity of additional COX-2-independent NSAID-regulated pathways can also impact apoptosis, as illustrated by NSAID induction of cell death in *COX-2* null cells [270–272] and by non-COX-inhibiting drugs [278]. NSAIDs can stimulate induction of multiple proapoptotic effectors, including 15-lipoxygenase-1 (15-LOX-1) [279, 280], NSAID-activated gene-1 (NAG-1) [281], and prostate apoptosis response 4 (Par-4) [282]. NSAID-mediated abrogation of antiapoptotic pathways via inhibition of PPARδ and 3-phosphoinositide-dependent protein kinase-1 (PDPK1) signaling has also been described [283, 284]. Several mechanisms have been advanced to explain PPARδ suppression, including disruption of PPARδ binding to target sequences, and down-regulated PPARδ expression, which can be a consequence of increased synthesis of the 15-LOX-1 product 13-S-HODE [283, 285–287]. NSAID-elicited suppression of apoptosis likely occurs through combination effects on these pathways. Intriguingly, a selective proapoptotic effect of sulindac on intestinal stem cells has been suggested, implying that NSAIDs could selectively impact the biology of cancer stem cells with activation of Wnt/β-catenin signaling [288], which could relate to the previously reported ability of PGE_2 to regulate stem cells [73–75, 289].

11.5.3 Cyclooxygenase-Independent NSAID Effects

The Wnt/β-catenin pathway has itself been identified as an NSAID target: suppression of β-catenin/TCF transcription can be elicited by multiple NSAIDs including aspirin, indomethacin, and sulindac [290–292]. Strikingly, reduced nuclear β-catenin is evident in adenomas from NSAID users [290, 293]. Aspirin-mediated inhibition of phosphatase 2A has been posited as one mechanism for attenuating Wnt/β-catenin signaling [291]. However, regulation of β-catenin signaling by PGE_2 via modulation of glycogen synthase kinase 3β has also been described [76, 294], suggesting that NSAIDs may regulate this pathway via both COX-dependent and COX-independent routes.

An emerging target of interest in the NSAID field is cGMP phosphodiesterase (PDE). Sulindac metabolites inhibit cGMP PDE, resulting in increased cGMP and protein kinase G activity in CRC and breast cancer cell lines [295–297]. Selective expression of cGMP PDE5 in cancer cell lines and tissues suggests this isoform as the target for sulindac and other NSAIDs [296, 297]. Parallel NSAID effects on cGMP PDE5 activity, apoptosis, and suppression of β-catenin/TCF signaling suggest that cGMP PDE5 regulates these latter responses, and causal relationships have been established using knockdown approaches.

The ability of COX inhibition to directly impact carcinogenesis through diminished COX-mediated activation of procarcinogens was discussed above (Sect. 11.2.3.6). However, NSAIDs can also regulate carcinogen metabolism independent of COX enzymes. For example, the major aspirin metabolite salicylic acid inhibits P phenolsulfotransferase, which mediates sulfation activation of carcinogens [298]. Additional COX-independent effects of NSAIDs of particular relevance with respect to obesity are described in the following section.

11.5.4 Key COX-Independent NSAID Effects in the Context of Obesity

Of the numerous "off-target" NSAID effects described to date, suppression of the NFκB pathway is strikingly relevant to obesity. Inhibition of NFκB transcriptional activity has been described for both aspirin and other NSAIDs, with a variety of proposed mechanisms that dictate NFκB protein localization such that it remains transcriptionally inactive [299–304]. NSAID-mediated suppression of NFκB signaling is likely to have profound consequences for obesity and neoplasia, given the key role of NFκB in both the upregulation and signaling of macrophage-derived inflammatory mediators in obese adipose.

Notably, salicylate treatment attenuates experimental insulin resistance in rodent models, and this has been attributed to NFκB suppression [265, 267]. Furthermore, selective COX-2 inhibitors (celecoxib, nimesulide) reduce parameters of metabolic disease in rats fed high-fat diets [305]. Clinically, improved insulin sensitivity and glycemia has been achieved through administration of aspirin, salicylate, salsalate (a non-acetylated prodrug of salicylate), and COXibs [306–310]. Intriguingly however, NSAID modulation of NFκB may only represent part of the explanation for these findings. Emerging data suggest probable involvement of AMPK as an additional mediator of aspirin/salicylate improvement of metabolic dysfunction. Salicylate has recently been shown to activate AMPK, and salicylate-induced decreases in plasma fatty acids and increased fat utilization are abrogated in *AMPK*-deficient mice [311, 312]. Thus, salicylate-elicited amelioration is likely to reflect modulation of both NFκB and AMPK.

Several other pathways that are altered in obesity-associated inflammation are also targeted by NSAIDs. Multiple lines of evidence support the ability of NSAIDs to attenuate PI3K/AKT signaling and growth control. NSAIDs can suppress PI3K/AKT signaling via COX inhibition, since PGE_2 induces activation of PI3K/AKT through G protein-β/γ subunits [76]. Additionally, NSAID-mediated inhibition of the upstream AKT activator 3-phosphoinositide-dependent protein kinase-1

(PDPK1) has been reported [284]. Reduced phosphorylation levels of ribosomal protein S6 and S6 kinase in rectal mucosa from aspirin-treated patients are consistent with attenuation of PI3K/AKT signaling [312], these proteins being key downstream mediators that regulate protein synthesis and hence cell growth. Furthermore, indomethacin activates the eIF2a kinase PKR in CRC cells, inducing a translation block [313]. Notably, a recent survey of cellular proteins acetylated by aspirin identified 33 substrates including metabolic enzymes, histones, and ribosomal and mitochondrial proteins [314]. These provocative findings suggest that aspirin-mediated acetylation of cellular proteins could have far-reaching metabolic consequences. Of note, aspirin-acetylated COX-2 synthesizes a novel class of lipid mediators, lipoxins, which actively promote resolution of acute inflammation [315], although the role of these lipid moieties in chronic inflammation is less well-defined.

The available evidence suggests that aspirin compounds can favorably impact pleiotropic pathways to improve metabolic health. Thus, NSAID-mediated attenuation of systemic inflammation might also be expected to translate into reduced risk of neoplasia through reducing the parameters of inflammation and thereby modulating the biology of the neoplastic cell.

11.6 Conclusions

Obesity as a determinant of increased cancer risk and poorer cancer outcome is now well established for cancers of multiple organ sites, including CRC and postmenopausal breast cancer. Obesity-associated systemic inflammation is likely to be a key driver of elevated neoplastic risk, with numerous plausible mechanisms consequent on elevated levels of inflammatory mediators and hormones. Both circulating and local tissue levels of these molecules can drive cancer progression and may also contribute to tumor initiation via increased DNA damage. Induction of numerous key protumorigenic signaling pathways may be elicited as a consequence of obesity-associated inflammation, including NFκB, Wnt/β-catenin, MAPK, JAK/STAT, as well as the PI3K/AKT/mTOR axis which is central to cell growth and homeostasis. Strikingly, many of these pathways are targeted by NSAIDs, and this drug class has demonstrated clinical efficacy for ameliorating metabolic dysfunction. Importantly, clinical findings support the antineoplastic activity of NSAIDs with respect to both colorectal and breast cancer, and consistent data have been obtained using animal models. To date there is limited epidemiologic evidence for selective protective effects of NSAIDs in obese individuals. Clinical evidence for a selective signal of NSAID utility for cancer prophylaxis in the context of obesity is mixed for CRC and completely absent for breast cancer. This may reflect the pleiotropic mechanisms, both COX-dependent and COX-independent, through which NSAIDs can attenuate the cancer process, which may confer protection in both lean and obese individuals. Nevertheless, NSAID-mediated suppression of COX/PG signaling, NFκB activity, and the PI3K/AKT/mTOR axis, as well as activation of AMPK signaling, are likely to have profound consequences for obesity-associated neoplasia. Furthermore, COX/PG-mediated upregulation of estrogen biosynthesis and signaling offers a plausible target for NSAID-mediated risk reduction with respect to breast and other hormone-sensitive cancers in the context of obesity.

References

1. Calle EE, Thun MJ (2004) Obesity and cancer. Oncogene 23(38):6365–6378. doi:10.1038/sj.onc.1207751 1207751 [pii]
2. Wolin KY, Carson K, Colditz GA (2010) Obesity and cancer. Oncologist 15(6):556–565. doi:theoncologist.2009-0285 [pii] 10.1634/theoncologist.2009-0285
3. World Cancer Research Fund/American Institute for Cancer Research (2007) Food nutrition, physical activity, and the prevention of cancer: a global perspective. World Cancer Research Fund/American Institute for Cancer Research, Washington, DC
4. Osborn O, Olefsky JM (2012) The cellular and signaling networks linking the immune system and metabolism in disease. Nat Med 18(3):363–374. doi:nm.2627 [pii] 10.1038/nm.2627
5. van Kruijsdijk RC, van der Wall E, Visseren FL (2009) Obesity and cancer: the role of dysfunctional adipose tissue. Cancer Epidemiol Biomarkers Prev 18(10):2569–2578. doi:1055-9965.EPI-09-0372 [pii] 10.1158/1055-9965.EPI-09-0372
6. Smith WL, Urade Y, Jakobsson PJ (2011) Enzymes of the cyclooxygenase pathways of prostanoid biosynthesis. Chem Rev 111(10):5821–5865. doi:10.1021/cr2002992
7. Herschman HR (1996) Prostaglandin synthase 2. Biochim Biophys Acta 1299(1):125–140
8. Brown JR, DuBois RN (2005) COX-2: a molecular target for colorectal cancer prevention. J Clin Oncol 23(12):2840–2855. doi:23/12/2840 [pii] 10.1200/JCO.2005.09.051
9. Howe LR, Subbaramaiah K, Brown AMC, Dannenberg AJ (2001) Cyclooxygenase-2: a target for the prevention and treatment of breast cancer. Endocr Relat Cancer 8(2):97–114
10. Eberhart CE, Coffey RJ, Radhika A, Giardiello FM, Ferrenbach S, DuBois RN (1994) Up-regulation of cyclooxygenase 2 gene expression in human colorectal adenomas and adenocarcinomas. Gastroenterology 107(4):1183–1188
11. Kargman SL, O'Neill GP, Vickers PJ, Evans JF, Mancini JA, Jothy S (1995) Expression of prostaglandin G/H synthase-1 and -2 protein in human colon cancer. Cancer Res 55(12):2556–2559
12. Sano H, Kawahito Y, Wilder RL, Hashiramoto A, Mukai S, Asai K, Kimura S, Kato H, Kondo M, Hla T (1995) Expression of cyclooxygenase-1 and -2 in human colorectal cancer. Cancer Res 55(17):3785–3789
13. Kutchera W, Jones DA, Matsunami N, Groden J, McIntyre TM, Zimmerman GA, White RL, Prescott SM (1996) Prostaglandin H synthase 2 is expressed abnormally in human colon cancer: evidence for a transcriptional effect. Proc Natl Acad Sci U S A 93(10):4816–4820
14. Boolbol SK, Dannenberg AJ, Chadburn A, Martucci C, Guo XJ, Ramonetti JT, Abreu-Goris M, Newmark HL, Lipkin ML, DeCosse JJ, Bertagnolli MM (1996) Cyclooxygenase-2 overexpression and tumor formation are blocked by sulindac in a murine model of familial adenomatous polyposis. Cancer Res 56(11):2556–2560
15. Oshima M, Dinchuk JE, Kargman SL, Oshima H, Hancock B, Kwong E, Trzaskos JM, Evans JF, Taketo MM (1996) Suppression of intestinal polyposis in $Apc^{\square 716}$ knockout mice by inhibition of cyclooxygenase 2 (COX-2). Cell 87(5):803–809
16. Williams CS, Luongo C, Radhika A, Zhang T, Lamps LW, Nanney LB, Beauchamp RD, DuBois RN (1996) Elevated cyclooxygenase-2 levels in Min mouse adenomas. Gastroenterology 111(4):1134–1140

17. DuBois RN, Radhika A, Reddy BS, Entingh AJ (1996) Increased cyclooxygenase-2 levels in carcinogen-induced rat colonic tumors. Gastroenterology 110(4):1259–1262
18. Singh J, Hamid R, Reddy BS (1997) Dietary fat and colon cancer: modulation of cyclooxygenase-2 by types and amount of dietary fat during the postinitiation stage of colon carcinogenesis. Cancer Res 57(16):3465–3470
19. Chapple KS, Cartwright EJ, Hawcroft G, Tisbury A, Bonifer C, Scott N, Windsor AC, Guillou PJ, Markham AF, Coletta PL, Hull MA (2000) Localization of cyclooxygenase-2 in human sporadic colorectal adenomas. Am J Pathol 156(2):545–553. doi:S0002-9440(10)64759-1 [pii] 10.1016/S0002-9440(10)64759-1
20. Hull MA, Booth JK, Tisbury A, Scott N, Bonifer C, Markham AF, Coletta PL (1999) Cyclooxygenase 2 is up-regulated and localized to macrophages in the intestine of Min mice. Br J Cancer 79(9–10):1399–1405. doi:10.1038/sj.bjc.6690224
21. Martinez ME, Heddens D, Earnest DL, Bogert CL, Roe D, Einspahr J, Marshall JR, Alberts DS (1999) Physical activity, body mass index, and prostaglandin E2 levels in rectal mucosa. J Natl Cancer Inst 91(11):950–953
22. Bennett A, Charlier EM, McDonald AM, Simpson JS, Stamford IF, Zebro T (1977) Prostaglandins and breast cancer. Lancet 2(8039):624–626. doi:S0140-6736(77)92496-5 [pii]
23. Rolland PH, Martin PM, Jacquemier J, Rolland AM, Toga M (1980) Prostaglandin in human breast cancer: evidence suggesting that an elevated prostaglandin production is a marker of high metastatic potential for neoplastic cells. J Natl Cancer Inst 64(5):1061–1070
24. Tan WC, Privett OS, Goldyne ME (1974) Studies of prostaglandins in rat mammary tumors induced by 7,12- dimethylbenz(a)anthracene. Cancer Res 34(12):3229–3231
25. Boland GP, Butt IS, Prasad R, Knox WF, Bundred NJ (2004) COX-2 expression is associated with an aggressive phenotype in ductal carcinoma in situ. Br J Cancer 90(2):423–429
26. Costa C, Soares R, Reis-Filho JS, Leitao D, Amendoeira I, Schmitt FC (2002) Cyclooxygenase 2 expression is associated with angiogenesis and lymph node metastasis in human breast cancer. J Clin Pathol 55(6):429–434
27. Davies G, Salter J, Hills M, Martin LA, Sacks N, Dowsett M (2003) Correlation between cyclooxygenase-2 expression and angiogenesis in human breast cancer. Clin Cancer Res 9(7):2651–2656
28. Denkert C, Winzer KJ, Muller BM, Weichert W, Pest S, Kobel M, Kristiansen G, Reles A, Siegert A, Guski H, Hauptmann S (2003) Elevated expression of cyclooxygenase-2 is a negative prognostic factor for disease free survival and overall survival in patients with breast carcinoma. Cancer 97(12):2978–2987
29. Half E, Tang XM, Gwyn K, Sahin A, Wathen K, Sinicrope FA (2002) Cyclooxygenase-2 expression in human breast cancers and adjacent ductal carcinoma in situ. Cancer Res 62(6):1676–1681
30. Kelly LM, Hill AD, Kennedy S, Connolly EM, Ramanath R, Teh S, Dijkstra B, Purcell R, McDermott EW, O'Higgins N (2003) Lack of prognostic effect of Cox-2 expression in primary breast cancer on short-term follow-up. Eur J Surg Oncol 29(9):707–710
31. Lim SC (2003) Role of COX-2, VEGF and cyclin D1 in mammary infiltrating duct carcinoma. Oncol Rep 10(5):1241–1249
32. Ristimaki A, Sivula A, Lundin J, Lundin M, Salminen T, Haglund C, Joensuu H, Isola J (2002) Prognostic significance of elevated cyclooxygenase-2 expression in breast cancer. Cancer Res 62(3):632–635
33. Shim JY, An HJ, Lee YH, Kim SK, Lee KP, Lee KS (2003) Overexpression of cyclooxygenase-2 is associated with breast carcinoma and its poor prognostic factors. Mod Pathol 16(12):1199–1204
34. Soslow RA, Dannenberg AJ, Rush D, Woerner BM, Khan KN, Masferrer J, Koki AT (2000) COX-2 is expressed in human pulmonary, colonic, and mammary tumors. Cancer 89(12):2637–2645
35. Spizzo G, Gastl G, Wolf D, Gunsilius E, Steurer M, Fong D, Amberger A, Margreiter R, Obrist P (2003) Correlation of COX-2 and Ep-CAM overexpression in human invasive breast cancer and its impact on survival. Br J Cancer 88(4):574–578

36. Watanabe O, Shimizu T, Imamura H, Kinoshita J, Utada Y, Okabe T, Kimura K, Hirano A, Yoshimatsu K, Aiba M, Ogawa K (2003) Expression of cyclooxygenase-2 in malignant and benign breast tumors. Anticancer Res 23(4):3215–3221
37. Wulfing P, Diallo R, Muller C, Wulfing C, Poremba C, Heinecke A, Rody A, Greb RR, Bocker W, Kiesel L (2003) Analysis of cyclooxygenase-2 expression in human breast cancer: high throughput tissue microarray analysis. J Cancer Res Clin Oncol 129(7):375–382
38. Yoshimura N, Sano H, Okamoto M, Akioka K, Ushigome H, Kadotani Y, Yoshimura R, Nobori S, Higuchi A, Ohmori Y, Nakamura K (2003) Expression of cyclooxygenase-1 and -2 in human breast cancer. Surg Today 33(11):805–811
39. Shim V, Gauthier ML, Sudilovsky D, Mantei K, Chew KL, Moore DH, Cha I, Tlsty TD, Esserman LJ (2003) Cyclooxygenase-2 expression is related to nuclear grade in ductal carcinoma in situ and is increased in its normal adjacent epithelium. Cancer Res 63(10): 2347–2350
40. Tan KB, Yong WP, Putti TC (2004) Cyclooxygenase-2 expression: a potential prognostic and predictive marker for high-grade ductal carcinoma in situ of the breast. Histopathology 44(1):24–28
41. Kirkpatrick K, Ogunkolade W, Elkak A, Bustin S, Jenkins P, Ghilchik M, Mokbel K (2002) The mRNA expression of cyclo-oxygenase-2 (COX-2) and vascular endothelial growth factor (VEGF) in human breast cancer. Curr Med Res Opin 18(4):237–241
42. Crawford YG, Gauthier ML, Joubel A, Mantei K, Kozakiewicz K, Afshari CA, Tlsty TD (2004) Histologically normal human mammary epithelia with silenced p16(INK4a) overexpress COX-2, promoting a premalignant program. Cancer Cell 5(3):263–273
43. Yang WT, Lewis MT, Hess K, Wong H, Tsimelzon A, Karadag N, Cairo M, Wei C, Meric-Bernstam F, Brown P, Arun B, Hortobagyi GN, Sahin A, Chang JC (2010) Decreased TGFbeta signaling and increased COX2 expression in high risk women with increased mammographic breast density. Breast Cancer Res Treat 119(2):305–314. doi:10.1007/s10549-009-0350-0
44. Holmes MD, Chen WY, Schnitt SJ, Collins L, Colditz GA, Hankinson SE, Tamimi RM (2011) COX-2 expression predicts worse breast cancer prognosis and does not modify the association with aspirin. Breast Cancer Res Treat 130(2):657–662. doi:10.1007/s10549-011-1651-7
45. Kerlikowske K, Molinaro AM, Gauthier ML, Berman HK, Waldman F, Bennington J, Sanchez H, Jimenez C, Stewart K, Chew K, Ljung BM, Tlsty TD (2010) Biomarker expression and risk of subsequent tumors after initial ductal carcinoma in situ diagnosis. J Natl Cancer Inst 102(9):627–637. doi:djq101 [pii] 10.1093/jnci/djq101
46. Subbaramaiah K, Norton L, Gerald W, Dannenberg AJ (2002) Cyclooxygenase-2 is overexpressed in HER-2/neu-positive breast cancer. Evidence for involvement of AP-1 and PEA3. J Biol Chem 277:18649–18657
47. Badawi AF, el-Sohemy A, Stephen LL, Ghoshal AK, Archer MC (1999) Modulation of the expression of the cyclooxygenase 1 and 2 genes in rat mammary glands: role of hormonal status and dietary fat. Adv Exp Med Biol 469:119–124
48. Hamid R, Singh J, Reddy BS, Cohen LA (1999) Inhibition by dietary menhaden oil of cyclooxygenase-1 and -2 in N-nitrosomethylurea-induced rat mammary tumors. Int J Oncol 14(3):523–528
49. Howe LR, Crawford HC, Subbaramaiah K, Hassell JA, Dannenberg AJ, Brown AMC (2001) PEA3 is upregulated in response to Wnt1 and activates the expression of cyclooxygenase-2. J Biol Chem 276(23):20108–20115
50. Howe LR, Subbaramaiah K, Patel J, Masferrer JL, Deora A, Hudis C, Thaler HT, Muller WJ, Du B, Brown AMC, Dannenberg AJ (2002) Celecoxib, a selective cyclooxygenase-2 inhibitor, protects against human epidermal growth factor receptor 2 (HER-2)/neu-induced breast cancer. Cancer Res 62:5405–5407
51. Nakatsugi S, Ohta T, Kawamori T, Mutoh M, Tanigawa T, Watanabe K, Sugie S, Sugimura T, Wakabayashi K (2000) Chemoprevention by nimesulide, a selective cyclooxygenase-2 inhib-

itor, of 2-amino-1-methyl-6-phenylimidazo[4,5-b]pyridine (PhIP)-induced mammary gland carcinogenesis in rats. Jpn J Cancer Res 91(9):886–892
52. Robertson FM, Parrett ML, Joarder FS, Ross M, Abou-Issa HM, Alshafie G, Harris RE (1998) Ibuprofen-induced inhibition of cyclooxygenase isoform gene expression and regression of rat mammary carcinomas. Cancer Lett 122(1–2):165–175
53. Subbaramaiah K, Howe LR, Bhardwaj P, Du B, Gravaghi C, Yantiss RK, Zhou XK, Blaho VA, Hla T, Yang P, Kopelovich L, Hudis CA, Dannenberg AJ (2011) Obesity is associated with inflammation and elevated aromatase expression in the mouse mammary gland. Cancer Prev Res (Phila) 4(3):329–346. doi:4/3/329 [pii] 10.1158/1940-6207.CAPR-10-0381
54. Subbaramaiah K, Morris PG, Zhou XK, Morrow M, Du B, Giri D, Kopelovich L, Hudis CA, Dannenberg AJ (2012) Increased levels of COX-2 and prostaglandin E2 contribute to elevated aromatase expression in inflamed breast tissue of obese women. Cancer Discov 2(4):356–365. doi:2159–8290.CD-11-0241 [pii] 10.1158/2159-8290.CD-11-0241
55. Cancello R, Henegar C, Viguerie N, Taleb S, Poitou C, Rouault C, Coupaye M, Pelloux V, Hugol D, Bouillot JL, Bouloumie A, Barbatelli G, Cinti S, Svensson PA, Barsh GS, Zucker JD, Basdevant A, Langin D, Clement K (2005) Reduction of macrophage infiltration and chemoattractant gene expression changes in white adipose tissue of morbidly obese subjects after surgery-induced weight loss. Diabetes 54(8):2277–2286. doi:54/8/2277 [pii]
56. Cinti S, Mitchell G, Barbatelli G, Murano I, Ceresi E, Faloia E, Wang S, Fortier M, Greenberg AS, Obin MS (2005) Adipocyte death defines macrophage localization and function in adipose tissue of obese mice and humans. J Lipid Res 46(11):2347–2355. doi:M500294-JLR200 [pii] 10.1194/jlr.M500294-JLR200
57. Murano I, Barbatelli G, Parisani V, Latini C, Muzzonigro G, Castellucci M, Cinti S (2008) Dead adipocytes, detected as crown-like structures, are prevalent in visceral fat depots of genetically obese mice. J Lipid Res 49(7):1562–1568. doi:M800019-JLR200 [pii] 10.1194/jlr.M800019-JLR200
58. Morris PG, Hudis CA, Giri D, Morrow M, Falcone DJ, Zhou XK, Du B, Brogi E, Crawford CB, Kopelovich L, Subbaramaiah K, Dannenberg AJ (2011) Inflammation and increased aromatase expression occur in the breast tissue of obese women with breast cancer. Cancer Prev Res (Phila) 4(7):1021–1029. doi:1940–6207.CAPR-11-0110 [pii] 10.1158/1940-6207.CAPR-11-0110
59. Sun X, Casbas-Hernandez P, Bigelow C, Makowski L, Joseph Jerry D, Smith Schneider S, Troester MA (2012) Normal breast tissue of obese women is enriched for macrophage markers and macrophage-associated gene expression. Breast Cancer Res Treat 131(3):1003–1012. doi:10.1007/s10549-011-1789-3 [doi]
60. Howe LR, Chang SH, Tolle KC, Dillon R, Young LJT, Cardiff RD, Newman RA, Yang P, Thaler HT, Muller WJ, Hudis C, Brown AMC, Hla T, Subbaramaiah K, Dannenberg AJ (2005) HER2/neu-induced mammary tumorigenesis and angiogenesis are reduced in cyclooxygenase-2 knockout mice. Cancer Res 65(21):10113–10119
61. Chulada PC, Thompson MB, Mahler JF, Doyle CM, Gaul BW, Lee C, Tiano HF, Morham SG, Smithies O, Langenbach R (2000) Genetic disruption of Ptgs-1, as well as Ptgs-2, reduces intestinal tumorigenesis in Min mice. Cancer Res 60(17):4705–4708
62. Tiano HF, Loftin CD, Akunda J, Lee CA, Spalding J, Sessoms A, Dunson DB, Rogan EG, Morham SG, Smart RC, Langenbach R (2002) Deficiency of either cyclooxygenase (COX)-1 or COX-2 alters epidermal differentiation and reduces mouse skin tumorigenesis. Cancer Res 62(12):3395–3401
63. Liu CH, Chang SH, Narko K, Trifan OC, Wu MT, Smith E, Haudenschild C, Lane TF, Hla T (2001) Overexpression of cyclooxygenase-2 is sufficient to induce tumorigenesis in transgenic mice. J Biol Chem 276:18563–18569
64. Chang SH, Liu CH, Conway R, Han DK, Nithipatikom K, Trifan OC, Lane TF, Hla T (2004) Role of prostaglandin E2-dependent angiogenic switch in cyclooxygenase 2-induced breast cancer progression. Proc Natl Acad Sci U S A 101(2):591–596

65. Muller-Decker K, Neufang G, Berger I, Neumann M, Marks F, Furstenberger G (2002) Transgenic cyclooxygenase-2 overexpression sensitizes mouse skin for carcinogenesis. Proc Natl Acad Sci U S A 99(19):12483–12488
66. Wang D, Dubois RN (2006) Prostaglandins and cancer. Gut 55(1):115–122. doi:gut.2004.047100 [pii] 10.1136/gut.2004.047100
67. Nakanishi M, Montrose DC, Clark P, Nambiar PR, Belinsky GS, Claffey KP, Xu D, Rosenberg DW (2008) Genetic deletion of mPGES-1 suppresses intestinal tumorigenesis. Cancer Res 68(9):3251–3259. doi:68/9/3251 [pii] 10.1158/0008-5472.CAN-07-6100
68. Hanahan D, Weinberg RA (2011) Hallmarks of cancer: the next generation. Cell 144(5): 646–674. doi:S0092-8674(11)00127-9 [pii] 10.1016/j.cell.2011.02.013
69. Bandyopadhyay GK, Imagawa W, Wallace D, Nandi S (1987) Linoleate metabolites enhance the in vitro proliferative response of mouse mammary epithelial cells to epidermal growth factor. J Biol Chem 262(6):2750–2756
70. Goin M, Pignataro O, Jimenez de Asua L (1993) Early cell cycle diacylglycerol (DAG) content and protein kinase C (PKC) activity enhancement potentiates prostaglandin F2 alpha (PGF2 alpha) induced mitogenesis in Swiss 3T3 cells. FEBS Lett 316(1):68–72
71. Nolan RD, Danilowicz RM, Eling TE (1988) Role of arachidonic acid metabolism in the mitogenic response of BALB/c 3T3 fibroblasts to epidermal growth factor. Mol Pharmacol 33(6):650–656
72. Greenhough A, Smartt HJ, Moore AE, Roberts HR, Williams AC, Paraskeva C, Kaidi A (2009) The COX-2/PGE2 pathway: key roles in the hallmarks of cancer and adaptation to the tumour microenvironment. Carcinogenesis 30(3):377–386. doi:bgp014 [pii] 10.1093/carcin/bgp014
73. North TE, Goessling W, Walkley CR, Lengerke C, Kopani KR, Lord AM, Weber GJ, Bowman TV, Jang IH, Grosser T, Fitzgerald GA, Daley GQ, Orkin SH, Zon LI (2007) Prostaglandin E2 regulates vertebrate haematopoietic stem cell homeostasis. Nature 447(7147):1007–1011. doi:nature05883 [pii] 10.1038/nature05883
74. Li HJ, Reinhardt F, Herschman HR, Weinberg RA (2012) Cancer-stimulated mesenchymal stem cells create a carcinoma stem cell niche via prostaglandin E2 signaling. Cancer Discov 2(9):840–855. doi:2159–8290.CD-12-0101 [pii] 10.1158/2159-8290.CD-12-0101
75. Rudnick JA, Arendt LM, Klebba I, Hinds JW, Iyer V, Gupta PB, Naber SP, Kuperwasser C (2011) Functional heterogeneity of breast fibroblasts is defined by a prostaglandin secretory phenotype that promotes expansion of cancer-stem like cells. PLoS One 6(9):e24605. doi:10.1371/journal.pone.0024605 PONE-D-11-11675 [pii]
76. Castellone MD, Teramoto H, Williams BO, Druey KM, Gutkind JS (2005) Prostaglandin E2 promotes colon cancer cell growth through a Gs-axin-beta-catenin signaling axis. Science 310(5753):1504–1510. doi:1116221 [pii] 10.1126/science.1116221
77. Buchanan FG, Wang D, Bargiacchi F, DuBois RN (2003) Prostaglandin E2 regulates cell migration via the intracellular activation of the epidermal growth factor receptor. J Biol Chem 278(37):35451–35457. doi:10.1074/jbc.M302474200 M302474200 [pii]
78. Pai R, Soreghan B, Szabo IL, Pavelka M, Baatar D, Tarnawski AS (2002) Prostaglandin E2 transactivates EGF receptor: a novel mechanism for promoting colon cancer growth and gastrointestinal hypertrophy. Nat Med 8(3):289–293. doi:10.1038/nm0302-289 nm0302-289 [pii]
79. Ben-Baruch A (2006) Inflammation-associated immune suppression in cancer: the roles played by cytokines, chemokines and additional mediators. Semin Cancer Biol 16(1):38–52. doi:S1044-579X(05)00059-3 [pii] 10.1016/j.semcancer.2005.07.006
80. Ostrand-Rosenberg S, Sinha P (2009) Myeloid-derived suppressor cells: linking inflammation and cancer. J Immunol 182(8):4499–4506. doi:182/8/4499 [pii] 10.4049/jimmunol.0802740
81. Chen EP, Smyth EM (2011) COX-2 and PGE2-dependent immunomodulation in breast cancer. Prostaglandins Other Lipid Mediat 96(1–4):14–20. doi:S1098-8823(11)00075-X [pii] 10.1016/j.prostaglandins.2011.08.005

82. Strassmann G, Patil-Koota V, Finkelman F, Fong M, Kambayashi T (1994) Evidence for the involvement of interleukin 10 in the differential deactivation of murine peritoneal macrophages by prostaglandin E2. J Exp Med 180(6):2365–2370
83. Markosyan N, Chen EP, Ndong VN, Yao Y, Sterner CJ, Chodosh LA, Lawson JA, Fitzgerald GA, Smyth EM (2011) Deletion of cyclooxygenase 2 in mouse mammary epithelial cells delays breast cancer onset through augmentation of type 1 immune responses in tumors. Carcinogenesis 32(10):1441–1449. doi:bgr134 [pii] 10.1093/carcin/bgr134
84. Balkwill FR, Mantovani A (2012) Cancer-related inflammation: common themes and therapeutic opportunities. Semin Cancer Biol 22(1):33–40. doi:S1044-579X(11)00106-4 [pii] 10.1016/j.semcancer.2011.12.005
85. Sombroek CC, Stam AG, Masterson AJ, Lougheed SM, Schakel MJ, Meijer CJ, Pinedo HM, van den Eertwegh AJ, Scheper RJ, de Gruijl TD (2002) Prostanoids play a major role in the primary tumor-induced inhibition of dendritic cell differentiation. J Immunol 168(9): 4333–4343
86. Stolina M, Sharma S, Zhu L, Dubinett SM (2000) Lung cancer cyclooxygenase-2 dependent inhibition of dendritic cells maturation and function. Proc Am Assoc Cancer Res 41:619
87. Tsujii M, DuBois RN (1995) Alterations in cellular adhesion and apoptosis in epithelial cells overexpressing prostaglandin endoperoxide synthase 2. Cell 83(3):493–501
88. Takahashi Y, Kawahara F, Noguchi M, Miwa K, Sato H, Seiki M, Inoue H, Tanabe T, Yoshimoto T (1999) Activation of matrix metalloproteinase-2 in human breast cancer cells overexpressing cyclooxygenase-1 or -2. FEBS Lett 460(1):145–148
89. Tsujii M, Kawano S, DuBois RN (1997) Cyclooxygenase-2 expression in human colon cancer cells increases metastatic potential. Proc Natl Acad Sci U S A 94(7):3336–3340
90. Connolly EM, Harmey JH, O'Grady T, Foley D, Roche-Nagle G, Kay E, Bouchier-Hayes DJ (2002) Cyclo-oxygenase inhibition reduces tumour growth and metastasis in an orthotopic model of breast cancer. Br J Cancer 87(2):231–237
91. Kundu N, Fulton AM (2002) Selective cyclooxygenase (COX)-1 or COX-2 inhibitors control metastatic disease in a murine model of breast cancer. Cancer Res 62(8):2343–2346
92. Roche-Nagle G, Connolly EM, Eng M, Bouchier-Hayes DJ, Harmey JH (2004) Antimetastatic activity of a cyclooxygenase-2 inhibitor. Br J Cancer 91(2):359–365
93. Miyaura C, Inada M, Suzawa T, Sugimoto Y, Ushikubi F, Ichikawa A, Narumiya S, Suda T (2000) Impaired bone resorption to prostaglandin E2 in prostaglandin E receptor EP4-knockout mice. J Biol Chem 275(26):19819–19823
94. Okada Y, Lorenzo JA, Freeman AM, Tomita M, Morham SG, Raisz LG, Pilbeam CC (2000) Prostaglandin G/H synthase-2 is required for maximal formation of osteoclast-like cells in culture. J Clin Invest 105(6):823–832
95. Ono K, Akatsu T, Murakami T, Nishikawa M, Yamamoto M, Kugai N, Motoyoshi K, Nagata N (1998) Important role of EP4, a subtype of prostaglandin (PG) E receptor, in osteoclast-like cell formation from mouse bone marrow cells induced by PGE2. J Endocrinol 158(3):R1–R5
96. Sabino MA, Ghilardi JR, Jongen JL, Keyser CP, Luger NM, Mach DB, Peters CM, Rogers SD, Schwei MJ, de Felipe C, Mantyh PW (2002) Simultaneous reduction in cancer pain, bone destruction, and tumor growth by selective inhibition of cyclooxygenase-2. Cancer Res 62(24):7343–7349
97. Powles TJ, Dowsett M, Easty GC, Easty DM, Neville AM (1976) Breast-cancer osteolysis, bone metastases, and anti-osteolytic effect of aspirin. Lancet 1(7960):608–610. doi:S0140-6736(76)90416-5 [pii]
98. Seno H, Oshima M, Ishikawa TO, Oshima H, Takaku K, Chiba T, Narumiya S, Taketo MM (2002) Cyclooxygenase 2- and prostaglandin E_2 receptor EP_2-dependent angiogenesis in $Apc^{\Delta 716}$ mouse intestinal polyps. Cancer Res 62(2):506–511
99. Takeda H, Sonoshita M, Oshima H, Sugihara K, Chulada PC, Langenbach R, Oshima M, Taketo MM (2003) Cooperation of cyclooxygenase 1 and cyclooxygenase 2 in intestinal polyposis. Cancer Res 63(16):4872–4877

100. Williams CS, Tsujii M, Reese J, Dey SK, DuBois RN (2000) Host cyclooxygenase-2 modulates carcinoma growth. J Clin Invest 105(11):1589–1594
101. Peterson HI (1983) Effects of prostaglandin synthesis inhibitors on tumor growth and vascularization. Experimental studies in the rat. Invasion Metastasis 3(3):151–159
102. Daniel TO, Liu H, Morrow JD, Crews BC, Marnett LJ (1999) Thromboxane A_2 is a mediator of cyclooxygenase-2-dependent endothelial migration and angiogenesis. Cancer Res 59(18):4574–4577
103. Leahy KM, Ornberg RL, Wang Y, Zweifel BS, Koki AT, Masferrer JL (2002) Cyclooxygenase-2 inhibition by celecoxib reduces proliferation and induces apoptosis in angiogenic endothelial cells in vivo. Cancer Res 62(3):625–631
104. Majima M, Isono M, Ikeda Y, Hayashi I, Hatanaka K, Harada Y, Katsumata O, Yamashina S, Katori M, Yamamoto S (1997) Significant roles of inducible cyclooxygenase (COX)-2 in angiogenesis in rat sponge implants. Jpn J Pharmacol 75(2):105–114
105. Masferrer JL, Leahy KM, Koki AT, Zweifel BS, Settle SL, Woerner BM, Edwards DA, Flickinger AG, Moore RJ, Seibert K (2000) Antiangiogenic and antitumor activities of cyclooxygenase-2 inhibitors. Cancer Res 60(5):1306–1311
106. Sawaoka H, Tsuji S, Tsujii M, Gunawan ES, Sasaki Y, Kawano S, Hori M (1999) Cyclooxygenase inhibitors suppress angiogenesis and reduce tumor growth in vivo. Lab Invest 79(12):1469–1477
107. Yamada M, Kawai M, Kawai Y, Mashima Y (1999) The effect of selective cyclooxygenase-2 inhibitor on corneal angiogenesis in the rat. Curr Eye Res 19(4):300–304
108. Jones MK, Wang H, Peskar BM, Levin E, Itani RM, Sarfeh IJ, Tarnawski AS (1999) Inhibition of angiogenesis by nonsteroidal anti-inflammatory drugs: insight into mechanisms and implications for cancer growth and ulcer healing. Nat Med 5(12):1418–1423
109. Tsujii M, Kawano S, Tsuji S, Sawaoka H, Hori M, DuBois RN (1998) Cyclooxygenase regulates angiogenesis induced by colon cancer cells. Cell 93(5):705–716. doi:S0092-8674(00)81433-6 [pii]
110. Gately S, Li WW (2004) Multiple roles of COX-2 in tumor angiogenesis: a target for antiangiogenic therapy. Semin Oncol 31(2 suppl 7):2–11
111. Leahy KM, Koki AT, Masferrer JL (2000) Role of cyclooxygenases in angiogenesis. Curr Med Chem 7(11):1163–1170
112. Miyazawa-Hoshimoto S, Takahashi K, Bujo H, Hashimoto N, Saito Y (2003) Elevated serum vascular endothelial growth factor is associated with visceral fat accumulation in human obese subjects. Diabetologia 46(11):1483–1488
113. Garcia de la Torre N, Rubio MA, Bordiu E, Cabrerizo L, Aparicio E, Hernandez C, Sanchez-Pernaute A, Diez-Valladares L, Torres AJ, Puente M, Charro AL (2008) Effects of weight loss after bariatric surgery for morbid obesity on vascular endothelial growth factor-A, adipocytokines, and insulin. J Clin Endocrinol Metab 93(11):4276–4281. doi:jc.2007-1370 [pii] 10.1210/jc.2007-1370
114. Silha JV, Krsek M, Sucharda P, Murphy LJ (2005) Angiogenic factors are elevated in overweight and obese individuals. Int J Obes (Lond) 29(11):1308–1314. doi:0802987 [pii] 10.1038/sj.ijo.0802987
115. Gu JW, Young E, Patterson SG, Makey KL, Wells J, Huang M, Tucker KB, Miele L (2011) Postmenopausal obesity promotes tumor angiogenesis and breast cancer progression in mice. Cancer Biol Ther 11(10):910–917. doi:15473 [pii]
116. Chen CT, Du Y, Yamaguchi H, Hsu JM, Kuo HP, Hortobagyi GN, Hung MC (2012) Targeting the IKKbeta/mTOR/VEGF signaling pathway as a potential therapeutic strategy for obesity-related breast cancer. Mol Cancer Ther 11(10):2212–2221. doi:1535–7163.MCT-12-0180 [pii] 10.1158/1535-7163.MCT-12-0180
117. Barnes NL, Warnberg F, Farnie G, White D, Jiang W, Anderson E, Bundred NJ (2007) Cyclooxygenase-2 inhibition: effects on tumour growth, cell cycling and lymphangiogenesis in a xenograft model of breast cancer. Br J Cancer 96(4):575–582. doi:6603593 [pii] 10.1038/sj.bjc.6603593

118. Timoshenko AV, Chakraborty C, Wagner GF, Lala PK (2006) COX-2-mediated stimulation of the lymphangiogenic factor VEGF-C in human breast cancer. Br J Cancer 94(8):1154–1163. doi:6603593 [pii] 10.1038/sj.bjc.6603593
119. Simpson ER, Clyne C, Rubin G, Boon WC, Robertson K, Britt K, Speed C, Jones M (2002) Aromatase—a brief overview. Annu Rev Physiol 64:93–127. doi:10.1146/annurev.physiol.64.081601.142703 64/1/93 [pii]
120. Bulun SE, Price TM, Aitken J, Mahendroo MS, Simpson ER (1993) A link between breast cancer and local estrogen biosynthesis suggested by quantification of breast adipose tissue aromatase cytochrome P450 transcripts using competitive polymerase chain reaction after reverse transcription. J Clin Endocrinol Metab 77(6):1622–1628
121. Lipton A, Santen RJ, Santner SJ, Harvey HA, Sanders SI, Matthews YL (1992) Prognostic value of breast cancer aromatase. Cancer 70(7):1951–1955
122. Miller WR, Anderson TJ, Jack WJ (1990) Relationship between tumour aromatase activity, tumour characteristics and response to therapy. J Steroid Biochem Mol Biol 37(6):1055–1059
123. Silva MC, Rowlands MG, Dowsett M, Gusterson B, McKinna JA, Fryatt I, Coombes RC (1989) Intratumoral aromatase as a prognostic factor in human breast carcinoma. Cancer Res 49(10):2588–2591
124. Agarwal VR, Bulun SE, Leitch M, Rohrich R, Simpson ER (1996) Use of alternative promoters to express the aromatase cytochrome P450 (CYP19) gene in breast adipose tissues of cancer-free and breast cancer patients. J Clin Endocrinol Metab 81(11):3843–3849
125. Chen S, Itoh T, Wu K, Zhou D, Yang C (2002) Transcriptional regulation of aromatase expression in human breast tissue. J Steroid Biochem Mol Biol 83(1–5):93–99
126. Mahendroo MS, Mendelson CR, Simpson ER (1993) Tissue-specific and hormonally controlled alternative promoters regulate aromatase cytochrome P450 gene expression in human adipose tissue. J Biol Chem 268(26):19463–19470
127. Zhou D, Chen S (1999) Identification and characterization of a cAMP-responsive element in the region upstream from promoter 1.3 of the human aromatase gene. Arch Biochem Biophys 371(2):179–190
128. Narumiya S, Sugimoto Y, Ushikubi F (1999) Prostanoid receptors: structures, properties, and functions. Physiol Rev 79(4):1193–1226
129. Brueggemeier RW, Richards JA, Joomprabutra S, Bhat AS, Whetstone JL (2001) Molecular pharmacology of aromatase and its regulation by endogenous and exogenous agents. J Steroid Biochem Mol Biol 79(1–5):75–84
130. Singh A, Purohit A, Ghilchik MW, Reed MJ (1999) The regulation of aromatase activity in breast fibroblasts: the role of interleukin-6 and prostaglandin E2. Endocr Relat Cancer 6(2):139–147
131. Zhao Y, Agarwal VR, Mendelson CR, Simpson ER (1996) Estrogen biosynthesis proximal to a breast tumor is stimulated by PGE_2 via cyclic AMP, leading to activation of promoter II of the CYP19 (aromatase) gene. Endocrinology 137(12):5739–5742
132. Subbaramaiah K, Hudis C, Chang SH, Hla T, Dannenberg AJ (2008) EP2 and EP4 receptors regulate aromatase expression in human adipocytes and breast cancer cells. Evidence of a BRCA1 and p300 exchange. J Biol Chem 283(6):3433–3444. doi:M705409200 [pii] 10.1074/jbc.M705409200
133. Prosperi JR, Robertson FM (2006) Cyclooxygenase-2 directly regulates gene expression of P450 Cyp19 aromatase promoter regions pII, pI.3 and pI.7 and estradiol production in human breast tumor cells. Prostaglandins Other Lipid Mediat 81(1–2):55–70. doi:S1098-8823(06)00115-8 [pii] 10.1016/j.prostaglandins.2006.07.003
134. Brodie AM, Lu Q, Long BJ, Fulton A, Chen T, Macpherson N, DeJong PC, Blankenstein MA, Nortier JW, Slee PH, van de Ven J, van Gorp JM, Elbers JR, Schipper ME, Blijham GH, Thijssen JH (2001) Aromatase and COX-2 expression in human breast cancers. J Steroid Biochem Mol Biol 79(1–5):41–47
135. Brueggemeier RW, Quinn AL, Parrett ML, Joarder FS, Harris RE, Robertson FM (1999) Correlation of aromatase and cyclooxygenase gene expression in human breast cancer specimens. Cancer Lett 140(1–2):27–35

136. Salhab M, Singh-Ranger G, Mokbel R, Jouhra F, Jiang WG, Mokbel K (2007) Cyclooxygenase-2 mRNA expression correlates with aromatase expression in human breast cancer. J Surg Oncol 96(5):424–428. doi:10.1002/jso.20740
137. Subbaramaiah K, Howe LR, Port ER, Brogi E, Fishman J, Liu CH, Hla T, Hudis C, Dannenberg AJ (2006) HER-2/neu status is a determinant of mammary aromatase activity in vivo: evidence for a cyclooxygenase-2-dependent mechanism. Cancer Res 66(10):5504–5511. doi:66/10/5504 [pii] 10.1158/0008-5472.CAN-05-4076
138. Gates MA, Tworoger SS, Eliassen AH, Missmer SA, Hankinson SE (2010) Analgesic use and sex steroid hormone concentrations in postmenopausal women. Cancer Epidemiol Biomarkers Prev 19(4):1033–1041. doi:1055-9965.EPI-09-0975 [pii] 10.1158/1055-9965.EPI-09-0975
139. Hudson AG, Gierach GL, Modugno F, Simpson J, Wilson JW, Evans RW, Vogel VG, Weissfeld JL (2008) Nonsteroidal anti-inflammatory drug use and serum total estradiol in postmenopausal women. Cancer Epidemiol Biomarkers Prev 17(3):680–687. doi:17/3/680 [pii] 10.1158/1055-9965.EPI-07-2739
140. Cauley JA, Gutai JP, Kuller LH, LeDonne D, Powell JG (1989) The epidemiology of serum sex hormones in postmenopausal women. Am J Epidemiol 129(6):1120–1131
141. Kaye SA, Folsom AR, Soler JT, Prineas RJ, Potter JD (1991) Associations of body mass and fat distribution with sex hormone concentrations in postmenopausal women. Int J Epidemiol 20(1):151–156
142. McTiernan A, Rajan KB, Tworoger SS, Irwin M, Bernstein L, Baumgartner R, Gilliland F, Stanczyk FZ, Yasui Y, Ballard-Barbash R (2003) Adiposity and sex hormones in postmenopausal breast cancer survivors. J Clin Oncol 21(10):1961–1966. doi:10.1200/JCO.2003.07.057 JCO.2003.07.057 [pii]
143. Wake DJ, Strand M, Rask E, Westerbacka J, Livingstone DE, Soderberg S, Andrew R, Yki-Jarvinen H, Olsson T, Walker BR (2007) Intra-adipose sex steroid metabolism and body fat distribution in idiopathic human obesity. Clin Endocrinol (Oxf) 66(3):440–446. doi:CEN2755 [pii] 10.1111/j.1365-2265.2007.02755.x
144. Marnett LJ (1992) Aspirin and the potential role of prostaglandins in colon cancer. Cancer Res 52(20):5575–5589
145. Corpet DE, Pierre F (2003) Point: from animal models to prevention of colon cancer. Systematic review of chemoprevention in min mice and choice of the model system. Cancer Epidemiol Biomarkers Prev 12(5):391–400
146. Rao CV, Reddy BS (2004) NSAIDs and chemoprevention. Curr Cancer Drug Targets 4(1):29–42
147. Howe LR (2005) Cyclooxygenase-2 and breast cancer. In: Yao AP (ed) Trends in breast cancer research, vol 9, Horizons in cancer research. Nova, New York, pp 1–38
148. Buchanan FG, Holla V, Katkuri S, Matta P, DuBois RN (2007) Targeting cyclooxygenase-2 and the epidermal growth factor receptor for the prevention and treatment of intestinal cancer. Cancer Res 67(19):9380–9388. doi:67/19/9380 [pii] 10.1158/0008-5472.CAN-07-0710
149. Torrance CJ, Jackson PE, Montgomery E, Kinzler KW, Vogelstein B, Wissner A, Nunes M, Frost P, Discafani CM (2000) Combinatorial chemoprevention of intestinal neoplasia. Nat Med 6(9):1024–1028. doi:10.1038/79534
150. Mann M, Sheng H, Shao J, Williams CS, Pisacane PI, Sliwkowski MX, DuBois RN (2001) Targeting cyclooxygenase 2 and HER-2/neu pathways inhibits colorectal carcinoma growth. Gastroenterology 120(7):1713–1719
151. Reddy BS, Patlolla JM, Simi B, Wang SH, Rao CV (2005) Prevention of colon cancer by low doses of celecoxib, a cyclooxygenase inhibitor, administered in diet rich in omega-3 polyunsaturated fatty acids. Cancer Res 65(17):8022–8027. doi:65/17/8022 [pii] 10.1158/0008-5472.CAN-05-0212
152. Reddy BS, Wang CX, Kong AN, Khor TO, Zheng X, Steele VE, Kopelovich L, Rao CV (2006) Prevention of azoxymethane-induced colon cancer by combination of low doses of atorvastatin, aspirin, and celecoxib in F 344 rats. Cancer Res 66(8):4542–4546. doi:66/8/4542 [pii] 10.1158/0008-5472.CAN-05-4428

153. Swamy MV, Patlolla JM, Steele VE, Kopelovich L, Reddy BS, Rao CV (2006) Chemoprevention of familial adenomatous polyposis by low doses of atorvastatin and celecoxib given individually and in combination to APCMin mice. Cancer Res 66(14): 7370–7377. doi:66/14/7370 [pii] 10.1158/0008-5472.CAN-05-4619
154. Ignatenko NA, Besselsen DG, Stringer DE, Blohm-Mangone KA, Cui H, Gerner EW (2008) Combination chemoprevention of intestinal carcinogenesis in a murine model of familial adenomatous polyposis. Nutr Cancer 60(suppl 1):30–35. doi:905311576 [pii] 10.1080/01635580802401317
155. Lanza-Jacoby S, Miller S, Flynn J, Gallatig K, Daskalakis C, Masferrer JL, Zweifel BS, Sembhi H, Russo IH (2003) The cyclooxygenase-2 inhibitor, celecoxib, prevents the development of mammary tumors in Her-2/neu mice. Cancer Epidemiol Biomarkers Prev 12(12):1486–1491
156. Woditschka S, Haag JD, Mau B, Lubet RA, Gould MN (2008) Chemopreventive effects of celecoxib are limited to hormonally responsive mammary carcinomas in the neu-induced retroviral rat model. Breast Cancer Res 10(1):R18. doi:bcr1864 [pii] 10.1186/bcr1864
157. Brown PH, Subbaramaiah K, Salmon AP, Baker R, Newman RA, Yang P, Zhou XK, Bissonnette RP, Dannenberg AJ, Howe LR (2008) Combination chemoprevention of HER2/neu-induced breast cancer using a cyclooxygenase-2 inhibitor and a retinoid X receptor-selective retinoid. Cancer Prev Res (Phila) 1(3):208–214. doi:1/3/208 [pii] 10.1158/1940-6207.CAPR-08-0021
158. Chan AT, Arber N, Burn J, Chia WK, Elwood P, Hull MA, Logan RF, Rothwell PM, Schror K, Baron JA (2012) Aspirin in the chemoprevention of colorectal neoplasia: an overview. Cancer Prev Res (Phila) 5(2):164–178. doi:1940–6207.CAPR-11-0391 [pii] 10.1158/1940-6207.CAPR-11-0391
159. Bosetti C, Rosato V, Gallus S, Cuzick J, La Vecchia C (2012) Aspirin and cancer risk: a quantitative review to 2011. Ann Oncol 23(6):1403–1415. doi:mds113 [pii] 10.1093/annonc/mds113
160. Reeves MJ, Newcomb PA, Trentham-Dietz A, Storer BE, Remington PL (1996) Nonsteroidal anti-inflammatory drug use and protection against colorectal cancer in women. Cancer Epidemiol Biomarkers Prev 5(12):955–960
161. Rosenberg L, Louik C, Shapiro S (1998) Nonsteroidal antiinflammatory drug use and reduced risk of large bowel carcinoma. Cancer 82(12):2326–2333. doi:10.1002/(SICI)1097-0142(19980615)82:12<2326::AID-CNCR5>3.0.CO;2-Q [pii]
162. Harris RE, Beebe-Donk J, Alshafie GA (2008) Similar reductions in the risk of human colon cancer by selective and nonselective cyclooxygenase-2 (COX-2) inhibitors. BMC Cancer 8:237. doi:1471-2407-8-237 [pii] 10.1186/1471-2407-8-237
163. Ruder EH, Laiyemo AO, Graubard BI, Hollenbeck AR, Schatzkin A, Cross AJ (2011) Non-steroidal anti-inflammatory drugs and colorectal cancer risk in a large, prospective cohort. Am J Gastroenterol 106(7):1340–1350. doi:ajg201138 [pii] 10.1038/ajg.2011.38
164. Rahme E, Barkun AN, Toubouti Y, Bardou M (2003) The cyclooxygenase-2-selective inhibitors rofecoxib and celecoxib prevent colorectal neoplasia occurrence and recurrence. Gastroenterology 125(2):404–412. doi:S0016508503008801 [pii]
165. Vinogradova Y, Hippisley-Cox J, Coupland C, Logan RF (2007) Risk of colorectal cancer in patients prescribed statins, nonsteroidal anti-inflammatory drugs, and cyclooxygenase-2 inhibitors: nested case–control study. Gastroenterology 133(2):393–402. doi:S0016-5085(07)01005-0 [pii] 10.1053/j.gastro.2007.05.023
166. Rothwell PM, Fowkes FG, Belch JF, Ogawa H, Warlow CP, Meade TW (2011) Effect of daily aspirin on long-term risk of death due to cancer: analysis of individual patient data from randomised trials. Lancet 377(9759):31–41. doi:S0140-6736(10)62110-1 [pii] 10.1016/S0140-6736(10)62110-1
167. Rothwell PM, Wilson M, Elwin CE, Norrving B, Algra A, Warlow CP, Meade TW (2010) Long-term effect of aspirin on colorectal cancer incidence and mortality: 20-year follow-up of five randomised trials. Lancet 376(9754):1741–1750. doi:S0140-6736(10)61543-7 [pii] 10.1016/S0140-6736(10)61543-7

168. Rothwell PM, Wilson M, Price JF, Belch JF, Meade TW, Mehta Z (2012) Effect of daily aspirin on risk of cancer metastasis: a study of incident cancers during randomised controlled trials. Lancet 379(9826):1591–1601. doi:S0140-6736(12)60209-8 [pii] 10.1016/S0140-6736(12)60209-8
169. Giardiello FM, Hamilton SR, Krush AJ, Piantadosi S, Hylind LM, Celano P, Booker SV, Robinson CR, Offerhaus GJ (1993) Treatment of colonic and rectal adenomas with sulindac in familial adenomatous polyposis. N Engl J Med 328(18):1313–1316
170. Giardiello FM, Spannhake EW, DuBois RN, Hylind LM, Robinson CR, Hubbard WC, Hamilton SR, Yang VW (1998) Prostaglandin levels in human colorectal mucosa: effects of sulindac in patients with familial adenomatous polyposis. Dig Dis Sci 43(2):311–316
171. Meyskens FL Jr, McLaren CE, Pelot D, Fujikawa-Brooks S, Carpenter PM, Hawk E, Kelloff G, Lawson MJ, Kidao J, McCracken J, Albers CG, Ahnen DJ, Turgeon DK, Goldschmid S, Lance P, Hagedorn CH, Gillen DL, Gerner EW (2008) Difluoromethylornithine plus sulindac for the prevention of sporadic colorectal adenomas: a randomized placebo-controlled, double-blind trial. Cancer Prev Res (Phila) 1(1):32–38. doi:10.1158/1940-6207.CAPR-08-0042
172. Baron JA, Cole BF, Sandler RS, Haile RW, Ahnen D, Bresalier R, McKeown-Eyssen G, Summers RW, Rothstein R, Burke CA, Snover DC, Church TR, Allen JI, Beach M, Beck GJ, Bond JH, Byers T, Greenberg ER, Mandel JS, Marcon N, Mott LA, Pearson L, Saibil F, van Stolk RU (2003) A randomized trial of aspirin to prevent colorectal adenomas. N Engl J Med 348(10):891–899
173. Logan RF, Grainge MJ, Shepherd VC, Armitage NC, Muir KR (2008) Aspirin and folic acid for the prevention of recurrent colorectal adenomas. Gastroenterology 134(1):29–38. doi:S0016-5085(07)01813-6 [pii] 10.1053/j.gastro.2007.10.014
174. Sandler RS, Halabi S, Baron JA, Budinger S, Paskett E, Keresztes R, Petrelli N, Pipas JM, Karp DD, Loprinzi CL, Steinbach G, Schilsky R (2003) A randomized trial of aspirin to prevent colorectal adenomas in patients with previous colorectal cancer. N Engl J Med 348(10):883–890
175. Benamouzig R, Deyra J, Martin A, Girard B, Jullian E, Piednoir B, Couturier D, Coste T, Little J, Chaussade S (2003) Daily soluble aspirin and prevention of colorectal adenoma recurrence: one-year results of the APACC trial. Gastroenterology 125(2):328–336. doi:S0016508503008874 [pii]
176. Cole BF, Logan RF, Halabi S, Benamouzig R, Sandler RS, Grainge MJ, Chaussade S, Baron JA (2009) Aspirin for the chemoprevention of colorectal adenomas: meta-analysis of the randomized trials. J Natl Cancer Inst 101(4):256–266. doi:djn485 [pii] 10.1093/jnci/djn485
177. Benamouzig R, Uzzan B, Deyra J, Martin A, Girard B, Little J, Chaussade S (2012) Prevention by daily soluble aspirin of colorectal adenoma recurrence: 4-year results of the APACC randomised trial. Gut 61(2):255–261. doi:gutjnl-2011-300113 [pii] 10.1136/gutjnl-2011-300113
178. Barnes CJ, Hamby-Mason RL, Hardman WE, Cameron IL, Speeg KV, Lee M (1999) Effect of aspirin on prostaglandin E2 formation and transforming growth factor alpha expression in human rectal mucosa from individuals with a history of adenomatous polyps of the colon. Cancer Epidemiol Biomarkers Prev 8(4 pt 1):311–315
179. Frommel TO, Dyavanapalli M, Oldham T, Kazi N, Lietz H, Liao Y, Mobarhan S (1997) Effect of aspirin on prostaglandin E2 and leukotriene B4 production in human colonic mucosa from cancer patients. Clin Cancer Res 3(2):209–213
180. Krishnan K, Ruffin MT, Normolle D, Shureiqi I, Burney K, Bailey J, Peters-Golden M, Rock CL, Boland CR, Brenner DE (2001) Colonic mucosal prostaglandin E2 and cyclooxygenase expression before and after low aspirin doses in subjects at high risk or at normal risk for colorectal cancer. Cancer Epidemiol Biomarkers Prev 10(5):447–453
181. Ruffin MT, Krishnan K, Rock CL, Normolle D, Vaerten MA, Peters-Golden M, Crowell J, Kelloff G, Boland CR, Brenner DE (1997) Suppression of human colorectal mucosal prostaglandins: determining the lowest effective aspirin dose. J Natl Cancer Inst 89(15):1152–1160

182. Sample D, Wargovich M, Fischer SM, Inamdar N, Schwartz P, Wang X, Do KA, Sinicrope FA (2002) A dose-finding study of aspirin for chemoprevention utilizing rectal mucosal prostaglandin E(2) levels as a biomarker. Cancer Epidemiol Biomarkers Prev 11(3):275–279
183. Burn J, Gerdes AM, Macrae F, Mecklin JP, Moeslein G, Olschwang S, Eccles D, Evans DG, Maher ER, Bertario L, Bisgaard ML, Dunlop MG, Ho JW, Hodgson SV, Lindblom A, Lubinski J, Morrison PJ, Murday V, Ramesar R, Side L, Scott RJ, Thomas HJ, Vasen HF, Barker G, Crawford G, Elliott F, Movahedi M, Pylvanainen K, Wijnen JT, Fodde R, Lynch HT, Mathers JC, Bishop DT (2011) Long-term effect of aspirin on cancer risk in carriers of hereditary colorectal cancer: an analysis from the CAPP2 randomised controlled trial. Lancet 378(9809):2081–2087. doi:S0140-6736(11)61049-0 [pii] 10.1016/S0140-6736(11)61049-0
184. Sturmer T, Glynn RJ, Lee IM, Manson JE, Buring JE, Hennekens CH (1998) Aspirin use and colorectal cancer: post-trial follow-up data from the Physicians' Health Study. Ann Intern Med 128(9):713–720
185. Cook NR, Lee IM, Gaziano JM, Gordon D, Ridker PM, Manson JE, Hennekens CH, Buring JE (2005) Low-dose aspirin in the primary prevention of cancer: the Women's Health Study: a randomized controlled trial. JAMA 294(1):47–55. doi:294/1/47 [pii] 10.1001/jama.294.1.47
186. Dube C, Rostom A, Lewin G, Tsertsvadze A, Barrowman N, Code C, Sampson M, Moher D (2007) The use of aspirin for primary prevention of colorectal cancer: a systematic review prepared for the U.S. Preventive Services Task Force. Ann Intern Med 146(5):365–375. doi:146/5/365 [pii]
187. Cuzick J, Otto F, Baron JA, Brown PH, Burn J, Greenwald P, Jankowski J, La Vecchia C, Meyskens F, Senn HJ, Thun M (2009) Aspirin and non-steroidal anti-inflammatory drugs for cancer prevention: an international consensus statement. Lancet Oncol 10(5):501–507. doi:S1470-2045(09)70035-X [pii] 10.1016/S1470-2045(09)70035-X
188. Chan AT, Giovannucci EL, Meyerhardt JA, Schernhammer ES, Curhan GC, Fuchs CS (2005) Long-term use of aspirin and nonsteroidal anti-inflammatory drugs and risk of colorectal cancer. JAMA 294(8):914–923. doi:294/8/914 [pii] 10.1001/jama.294.8.914
189. Chan AT, Giovannucci EL, Meyerhardt JA, Schernhammer ES, Wu K, Fuchs CS (2008) Aspirin dose and duration of use and risk of colorectal cancer in men. Gastroenterology 134(1):21–28. doi:S0016-5085(07)01745-3 [pii] 10.1053/j.gastro.2007.09.035
190. Chan AT, Manson JE, Feskanich D, Stampfer MJ, Colditz GA, Fuchs CS (2007) Long-term aspirin use and mortality in women. Arch Intern Med 167(6):562–572. doi:167/6/562 [pii] 10.1001/archinte.167.6.562
191. Coghill AE, Newcomb PA, Chia VM, Zheng Y, Wernli KJ, Passarelli MN, Potter JD (2011) Pre-diagnostic NSAID use but not hormone therapy is associated with improved colorectal cancer survival in women. Br J Cancer 104(5):763–768. doi:6606041 [pii] 10.1038/sj.bjc.6606041
192. Zell JA, Ziogas A, Bernstein L, Clarke CA, Deapen D, Largent JA, Neuhausen SL, Stram DO, Ursin G, Anton-Culver H (2009) Nonsteroidal anti-inflammatory drugs: effects on mortality after colorectal cancer diagnosis. Cancer 115(24):5662–5671. doi:10.1002/cncr.24705
193. Din FV, Theodoratou E, Farrington SM, Tenesa A, Barnetson RA, Cetnarskyj R, Stark L, Porteous ME, Campbell H, Dunlop MG (2010) Effect of aspirin and NSAIDs on risk and survival from colorectal cancer. Gut 59(12):1670–1679. doi:gut.2009.203000 [pii] 10.1136/gut.2009.203000
194. Chan AT, Ogino S, Fuchs CS (2009) Aspirin use and survival after diagnosis of colorectal cancer. JAMA 302(6):649–658. doi:302/6/649 [pii] 10.1001/jama.2009.1112
195. Bastiaannet E, Sampieri K, Dekkers OM, de Craen AJ, van Herk-Sukel MP, Lemmens V, van den Broek CB, Coebergh JW, Herings RM, van de Velde CJ, Fodde R, Liefers GJ (2012) Use of aspirin postdiagnosis improves survival for colon cancer patients. Br J Cancer 106(9):1564–1570. doi:bjc2012101 [pii] 10.1038/bjc.2012.101
196. Walker AJ, Grainge MJ, Card TR (2012) Aspirin and other non-steroidal anti-inflammatory drug use and colorectal cancer survival: a cohort study. Br J Cancer 107(9):1602–1607. doi:bjc2012427 [pii] 10.1038/bjc.2012.427

197. Chan AT, Ogino S, Fuchs CS (2007) Aspirin and the risk of colorectal cancer in relation to the expression of COX-2. N Engl J Med 356(21):2131–2142. doi:356/21/2131 [pii] 10.1056/NEJMoa067208
198. Benamouzig R, Uzzan B, Martin A, Deyra J, Little J, Girard B, Chaussade S (2010) Cyclooxygenase-2 expression and recurrence of colorectal adenomas: effect of aspirin chemoprevention. Gut 59(5):622–629. doi:59/5/622 [pii] 10.1136/gut.2008.175406
199. Liao X, Lochhead P, Nishihara R, Morikawa T, Kuchiba A, Yamauchi M, Imamura Y, Qian ZR, Baba Y, Shima K, Sun R, Nosho K, Meyerhardt JA, Giovannucci E, Fuchs CS, Chan AT, Ogino S (2012) Aspirin use, tumor PIK3CA mutation, and colorectal-cancer survival. N Engl J Med 367(17):1596–1606. doi:10.1056/NEJMoa1207756
200. Johnson CC, Hayes RB, Schoen RE, Gunter MJ, Huang WY (2010) Non-steroidal anti-inflammatory drug use and colorectal polyps in the Prostate, Lung, Colorectal, and Ovarian Cancer Screening Trial. Am J Gastroenterol 105(12):2646–2655. doi:ajg2010349 [pii] 10.1038/ajg.2010.349
201. Zhang X, Smith-Warner SA, Chan AT, Wu K, Spiegelman D, Fuchs CS, Willett WC, Giovannucci EL (2011) Aspirin use, body mass index, physical activity, plasma C-peptide, and colon cancer risk in US health professionals. Am J Epidemiol 174(4):459–467. doi:kwr115 [pii] 10.1093/aje/kwr115
202. Kim S, Baron JA, Mott LA, Burke CA, Church TR, McKeown-Eyssen GE, Cole BF, Haile RW, Sandler RS (2006) Aspirin may be more effective in preventing colorectal adenomas in patients with higher BMI (United States). Cancer Causes Control 17(10):1299–1304. doi:10.1007/s10552-006-0075-x
203. Steinbach G, Lynch PM, Phillips RK, Wallace MH, Hawk E, Gordon GB, Wakabayashi N, Saunders B, Shen Y, Fujimura T, Su LK, Levin B (2000) The effect of celecoxib, a cyclooxygenase-2 inhibitor, in familial adenomatous polyposis. N Engl J Med 342(26):1946–1952
204. Arber N, Eagle CJ, Spicak J, Racz I, Dite P, Hajer J, Zavoral M, Lechuga MJ, Gerletti P, Tang J, Rosenstein RB, Macdonald K, Bhadra P, Fowler R, Wittes J, Zauber AG, Solomon SD, Levin B (2006) Celecoxib for the prevention of colorectal adenomatous polyps. N Engl J Med 355(9):885–895. doi:355/9/885 [pii] 10.1056/NEJMoa061652
205. Baron JA, Sandler RS, Bresalier RS, Quan H, Riddell R, Lanas A, Bolognese JA, Oxenius B, Horgan K, Loftus S, Morton DG (2006) A randomized trial of rofecoxib for the chemoprevention of colorectal adenomas. Gastroenterology 131(6):1674–1682. doi:S0016-5085(06)01994-9 [pii] 10.1053/j.gastro.2006.08.079
206. Bertagnolli MM, Eagle CJ, Zauber AG, Redston M, Solomon SD, Kim K, Tang J, Rosenstein RB, Wittes J, Corle D, Hess TM, Woloj GM, Boisserie F, Anderson WF, Viner JL, Bagheri D, Burn J, Chung DC, Dewar T, Foley TR, Hoffman N, Macrae F, Pruitt RE, Saltzman JR, Salzberg B, Sylwestrowicz T, Gordon GB, Hawk ET (2006) Celecoxib for the prevention of sporadic colorectal adenomas. N Engl J Med 355(9):873–884
207. Yan M, Myung SJ, Fink SP, Lawrence E, Lutterbaugh J, Yang P, Zhou X, Liu D, Rerko RM, Willis J, Dawson D, Tai HH, Barnholtz-Sloan JS, Newman RA, Bertagnolli MM, Markowitz SD (2009) 15-Hydroxyprostaglandin dehydrogenase inactivation as a mechanism of resistance to celecoxib chemoprevention of colon tumors. Proc Natl Acad Sci U S A 106(23): 9409–9413. doi:0902367106 [pii] 10.1073/pnas.0902367106
208. Tai HH, Chi X, Tong M (2011) Regulation of 15-hydroxyprostaglandin dehydrogenase (15-PGDH) by non-steroidal anti-inflammatory drugs (NSAIDs). Prostaglandins Other Lipid Mediat 96(1–4):37–40. doi:S1098-8823(11)00045-1 [pii] 10.1016/j.prostaglandins.2011.06.005
209. Baron JA, Sandler RS, Bresalier RS, Lanas A, Morton DG, Riddell R, Iverson ER, Demets DL (2008) Cardiovascular events associated with rofecoxib: final analysis of the APPROVe trial. Lancet 372(9651):1756–1764. doi:S0140-6736(08)61490-7 [pii] 10.1016/S0140-6736(08)61490-7
210. Bresalier RS, Sandler RS, Quan H, Bolognese JA, Oxenius B, Horgan K, Lines C, Riddell R, Morton D, Lanas A, Konstam MA, Baron JA (2005) Cardiovascular events associated with rofecoxib in a colorectal adenoma chemoprevention trial. N Engl J Med 352(11):1092–1102. doi:NEJMoa050493 [pii] 10.1056/NEJMoa050493

211. Solomon SD, McMurray JJ, Pfeffer MA, Wittes J, Fowler R, Finn P, Anderson WF, Zauber A, Hawk E, Bertagnolli M (2005) Cardiovascular risk associated with celecoxib in a clinical trial for colorectal adenoma prevention. N Engl J Med 352(11):1071–1080. doi:NEJMoa050405 [pii] 10.1056/NEJMoa050405
212. Solomon SD, Pfeffer MA, McMurray JJ, Fowler R, Finn P, Levin B, Eagle C, Hawk E, Lechuga M, Zauber AG, Bertagnolli MM, Arber N, Wittes J (2006) Effect of celecoxib on cardiovascular events and blood pressure in two trials for the prevention of colorectal adenomas. Circulation 114(10):1028–1035. doi:CIRCULATIONAHA.106.636746 [pii] 10.1161/CIRCULATIONAHA.106.636746
213. Kerr DJ, Dunn JA, Langman MJ, Smith JL, Midgley RS, Stanley A, Stokes JC, Julier P, Iveson C, Duvvuri R, McConkey CC (2007) Rofecoxib and cardiovascular adverse events in adjuvant treatment of colorectal cancer. N Engl J Med 357(4):360–369. doi:357/4/360 [pii] 10.1056/NEJMoa071841
214. Rostom A, Dube C, Lewin G, Tsertsvadze A, Barrowman N, Code C, Sampson M, Moher D (2007) Nonsteroidal anti-inflammatory drugs and cyclooxygenase-2 inhibitors for primary prevention of colorectal cancer: a systematic review prepared for the U.S. Preventive Services Task Force. Ann Intern Med 146(5):376–389. doi:146/5/376 [pii]
215. Chan AT, Sima CS, Zauber AG, Ridker PM, Hawk ET, Bertagnolli MM (2011) C-reactive protein and risk of colorectal adenoma according to celecoxib treatment. Cancer Prev Res (Phila) 4(8):1172–1180. doi:4/8/1172 [pii] 10.1158/1940-6207.CAPR-10-0403
216. Luo T, Yan HM, He P, Luo Y, Yang YF, Zheng H (2012) Aspirin use and breast cancer risk: a meta-analysis. Breast Cancer Res Treat 131(2):581–587. doi:10.1007/s10549-011-1747-0
217. Takkouche B, Regueira-Mendez C, Etminan M (2008) Breast cancer and use of nonsteroidal anti-inflammatory drugs: a meta-analysis. J Natl Cancer Inst 100(20):1439–1447. doi:djn324 [pii] 10.1093/jnci/djn324
218. Ashok V, Dash C, Rohan TE, Sprafka JM, Terry PD (2011) Selective cyclooxygenase-2 (COX-2) inhibitors and breast cancer risk. Breast 20(1):66–70. doi:S0960-9776(10)00176-1 [pii] 10.1016/j.breast.2010.07.004
219. Brasky TM, Bonner MR, Moysich KB, Ambrosone CB, Nie J, Tao MH, Edge SB, Kallakury BV, Marian C, Trevisan M, Shields PG, Freudenheim JL (2010) Non-steroidal anti-inflammatory drug (NSAID) use and breast cancer risk in the Western New York Exposures and Breast Cancer (WEB) Study. Cancer Causes Control 21(9):1503–1512. doi:10.1007/s10552-010-9579-5
220. Harris RE, Beebe-Donk J, Alshafie GA (2006) Reduction in the risk of human breast cancer by selective cyclooxygenase-2 (COX-2) inhibitors. BMC Cancer 6:27. doi:1471-2407-6-27 [pii] 10.1186/1471-2407-6-27
221. Harris RE, Chlebowski RT, Jackson RD, Frid DJ, Ascenseo JL, Anderson G, Loar A, Rodabough RJ, White E, McTiernan A (2003) Breast cancer and nonsteroidal anti-inflammatory drugs: prospective results from the Women's Health Initiative. Cancer Res 63(18):6096–6101
222. Terry MB, Gammon MD, Zhang FF, Tawfik H, Teitelbaum SL, Britton JA, Subbaramaiah K, Dannenberg AJ, Neugut AI (2004) Association of frequency and duration of aspirin use and hormone receptor status with breast cancer risk. JAMA 291(20):2433–2440
223. Bosco JL, Palmer JR, Boggs DA, Hatch EE, Rosenberg L (2011) Regular aspirin use and breast cancer risk in US Black women. Cancer Causes Control 22(11):1553–1561. doi:10.1007/s10552-011-9832-6
224. Garcia Rodriguez LA, Gonzalez-Perez A (2004) Risk of breast cancer among users of aspirin and other anti-inflammatory drugs. Br J Cancer 91(3):525–529. doi:10.1038/sj.bjc.6602003 6602003 [pii]
225. Rahme E, Ghosn J, Dasgupta K, Rajan R, Hudson M (2005) Association between frequent use of nonsteroidal anti-inflammatory drugs and breast cancer. BMC Cancer 5:159. doi:1471-2407-5-159 [pii] 10.1186/1471-2407-5-159
226. Vinogradova Y, Coupland C, Hippisley-Cox J (2011) Exposure to cyclooxygenase-2 inhibitors and risk of cancer: nested case–control studies. Br J Cancer 105(3):452–459. doi:bjc2011252 [pii] 10.1038/bjc.2011.252

227. Valsecchi ME, Pomerantz SC, Jaslow R, Tester W (2009) Reduced risk of bone metastasis for patients with breast cancer who use COX-2 inhibitors. Clin Breast Cancer 9(4):225–230. doi:S1526-8209(11)70672-3 [pii] 10.3816/CBC.2009.n.038
228. Brasky TM, Bonner MR, Moysich KB, Ambrosone CB, Nie J, Tao MH, Edge SB, Kallakury BV, Marian C, Goerlitz DS, Trevisan M, Shields PG, Freudenheim JL (2011) Non-steroidal anti-inflammatory drugs (NSAIDs) and breast cancer risk: differences by molecular subtype. Cancer Causes Control 22(7):965–975. doi:10.1007/s10552-011-9769-9
229. Friis S, Thomassen L, Sorensen HT, Tjonneland A, Overvad K, Cronin-Fenton DP, Vogel U, McLaughlin JK, Blot WJ, Olsen JH (2008) Nonsteroidal anti-inflammatory drug use and breast cancer risk: a Danish cohort study. Eur J Cancer Prev 17(2):88–96. doi:10.1097/CEJ.0b013e3282b6fd55 00008469-200804000-00004 [pii]
230. Ready A, Velicer CM, McTiernan A, White E (2008) NSAID use and breast cancer risk in the VITAL cohort. Breast Cancer Res Treat 109(3):533–543. doi:10.1007/s10549-007-9665-x
231. Cronin-Fenton DP, Pedersen L, Lash TL, Friis S, Baron JA, Sorensen HT (2010) Prescriptions for selective cyclooxygenase-2 inhibitors, non-selective non-steroidal anti-inflammatory drugs, and risk of breast cancer in a population-based case–control study. Breast Cancer Res 12(2):R15. doi:bcr2482 [pii] 10.1186/bcr2482
232. Egan KM, Stampfer MJ, Giovannucci E, Rosner BA, Colditz GA (1996) Prospective study of regular aspirin use and the risk of breast cancer. J Natl Cancer Inst 88(14):988–993
233. Eliassen AH, Chen WY, Spiegelman D, Willett WC, Hunter DJ, Hankinson SE (2009) Use of aspirin, other nonsteroidal anti-inflammatory drugs, and acetaminophen and risk of breast cancer among premenopausal women in the Nurses' Health Study II. Arch Intern Med 169(2):115–121; discussion 121. doi:169/2/115 [pii] 10.1001/archinternmed.2008.537
234. Zhang X, Smith-Warner SA, Collins LC, Rosner B, Willett WC, Hankinson SE (2012) Use of aspirin, other nonsteroidal anti-inflammatory drugs, and acetaminophen and postmenopausal breast cancer incidence. J Clin Oncol 30(28):3468–3477. doi:JCO.2012.42.2006 [pii] 10.1200/JCO.2012.42.2006
235. Jacobs EJ, Thun MJ, Connell CJ, Rodriguez C, Henley SJ, Feigelson HS, Patel AV, Flanders WD, Calle EE (2005) Aspirin and other nonsteroidal anti-inflammatory drugs and breast cancer incidence in a large U.S. cohort. Cancer Epidemiol Biomarkers Prev 14(1):261–264
236. Gierach GL, Lacey JV Jr, Schatzkin A, Leitzmann MF, Richesson D, Hollenbeck AR, Brinton LA (2008) Nonsteroidal anti-inflammatory drugs and breast cancer risk in the National Institutes of Health-AARP Diet and Health Study. Breast Cancer Res 10(2):R38
237. Marshall SF, Bernstein L, Anton-Culver H, Deapen D, Horn-Ross PL, Mohrenweiser H, Peel D, Pinder R, Purdie DM, Reynolds P, Stram D, West D, Wright WE, Ziogas A, Ross RK (2005) Nonsteroidal anti-inflammatory drug use and breast cancer risk by stage and hormone receptor status. J Natl Cancer Inst 97(11):805–812
238. Gill JK, Maskarinec G, Wilkens LR, Pike MC, Henderson BE, Kolonel LN (2007) Nonsteroidal antiinflammatory drugs and breast cancer risk: the multiethnic cohort. Am J Epidemiol 166(10):1150–1158. doi:kwm195 [pii] 10.1093/aje/kwm195
239. Zhang Y, Coogan PF, Palmer JR, Strom BL, Rosenberg L (2005) Use of nonsteroidal antiinflammatory drugs and risk of breast cancer: the Case–Control Surveillance Study revisited. Am J Epidemiol 162(2):165–170. doi:kwi182 [pii] 10.1093/aje/kwi182
240. Gallicchio L, Visvanathan K, Burke A, Hoffman SC, Helzlsouer KJ (2007) Nonsteroidal anti-inflammatory drugs and the risk of developing breast cancer in a population-based prospective cohort study in Washington County, MD. Int J Cancer 121(1):211–215. doi:10.1002/ijc.22656
241. Kirsh VA, Kreiger N, Cotterchio M, Sloan M, Theis B (2007) Nonsteroidal antiinflammatory drug use and breast cancer risk: subgroup findings. Am J Epidemiol 166(6):709–716. doi:kwm216 [pii] 10.1093/aje/kwm216
242. Bardia A, Olson JE, Vachon CM, Lazovich D, Vierkant RA, Wang AH, Limburg PJ, Anderson KE, Cerhan JR (2011) Effect of aspirin and other NSAIDs on postmenopausal breast cancer incidence by hormone receptor status: results from a prospective cohort study. Breast Cancer Res Treat 126(1):149–155. doi:10.1007/s10549-010-1074-x

243. Blair CK, Sweeney C, Anderson KE, Folsom AR (2007) NSAID use and survival after breast cancer diagnosis in post-menopausal women. Breast Cancer Res Treat 101(2):191–197. doi:10.1007/s10549-006-9277-x
244. Holmes MD, Chen WY, Li L, Hertzmark E, Spiegelman D, Hankinson SE (2010) Aspirin intake and survival after breast cancer. J Clin Oncol 28(9):1467–1472. doi:JCO.2009.22.7918 [pii] 10.1200/JCO.2009.22.7918
245. Kwan ML, Habel LA, Slattery ML, Caan B (2007) NSAIDs and breast cancer recurrence in a prospective cohort study. Cancer Causes Control 18(6):613–620. doi:10.1007/s10552-007-9003-y
246. Li Y, Brasky TM, Nie J, Ambrosone CB, McCann SE, Shields PG, Trevisan M, Edge SB, Freudenheim JL (2012) Use of nonsteroidal anti-inflammatory drugs and survival following breast cancer diagnosis. Cancer Epidemiol Biomarkers Prev 21(1):239–242. doi:1055–9965. EPI-11-1012 [pii] 10.1158/1055-9965.EPI-11-1012
247. Wernli KJ, Hampton JM, Trentham-Dietz A, Newcomb PA (2011) Use of antidepressants and NSAIDs in relation to mortality in long-term breast cancer survivors. Pharmacoepidemiol Drug Saf 20(2):131–137. doi:10.1002/pds.2064
248. Canney PA, Machin MA, Curto J (2006) A feasibility study of the efficacy and tolerability of the combination of Exemestane with the COX-2 inhibitor Celecoxib in post-menopausal patients with advanced breast cancer. Eur J Cancer 42(16):2751–2756. doi:S0959-8049(06)00723-4 [pii] 10.1016/j.ejca.2006.08.014
249. Dang CT, Dannenberg AJ, Subbaramaiah K, Dickler MN, Moasser MM, Seidman AD, D'Andrea GM, Theodoulou M, Panageas KS, Norton L, Hudis CA (2004) Phase II study of celecoxib and trastuzumab in metastatic breast cancer patients who have progressed after prior trastuzumab-based treatments. Clin Cancer Res 10(12 pt 1):4062–4067. doi:10.1158/1078-0432.CCR-03-0463 10/12/4062 [pii]
250. Dirix LY, Ignacio J, Nag S, Bapsy P, Gomez H, Raghunadharao D, Paridaens R, Jones S, Falcon S, Carpentieri M, Abbattista A, Lobelle JP (2008) Treatment of advanced hormone-sensitive breast cancer in postmenopausal women with exemestane alone or in combination with celecoxib. J Clin Oncol 26(8):1253–1259. doi:26/8/1253 [pii] 10.1200/JCO.2007.13.3744
251. Fabi A, Metro G, Papaldo P, Mottolese M, Melucci E, Carlini P, Sperduti I, Russillo M, Gelibter A, Ferretti G, Tomao S, Milella M, Cognetti F (2008) Impact of celecoxib on capecitabine tolerability and activity in pretreated metastatic breast cancer: results of a phase II study with biomarker evaluation. Cancer Chemother Pharmacol 62(4):717–725. doi:10.1007/s00280-007-0650-1
252. Chow LW, Yip AY, Loo WT, Lam CK, Toi M (2008) Celecoxib anti-aromatase neoadjuvant (CAAN) trial for locally advanced breast cancer. J Steroid Biochem Mol Biol 111(1–2):13–17. doi:S0960-0760(08)00089-7 [pii] 10.1016/j.jsbmb.2008.04.004
253. Sauter ER, Qin W, Schlatter L, Hewett JE, Flynn JT (2006) Celecoxib decreases prostaglandin E2 concentrations in nipple aspirate fluid from high risk postmenopausal women and women with breast cancer. BMC Cancer 6:248. doi:1471-2407-6-248 [pii] 10.1186/1471-2407-6-248
254. Bundred NJ, Cramer A, Morris J, Renshaw L, Cheung KL, Flint P, Johnson R, Young O, Landberg G, Grassby S, Turner L, Baildam A, Barr L, Dixon JM (2010) Cyclooxygenase-2 inhibition does not improve the reduction in ductal carcinoma in situ proliferation with aromatase inhibitor therapy: results of the ERISAC randomized placebo-controlled trial. Clin Cancer Res 16(5):1605–1612. doi:1078–0432.CCR-09-1623 [pii] 10.1158/1078-0432.CCR-09-1623
255. Martin LA, Davies GL, Weigel MT, Betambeau N, Hills MJ, Salter J, Walsh G, A'Hern R, Dowsett M (2010) Pre-surgical study of the biological effects of the selective cyclo-oxygenase-2 inhibitor celecoxib in patients with primary breast cancer. Breast Cancer Res Treat 123(3):829–836. doi:10.1007/s10549-010-1100-z
256. Lustberg MB, Povoski SP, Zhao W, Ziegler RM, Sugimoto Y, Ruppert AS, Lehman AM, Shiels DR, Mrozek E, Ramaswamy B, Layman RM, Brueggemeier RW, Shapiro CL (2011) Phase II trial of neoadjuvant exemestane in combination with celecoxib in postmenopausal women who have breast cancer. Clin Breast Cancer 11(4):221–227. doi:S1526-8209(11)00035-8 [pii] 10.1016/j.clbc.2011.03.022

257. Balkwill F, Mantovani A (2001) Inflammation and cancer: back to Virchow? Lancet 357(9255):539–545. doi:S0140-6736(00)04046-0 [pii] 10.1016/S0140-6736(00)04046-0
258. Harvey AE, Lashinger LM, Hursting SD (2011) The growing challenge of obesity and cancer: an inflammatory issue. Ann N Y Acad Sci 1229:45–52. doi:10.1111/j.1749-6632.2011.06096.x
259. Pollak M (2008) Insulin and insulin-like growth factor signalling in neoplasia. Nat Rev Cancer 8(12):915–928. doi:nrc2536 [pii] 10.1038/nrc2536
260. Wong KK, Engelman JA, Cantley LC (2010) Targeting the PI3K signaling pathway in cancer. Curr Opin Genet Dev 20(1):87–90. doi:S0959-437X(09)00178-6 [pii] 10.1016/j.gde.2009.11.002
261. Desbois-Mouthon C, Cadoret A, Blivet-Van Eggelpoel MJ, Bertrand F, Cherqui G, Perret C, Capeau J (2001) Insulin and IGF-1 stimulate the beta-catenin pathway through two signalling cascades involving GSK-3beta inhibition and Ras activation. Oncogene 20(2):252–259. doi:10.1038/sj.onc.1204064
262. Giles RH, van Es JH, Clevers H (2003) Caught up in a Wnt storm: Wnt signaling in cancer. Biochim Biophys Acta 1653(1):1–24
263. Cleary MP, Grossmann ME, Ray A (2010) Effect of obesity on breast cancer development. Vet Pathol 47(2):202–213. doi:0300985809357753 [pii] 10.1177/0300985809357753
264. Arkan MC, Hevener AL, Greten FR, Maeda S, Li ZW, Long JM, Wynshaw-Boris A, Poli G, Olefsky J, Karin M (2005) IKK-beta links inflammation to obesity-induced insulin resistance. Nat Med 11(2):191–198. doi:nm1185 [pii] 10.1038/nm1185
265. Cai D, Yuan M, Frantz DF, Melendez PA, Hansen L, Lee J, Shoelson SE (2005) Local and systemic insulin resistance resulting from hepatic activation of IKK-beta and NF-kappaB. Nat Med 11(2):183–190. doi:nm1166 [pii] 10.1038/nm1166
266. Chiang SH, Bazuine M, Lumeng CN, Geletka LM, Mowers J, White NM, Ma JT, Zhou J, Qi N, Westcott D, Delproposto JB, Blackwell TS, Yull FE, Saltiel AR (2009) The protein kinase IKKepsilon regulates energy balance in obese mice. Cell 138(5):961–975. doi:S0092-8674(09)00793-4 [pii] 10.1016/j.cell.2009.06.046
267. Yuan M, Konstantopoulos N, Lee J, Hansen L, Li ZW, Karin M, Shoelson SE (2001) Reversal of obesity- and diet-induced insulin resistance with salicylates or targeted disruption of Ikkbeta. Science 293(5535):1673–1677. doi:10.1126/science.1061620 293/5535/1673 [pii]
268. Laplante M, Sabatini DM (2012) mTOR signaling in growth control and disease. Cell 149(2):274–293. doi:S0092-8674(12)00351-0 [pii] 10.1016/j.cell.2012.03.017
269. Piazza GA, Alberts DS, Hixson LJ, Paranka NS, Li H, Finn T, Bogert C, Guillen JM, Brendel K, Gross PH, Sperl G, Ritchie J, Burt RW, Ellsworth L, Ahnen DJ, Pamukcu R (1997) Sulindac sulfone inhibits azoxymethane-induced colon carcinogenesis in rats without reducing prostaglandin levels. Cancer Res 57(14):2909–2915
270. Elder DJ, Halton DE, Hague A, Paraskeva C (1997) Induction of apoptotic cell death in human colorectal carcinoma cell lines by a cyclooxygenase-2 (COX-2)-selective nonsteroidal anti- inflammatory drug: independence from COX-2 protein expression. Clin Cancer Res 3(10):1679–1683
271. Hanif R, Pittas A, Feng Y, Koutsos MI, Qiao L, Staiano-Coico L, Shiff SI, Rigas B (1996) Effects of nonsteroidal anti-inflammatory drugs on proliferation and on induction of apoptosis in colon cancer cells by a prostaglandin-independent pathway. Biochem Pharmacol 52(2):237–245
272. Zhang X, Morham SG, Langenbach R, Young DA (1999) Malignant transformation and antineoplastic actions of nonsteroidal antiinflammatory drugs (NSAIDs) on cyclooxygenase-null embryo fibroblasts. J Exp Med 190(4):451–459
273. Sheng H, Shao J, Morrow JD, Beauchamp RD, DuBois RN (1998) Modulation of apoptosis and Bcl-2 expression by prostaglandin E2 in human colon cancer cells. Cancer Res 58(2):362–366
274. Chan TA, Morin PJ, Vogelstein B, Kinzler KW (1998) Mechanisms underlying nonsteroidal antiinflammatory drug-mediated apoptosis. Proc Natl Acad Sci U S A 95(2):681–686

275. Cao Y, Pearman AT, Zimmerman GA, McIntyre TM, Prescott SM (2000) Intracellular unesterified arachidonic acid signals apoptosis. Proc Natl Acad Sci U S A 97(21):11280–11285
276. Gupta RA, Tan J, Krause WF, Geraci MW, Willson TM, Dey SK, DuBois RN (2000) Prostacyclin-mediated activation of peroxisome proliferator-activated receptor delta in colorectal cancer. Proc Natl Acad Sci U S A 97(24):13275–13280. doi:10.1073/pnas.97.24.13275 97/24/13275 [pii]
277. Cutler NS, Graves-Deal R, LaFleur BJ, Gao Z, Boman BM, Whitehead RH, Terry E, Morrow JD, Coffey RJ (2003) Stromal production of prostacyclin confers an antiapoptotic effect to colonic epithelial cells. Cancer Res 63(8):1748–1751
278. Piazza GA, Rahm AL, Krutzsch M, Sperl G, Paranka NS, Gross PH, Brendel K, Burt RW, Alberts DS, Pamukcu R, Ahnen DJ (1995) Antineoplastic drugs sulindac sulfide and sulfone inhibit cell growth by inducing apoptosis. Cancer Res 55(14):3110–3116
279. Shureiqi I, Chen D, Lee JJ, Yang P, Newman RA, Brenner DE, Lotan R, Fischer SM, Lippman SM (2000) 15-LOX-1: a novel molecular target of nonsteroidal anti-inflammatory drug-induced apoptosis in colorectal cancer cells. J Natl Cancer Inst 92(14):1136–1142
280. Wu J, Xia HH, Tu SP, Fan DM, Lin MC, Kung HF, Lam SK, Wong BC (2003) 15-Lipoxygenase-1 mediates cyclooxygenase-2 inhibitor-induced apoptosis in gastric cancer. Carcinogenesis 24(2):243–247
281. Baek SJ, Kim KS, Nixon JB, Wilson LC, Eling TE (2001) Cyclooxygenase inhibitors regulate the expression of a TGF-beta superfamily member that has proapoptotic and antitumorigenic activities. Mol Pharmacol 59(4):901–908
282. Zhang Z, DuBois RN (2000) Par-4, a proapoptotic gene, is regulated by NSAIDs in human colon carcinoma cells. Gastroenterology 118(6):1012–1017. doi:S0016508500948704 [pii]
283. He TC, Chan TA, Vogelstein B, Kinzler KW (1999) PPARdelta is an APC-regulated target of nonsteroidal anti-inflammatory drugs. Cell 99(3):335–345. doi:S0092-8674(00)81664-5 [pii]
284. Arico S, Pattingre S, Bauvy C, Gane P, Barbat A, Codogno P, Ogier-Denis E (2002) Celecoxib induces apoptosis by inhibiting 3-phosphoinositide-dependent protein kinase-1 activity in the human colon cancer HT-29 cell line. J Biol Chem 277(31):27613–27621. doi:10.1074/jbc.M201119200 M201119200 [pii]
285. Liou JY, Ghelani D, Yeh S, Wu KK (2007) Nonsteroidal anti-inflammatory drugs induce colorectal cancer cell apoptosis by suppressing 14-3-3epsilon. Cancer Res 67(7):3185–3191. doi:67/7/3185 [pii] 10.1158/0008-5472.CAN-06-3431
286. Ouyang N, Williams JL, Rigas B (2006) NO-donating aspirin isomers downregulate peroxisome proliferator-activated receptor (PPAR)delta expression in APC(min/+) mice proportionally to their tumor inhibitory effect: implications for the role of PPARdelta in carcinogenesis. Carcinogenesis 27(2):232–239. doi:bgi221 [pii] 10.1093/carcin/bgi221
287. Shureiqi I, Jiang W, Zuo X, Wu Y, Stimmel JB, Leesnitzer LM, Morris JS, Fan HZ, Fischer SM, Lippman SM (2003) The 15-lipoxygenase-1 product 13-S-hydroxyoctadecadienoic acid down-regulates PPAR-delta to induce apoptosis in colorectal cancer cells. Proc Natl Acad Sci U S A 100(17):9968–9973. doi:10.1073/pnas.1631086100 1631086100 [pii]
288. Qiu W, Wang X, Leibowitz B, Liu H, Barker N, Okada H, Oue N, Yasui W, Clevers H, Schoen RE, Yu J, Zhang L (2010) Chemoprevention by nonsteroidal anti-inflammatory drugs eliminates oncogenic intestinal stem cells via SMAC-dependent apoptosis. Proc Natl Acad Sci U S A 107(46):20027–20032. doi:1010430107 [pii] 10.1073/pnas.1010430107
289. Liou JY, Ellent DP, Lee S, Goldsby J, Ko BS, Matijevic N, Huang JC, Wu KK (2007) Cyclooxygenase-2-derived prostaglandin e2 protects mouse embryonic stem cells from apoptosis. Stem Cells 25(5):1096–1103. doi:2006–0505 [pii] 10.1634/stemcells.2006-0505
290. Boon EM, Keller JJ, Wormhoudt TA, Giardiello FM, Offerhaus GJ, van der Neut R, Pals ST (2004) Sulindac targets nuclear beta-catenin accumulation and Wnt signalling in adenomas of patients with familial adenomatous polyposis and in human colorectal cancer cell lines. Br J Cancer 90(1):224–229. doi:10.1038/sj.bjc.6601505 6601505 [pii]

291. Bos CL, Kodach LL, van den Brink GR, Diks SH, van Santen MM, Richel DJ, Peppelenbosch MP, Hardwick JC (2006) Effect of aspirin on the Wnt/beta-catenin pathway is mediated via protein phosphatase 2A. Oncogene 25(49):6447–6456. doi:1209658 [pii] 10.1038/sj.onc.1209658
292. Dihlmann S, Siermann A, von Knebel Doeberitz M (2001) The nonsteroidal anti-inflammatory drugs aspirin and indomethacin attenuate beta-catenin/TCF-4 signaling. Oncogene 20(5):645–653. doi:10.1038/sj.onc.1204123
293. Greenspan EJ, Madigan JP, Boardman LA, Rosenberg DW (2011) Ibuprofen inhibits activation of nuclear {beta}-catenin in human colon adenomas and induces the phosphorylation of GSK-3{beta}. Cancer Prev Res (Phila) 4(1):161–171. doi:4/1/161 [pii] 10.1158/1940-6207.CAPR-10-0021
294. Shao J, Jung C, Liu C, Sheng H (2005) Prostaglandin E2 stimulates the beta-catenin/T cell factor-dependent transcription in colon cancer. J Biol Chem 280(28):26565–26572. doi:M413056200 [pii] 10.1074/jbc.M413056200
295. Thompson WJ, Piazza GA, Li H, Liu L, Fetter J, Zhu B, Sperl G, Ahnen D, Pamukcu R (2000) Exisulind induction of apoptosis involves guanosine 3′,5′-cyclic monophosphate phosphodiesterase inhibition, protein kinase G activation, and attenuated beta-catenin. Cancer Res 60(13):3338–3342
296. Tinsley HN, Gary BD, Keeton AB, Zhang W, Abadi AH, Reynolds RC, Piazza GA (2009) Sulindac sulfide selectively inhibits growth and induces apoptosis of human breast tumor cells by phosphodiesterase 5 inhibition, elevation of cyclic GMP, and activation of protein kinase G. Mol Cancer Ther 8(12):3331–3340. doi:1535–7163.MCT-09-0758 [pii] 10.1158/1535-7163.MCT-09-0758
297. Tinsley HN, Gary BD, Thaiparambil J, Li N, Lu W, Li Y, Maxuitenko YY, Keeton AB, Piazza GA (2010) Colon tumor cell growth-inhibitory activity of sulindac sulfide and other nonsteroidal anti-inflammatory drugs is associated with phosphodiesterase 5 inhibition. Cancer Prev Res (Phila) 3(10):1303–1313. doi:1940–6207.CAPR-10-0030 [pii] 10.1158/1940-6207.CAPR-10-0030
298. Harris RM, Hawker RJ, Langman MJ, Singh S, Waring RH (1998) Inhibition of phenolsulphotransferase by salicylic acid: a possible mechanism by which aspirin may reduce carcinogenesis. Gut 42(2):272–275
299. Kopp E, Ghosh S (1994) Inhibition of NF-kappa B by sodium salicylate and aspirin. Science 265(5174):956–959
300. Grilli M, Pizzi M, Memo M, Spano P (1996) Neuroprotection by aspirin and sodium salicylate through blockade of NF-kappaB activation. Science 274(5291):1383–1385
301. Yamamoto Y, Yin MJ, Lin KM, Gaynor RB (1999) Sulindac inhibits activation of the NF-kappaB pathway. J Biol Chem 274(38):27307–27314
302. Yin MJ, Yamamoto Y, Gaynor RB (1998) The anti-inflammatory agents aspirin and salicylate inhibit the activity of I(kappa)B kinase-beta. Nature 396(6706):77–80. doi:10.1038/23948
303. Din FV, Stark LA, Dunlop MG (2005) Aspirin-induced nuclear translocation of NFkappaB and apoptosis in colorectal cancer is independent of p53 status and DNA mismatch repair proficiency. Br J Cancer 92(6):1137–1143. doi:6602455 [pii] 10.1038/sj.bjc.6602455
304. Loveridge CJ, MacDonald AD, Thoms HC, Dunlop MG, Stark LA (2008) The proapoptotic effects of sulindac, sulindac sulfone and indomethacin are mediated by nucleolar translocation of the RelA(p65) subunit of NF-kappaB. Oncogene 27(18):2648–2655. doi:1210891 [pii] 10.1038/sj.onc.1210891
305. Hsieh PS, Jin JS, Chiang CF, Chan PC, Chen CH, Shih KC (2009) COX-2-mediated inflammation in fat is crucial for obesity-linked insulin resistance and fatty liver. Obesity (Silver Spring) 17(6):1150–1157. doi:oby2008674 [pii] 10.1038/oby.2008.674
306. Gonzalez-Ortiz M, Martinez-Abundis E, Balcazar-Munoz BR, Robles-Cervantes JA (2001) Inhibition of cyclooxygenase-1 or -2 on insulin sensitivity in healthy subjects. Horm Metab Res 33(4):250–253

307. Hundal RS, Petersen KF, Mayerson AB, Randhawa PS, Inzucchi S, Shoelson SE, Shulman GI (2002) Mechanism by which high-dose aspirin improves glucose metabolism in type 2 diabetes. J Clin Invest 109(10):1321–1326. doi:10.1172/JCI14955
308. Fleischman A, Shoelson SE, Bernier R, Goldfine AB (2008) Salsalate improves glycemia and inflammatory parameters in obese young adults. Diabetes Care 31(2):289–294. doi:dc07-1338 [pii] 10.2337/dc07-1338
309. Shoelson SE, Lee J, Yuan M (2003) Inflammation and the IKK beta/I kappa B/NF-kappa B axis in obesity- and diet-induced insulin resistance. Int J Obes Relat Metab Disord 27(suppl 3):S49–S52. doi:10.1038/sj.ijo.0802501 0802501 [pii]
310. Goldfine AB, Silver R, Aldhahi W, Cai D, Tatro E, Lee J, Shoelson SE (2008) Use of salsalate to target inflammation in the treatment of insulin resistance and type 2 diabetes. Clin Transl Sci 1(1):36–43. doi:10.1111/j.1752-8062.2008.00026.x
311. Hawley SA, Fullerton MD, Ross FA, Schertzer JD, Chevtzoff C, Walker KJ, Peggie MW, Zibrova D, Green KA, Mustard KJ, Kemp BE, Sakamoto K, Steinberg GR, Hardie DG (2012) The ancient drug salicylate directly activates AMP-activated protein kinase. Science 336(6083):918–922. doi:science.1215327 [pii] 10.1126/science.1215327
312. Din FV, Valanciute A, Houde VP, Zibrova D, Green KA, Sakamoto K, Alessi DR, Dunlop MG (2012) Aspirin inhibits mTOR signaling, activates AMP-activated protein kinase, and induces autophagy in colorectal cancer cells. Gastroenterology 142(7):1504–1515 e1503. doi:S0016-5085(12)00313-7 [pii] 10.1053/j.gastro.2012.02.05
313. Brunelli C, Amici C, Angelini M, Fracassi C, Belardo G, Santoro MG (2012) The non-steroidal anti-inflammatory drug indomethacin activates the eIF2alpha kinase PKR, causing a translational block in human colorectal cancer cells. Biochem J 443(2):379–386. doi:BJ20111236 [pii] 10.1042/BJ20111236
314. Marimuthu S, Chivukula RS, Alfonso LF, Moridani M, Hagen FK, Bhat GJ (2011) Aspirin acetylates multiple cellular proteins in HCT-116 colon cancer cells: identification of novel targets. Int J Oncol 39(5):1273–1283. doi:10.3892/ijo.2011.1113
315. Claria J, Serhan CN (1995) Aspirin triggers previously undescribed bioactive eicosanoids by human endothelial cell-leukocyte interactions. Proc Natl Acad Sci U S A 92(21):9475–9479
316. Ruschoff J, Wallinger S, Dietmaier W, Bocker T, Brockhoff G, Hofstadter F, Fishel R (1998) Aspirin suppresses the mutator phenotype associated with hereditary nonpolyposis colorectal cancer by genetic selection. Proc Natl Acad Sci U S A 95(19):11301–11306
317. He H, Xia HH, Wang JD, Gu Q, Lin MC, Zou B, Lam SK, Chan AO, Yuen MF, Kung HF, Wong BC (2006) Inhibition of human telomerase reverse transcriptase by nonsteroidal antiinflammatory drugs in colon carcinoma. Cancer 106(6):1243–1249. doi:10.1002/cncr.21694

Chapter 12
Omega-3 Fatty Acids in Cancer Prevention and Control: A Membrane Perspective

Harmony F. Turk, Jennifer M. Monk, Tim Y. Hou, and Robert S. Chapkin

Abstract Long-chain n-3 polyunsaturated fatty acids (PUFA) have been shown to provide health benefits in a number of diseases including several forms of cancer. In this chapter we will discuss in detail some of the prominent mechanisms through which n-3 PUFA and its metabolites are believed to function in the prevention of colon tumorigenesis. At the plasma membrane, n-3 PUFA antagonize the production of inflammatory and procarcinogenic n-6 PUFA (i.e., arachidonic acid)-derived metabolites. Additionally, the highly unsaturated nature of n-3 PUFA impacts cell membrane properties and dynamics thereby altering numerous cellular functions, including intracellular signaling, cell growth, survival, and proliferation. Due to the sterically incompatible relationship between docosahexaenoic acid, sphingolipids, and cholesterol, the major constituents of lipid rafts, n-3 PUFA modulate these crucial membrane microdomains and perturb efficient signal transduction thereby eliciting the same biological effects exploited by some anti-cancer therapies. Moreover, we discuss how alterations in lipid rafts and downstream signaling impact both epithelial cells and the activation of immune cells. This is noteworthy, because chronic inflammation plays a critical role in tumorigenesis. Therefore, we also present an overview regarding the anti-inflammatory and immunomodulatory mechanisms

H.F. Turk, Ph.D. • J.M. Monk, Ph.D. • T.Y. Hou, B.S.
Program in Integrative Nutrition and Complex Diseases
and the Center for Environmental and Rural Health Texas A&M University,
College Station, TX 77843, USA
e-mail: hfturk@tamu.edu; monkjpdf@gmail.com; tim.hou@tamu.edu

R.S. Chapkin, Ph.D. (✉)
Kleberg Center, Room 218B, TAMU 2253, Texas A&M University,
College Station, TX 77843-2253, USA
e-mail: r-chapkin@tamu.edu

A.J. Dannenberg and N.A. Berger (eds.), *Obesity, Inflammation and Cancer*,
Energy Balance and Cancer 7, DOI 10.1007/978-1-4614-6819-6_12,

through which n-3 PUFA perturb the tumor microenvironment and downregulate the activation of critical transcription factors and target genes with an established role in cancer development. Finally, we discuss recent evidence suggesting that n-3 PUFA in combination with other dietary bioactive nutrients, such as soluble fiber and curcumin, could be beneficial in cancer prevention. Collectively, we demonstrate that dietary n-3 PUFA have utility in the prevention of cancer development through mechanisms centered at both the molecular, cellular (plasma membrane), and tissue level.

12.1 n-3 PUFA

Numerous health benefits have been attributed to fish oil. The most notable bioactive components of fish oil are the long-chain omega-3 polyunsaturated fatty acids (n-3 PUFA), eicosapentaenoic acid (EPA, $20{:}5^{\Delta5,8,11,14,17}$), and docosahexaenoic acid (DHA, $22{:}6^{\Delta4,7,10,13,16,19}$). Fatty acids that have the last double-bond three carbons from the terminal methyl group are characterized as n-3 fatty acids. Mammals lack the desaturases that are required to synthesize n-3 PUFA, so they must be acquired through the diet. The major dietary source of long-chain n-3 PUFA is fatty, cold-water fish, which ingest zooplankton and phytoplankton rich in EPA and DHA [1]. Another shorter-chain n-3 PUFA, α-linolenic acid (α-LNA, $18{:}3^{\Delta9,12,15}$), can be obtained from plant sources, including walnuts, seeds (flaxseed), dark green leafy vegetables, and soybeans. Humans express the enzymes required for conversion of α-LNA to EPA and EPA to DHA; however, conversion rates are extremely low. Kinetic analyses suggest that conversion of α-LNA to EPA to be as low as 0.2 %. Further conversion to DHA is calculated to be as low as 0.05 % [2]. Therefore, the optimal way to increase tissue levels of long-chain n-3 PUFA is by directly ingesting foods or supplements high in both EPA and DHA.

12.2 n-3 PUFA in Cancer

As mentioned above, EPA and DHA have been shown to provide health benefits in a number of diseases. Of interest, there is growing evidence that n-3 PUFA may prevent multiple forms of cancer, including some of the top causes of cancer mortality: breast, lung, prostate, and colon cancer [3–6]. In this chapter, we will focus on the evidence for the role of n-3 PUFA in the prevention and treatment of colon cancer. Importantly, many of the mechanisms of action of n-3 PUFA that we will discuss are common between multiple cancer types.

12.3 Human Data

12.3.1 Observational Studies

It is important to first appreciate the mounting human data documenting the role on n-3 PUFA in colon cancer prevention. Numerous epidemiological studies have demonstrated distinct trends in which an increase in consumption of n-3 PUFA is associated with reduced risk of colon cancer. Studies have established a significant risk reduction of colorectal cancer (CRC) with increased consumption of fish, n-3 PUFA, and DHA or EPA [7–9]. A recent study found that dietary intake of marine n-3 PUFA was associated with a decreased risk of adenomatous polyps in women [10], and another described an inverse relationship between the intake of n-3 PUFA and the risk of colon cancer in the proximal colon in both men and women [11]. Additionally, a Scottish study described a dose-dependent reduction in colon cancer risk with increased intake of total n-3 PUFA, EPA alone, and DHA alone [12]. A 22-year prospective study established that intake of fish and fatty acids from fish may decrease the risk of colon cancer [13]. Similarly, a meta-analysis of prospective cohort studies additionally found that fish consumption reduces colon cancer risk [14]. Furthermore, data from 24 European countries established an inverse correlation between fish and fish oil consumption with CRC mortality [4]. Other studies have shown that individuals with the highest percentage of circulating n-3 PUFA display the lowest risk of colon cancer [15–17]. Historically, a very low prevalence of colon cancer existed in Asian countries, which was attributed to a high consumption of fish. However, recent lifestyle changes of these populations have resulted in similar rates of colon cancer to the population of the United States [18]. These trends indicate a role of diet in colon cancer risk.

12.3.2 Clinical Studies

Due to the abundance of data suggesting a protective role of n-3 PUFA, numerous clinical trials have been undertaken to further clarify these effects. A short-term trial demonstrated that fish oil reduces mucosal proliferation, which may protect high-risk subjects from colon cancer [19]. Additionally, a double-blind, cross-over trial further indicated a reduction in mucosal proliferation as well as reduced PGE_2 synthesis in test subjects receiving fish oil [20]. Another study conducted by Anti et al. demonstrated that fish oil supplementation has both short-term and long-term normalizing effects on the abnormal proliferation patterns associated with increased colon cancer risk [21]. Very few studies have focused specifically on purified DHA or EPA. However, two studies have shown a positive effect of purified EPA. One of these investigations found that highly purified EPA reduces crypt cell proliferation and increases apoptosis in subjects with a history of colorectal adenomas [22]. In a complementary study, the administration of EPA to a population of genetically predisposed subjects for colon cancer exhibited a reduction in the number and diameter of colonic polyps, which can serve as precursors to colon cancer [23]. Overall, human clinical trials have demonstrated a potential use of n-3 PUFA in colon cancer prevention.

12.3.3 Nutrigenetics

Although many studies have demonstrated the beneficial effect of n-3 PUFA in regard to colon tumorigenesis, a subset of studies have failed to show any effect. The answer to some of these differences may lie in the field of nutrigenetics. Nutrigenetics is the study of how certain genes are affected by nutrients, especially with regard to disease. Recently, it has been reported that a single nucleotide polymorphisms (SNPs) within a DNA repair gene modified the effect of marine n-3 PUFA on colon cancer risk [24]. Further developments in this field of research will likely enhance our understanding of the role of n-3 PUFA in colon cancer prevention.

In cancer, many different abnormalities can contribute to the formation of a tumor. Therefore, not all patients will respond to the same type of therapy. If a nutrient is going to be effective for prevention or treatment of cancer, the mechanism of action will elucidate the patients most likely to respond to a given therapy. Therefore, we will now discuss the mechanisms by which n-3 PUFA are believed to function in the prevention of colon tumorigenesis.

12.4 Membrane Effects of n-3 PUFA

Multiple studies have been undertaken to elucidate the mechanisms of action of n-3 PUFA, and the current literature contains evidence of various interactions to explain the antineoplastic activity of DHA and EPA. Of interest, long-chain n-3 PUFA have been shown to alter membrane dynamics, which could explain the pleiotropic effects of these fatty acids. DHA and EPA are both long, highly unsaturated fatty acids, which can directly affect many of the properties of the plasma membrane. Modification to cell membranes can extensively alter numerous cellular functions, including cellular signaling, growth, survival, and proliferation. For example, both EPA and DHA have been shown to increase membrane fluidity [25, 26], and membrane fluidity can directly affect protein diffusion within the membrane. This can then modify multiple cell signaling events which require specific spatiotemporal regulation of protein–protein interactions. n-3 PUFA have additionally been shown to alter permeability, phase behavior, fusion, and flip-flop of the plasma membrane [26, 27]. These effects on the membrane directly contribute to the ability of n-3 PUFA to alter important membrane microdomains. In fact, recent evidence clearly demonstrates that EPA and DHA can perturb highly crucial microdomains known as lipid rafts [28–31].

Lipid rafts are extremely small membrane microdomains that can vary between 5 and 200 nm [32]. Lipid rafts are further characterized as dynamic plasma membrane domains enriched with cholesterol and sphingolipids [33]. These domains contain highly ordered lipid assemblies and exhibit liquid-ordered characteristics due to the interaction between cholesterol and sphingolipids. Furthermore, lipids that are localized to these domains comprise mostly saturated, long hydrocarbon chains and hydroxylated ceramide backbones [34, 35]. The composition of lipid rafts imparts two distinguishing characteristics: detergent-resistance and low-buoyant density [36]. These qualities have been exploited by researchers to explore the protein composition of lipid rafts. It is important to note that lipid rafts are highly heterogeneous [37]. Additionally, the structure of lipid rafts is decidedly dynamic, and lipid rafts constantly alter their composition of both lipids and proteins. These properties of rafts facilitate the ability of cells to respond to external stimuli and activate intracellular signaling pathways that originate at the cell surface.

The dynamics and heterogeneity of lipid rafts allow for a large number of signals to be transduced from the extracellular environment to intracellular domains. Membrane rafts play a fundamental role in mediating multiple cell functions, including signal transduction [38–42]. Many proteins that function as signaling partners accumulate in lipid rafts, which serve as a platform to support highly efficient signal transduction. Although lipid rafts are small, they can be stimulated to coalesce and form large, stable platforms for signaling [43, 44]. Many important signaling proteins have been shown to be highly enriched in lipid raft domains, including epidermal growth factor receptor (EGFR), G-protein-coupled receptors (GPCR), platelet-derived growth factor receptor (PDGFR), nerve growth factor receptor (NGFR), vascular endothelial growth factor receptor (VEGFR), H-Ras, and many more [33, 45–47]. In many cases, the function of proteins depends greatly on their localization within lipid rafts [48]. These lipid raft-mediated signaling events are essential for a plethora of cell functions.

Lipid rafts have been found to participate in a variety of cellular events. For instance, extensive studies have been performed to elucidate the role of lipid rafts in the activation of T lymphocytes, which requires clustering of signaling components in a large, stable lipid raft domain at the immunological synapse (IS, [49, 50]). Furthermore, extrinsic induction of apoptosis relies on signaling through lipid rafts, and the composition of lipid rafts has been shown to affect apoptotic responses [48, 51, 52]. Of interest, recent evidence suggests that lipid rafts may modulate the malignant transformation process. In fact, the levels of lipid rafts are increased in many types of cancer [53–55]. Additionally, lipid rafts mediate many of the cell signaling events that are often constitutively or hyperactivated in cancer [56–61], including EGFR, Akt, Ras, Src, and HER2. Research suggests that disruption of lipid rafts may enhance responsiveness to anti-cancer therapies [62]. Furthermore, some anti-cancer drugs have beneficial effects through alteration of the protein content of lipid rafts [48]. In colon cancer, lipid rafts have been shown to function in cell death-mediated signaling [63, 64], entry of bioactive compounds [65], and localization of key proteins involved in immune response [66].

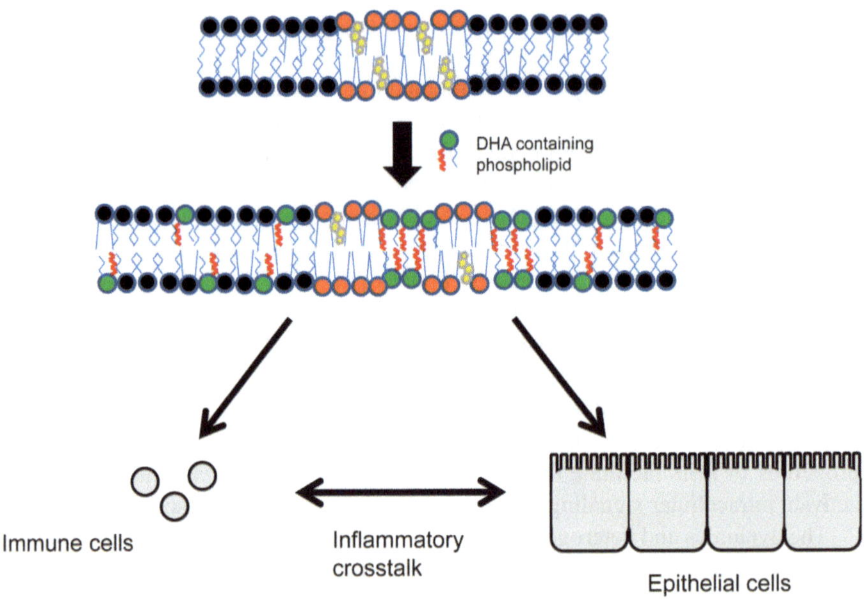

Fig. 12.1 Cross-talk between immune and epithelial cells in inflammation and cancer. Immune cells, e.g., T lymphocytes and epithelial cells engage in constant cross-talk to monitor and survey the physiological environment for foreign antigens and protect the host. Since EPA- and DHA-containing phospholipids are incorporated into membranes of many cell types and affect their signaling, understanding the molecular mechanisms by which n-3 PUFA impact signaling in both immune and epithelial cells will serve to elucidate the effects of n-3 PUFA on inflammation and carcinogenesis

Extensive studies have demonstrated that DHA, due to its high degree of unsaturation, is sterically incompatible with cholesterol [27], a major constituent of rafts. Additionally, DHA displays nonideal mixing with sphingomyelin, another major component of lipid rafts, in model membranes [31]. Conceivably, disruption of lipid rafts could be caused by the lack of affinity of n-3 PUFA for lipid raft components. As mentioned above, several critical processes involve lipid rafts, including T cell activation, signal transduction, and protein and lipid trafficking [33]. Many of these lipid raft-mediated processes play an integral role in colon tumorigenesis. Signaling pathways emanating from lipid rafts, frequently exacerbated in cancer, mediate a variety of tumor-promoting activities, including cell proliferation, migration, and invasion [55]. Additionally, chronic inflammation, central to the process of tumorigenesis [67], involves excessive lipid raft-mediated T cell activation. Numerous recent discoveries highlight the role of n-3 PUFA in the regulation of lipid rafts and lipid raft-mediated signaling in both immune cells and epithelial cells. Due to the extensive cross-talk between immune and epithelial cells, the ability of n-3 PUFA to alter signaling in both cell types likely contributes to a greater effect with respect to cancer prevention than targeting either cell type on its own (Fig. 12.1).

12.5 Effect of n-3 PUFA on Epithelial Cell Lipid Rafts

12.5.1 Epithelial Raft Size

Lipid raft size is an important feature in lipid raft function [68] and has been shown to be altered by n-3 PUFA [28]. Lipid raft size is integral for dynamic lateral segregation of signaling proteins into microdomains. Partitioning of proteins into rafts can increase specific protein–protein collision rates to facilitate efficient signaling. However, to maximize this essential, biologically relevant function, rafts must be mobile and small, with a diameter up to 14 nm [68]. Therefore, altering the size of a lipid raft would likely have adverse effects on signaling. In HeLa cells and T lymphocytes, treatment with DHA resulted in enhanced clustering of lipid raft domains compared to untreated cells [28]. Additionally, DHA treatment was found to increase the height of lipid rafts, while reducing the overall number of lipid rafts [69]. Altering lipid raft size has implications with regard to the regulation of multiple signaling events associated with these microdomains.

12.5.2 Lipid Raft Composition

In addition to the size of lipid rafts, the composition of rafts is important for their function. Interestingly, n-3 PUFA have been found to alter the composition of lipid rafts. Upon feeding mice a diet enriched in n-3 PUFA, the cholesterol content of lipid rafts in colonocytes was reduced by 46 % compared to mice fed a diet enriched in n-6 PUFA [70]. Furthermore, treatment of both endothelial and breast cancer cells with DHA was also found to reduce raft sphingomyelin and cholesterol content [71, 72]. These n-3 PUFA-induced modifications of lipid raft composition are significant because cholesterol and sphingomyelin are major building blocks of lipid rafts that promote the formation of hydrophobic liquid-ordered molecular packing.

In addition to altering the lipid composition of rafts, n-3 PUFA have been shown to modulate raft protein composition. This effect of DHA on Ras is highly important due to the central role that Ras signaling plays in colon carcinogenesis [73]. Feeding mice a diet enriched in n-3 PUFA was found to reduce the localization of H-Ras to colonocyte lipid rafts [70]. Moreover, treating immortalized young adult mouse colonocytes (YAMC) with DHA was also found to decrease the localization of H-Ras to lipid raft domains [70]. DHA was then further shown to not only reduce the localization of H-Ras to lipid rafts but also to inhibit the plasma membrane targeting of H-Ras [74].

EGFR is a transmembrane receptor tyrosine kinase that mediates multiple oncogenic signaling pathways and is enriched in lipid raft domains. Localization of EGFR to lipid rafts is essential for efficient signal transduction. Interestingly, treatment of breast, lung, and colon cell lines with DHA or EPA was found to reduce the localization of EGFR to lipid rafts [71, 75, 76]. In addition to Ras, Src and Fyn are two other important lipid raft localized signaling mediators. Treatment of the colon cancer cell line, Caco-2, with DHA led to a delocalization of these proteins from rafts [77]. The effect of n-3 PUFA on localization of these signaling proteins to lipid rafts directly altered cell signaling.

12.5.3 *Lipid Raft-Mediated Cell Functions*

The n-3 PUFA-induced alterations to lipid raft size and composition culminate in the modification to lipid raft-mediated cell functions. For example, n-3 PUFA have been found to alter protein function and signaling. In the case of EGFR, n-3 PUFA increase EGFR phosphorylation [71, 75, 76]. EGFR phosphorylation is a canonical precursor to EGFR signaling. However, n-3 PUFA have been found to paradoxically disrupt EGFR signal transduction [75, 76]. n-3 PUFA have also been shown to reduce activation of Ras [76, 78], which is a lipid raft-mediated process [79–81]. Lipid rafts have also been shown to play a central role in the activation of STAT3 [82], another important signaling mediator that is often hyperactivated in colon cancer. Both in vivo and in vitro activation of STAT3 have been shown to be significantly reduced in colonocytes by n-3 PUFA [76, 83]. Collectively, these signaling events are important for mediating cell proliferation and apoptosis. The altered lipid raft-mediated cell signaling caused by n-3 PUFA has been shown to reduce cell proliferation and increase apoptosis in multiple epithelial cell lines [69, 76, 84]. Interestingly, calcium influx is also a lipid raft-mediated process. It has been shown that DHA decreases oxidative-stress-induced calcium influx [72]. These data signify the implicit role of n-3 PUFA in the regulation of lipid raft-mediated cell processes.

12.6 Additional Mechanisms of Action

Although there is significant evidence for the role that n-3 PUFA plays in altering the plasma membrane, there are likely multiple mechanisms by which these fatty acids function. Research has identified additional mechanisms of n-3 PUFA action. These mechanisms may be dependent or independent of the effects of n-3 PUFA on lipid rafts (described below).

12.7 The Role of Inflammation and Cancer

Inflammation plays a critical role in tumorigenesis, and an inflammatory microenvironment is an essential component of all tumors, including those in which a direct causal relationship with inflammation has not been elucidated [85]. It is well accepted that inflammation facilitates the initiation and progression of normal cells to malignancy through production of inflammatory cytokines (i.e., IL-1, TNFα, and IL-6) and an array of reactive oxygen and nitrogen species [86, 87]. Downstream, these mediators subsequently activate transcription factors (i.e., NF-κB), inducible nitrogen synthase, and cyclooxygenase-2-related signaling pathways, which generally delay or suppress apoptosis in intestinal epithelial cells and modulate angiogenesis and drug-metabolizing enzymes, especially phase II enzyme induction [87–89]. Additionally, arachidonic acid metabolites are also capable of stimulating oncogenic pathways which favor tumor growth, invasiveness, and metastasis. Therefore, anti-inflammatory agents hold promise for decreasing the incidence of colon and other epithelial cancers [90]. A prominent hallmark of cancer-related inflammation is the presence of various types of immune cells and inflammatory mediators (cytokines, chemokines, growth factors, eicosanoids, reactive oxygen, and nitrogen species) which interact with each other by direct contact or by cytokine and chemokine production (autocrine or paracrine signaling), ultimately promoting tumor cell growth, progression, and metastasis [90].

The interrelationship between chronic inflammation and carcinogenesis is perhaps best documented in the colon. CRC is the third most common malignancy and fourth most common cause of cancer mortality worldwide [91]. Approximately only 20 % of CRC cases have a familial basis, i.e., familial adenomatous polyposis and hereditary nonpolyposis CRC [92]; therefore, environmental causes rather than heritable genetic changes represent the largest contributor and modifiable risk factor for CRC development. Chronic intestinal inflammation both precedes and promotes tumor development. Chronic inflammatory conditions such as inflammatory bowel disease (IBD) are associated with highly enhanced colon mucosa carcinogenesis [93]. Colitis-associated cancer represents a CRC subtype that is associated with a high mortality rate [94] and develops over time in more than 20 % of IBD patients following disease onset [95]. Typically colitis-associated chronic inflammation promotes cancer development by inducing oxidative damage to DNA, contributing to p53 mutations in both tumor cells and the inflamed but nondysplastic epithelium [96, 97]. Experimentally, this condition is recapitulated by the combination of a procarcinogen (azoxymethane, AOM) preceding repeated cycles of dextran sodium sulfate (DSS) administration to induce chronic colonic inflammation. Importantly, n-3 PUFA exert anti-inflammatory and immunomodulatory actions through several molecular pathways crucial for the development of inflammatory processes [93].

12.7.1 COX-2-Mediated Effects

A number of studies have addressed the role that n-3 PUFA plays in regulating cyclooxygenase (COX)-2, a key enzyme involved in production of prostaglandins. Arachidonic acid (AA; 20:4 n-6) serves as the precursor to 2-series prostaglandins which are generally considered to be proinflammatory. Importantly, PGE_2, one of the most well-studied prostaglandins synthesized from AA, is directly linked to colon cancer risk (described below). Besides antagonizing AA, n-3 PUFA can serve as precursors for prostaglandin synthesis; however, the 3-series prostaglandins are anti-inflammatory [98–100]. For example, PGE_3 produced from EPA mediates multiple anti-inflammatory actions. As mentioned above, inflammation is intimately linked to colon cancer risk; therefore, formation of anti-inflammatory compounds is of critical importance in colon cancer prevention. Consumption and subsequent incorporation of fish oil (FO)-derived n-3 PUFA into inflammatory cell phospholipids occur at the expense of n-6 PUFA, thereby decreasing the amount of AA substrate available for synthesis of AA-derived eicosanoids resulting in an altered eicosanoid profile with reduced levels of 2- and 4-series prostaglandins and leukotrienes in inflammatory/immune cells [101–105]. Therefore, one anti-inflammatory mechanism of n-3 PUFA is to inhibit the production of AA-derived eicosanoids.

In several types of cancer, including most colon tumors, COX-2 is overexpressed, and subsequently nonsteroidal anti-inflammatory drugs (NSAIDs) and selective inhibitors of COX-2 (celecoxib) have been demonstrated to be highly effective colon cancer chemopreventive agents [106, 107]. Aberrant expression of COX-2 leads to an increase in PGE_2 levels, which activates the *Wnt* signaling pathway in CRC and plays a well-established role by enhancing intestinal cell proliferation, angiogenesis, cell migration, and invasion while inhibiting protective apoptotic signaling pathways [108–111]. Additionally, PGE_2 levels are elevated in human colon carcinomas compared to normal mucosa [112] and in the colonic mucosa of rats injected with AOM during the initiation and post initiation stages of carcinogenesis [113]. Interestingly, a novel null mouse model, in which tissue AA mass is reduced to near undetectable levels via systemic disruption of the *Fads1* (Δ5 desaturase) gene, exhibits reduced colon cell proliferation and suppressed PGE_2 levels in the colonic mucosa in null mice compared to wild-type littermates [114]. Future work is needed in order to determine if these mice are resistant to the induction of colon cancer.

n-3 PUFA decrease expression of COX-2 both in vitro and in vivo. In rodents, dietary n-3 PUFA has been shown to reduce COX-2 expression in the colonic mucosa [115, 116]. Moreover, n-3 PUFA have been shown to suppress inflammatory eicosanoid production by suppressing COX-2 activity [117, 118]. Additionally, *Fat-1* (omega-3 desaturase) transgenic mice have lower expression of COX-2 than wild-type littermates [119]. In two distinct human colon cancer cell lines, EPA and DHA have been shown to reduce COX-2 expression [120, 121]. Furthermore, nude mice injected with human colon cancer cells and fed either DHA or EPA exhibit a decrease in both COX-2 expression, PGE_2 synthesis, and tumor growth [120]. Additionally, a combination of DHA with celecoxib, a selective COX-2 inhibitor, was shown to decrease COX-2 expression in a human colon cancer cell line, and the combination was more effective than either treatment alone [122]. COX-2 inhibitors have been found to have beneficial effects for colon cancer, but they can have detrimental side effects that limit their applicability. Therefore, utilization of n-3 PUFA, i.e., innocuous dietary lipids with very minor side effects, could provide the benefits of COX-2 inhibitors without the medical complications.

12.7.2 Effects of n-3 PUFA Metabolites

In addition to the anti-inflammatory metabolites produced by COX-2-mediated metabolism, other enzymes can produce additional anti-inflammatory and pro-resolving compounds from n-3 PUFA. Included among these anti-inflammation and pro-resolution metabolites are resolvins and protectins, which have been shown to elicit potent anti-inflammatory effects on inflammatory/immune cells [123–125]. DHA is the precursor for the D-series resolvins and protectin D1 (PD1), whereas EPA serves as the origin for resolvins E1 (RvE1) and E2 (RvE2) [125–127]. These novel DHA- and EPA-derived metabolites were first observed during the resolution phase of acute inflammation [123, 125]. The resolvins can stimulate the resolution phase of inflammation to begin at an earlier time point, which reduces the overall exposure to inflammation [128]. Generally, resolvins actively contribute to the resolution of inflammation via the removal of inflammatory cells and the restoration of tissue integrity, and these processes can be further facilitated by the presence of aspirin [90]. RvE1 has been found to reduce inflammation, dendritic cell migration, and the production of inflammatory cytokines [129]. Furthermore, PD1 protects tissues from oxidative damage [128]. PD1 also exhibits anti-inflammatory properties by reducing neutrophil infiltration and production of proinflammatory cytokines, specifically during the resolution phase of inflammation [130]. These recently discovered metabolites of DHA and EPA serve as an additional mechanism by which dietary fatty acids can facilitate reduction of inflammation, which is an important aspect of cancer prevention.

12.8 n-3 PUFA and Transcriptional Regulation

Since the majority of factors perturbed in cancer are signaling molecules, their downstream effectors, transcription factors, hold promise for cancer therapy [131]. Moreover, oncogenic transcription factors represent a useful class of therapeutic targets since they often have increased expression and activity in a variety of cancer types. One mechanism through which n-3 PUFA exert their anti-inflammatory and anti-cancer effects is through modulation of transcription factor activity and/or expression, thereby altering the downstream gene expression profile as discussed below. Aberrant NF-κB activation has been detected in more than 50 % of colorectal and colitis-associated tumors using animal models, establishing a role for NF-κB in CRC development [132, 133]. NF-κB activation promotes tumorigenesis by inhibiting apoptosis (via activation of Bcl2, Bcl-xL cFLIP, and other genes), increasing cell proliferation, angiogenesis, and promoting cell invasion and metastasis [134, 135]. Moreover, aberrant NF-κB signaling has been proposed to be one of the mechanisms through which chronic inflammation leads to cancer [135].

Most tumor-promoting cytokines are activated by NF-κB directly or in conjunction with other inflammatory stimuli in premalignant cells and immune/inflammatory cells [136]. In this context, deletion of IKKβ within myeloid cells (i.e., dendritic cells, macrophages, and neutrophils) resulted in a 50 % decrease in tumor incidence and a substantial reduction in tumor size [137], thereby inhibiting inflammation-induced enterocyte proliferation and inflammation-stimulated growth of colitis-associated tumors. Therefore, activation of NF-κB in lamina propria-residing myeloid cells results in the production of cytokines that can act as growth factors for premalignant enterocytes, since several proinflammatory cytokines and chemokines (such as TNF, IL-1, IL-6, and CXCL8) are encoded by target genes of the IKKβ-dependent NF-κB activation pathway and are associated with tumor development and progression [132]. Inflammatory cytokines play crucial roles in colon cancer. For example, TNFα and IL-6 serum levels are associated with increased risk of colorectal adenoma development [138]. Additionally, injection of neutralizing antibodies specific for the IL-6 receptor has been shown to decrease both tumor number and size, thereby demonstrating the critical role of the IL-6 signaling pathway in colon cancer development [139].

In vitro studies have shown that n-3 PUFA exert inhibitory effects on NF-κB by decreasing IκB phosphorylation and activation, subsequently reducing inflammatory cytokine production (IL-1β, IL-6, and TNFα) [140–143]. In a colitis-associated cancer model, *Fat-1* transgenic mice, which endogenously produce long-chain n-3 PUFA de novo, exhibited a suppression of tumorigenesis compared to wild-type mice. This was associated with a reduction in tumor incidence and coincided with lower colonic NF-κB activity and increased mRNA expression of the antiproliferative cytokine TGF-β [144]. The suppression of NF-κB in the chronic colitis/malignant transformation model mirrored the anti-inflammatory effect of n-3 PUFA during acute colitis, wherein NF-κB activation and mRNA expression of key inflammatory makers (inducible nitric oxide synthase [iNOS], IL-1β, and TNFα) were suppressed in the *Fat-1* mouse [145].

Following splenic $CD4^+$ T cell activation, either dietary FO or a diet enriched in DHA ethyl esters inhibited the DNA-binding activity of both NF-κB and AP-1 [146]. Similarly, both NF-κB and AP-1 DNA-binding activity were modestly affected in human fibroblasts by EPA treatment in comparison to AA [147]. AP-1 is overexpressed and has been shown to have increased activity in several types of cancer including breast, ovarian, cervical, and colorectal, consistent with its fundamental role in oncogenesis [148–151]. Functionally, AP-1 is implicated in biological processes integral to cellular transformation including proliferation, apoptosis, differentiation, invasion, and motility [152]. Collectively, these data suggest that n-3 PUFA can potentially impact AP-1-mediated early and late events in carcinogenesis [153].

Another inflammatory transcription factor whose colonic activation and expression is reduced by n-3 PUFA is STAT3 [83, 154]. STAT3 is constitutively activated in biopsies from cancer patients and cell lines. Numerous cytokines, growth factors, and oncogenic proteins activate STAT3, leading to cell transformation and tumorigenesis, as STAT3 target genes are involved in multiple steps of the metastatic process, including invasion, cell survival, self-renewal, angiogenesis, and tumor cell immune evasion [155, 156]. Moreover, STAT3 activation functionally links inflammation and cancer wherein aberrant phosphorylated STAT3 expression in inflammatory cells residing within the tumor microenvironment triggers an epithelial cell survival and growth response that promotes the overgrowth of neoplastic cells [156]. In summary, STAT3-mediated reciprocal interactions between tumor, inflammatory, and stromal cells collectively make up and prime the inflammatory tumor cellular microenvironment. Therefore, suppressed STAT3 activation by n-3 PUFA represents a potential therapeutic target for the purpose of transcriptional repression.

12.8.1 Role of n-3 PUFA as Receptor Ligands

DHA and EPA have been shown to act as ligands for peroxisome proliferator-activated receptors (PPARs) and retinoid X receptor alpha, which are both nuclear receptors [157, 158]. Binding of ligand to these nuclear receptors stimulates their activity as transcription factors, which alters gene expression. Collectively, these outcomes mediate numerous biological functions, including lipid metabolism, cell differentiation, and cell death. In this context, PPARγ, one of the isoforms of PPAR, was found to mediate DHA-induced apoptosis [159]. The ability of PPARγ to promote differentiation and maturation of epithelial cells has led to studies of its potential role in the cause of CRC, and animal studies suggest both pro- and anti-cancer properties in the colon [160–162]; however, the majority of studies point to PPARγ ligands as chemopreventive agents [163].

PPARs are also involved in the regulation of the production of inflammatory mediators, thereby exerting anti-inflammatory effects [164]. PPARs can physically interact and thereby interfere with other transcription factors involved in proinflammatory signal transduction pathways including AP-1, STAT-1, NF-κB, and NFAT [164, 165]. The ability of activated PPARs to transrepress inflammatory responses mediated by NF-κB represents an important functional mechanism for n-3 PUFA, since NF-κB links inflammation and immunity to cancer development and progression [132, 166, 167]. PPARs further regulate inflammatory processes by suppressing the expression of iNOS thereby inhibiting NO production, inflammatory cytokines (TNFα, IL-6, IL-1β, and IL-12), reducing macrophage recruitment to inflammatory sites by repressing transcription of MCP-1 and its receptor CCR2, inducing T cell anergy, and increasing the formation and suppressive function of regulatory T cells [164, 165]. Therefore, it has been proposed that n-3 PUFA exert their anti-inflammatory and immune suppressive effects through binding to PPARs [168–170]. However, with respect to ligand-binding specificity, this class of nuclear receptor binds n-3 and n-6 PUFA with equal affinity and lacks fatty acid class (n-3 vs. n-6) specificity [157, 171–173]. Therefore, the anti-inflammatory/chemoprotective effects of n-3 PUFA are likely not mediated directly via PPARs alone.

n-3 fatty acids have additionally been found to serve as ligands for the G-protein-coupled receptor GPR120. Binding of n-3 PUFA to GPR120 results in multiple anti-inflammatory effects suppressing NF-κB activation and reducing inflammatory cytokine production in monocytes and macrophages [174]. Moreover, in GPR120 knockdown cells, the inhibitory effect of DHA on the response to endotoxin was abolished, thereby indicating that the inhibition of NF-κB by n-3 PUFA occurs via a GPR120-dependent mechanism [174]. However, there is concern that the effects of fatty acids on GPR120 are not restricted to n-3 PUFA [175]. Therefore, its activation in vivo using biologically relevant delivery models requires further investigation.

Metabolites of DHA and EPA have been found to function as ligands for some receptors. The PGE_2 EP4 receptor is involved in signaling events that are integral in colonic tumorigenesis [176, 177]. Interestingly, PGE_3 (an EPA-derived metabolite) can also bind to this receptor but with reduced affinity and efficacy compared to PGE_2 [178]. Furthermore, PGE_3 can act as an antagonist to EP4 signaling in human CRC cells [178], and DHA and EPA can reduce EP4 signaling due to the reduced production of COX-2-derived PGE_2 as discussed above. Likewise, RvE1 can serve as a ligand for the ChemR23 receptor. Binding of RvE1 to this receptor has been shown to attenuate NF-κB [129]. RvE1 can also bind to the leukotriene B4 receptor 1, where it serves as a partial agonist and hinders receptor signaling [179]. RvD1 can interact with the lipoxin A_4 receptor ALX and the orphan receptor GPR32 on phagocytes [180]. This interaction stimulates phagocytosis to aid in the resolution of acute inflammation. Overall, n-3 PUFA can directly serve as ligands, as well as precursors of ligands (DHA- and EPA-derived protectins and resolvins), which can directly regulate cell signaling and inflammation biology.

12.8.2 n-3 PUFA and T Cell Activation

Understanding how T cell activation is modulated by n-3 PUFA, specifically with respect to the dampening of uncontrolled T cell-mediated inflammatory responses, could beneficially impact the carcinogenic process. T cells isolated from mice fed an n-3 PUFA-enriched fish oil diet exhibit an altered phospholipid distribution in the liquid-ordered phase compared to mice fed an n-6 PUFA-enriched control diet. Specifically, the percentage of sphingomyelin in the liquid-ordered phase of T cells isolated from the n-3 PUFA-enriched fish oil group exhibited a 50 % decrease compared to the n-6 PUFA-enriched corn oil group. Furthermore, n-3 PUFA such as EPA and DHA are readily incorporated into the phosphatidylserine and glycerophosphoethanolamine in the liquid-ordered phase [181]. These results are consistent with fatty acid changes seen in the aforementioned cell types. Since lipid–lipid interaction and the formation of liquid-ordered and disordered phases are dependent on the composition of not only the fatty acid species but also the phospholipid composition [182], the perturbation of plasma membrane composition by n-3 PUFA could directly influence T cell activation.

T cell activation depends on the formation of an immunological synapse (IS) which requires lipid raft (i.e., cholesterol, sphingomyelin, and saturated phosphatidylcholine) accumulation at the IS [183]. Interestingly, disruption of the liquid-ordered phase of the plasma membrane by oxysterol 7-ketocholesterol (7KC) resulted in suppressed lipid condensation upon T cell activation [184]. Since dietary n-3 PUFA such as EPA and DHA can be incorporated and disrupt lipid-ordered phase of the plasma membrane [185], it is not surprising that n-3 PUFA impact the lipidomic rearrangement upon $CD4^+$ T cell activation. In $CD4^+$ T cells isolated from transgenic mice that generate n-3 fatty acids de novo (*Fat-1* mice), the lipid-ordered phase at the IS was increased [29], demonstrating how n-3 PUFA are capable of disrupting the lipidomic rearrangement of the IS, impacting downstream T cell signaling and activation.

Another lipidomic impact exerted by n-3 PUFA is observed at the level of phosphatidylinositol-(4,5)-bisphosphate [$PI(4,5)P_2$], a "phospholipid" in the inner leaflet of the plasma membrane [186]. Hydrolysis by phospholipase C generates the second messengers inositol-1,4,5-trisphosphate (IP_3) and diacylglycerol (DAG), which mediate calcium signaling and protein kinase C (PKC) recruitment, respectively. Recent research has focused on the role of $PI(4,5)P_2$ as a second messenger itself, recruiting and/or activating effector proteins to various $PI(4,5)P_2$-enriched compartments (i.e., plasma membrane, nucleus). The presence of n-3 PUFA such as EPA and DHA is correlated to a decrease in the n-6 PUFA arachidonic acid ($20{:}4^{\Delta 5,8,11,14}$) content, which is the predominant fatty acid species at the *sn-2* position of $PI(4,5)P_2$ [187]. Thus, it is expected that n-3 PUFA can antagonize the synthesis of $PI(4,5)P_2$. Indeed, the platelet 1-acyl-glycero 3-phosphoinositol acyltransferase that remodels $PI(4,5)P_2$ at the *sn*-2 position prefers n-6 PUFA over n-3 PUFA, suggesting a competitive inhibition by EPA and DHA [188]. Consistent with the acyltransferase specificity, in the *Fat-1* $CD4^+$ T cells, $PI(4,5)P_2$ was decreased by 50 % compared to wild-type control, leading to defects in $PI(4,5)P_2$ metabolism upon T cell activation [189]. This phenotype was recapitulated in a dietary model, in which $CD4^+$ T cells isolated from animals fed a 4 % DHA-triglyceride diet exhibited a 25 % decrease in basal $PI(4,5)P_2$. Defects in $PI(4,5)P_2$ metabolism were also observed in DHA-triglyceride-enriched $CD4^+$ T cells upon activation. Interestingly, the kinetics of DAG, the hydrolysis product of $PI(4,5)P_2$, was suppressed upon activation by concanavalin A in lymphocyte population from mice fed an EPA- or DHA-enriched diet [190].

Proteomic rearrangements and their contribution to the formation of the IS begin with the activation of the tyrosine kinases lymphocyte-specific protein tyrosine kinase (LCK) and zeta-chain-associated protein kinase 70 (ZAP70). Their activation results in the phosphorylation of linker for activation of T cells (LAT) and the assembly of the signalsome composed of effector proteins such as GADS, SLP76, NCK, ITK, VAV, PAK, and PLC-γ1 [191, 192]. Protein localization to specific membrane compartments (i.e., lipid raft-enriched IS) can be affected by two non-mutually exclusive mechanisms: (1) alterations in posttranslational modifications and (2) disruption of membrane microdomains. Proteins may be posttranslationally modified to incorporate fatty acids into various amino acid residues to stabilize their association to specific membrane microdomains. Therefore, it is of interest to note that in Jurkat T cells incubated with EPA, src kinase Fyn failed to localize to the detergent-resistant membrane by inhibiting its palmitoylation [193]. This was correlated with the presence of posttranslational modification by monounsaturated and polyunsaturated fatty acids [194], demonstrating that n-3 PUFA are capable of modulating posttranslational fatty acid modification and therefore displace proteins from appropriate membrane compartments. Disruption of membrane microdomains by n-3 PUFA can also affect protein localization. Indeed, LAT, an important protein recruited early to the signalsome, was displaced from the detergent-resistant membrane after treatment with n-3 PUFA [195]. This appeared to be the first proteomic perturbation induced by n-3 PUFA, as PUFA had no effect on the phosphorylation of the CD3ζ by LCK or the binding and phosphorylation of ZAP-70 [196]. The localization of PLC-γ1 and subsequent phosphorylation was also suppressed in $CD4^+$ T cells enriched in n-3 PUFA, demonstrating additional proteomic effects induced by n-3 PUFA [29].

Actin reorganization is required for trafficking of the T cell receptor to the site of contact with antigen-presenting cells in order to survey antigens presented in MHCII. Upon antigen recognition by the TCR, actin stabilizes the IS to allow for protein complex formation and downstream signaling [197–199]. Actin-regulatory proteins that are important for T cell activation and are also known to be regulated by $PI(4,5)P_2$ include the Wiskoff-Aldrich syndrome protein (WASP), ERM proteins, talin, WAVE, and ADF/cofilin [200–204]. The decrease of $PI(4,5)P_2$ by n-3 PUFA has been correlated with suppressed actin remodeling upon T cell activation in n-3 PUFA-enriched $CD4^+$ T cells [29, 189]. It was further shown that the recruitment of WASP to the IS was suppressed in *Fat-1* $CD4^+$ T cells, elucidating one pathway in which the effects of n-3 PUFA on actin remodeling could be mediated. These results indicate that n-3 PUFA such as DHA suppress early T cell activation events by antagonizing $PI(4,5)P_2$ and subsequent $PI(4,5)P_2$-dependent signaling processes such as actin remodeling [189]. This is consistent with results seen in the human Jurkat T cell line E6-1 incubated with either EPA or DHA, where the localization of F-actin upon IS formation was suppressed in cells exposed to individual n-3 PUFA [205]. Actin cytoskeleton can also be regulated by the nucleotide exchange factor VAV, which is recruited to the IS upon T cell activation. EPA treatment of T cells showed an inhibition of VAV phosphorylation, leading to instability of the IS and antigen-presenting T cell conjugates [205]. These observations demonstrate the immunosuppressive pleiotropic effects of n-3 PUFA with respect to the actin cytoskeleton, cell lipidomics, and proteomics.

The assembly of the signalsome allows for further downstream signaling and the activation of transcription factors for sustained T cell activation. Signal transduction from the plasma membrane to the nucleus includes the translocation and activation of PKCθ [206], translocation of the mitochondria to the IS [207, 208], Ca^{2+} signaling [209], and ultimately the activation of transcription factors such as NFAT, NF-κB, and AP-1 to induce specific gene expression required for cell proliferation and differentiation [210–212]. Similar to LAT, which was displaced from detergent-resistant membranes, the colocalization of PKCθ to GM1-enriched lipid rafts was decreased in splenic T cells isolated from mice fed an n-3 PUFA-enriched diet, leading to the suppression of PKCθ to the IS [146]. In parallel, downstream signaling events such as mitochondrial translocation and Ca^{2+} signaling were also suppressed

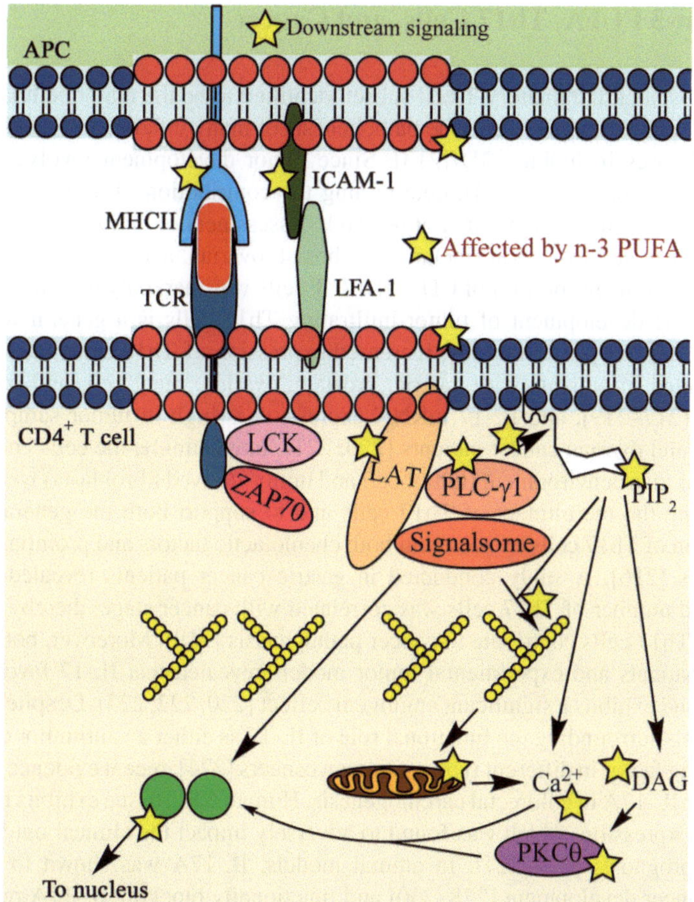

Fig. 12.2 Pleiotropic effects of n-3 PUFA on CD4+ T cell and APC activation. Upon the formation of the immunological synapse, several early signaling events are perturbed in both the CD4+ T cell and the APC. For the APC, n-3 PUFA perturbs the organization of the lipid microdomains, leading to changes in the conformation, lateral organization, and trafficking of MHCII and adhesion proteins such as ICAM-1. This results in alterations in the downstream signaling of APC. For the CD4+ T cells, incorporation of n-3 PUFA leads to changes in the lipidomic (i.e., suppression of sphingomyelin and $PI(4,5)P_2$, changes in lipid microdomains) and proteomic (i.e., decreased LAT and PLC-γ1 localization) rearrangements that occur upon immunological synapse formation. These effects lead to decreased actin remodeling and downstream events such as (1) mitochondrial translocation and Ca^{2+} signaling, (2) DAG metabolism and PKCθ recruitment, (3) activation of transcription factors, and (4) cellular processes such as proliferation, differentiation, and inflammatory cytokine secretion

in CD4+ T cells enriched in n-3 PUFA. Importantly, the upstream suppression by n-3 PUFA has been linked to a decrease in AP-1 and NF-κB activation and ultimately the IL-2 secretion and proliferation of CD4+ T cells following activation [29, 146] (Fig. 12.2).

12.9 n-3 PUFA, Th17 Cells, and Cancer

Th17 cells, an inflammatory T cell subset identified to be the major cellular source of IL-17, play a critical role in the pathogenesis of many inflammatory and autoimmune diseases including IBD [213]. Since tumor development involves chronic inflammatory processes [214], determining the contribution of Th17 cells to this process represents an active area of research. Assessment of tumor-infiltrating lymphocytes isolated from skin (melanoma), breast, ovarian, and colon cancer tumors revealed that the proportion of $CD4^+$ $IL17^+$ T cells was markedly elevated, suggesting that the development of tumor-infiltrating Th17 cells is a general feature of cancer [215, 216]. Additionally, increased expression of Th17 cells has also been documented in patients with gastric, prostate, ovarian, renal cell, and pancreatic cancers [217–219], and IL-17 mRNA expression is high in tumor samples from prostate and ovarian cancer patients [220, 221]. Interestingly, the cells comprising the tumor microenvironment (tumor cell and tumor-derived fibroblasts) were shown to mediate the recruitment of Th17 cells and to support both the generation and expansion of Th17 cells by secreting both chemotactic factors and proinflammatory cytokines [216]. A study conducted in gastric cancer patients revealed that the increased number of Th17 cells was correlated with cancer stage, thereby suggesting that Th17 cells contribute to cancer pathogenesis [219]. Moreover, both human cancer patients and experimental tumor models revealed that IL-17 favors tumor growth and exhibits a significant angiogenic effect [220, 222, 223]. Despite conflicting reports surrounding the functional role of IL-17 as either an antitumor or tumor-promoting factor in different types of human cancers [224], recent evidence supports a role for IL-17A in colorectal carcinogenesis. Human CRC tissue exhibits increased IL-17A expression, which was found to adversely impact the clinical outcome and patient prognosis [225–227]. In animal models, IL-17A was shown to promote colon cancer development [228–230] and functionally blocking IL-17A resulted in decreased formation of hyperplasia and cancer [228]. Overexpression of IL-17 in tumors leads to increased angiogenesis and tumor growth [222], and $IL\text{-}17^{-/-}$ and $IL\text{-}17R^{-/-}$ mice exhibit reduced tumor growth [231, 232]. Additionally, in a mouse model of colon cancer (AOM + DSS), both tumor size and number of colonic tumors per mouse were reduced in IL-17A-deficient mice [233]. Interestingly, in the same cancer model, dietary fish oil reduced the total number of colonic tumor entities (adenomas + adenocarcinomas) per mouse and exhibited a greater percentage of animals that were nonresponsive in terms of tumor formation (48 % vs. 18 %), as compared to mice fed a control diet (5 % corn oil) [83]. Moreover, colonic mucosal mRNA expression of both the Th17 cell signature cytokines (IL-17A and IL-17F) and inflammatory cytokines supporting Th17 cell differentiation (IL-6) and phenotype maintenance (IL-23) was significantly reduced by dietary FO [83]. Lastly, activated (i.e., phosphorylated) STAT3 colonic protein levels were reduced by dietary FO [83], which is significant because T cell expression of STAT3 plays a critical role in Th17 cell differentiation [234] and promotes a procarcinogenic Th17 cell response [230, 235]. Interestingly, the tumorigenic effects of IL-17 are mediated, at least in part, by IL-6 via a STAT3-dependent mechanism [232], and all three of these mediators were reduced in the colon by dietary fish oil [83]. n-3 PUFA also exhibited a suppressive effect on colonic IL-23 and IL-23R mRNA expression, a key cytokine that not only promotes pathogenic Th1 and Th17 responses in the intestine [236] but also represents an important molecular link between chronic inflammation and carcinogenesis as demonstrated by IL-23p19$^{-/-}$ mice which are resistant to tumor induction [237]. Therefore, this research demonstrates that within the colonic tumor microenvironment, expression of critical factors impacting aspects of Th17 cell polarization and function, which have independent procarcinogenic roles, was collectively suppressed by dietary fish oil, resulting in reduced colonic tumor formation. Although the molecular mechanisms of n-3 PUFA action on Th17 cells have not been elucidated, these data provide insight into the underlying anti-inflammatory and chemopreventive actions of n-3 PUFA.

12.9.1 Oxidative Stress and the Induction of Apoptosis

Studies in several types of cancer, including breast, prostate, and colon, have established that oxidative stress modulates both cancer initiation and progression [238]. In cancer cells, high levels of reactive oxygen species (ROS) can result from increased metabolic activity, mitochondrial dysfunction due to hypoxia or mitophagy, peroxisome activity, uncontrolled growth factor or cytokine signaling, and oncogene activity in addition to enhanced activity of known ROS sources (e.g., NADPH oxidase, COX, and LOX) [238–241]. Moreover, ROS can promote many aspects of tumor onset and progression towards a malignant phenotype [238]. Thus, if the cellular oxidative state increases, ROS form and can cause both cellular and DNA damage. n-3 PUFA, due to their high levels of unsaturation, are oxidatively susceptible lipids [242, 243]. Interestingly, formation of lipid peroxidation products can activate the "intrinsic" mitochondrial apoptosis pathway, which serves as a sensor for damage and oxidative stress [244]. This is noteworthy, because the balance between proliferation and apoptosis is critical for the maintenance of steady-state number for cell populations, particularly in the colon. Feeding a diet containing fish oil can increase oxidative stress and activate apoptosis [243]. In cell culture, treatment with DHA increased lipid oxidation and reduced mitochondrial membrane potential, which resulted in increased apoptosis [245]. Furthermore, in genetically oxidatively stressed mice, feeding fish oil enhanced apoptosis and altered mitochondrial metabolic activity [246]. These findings support the contention that n-3 PUFA alters cellular oxidation to enhance apoptosis and thereby prevent colon cancer.

12.10 Combined Bioactive Nutrients

There is also a wealth of data suggesting a synergistic effect of n-3 PUFA with other dietary compounds. It is important to note that no nutrient is ever consumed in isolation. Therefore, it is imperative to understand how nutrients function in combination with each other.

12.10.1 n-3 PUFA and Fiber

In addition to n-3 PUFA, fiber is another commonly studied nutrient in the context of colonic health. In colonocytes and colon cancer cells, treatment with the combination of DHA and butyrate, a fiber fermentation product, has been shown to induce apoptosis and alter mitochondrial-to-cytosolic Ca^{2+} levels [247, 248]. This effect was specific to the combination of DHA and butyrate, i.e., other fatty acids combined with butyrate did not exert the same effects. DHA and butyrate have also been shown to synergistically increase lipid oxidation in colonocytes [245]. Feeding carcinogen-injected rats a diet of fish oil in combination with enterically coated butyrate, elevated apoptosis and decreased the formation of aberrant crypts foci, which can serve as precursors for colonic tumors [249]. Combination of fish oil and the fermentable fiber pectin was additionally found to reduce oxidative DNA damage and increase apoptosis [250]. Overall, these data indicate that the combination of fish oil and fiber could provide more protection against colon cancer than either nutrient alone.

12.10.2 n-3 PUFA and Curcumin

Another nutrient that has attracted a great deal of interest is curcumin. Curcumin is a curcumoid found in the Indian spice turmeric. Treatment of macrophages with a combination of curcumin and DHA or EPA was found to synergistically suppress nitric oxide levels, PGE_2 synthesis, and the levels of enzymes involved in formation of proinflammatory mediators [251]. The combination of fish oil and curcumin has been shown to modulate colonocyte gene expression and enhance resolution of chronic inflammation, in part, by suppressing a key inflammatory mediator, NF-κB [154]. The combination of fish oil and curcumin was also found to reduce proliferation of $CD4^+$ T cells to favorably modulate inflammation [252]. Together, these studies suggest that the combination of fish oil and curcumin could have potent anti-inflammatory effects, which is beneficial for colon cancer prevention.

12.10.3 Ratio of n-6 to n-3 PUFA

In a pivotal study, Simopoulos identified the ratio of n-6 to n-3 PUFA as an important indicator of the overall health outcome [253]. Currently, the n-6:n-3 ratio in the average Western diet is calculated to be between 10:1 and 20:1 [254–256]. One of the reasons that the ratio of these two classes of dietary lipid is important is the competition between n-3 and n-6 PUFA for many enzymes. These fatty acids utilize many of the same elongases and desaturases required for production of the longer-chain fatty acids in each series. Therefore, if levels of α-LNA are high enough, this essential fatty acid can compete with linoleic acid (LA 18:2 n-6) and thereby reduce the formation of AA. Likewise, as mentioned above, both EPA and AA can be metabolized by COX-2 to produce anti-inflammatory and proinflammatory prostaglandins, respectively. Therefore, if levels of n-6 PUFA are substantially higher than n-3 PUFA, more prostaglandins will be produced from AA as opposed to EPA. This could then hinder the anti-inflammatory activities of EPA and DHA. Additionally, these PUFA will compete for esterification into phospholipids and, therefore, insertion into the plasma membrane. Due to the important actions of n-3 PUFA in the plasma membrane, it is imperative that the ratio of n-6 to n-3 PUFA be sufficiently low in order for fatty acyl incorporation to occur. Furthermore, whereas n-3 PUFA is associated with reduced colon cancer risk, n-6 PUFA has been found to increase the risk of colorectal adenomas [17]. Therefore, understanding the roles of all dietary lipids, as well as the ways they interact and compete with each, is required in order to make informed recommendations for overall human health.

12.11 Differential Effects of DHA and EPA

A significant proportion of the published studies describing the effects of n-3 PUFA on cellular functions utilizes either fish oil, purified DHA, or a combination of EPA and DHA [28, 71, 146]. This can likely be a source of some of the inconsistencies in the literature. Specifically, although many commercially available fish oils contain approximately a 2:1 ratio of EPA to DHA, fish oils from different sources contain variable mixtures of EPA and DHA. This can make it difficult or impossible to compare results from different studies. Furthermore, whether the effects of n-3 PUFA supplementation are due to EPA, DHA, or both is often undetermined and unappreciated. EPA is both two carbons shorter and less unsaturated than DHA, and the structural differences between these two fatty acids are enough to result in functional differences [69]. A recent biophysical study in model membranes demonstrated differential efficacies of DHA and EPA to modify lipid raft composition and organization [185]. Studies on the effects of DHA compared to EPA on immune cell lipid rafts have observed differential effects of these fatty acids on membrane order [30]. Comparisons of the anti-inflammatory profiles of macrophages treated with DHA or EPA indicated that DHA may be more effective than EPA in alleviating inflammation [142]. Additionally, a study compared the effects of DHA or EPA on EGFR signaling found that DHA but not EPA altered EGFR activation status and signaling [76]. These data highlight the need to better understand the mechanisms by which EPA and DHA differentially modulate cell membranes.

In summary, cogent evidence indicates that n-3 PUFA confer protection against several forms of cancer. We propose that n-3 fatty acids represent a class of *novel innocuous dietary bioactives for lipid raft targeted therapy* and cancer prevention. So why have clinical practitioners currently searching for toxicologically *innocuous cancer chemoprevention* approaches not embraced n-3 PUFA? Perhaps many studies looking for n-3 PUFA-related effects in cancer populations have errored by intervening too late in the disease process [257]. Undoubtedly, additional work is needed in order to fully elucidate the chemopreventive actions of n-3 PUFA at the molecular, cellular, animal model, and clinical level.

We propose that it is time to get serious about cancer prevention. Unfortunately, only a small fraction of current funding in the USA for cancer research is targeted to early detection and prevention. In addition, less than 1.5 % of total biomedical research funding is devoted to implementation of cancer prevention programs [258]. According to Vogelstein and colleagues [259], the only sure way to reduce cancer risk is through screening and a healthy lifestyle.

Acknowledgments This work was supported by National Institute of Health grants CA59034, CA129444, CA168312 and by USDA 2010-34402-20875 and the Cancer Prevention and Research Institute of Texas grant RP100473.

References

1. Sargent JR (1997) Fish oils and human diet. Br J Nutr 78(suppl 1):S5–S13
2. Pawlosky RJ, Hibbeln JR, Novotny JA, Salem N Jr (2001) Physiological compartmental analysis of alpha-linolenic acid metabolism in adult humans. J Lipid Res 42:1257–1265
3. Berquin IM, Edwards IJ, Chen YQ (2008) Multi-targeted therapy of cancer by omega-3 fatty acids. Cancer Lett 269:363–377
4. Caygill CP, Charlett A, Hill MJ (1996) Fat, fish, fish oil and cancer. Br J Cancer 74:159–164
5. Gerber M (2012) Omega-3 fatty acids and cancers: a systematic update review of epidemiological studies. Br J Nutr 107(suppl 2):S228–S239
6. Pauwels EK, Kairemo K (2008) Fatty acid facts, part II: role in the prevention of carcinogenesis, or, more fish on the dish? Drug News Perspect 21:504–510
7. Gerber M (2009) Background review paper on total fat, fatty acid intake and cancers. Ann Nutr Metab 55:140–161
8. Kim S, Sandler DP, Galanko J, Martin C, Sandler RS (2010) Intake of polyunsaturated fatty acids and distal large bowel cancer risk in whites and African Americans. Am J Epidemiol 171:969–979
9. Kimura Y, Kono S, Toyomura K, Nagano J, Mizoue T, Moore MA, Mibu R, Tanaka M, Kakeji Y, Maehara Y, Okamura T, Ikejiri K, Futami K, Yasunami Y, Maekawa T, Takenaka K, Ichimiya H, Imaizumi N (2007) Meat, fish and fat intake in relation to subsite-specific risk of colorectal cancer: The Fukuoka Colorectal Cancer Study. Cancer Sci 98:590–597
10. Murff HJ, Shrubsole MJ, Cai Q, Smalley WE, Dai Q, Milne GL, Ness RM, Zheng W (2012) Dietary intake of PUFA and colorectal polyp risk. Am J Clin Nutr 95:703–712
11. Sasazuki S, Inoue M, Iwasaki M, Sawada N, Shimazu T, Yamaji T, Takachi R, Tsugane S (2011) Intake of n-3 and n-6 polyunsaturated fatty acids and development of colorectal cancer by subsite: Japan Public Health Center-based prospective study. Int J Cancer 129:1718–1729
12. Theodoratou E, McNeill G, Cetnarskyj R, Farrington SM, Tenesa A, Barnetson R, Porteous M, Dunlop M, Campbell H (2007) Dietary fatty acids and colorectal cancer: a case–control study. Am J Epidemiol 166:181–195
13. Hall MN, Chavarro JE, Lee IM, Willett WC, Ma J (2008) A 22-year prospective study of fish, n-3 fatty acid intake, and colorectal cancer risk in men. Cancer Epidemiol Biomarkers Prev 17:1136–1143

14. Geelen A, Schouten JM, Kamphuis C, Stam BE, Burema J, Renkema JM, Bakker EJ, van't Veer P, Kampman E (2007) Fish consumption, n-3 fatty acids, and colorectal cancer: a meta-analysis of prospective cohort studies. Am J Epidemiol 166:1116–1125
15. Hall MN, Campos H, Li H, Sesso HD, Stampfer MJ, Willett WC, Ma J (2007) Blood levels of long-chain polyunsaturated fatty acids, aspirin, and the risk of colorectal cancer. Cancer Epidemiol Biomarkers Prev 16:314–321
16. Kojima M, Wakai K, Tokudome S, Suzuki K, Tamakoshi K, Watanabe Y, Kawado M, Hashimoto S, Hayakawa N, Ozasa K, Toyoshima H, Suzuki S, Ito Y, Tamakoshi A, J.S. Group (2005) Serum levels of polyunsaturated fatty acids and risk of colorectal cancer: a prospective study. Am J Epidemiol 161:462–471
17. Pot GK, Geelen A, van Heijningen EM, Siezen CL, van Kranen HJ, Kampman E (2008) Opposing associations of serum n-3 and n-6 polyunsaturated fatty acids with colorectal adenoma risk: an endoscopy-based case–control study. Int J Cancer 123:1974–1977
18. Jemal A, Bray F, Center MM, Ferlay J, Ward E, Forman D (2011) Global cancer statistics. CA Cancer J Clin 61:69–90
19. Anti M, Marra G, Armelao F, Bartoli GM, Ficarelli R, Percesepe A, De Vitis I, Maria G, Sofo L, Rapaccini GL et al (1992) Effect of omega-3 fatty acids on rectal mucosal cell proliferation in subjects at risk for colon cancer. Gastroenterology 103:883–891
20. Bartram HP, Gostner A, Scheppach W, Reddy BS, Rao CV, Dusel G, Richter F, Richter A, Kasper H (1993) Effects of fish oil on rectal cell proliferation, mucosal fatty acids, and prostaglandin E2 release in healthy subjects. Gastroenterology 105:1317–1322
21. Anti M, Armelao F, Marra G, Percesepe A, Bartoli GM, Palozza P, Parrella P, Canetta C, Gentiloni N, De Vitis I et al (1994) Effects of different doses of fish oil on rectal cell proliferation in patients with sporadic colonic adenomas. Gastroenterology 107:1709–1718
22. Courtney ED, Matthews S, Finlayson C, Di Pierro D, Belluzzi A, Roda E, Kang JY, Leicester RJ (2007) Eicosapentaenoic acid (EPA) reduces crypt cell proliferation and increases apoptosis in normal colonic mucosa in subjects with a history of colorectal adenomas. Int J Colorectal Dis 22:765–776
23. West NJ, Clark SK, Phillips RK, Hutchinson JM, Leicester RJ, Belluzzi A, Hull MA (2010) Eicosapentaenoic acid reduces rectal polyp number and size in familial adenomatous polyposis. Gut 59:918–925
24. Stern MC, Butler LM, Corral R, Joshi AD, Yuan JM, Koh WP, Yu MC (2009) Polyunsaturated fatty acids, DNA repair single nucleotide polymorphisms and colorectal cancer in the Singapore Chinese Health Study. J Nutrigenet Nutrigenomics 2:273–279
25. Hashimoto M, Hossain S, Yamasaki H, Yazawa K, Masumura S (1999) Effects of eicosapentaenoic acid and docosahexaenoic acid on plasma membrane fluidity of aortic endothelial cells. Lipids 34:1297–1304
26. Stillwell W, Wassall SR (2003) Docosahexaenoic acid: membrane properties of a unique fatty acid. Chem Phys Lipids 126:1–27
27. Wassall SR, Brzustowicz MR, Shaikh SR, Cherezov V, Caffrey M, Stillwell W (2004) Order from disorder, corralling cholesterol with chaotic lipids. The role of polyunsaturated lipids in membrane raft formation. Chem Phys Lipids 132:79–88
28. Chapkin RS, Wang N, Fan YY, Lupton JR, Prior IA (2008) Docosahexaenoic acid alters the size and distribution of cell surface microdomains. Biochim Biophys Acta 1778:466–471
29. Kim W, Fan YY, Barhoumi R, Smith R, McMurray DN, Chapkin RS (2008) n-3 polyunsaturated fatty acids suppress the localization and activation of signaling proteins at the immunological synapse in murine CD4+ T cells by affecting lipid raft formation. J Immunol 181:6236–6243
30. Rockett BD, Teague H, Harris M, Melton M, Williams J, Wassall SR, Shaikh SR (2012) Fish oil increases raft size and membrane order of B cells accompanied by differential effects on function. J Lipid Res 53:674–685
31. Shaikh SR, Locascio DS, Soni SP, Wassall SR, Stillwell W (2009) Oleic- and docosahexaenoic acid-containing phosphatidylethanolamines differentially phase separate from sphingomyelin. Biochim Biophys Acta 1788:2421–2426

32. Pike LJ (2006) Rafts defined: a report on the keystone symposium on lipid rafts and cell function. J Lipid Res 47:1597–1598
33. Pike LJ (2005) Growth factor receptors, lipid rafts and caveolae: an evolving story. Biochim Biophys Acta 1746:260–273
34. Brown DA, London E (2000) Structure and function of sphingolipid- and cholesterol-rich membrane rafts. J Biol Chem 275:17221–17224
35. Simons K, Sampaio JL (2011) Membrane organization and lipid rafts. Cold Spring Harb Perspect Biol 3:a004697
36. Jacobson K, Mouritsen OG, Anderson RG (2007) Lipid rafts: at a crossroad between cell biology and physics. Nat Cell Biol 9:7–14
37. Pike LJ (2004) Lipid rafts: heterogeneity on the high seas. Biochem J 378:281–292
38. Deans JP, Robbins SM, Polyak MJ, Savage JA (1998) Rapid redistribution of CD20 to a low density detergent-insoluble membrane compartment. J Biol Chem 273:344–348
39. Field KA, Holowka D, Baird B (1997) Compartmentalized activation of the high affinity immunoglobulin E receptor within membrane domains. J Biol Chem 272:4276–4280
40. Simons K, Toomre D (2000) Lipid rafts and signal transduction. Nat Rev Mol Cell Biol 1:31–39
41. Smart EJ, Graf GA, McNiven MA, Sessa WC, Engelman JA, Scherer PE, Okamoto T, Lisanti MP (1999) Caveolins, liquid-ordered domains, and signal transduction. Mol Cell Biol 19:7289–7304
42. Staubach S, Hanisch FG (2011) Lipid rafts: signaling and sorting platforms of cells and their roles in cancer. Expert Rev Proteomics 8:263–277
43. Hofman EG, Ruonala MO, Bader AN, van den Heuvel D, Voortman J, Roovers RC, Verkleij AJ, Gerritsen HC, van Bergen En Henegouwen PM (2008) EGF induces coalescence of different lipid rafts. J Cell Sci 121:2519–2528
44. Kono H, Suzuki T, Yamamoto K, Okada M, Yamamoto T, Honda Z (2002) Spatial raft coalescence represents an initial step in Fc gamma R signaling. J Immunol 169:193–203
45. Patel HH, Murray F, Insel PA (2008) G-protein-coupled receptor-signaling components in membrane raft and caveolae microdomains. Handb Exp Pharmacol 186:167–184
46. Pike LJ, Han X, Gross RW (2005) Epidermal growth factor receptors are localized to lipid rafts that contain a balance of inner and outer leaflet lipids: a shotgun lipidomics study. J Biol Chem 280:26796–26804
47. Prior IA, Muncke C, Parton RG, Hancock JF (2003) Direct visualization of Ras proteins in spatially distinct cell surface microdomains. J Cell Biol 160:165–170
48. George KS, Wu S (2012) Lipid raft: a floating island of death or survival. Toxicol Appl Pharmacol 259:311–319
49. Alonso MA, Millan J (2001) The role of lipid rafts in signalling and membrane trafficking in T lymphocytes. J Cell Sci 114:3957–3965
50. Drbal K, Moertelmaier M, Holzhauser C, Muhammad A, Fuertbauer E, Howorka S, Hinterberger M, Stockinger H, Schutz GJ (2007) Single-molecule microscopy reveals heterogeneous dynamics of lipid raft components upon TCR engagement. Int Immunol 19:675–684
51. Algeciras-Schimnich A, Shen L, Barnhart BC, Murmann AE, Burkhardt JK, Peter ME (2002) Molecular ordering of the initial signaling events of CD95. Mol Cell Biol 22:207–220
52. Bang B, Gniadecki R, Gajkowska B (2005) Disruption of lipid rafts causes apoptotic cell death in HaCaT keratinocytes. Exp Dermatol 14:266–272
53. Hazarika P, McCarty MF, Prieto VG, George S, Babu D, Koul D, Bar-Eli M, Duvic M (2004) Up-regulation of Flotillin-2 is associated with melanoma progression and modulates expression of the thrombin receptor protease activated receptor 1. Cancer Res 64:7361–7369
54. Li YC, Park MJ, Ye SK, Kim CW, Kim YN (2006) Elevated levels of cholesterol-rich lipid rafts in cancer cells are correlated with apoptosis sensitivity induced by cholesterol-depleting agents. Am J Pathol 168:1107–1118; quiz 1404–1405
55. Patra SK (2008) Dissecting lipid raft facilitated cell signaling pathways in cancer. Biochim Biophys Acta 1785:182–206

56. Fedida-Metula S, Feldman B, Koshelev V, Levin-Gromiko U, Voronov E, Fishman D (2012) Lipid rafts couple store-operated Ca2+ entry to constitutive activation of PKB/Akt in a Ca2+/calmodulin-, Src- and PP2A-mediated pathway and promote melanoma tumor growth. Carcinogenesis 33:740–750
57. Irwin ME, Bohin N, Boerner JL (2011) Src family kinases mediate epidermal growth factor receptor signaling from lipid rafts in breast cancer cells. Cancer Biol Ther 12:718–726
58. Lasserre R, Guo XJ, Conchonaud F, Hamon Y, Hawchar O, Bernard AM, Soudja SM, Lenne PF, Rigneault H, Olive D, Bismuth G, Nunes JA, Payrastre B, Marguet D, He HT (2008) Raft nanodomains contribute to Akt/PKB plasma membrane recruitment and activation. Nat Chem Biol 4:538–547
59. Orr G, Hu D, Ozcelik S, Opresko LK, Wiley HS, Colson SD (2005) Cholesterol dictates the freedom of EGF receptors and HER2 in the plane of the membrane. Biophys J 89:1362–1373
60. Ringerike T, Blystad FD, Levy FO, Madshus IH, Stang E (2002) Cholesterol is important in control of EGF receptor kinase activity but EGF receptors are not concentrated in caveolae. J Cell Sci 115:1331–1340
61. Roy UK, Rial NS, Kachel KL, Gerner EW (2008) Activated K-RAS increases polyamine uptake in human colon cancer cells through modulation of caveolar endocytosis. Mol Carcinog 47:538–553
62. Irwin ME, Mueller KL, Bohin N, Ge Y, Boerner JL (2011) Lipid raft localization of EGFR alters the response of cancer cells to the EGFR tyrosine kinase inhibitor gefitinib. J Cell Physiol 226(9):2316–2328
63. Lacour S, Hammann A, Grazide S, Lagadic-Gossmann D, Athias A, Sergent O, Laurent G, Gambert P, Solary E, Dimanche-Boitrel MT (2004) Cisplatin-induced CD95 redistribution into membrane lipid rafts of HT29 human colon cancer cells. Cancer Res 64:3593–3598
64. Rebillard A, Tekpli X, Meurette O, Sergent O, LeMoigne-Muller G, Vernhet L, Gorria M, Chevanne M, Christmann M, Kaina B, Counillon L, Gulbins E, Lagadic-Gossmann D, Dimanche-Boitrel MT (2007) Cisplatin-induced apoptosis involves membrane fluidification via inhibition of NHE1 in human colon cancer cells. Cancer Res 67:7865–7874
65. Adachi S, Nagao T, Ingolfsson HI, Maxfield FR, Andersen OS, Kopelovich L, Weinstein IB (2007) The inhibitory effect of (−)-epigallocatechin gallate on activation of the epidermal growth factor receptor is associated with altered lipid order in HT29 colon cancer cells. Cancer Res 67:6493–6501
66. Bacso Z, Bene L, Damjanovich L, Damjanovich S (2002) INF-gamma rearranges membrane topography of MHC-I and ICAM-1 in colon carcinoma cells. Biochem Biophys Res Commun 290:635–640
67. Ullman TA, Itzkowitz SH (2011) Intestinal inflammation and cancer. Gastroenterology 140:1807–1816
68. Nicolau DV Jr, Burrage K, Parton RG, Hancock JF (2006) Identifying optimal lipid raft characteristics required to promote nanoscale protein-protein interactions on the plasma membrane. Mol Cell Biol 26:313–323
69. Corsetto PA, Cremona A, Montorfano G, Jovenitti IE, Orsini F, Arosio P, Rizzo AM (2012) Chemical-physical changes in cell membrane microdomains of breast cancer cells after omega-3 PUFA incorporation. Cell Biochem Biophys 64(1):45–59
70. Ma DW, Seo J, Davidson LA, Callaway ES, Fan YY, Lupton JR, Chapkin RS (2004) n-3 PUFA alter caveolae lipid composition and resident protein localization in mouse colon. FASEB J 18:1040–1042
71. Schley PD, Brindley DN, Field CJ (2007) (n-3) PUFA alter raft lipid composition and decrease epidermal growth factor receptor levels in lipid rafts of human breast cancer cells. J Nutr 137:548–553
72. Ye S, Tan L, Ma J, Shi Q, Li J (2010) Polyunsaturated docosahexaenoic acid suppresses oxidative stress induced endothelial cell calcium influx by altering lipid composition in membrane caveolar rafts. Prostaglandins Leukot Essent Fatty Acids 83:37–43

73. Poulogiannis G, Luo F, Arends MJ (2012) RAS signalling in the colorectum in health and disease. Cell Commun Adhes 19:1–9
74. Seo J, Barhoumi R, Johnson AE, Lupton JR, Chapkin RS (2006) Docosahexaenoic acid selectively inhibits plasma membrane targeting of lipidated proteins. FASEB J 20:770–772
75. Rogers KR, Kikawa KD, Mouradian M, Hernandez K, McKinnon KM, Ahwah SM, Pardini RS (2010) Docosahexaenoic acid alters epidermal growth factor receptor-related signaling by disrupting its lipid raft association. Carcinogenesis 31:1523–1530
76. Turk HF, Barhoumi R, Chapkin RS (2012) Alteration of EGFR spatiotemporal dynamics suppresses signal transduction. PLoS One 7:e39682
77. Duraisamy Y, Lambert D, O'Neill CA, Padfield PJ (2007) Differential incorporation of docosahexaenoic acid into distinct cholesterol-rich membrane raft domains. Biochem Biophys Res Commun 360:885–890
78. Collett ED, Davidson LA, Fan YY, Lupton JR, Chapkin RS (2001) n-6 and n-3 polyunsaturated fatty acids differentially modulate oncogenic Ras activation in colonocytes. Am J Physiol Cell Physiol 280:C1066–C1075
79. Eisenberg S, Beckett AJ, Prior IA, Dekker FJ, Hedberg C, Waldmann H, Ehrlich M, Henis YI (2011) Raft protein clustering alters N-Ras membrane interactions and activation pattern. Mol Cell Biol 31:3938–3952
80. Eisenberg S, Henis YI (2008) Interactions of Ras proteins with the plasma membrane and their roles in signaling. Cell Signal 20:31–39
81. Prior IA, Hancock JF (2012) Ras trafficking, localization and compartmentalized signalling. Semin Cell Dev Biol 23(2):145–153
82. Sehgal PB, Guo GG, Shah M, Kumar V, Patel K (2002) Cytokine signaling: STATS in plasma membrane rafts. J Biol Chem 277:12067–12074
83. Monk JM, Kim W, Callaway E, Turk HF, Foreman JE, Peters JM, He W, Weeks B, Alaniz RC, McMurray DN, Chapkin RS (2012) Immunomodulatory action of dietary fish oil and targeted deletion of intestinal epithelial cell PPARdelta in inflammation-induced colon carcinogenesis. Am J Physiol Gastrointest Liver Physiol 302:G153–G167
84. Ravacci GR, Brentani MM, Tortelli T Jr, Torrinhas RS, Saldanha T, Torres EA, Waitzberg DL (2012) Lipid raft disruption by docosahexaenoic acid induces apoptosis in transformed human mammary luminal epithelial cells harboring HER-2 overexpression. J Nutr Biochem 24:505–515
85. Mantovani A, Allavena P, Sica A, Balkwill F (2008) Cancer-related inflammation. Nature 454:436–444
86. Chen R, Rabinovitch PS, Crispin DA, Emond MJ, Bronner MP, Brentnall TA (2005) The initiation of colon cancer in a chronic inflammatory setting. Carcinogenesis 26:1513–1519
87. Ohshima H, Tazawa H, Sylla BS, Sawa T (2005) Prevention of human cancer by modulation of chronic inflammatory processes. Mutat Res 591:110–122
88. Shishodia S, Aggarwal BB (2004) Nuclear factor-kappaB: a friend or a foe in cancer? Biochem Pharmacol 68:1071–1080
89. Yan SR, Joseph RR, Rosen K, Reginato MJ, Jackson A, Allaire N, Brugge JS, Jobin C, Stadnyk AW (2005) Activation of NF-kappaB following detachment delays apoptosis in intestinal epithelial cells. Oncogene 24:6482–6491
90. Janakiram NB, Mohammed A, Rao CV (2011) Role of lipoxins, resolvins, and other bioactive lipids in colon and pancreatic cancer. Cancer Metastasis Rev 30:507–523
91. Tenesa A, Dunlop MG (2009) New insights into the aetiology of colorectal cancer from genome-wide association studies. Nat Rev Genet 10:353–358
92. Rustgi AK (2007) The genetics of hereditary colon cancer. Genes Dev 21:2525–2538
93. Chapkin RS, McMurray DN, Lupton JR (2007) Colon cancer, fatty acids and anti-inflammatory compounds. Curr Opin Gastroenterol 23:48–54
94. Feagins LA, Souza RF, Spechler SJ (2009) Carcinogenesis in IBD: potential targets for the prevention of colorectal cancer. Nat Rev Gastroenterol Hepatol 6:297–305

95. Lakatos PL, Lakatos L (2008) Risk for colorectal cancer in ulcerative colitis: changes, causes and management strategies. World J Gastroenterol 14:3937–3947
96. Choi J, Yoon SH, Kim JE, Rhee KH, Youn HS, Chung MH (2002) Gene-specific oxidative DNA damage in Helicobacter pylori-infected human gastric mucosa. Int J Cancer 99:485–490
97. Kraus S, Arber N (2009) Inflammation and colorectal cancer. Curr Opin Pharmacol 9:405–410
98. Bagga D, Wang L, Farias-Eisner R, Glaspy JA, Reddy ST (2003) Differential effects of prostaglandin derived from omega-6 and omega-3 polyunsaturated fatty acids on COX-2 expression and IL-6 secretion. Proc Natl Acad Sci USA 100:1751–1756
99. Dooper MM, Wassink L, M'Rabet L, Graus YM (2002) The modulatory effects of prostaglandin-E on cytokine production by human peripheral blood mononuclear cells are independent of the prostaglandin subtype. Immunology 107:152–159
100. Miles EA, Allen E, Calder PC (2002) In vitro effects of eicosanoids derived from different 20-carbon fatty acids on production of monocyte-derived cytokines in human whole blood cultures. Cytokine 20:215–223
101. Kelley DS, Taylor PC, Nelson GJ, Schmidt PC, Ferretti A, Erickson KL, Yu R, Chandra RK, Mackey BE (1999) Docosahexaenoic acid ingestion inhibits natural killer cell activity and production of inflammatory mediators in young healthy men. Lipids 34:317–324
102. Meydani SN, Endres S, Woods MM, Goldin BR, Soo C, Morrill-Labrode A, Dinarello CA, Gorbach SL (1991) Oral (n-3) fatty acid supplementation suppresses cytokine production and lymphocyte proliferation: comparison between young and older women. J Nutr 121:547–555
103. Rees D, Miles EA, Banerjee T, Wells SJ, Roynette CE, Wahle KW, Calder PC (2006) Dose-related effects of eicosapentaenoic acid on innate immune function in healthy humans: a comparison of young and older men. Am J Clin Nutr 83:331–342
104. Trebble TM, Wootton SA, Miles EA, Mullee M, Arden NK, Ballinger AB, Stroud MA, Burdge GC, Calder PC (2003) Prostaglandin E2 production and T cell function after fish-oil supplementation: response to antioxidant cosupplementation. Am J Clin Nutr 78:376–382
105. von Schacky C, Kiefl R, Jendraschak E, Kaminski WE (1993) n-3 fatty acids and cysteinyl-leukotriene formation in humans in vitro, ex vivo, and in vivo. J Lab Clin Med 121:302–309
106. Maitra A, Ashfaq R, Gunn CR, Rahman A, Yeo CJ, Sohn TA, Cameron JL, Hruban RH, Wilentz RE (2002) Cyclooxygenase 2 expression in pancreatic adenocarcinoma and pancreatic intraepithelial neoplasia: an immunohistochemical analysis with automated cellular imaging. Am J Clin Pathol 118:194–201
107. Rao CV, Reddy BS (2004) NSAIDs and chemoprevention. Curr Cancer Drug Targets 4:29–42
108. Buchanan FG, DuBois RN (2006) Connecting COX-2 and Wnt in cancer. Cancer Cell 9:6–8
109. Greenhough A, Wallam CA, Hicks DJ, Moorghen M, Williams AC, Paraskeva C (2010) The proapoptotic BH3-only protein Bim is downregulated in a subset of colorectal cancers and is repressed by antiapoptotic COX-2/PGE(2) signalling in colorectal adenoma cells. Oncogene 29:3398–3410
110. Park SW, Kim HS, Choi MS, Jeong WJ, Heo DS, Kim KH, Sung MW (2011) The effects of the stromal cell-derived cyclooxygenase-2 metabolite prostaglandin E2 on the proliferation of colon cancer cells. J Pharmacol Exp Ther 336:516–523
111. Wang D, Dubois RN (2010) The role of COX-2 in intestinal inflammation and colorectal cancer. Oncogene 29:781–788
112. Eberhart CE, Coffey RJ, Radhika A, Giardiello FM, Ferrenbach S, DuBois RN (1994) Up-regulation of cyclooxygenase 2 gene expression in human colorectal adenomas and adenocarcinomas. Gastroenterology 107:1183–1188
113. Kulkarni N, Zang E, Kelloff G, Reddy BS (1992) Effect of the chemopreventive agents piroxicam and D, L-alpha-difluoromethylornithine on intermediate biomarkers of colon carcinogenesis. Carcinogenesis 13:995–1000

114. Fan YY, Monk JM, Hou TY, Callway E, Vincent L, Weeks B, Yang P, Chapkin RS (2012) Characterization of an arachidonic acid-deficient (Fads1 knockout) mouse model. J Lipid Res 53:1287–1295
115. Badawi AF, El-Sohemy A, Stephen LL, Ghoshal AK, Archer MC (1998) The effect of dietary n-3 and n-6 polyunsaturated fatty acids on the expression of cyclooxygenase 1 and 2 and levels of p21ras in rat mammary glands. Carcinogenesis 19:905–910
116. Hamid R, Singh J, Reddy BS, Cohen LA (1999) Inhibition by dietary menhaden oil of cyclooxygenase-1 and -2 in N-nitrosomethylurea-induced rat mammary tumors. Int J Oncol 14:523–528
117. Ringbom T, Huss U, Stenholm A, Flock S, Skattebol L, Perera P, Bohlin L (2001) Cox-2 inhibitory effects of naturally occurring and modified fatty acids. J Nat Prod 64:745–749
118. Singh J, Hamid R, Reddy BS (1997) Dietary fat and colon cancer: modulation of cyclooxygenase-2 by types and amount of dietary fat during the postinitiation stage of colon carcinogenesis. Cancer Res 57:3465–3470
119. Gravaghi C, La Perle KM, Ogrodwski P, Kang JX, Quimby F, Lipkin M, Lamprecht SA (2011) Cox-2 expression, PGE(2) and cytokines production are inhibited by endogenously synthesized n-3 PUFA in inflamed colon of fat-1 mice. J Nutr Biochem 22:360–365
120. Calviello G, Di Nicuolo F, Gragnoli S, Piccioni E, Serini S, Maggiano N, Tringali G, Navarra P, Ranelletti FO, Palozza P (2004) n-3 PUFA reduce VEGF expression in human colon cancer cells modulating the COX-2/PGE2 induced ERK-1 and -2 and HIF-1alpha induction pathway. Carcinogenesis 25:2303–2310
121. Habbel P, Weylandt KH, Lichopoj K, Nowak J, Purschke M, Wang JD, He CW, Baumgart DC, Kang JX (2009) Docosahexaenoic acid suppresses arachidonic acid-induced proliferation of LS-174T human colon carcinoma cells. World J Gastroenterol 15:1079–1084
122. Swamy MV, Cooma I, Patlolla JM, Simi B, Reddy BS, Rao CV (2004) Modulation of cyclooxygenase-2 activities by the combined action of celecoxib and decosahexaenoic acid: novel strategies for colon cancer prevention and treatment. Mol Cancer Ther 3:215–221
123. Hong S, Gronert K, Devchand PR, Moussignac RL, Serhan CN (2003) Novel docosatrienes and 17S-resolvins generated from docosahexaenoic acid in murine brain, human blood, and glial cells. Autacoids in anti-inflammation. J Biol Chem 278:14677–14687
124. Serhan CN, Chiang N, Van Dyke TE (2008) Resolving inflammation: dual anti-inflammatory and pro-resolution lipid mediators. Nat Rev Immunol 8:349–361
125. Serhan CN, Clish CB, Brannon J, Colgan SP, Chiang N, Gronert K (2000) Novel functional sets of lipid-derived mediators with antiinflammatory actions generated from omega-3 fatty acids via cyclooxygenase 2-nonsteroidal antiinflammatory drugs and transcellular processing. J Exp Med 192:1197–1204
126. Mukherjee PK, Marcheselli VL, Serhan CN, Bazan NG (2004) Neuroprotectin D1: a docosahexaenoic acid-derived docosatriene protects human retinal pigment epithelial cells from oxidative stress. Proc Natl Acad Sci USA 101:8491–8496
127. Serhan CN, Hong S, Gronert K, Colgan SP, Devchand PR, Mirick G, Moussignac RL (2002) Resolvins: a family of bioactive products of omega-3 fatty acid transformation circuits initiated by aspirin treatment that counter proinflammation signals. J Exp Med 196:1025–1037
128. Bannenberg G, Arita M, Serhan CN (2007) Endogenous receptor agonists: resolving inflammation. ScientificWorldJournal 7:1440–1462
129. Arita M, Bianchini F, Aliberti J, Sher A, Chiang N, Hong S, Yang R, Petasis NA, Serhan CN (2005) Stereochemical assignment, antiinflammatory properties, and receptor for the omega-3 lipid mediator resolvin E1. J Exp Med 201:713–722
130. Bannenberg GL, Chiang N, Ariel A, Arita M, Tjonahen E, Gotlinger KH, Hong S, Serhan CN (2005) Molecular circuits of resolution: formation and actions of resolvins and protectins. J Immunol 174:4345–4355
131. Darnell JE Jr (2002) Transcription factors as targets for cancer therapy. Nat Rev Cancer 2:740–749
132. Karin M, Greten FR (2005) NF-kappaB: linking inflammation and immunity to cancer development and progression. Nat Rev Immunol 5:749–759

133. Kojima M, Morisaki T, Sasaki N, Nakano K, Mibu R, Tanaka M, Katano M (2004) Increased nuclear factor-kB activation in human colorectal carcinoma and its correlation with tumor progression. Anti-cancer Res 24:675–681
134. Meylan E, Dooley AL, Feldser DM, Shen L, Turk E, Ouyang C, Jacks T (2009) Requirement for NF-kappaB signalling in a mouse model of lung adenocarcinoma. Nature 462:104–107
135. Naugler WE, Karin M (2008) NF-kappaB and cancer-identifying targets and mechanisms. Curr Opin Genet Dev 18:19–26
136. Karin M (2006) Nuclear factor-kappaB in cancer development and progression. Nature 441:431–436
137. Greten FR, Eckmann L, Greten TF, Park JM, Li ZW, Egan LJ, Kagnoff MF, Karin M (2004) IKKbeta links inflammation and tumorigenesis in a mouse model of colitis-associated cancer. Cell 118:285–296
138. Kim S, Keku TO, Martin C, Galanko J, Woosley JT, Schroeder JC, Satia JA, Halabi S, Sandler RS (2008) Circulating levels of inflammatory cytokines and risk of colorectal adenomas. Cancer Res 68:323–328
139. Becker C, Fantini MC, Schramm C, Lehr HA, Wirtz S, Nikolaev A, Burg J, Strand S, Kiesslich R, Huber S, Ito H, Nishimoto N, Yoshizaki K, Kishimoto T, Galle PR, Blessing M, Rose-John S, Neurath MF (2004) TGF-beta suppresses tumor progression in colon cancer by inhibition of IL-6 trans-signaling. Immunity 21:491–501
140. Lee JY, Sohn KH, Rhee SH, Hwang D (2001) Saturated fatty acids, but not unsaturated fatty acids, induce the expression of cyclooxygenase-2 mediated through Toll-like receptor 4. J Biol Chem 276:16683–16689
141. Novak TE, Babcock TA, Jho DH, Helton WS, Espat NJ (2003) NF-kappa B inhibition by omega-3 fatty acids modulates LPS-stimulated macrophage TNF-alpha transcription. Am J Physiol Lung Cell Mol Physiol 284:L84–L89
142. Weldon SM, Mullen AC, Loscher CE, Hurley LA, Roche HM (2007) Docosahexaenoic acid induces an anti-inflammatory profile in lipopolysaccharide-stimulated human THP-1 macrophages more effectively than eicosapentaenoic acid. J Nutr Biochem 18:250–258
143. Zhao Y, Joshi-Barve S, Barve S, Chen LH (2004) Eicosapentaenoic acid prevents LPS-induced TNF-alpha expression by preventing NF-kappaB activation. J Am Coll Nutr 23:71–78
144. Nowak J, Weylandt KH, Habbel P, Wang J, Dignass A, Glickman JN, Kang JX (2007) Colitis-associated colon tumorigenesis is suppressed in transgenic mice rich in endogenous n-3 fatty acids. Carcinogenesis 28:1991–1995
145. Hudert CA, Weylandt KH, Lu Y, Wang J, Hong S, Dignass A, Serhan CN, Kang JX (2006) Transgenic mice rich in endogenous omega-3 fatty acids are protected from colitis. Proc Natl Acad Sci USA 103:11276–11281
146. Fan YY, Ly LH, Barhoumi R, McMurray DN, Chapkin RS (2004) Dietary docosahexaenoic acid suppresses T cell protein kinase C theta lipid raft recruitment and IL-2 production. J Immunol 173:6151–6160
147. Maziere C, Conte MA, Degonville J, Ali D, Maziere JC (1999) Cellular enrichment with polyunsaturated fatty acids induces an oxidative stress and activates the transcription factors AP1 and NFkappaB. Biochem Biophys Res Commun 265:116–122
148. Gee JM, Robertson JF, Ellis IO, Nicholson RI (2000) Impact of activation of MAP kinase family members on endocrine response and survival in clinical breast cancer. Eur J Cancer 36(suppl 4):105
149. Neyns B, Teugels E, Bourgain C, Birrerand M, De Greve J (1999) Alteration of jun proto-oncogene status by plasmid transfection affects growth of human ovarian cancer cells. Int J Cancer 82:687–693
150. Prusty BK, Das BC (2005) Constitutive activation of transcription factor AP-1 in cervical cancer and suppression of human papillomavirus (HPV) transcription and AP-1 activity in HeLa cells by curcumin. Int J Cancer 113:951–960
151. Wang H, Birkenbach M, Hart J (2000) Expression of Jun family members in human colorectal adenocarcinoma. Carcinogenesis 21:1313–1317

152. Ashida R, Tominaga K, Sasaki E, Watanabe T, Fujiwara Y, Oshitani N, Higuchi K, Mitsuyama S, Iwao H, Arakawa T (2005) AP-1 and colorectal cancer. Inflammopharmacology 13:113–125
153. Eferl R, Wagner EF (2003) AP-1: a double-edged sword in tumorigenesis. Nat Rev Cancer 3:859–868
154. Jia Q, Ivanov I, Zlatev ZZ, Alaniz RC, Weeks BR, Callaway ES, Goldsby JS, Davidson LA, Fan YY, Zhou L, Lupton JR, McMurray DN, Chapkin RS (2011) Dietary fish oil and curcumin combine to modulate colonic cytokinetics and gene expression in dextran sodium sulphate-treated mice. Br J Nutr 106:519–529
155. Devarajan E, Huang S (2009) STAT3 as a central regulator of tumor metastases. Curr Mol Med 9:626–633
156. Jarnicki A, Putoczki T, Ernst M (2010) Stat3: linking inflammation to epithelial cancer—more than a "gut" feeling? Cell Div 5:14
157. Fan YY, Spencer TE, Wang N, Moyer MP, Chapkin RS (2003) Chemopreventive n-3 fatty acids activate RXRalpha in colonocytes. Carcinogenesis 24:1541–1548
158. Jump DB (2002) The biochemistry of n-3 polyunsaturated fatty acids. J Biol Chem 277:8755–8758
159. Zand H, Rhimipour A, Bakhshayesh M, Shafiee M, Nour Mohammadi I, Salimi S (2007) Involvement of PPAR-gamma and p53 in DHA-induced apoptosis in Reh cells. Mol Cell Biochem 304:71–77
160. Carter AB, Misyak SA, Hontecillas R, Bassaganya-Riera J (2009) Dietary modulation of inflammation-induced colorectal cancer through PPARgamma. PPAR Res 2009:498352
161. Lefebvre M, Paulweber B, Fajas L, Woods J, McCrary C, Colombel JF, Najib J, Fruchart JC, Datz C, Vidal H, Desreumaux P, Auwerx J (1999) Peroxisome proliferator-activated receptor gamma is induced during differentiation of colon epithelium cells. J Endocrinol 162:331–340
162. Saez E, Tontonoz P, Nelson MC, Alvarez JG, Ming UT, Baird SM, Thomazy VA, Evans RM (1998) Activators of the nuclear receptor PPARgamma enhance colon polyp formation. Nat Med 4:1058–1061
163. Voutsadakis IA (2007) Peroxisome proliferator-activated receptor gamma (PPARgamma) and colorectal carcinogenesis. J Cancer Res Clin Oncol 133:917–928
164. Szanto A, Nagy L (2008) The many faces of PPARgamma: anti-inflammatory by any means? Immunobiology 213:789–803
165. Thompson EA (2007) PPARgamma physiology and pathology in gastrointestinal epithelial cells. Mol Cells 24:167–176
166. Karrasch T, Jobin C (2008) NF-kappaB and the intestine: friend or foe? Inflamm Bowel Dis 14:114–124
167. Pascual G, Fong AL, Ogawa S, Gamliel A, Li AC, Perissi V, Rose DW, Willson TM, Rosenfeld MG, Glass CK (2005) A SUMOylation-dependent pathway mediates transrepression of inflammatory response genes by PPAR-gamma. Nature 437:759–763
168. Kong W, Yen JH, Vassiliou E, Adhikary S, Toscano MG, Ganea D (2010) Docosahexaenoic acid prevents dendritic cell maturation and in vitro and in vivo expression of the IL-12 cytokine family. Lipids Health Dis 9:12
169. Zapata-Gonzalez F, Rueda F, Petriz J, Domingo P, Villarroya F, Diaz-Delfin J, de Madariaga MA, Domingo JC (2008) Human dendritic cell activities are modulated by the omega-3 fatty acid, docosahexaenoic acid, mainly through PPAR(gamma):RXR heterodimers: comparison with other polyunsaturated fatty acids. J Leukoc Biol 84:1172–1182
170. Zhao P, Xia HR, Lu GW, Li LX, Meng XL (2004) [The analysis for Raman spectra of Nd:YVO4]. Guang Pu Xue Yu Guang Pu Fen Xi 24:71–73
171. Forman BM, Chen J, Evans RM (1997) Hypolipidemic drugs, polyunsaturated fatty acids, and eicosanoids are ligands for peroxisome proliferator-activated receptors alpha and delta. Proc Natl Acad Sci USA 94:4312–4317
172. Kliewer SA, Sundseth SS, Jones SA, Brown PJ, Wisely GB, Koble CS, Devchand P, Wahli W, Willson TM, Lenhard JM, Lehmann JM (1997) Fatty acids and eicosanoids regulate gene expression through direct interactions with peroxisome proliferator-activated receptors alpha and gamma. Proc Natl Acad Sci USA 94:4318–4323

173. Xu HE, Lambert MH, Montana VG, Parks DJ, Blanchard SG, Brown PJ, Sternbach DD, Lehmann JM, Wisely GB, Willson TM, Kliewer SA, Milburn MV (1999) Molecular recognition of fatty acids by peroxisome proliferator-activated receptors. Mol Cell 3:397–403
174. Oh DY, Talukdar S, Bae EJ, Imamura T, Morinaga H, Fan W, Li P, Lu WJ, Watkins SM, Olefsky JM (2010) GPR120 is an omega-3 fatty acid receptor mediating potent anti-inflammatory and insulin-sensitizing effects. Cell 142:687–698
175. Im DS (2012) Omega-3 fatty acids in anti-inflammation (pro-resolution) and GPCRs. Prog Lipid Res 51:232–237
176. Chandramouli A, Mercado-Pimentel ME, Hutchinson A, Gibadulinova A, Olson ER, Dickinson S, Shanas R, Davenport J, Owens J, Bhattacharyya AK, Regan JW, Pastorekova S, Arumugam T, Logsdon CD, Nelson MA (2010) The induction of S100p expression by the prostaglandin E(2) (PGE(2))/EP4 receptor signaling pathway in colon cancer cells. Cancer Biol Ther 10:1056–1066
177. Cherukuri DP, Chen XB, Goulet AC, Young RN, Han Y, Heimark RL, Regan JW, Meuillet E, Nelson MA (2007) The EP4 receptor antagonist, L-161,982, blocks prostaglandin E2-induced signal transduction and cell proliferation in HCA-7 colon cancer cells. Exp Cell Res 313:2969–2979
178. Hawcroft G, Loadman PM, Belluzzi A, Hull MA (2010) Effect of eicosapentaenoic acid on E-type prostaglandin synthesis and EP4 receptor signaling in human colorectal cancer cells. Neoplasia 12:618–627
179. Arita M, Ohira T, Sun YP, Elangovan S, Chiang N, Serhan CN (2007) Resolvin E1 selectively interacts with leukotriene B4 receptor BLT1 and ChemR23 to regulate inflammation. J Immunol 178:3912–3917
180. Krishnamoorthy S, Recchiuti A, Chiang N, Yacoubian S, Lee CH, Yang R, Petasis NA, Serhan CN (2010) Resolvin D1 binds human phagocytes with evidence for proresolving receptors. Proc Natl Acad Sci USA 107:1660–1665
181. Fan YY, McMurray DN, Ly LH, Chapkin RS (2003) Dietary (n-3) polyunsaturated fatty acids remodel mouse T-cell lipid rafts. J Nutr 133:1913–1920
182. Simons K, Vaz WL (2004) Model systems, lipid rafts, and cell membranes. Annu Rev Biophys Biomol Struct 33:269–295
183. Zech T, Ejsing CS, Gaus K, de Wet B, Shevchenko A, Simons K, Harder T (2009) Accumulation of raft lipids in T-cell plasma membrane domains engaged in TCR signalling. EMBO J 28:466–476
184. Rentero C, Zech T, Quinn CM, Engelhardt K, Williamson D, Grewal T, Jessup W, Harder T, Gaus K (2008) Functional implications of plasma membrane condensation for T cell activation. PLoS One 3:e2262
185. Williams JA, Batten SE, Harris M, Rockett BD, Shaikh SR, Stillwell W, Wassall SR (2012) Docosahexaenoic and eicosapentaenoic acids segregate differently between raft and nonraft domains. Biophys J 103:228–237
186. Kwiatkowska K (2010) One lipid, multiple functions: how various pools of PI(4,5)P(2) are created in the plasma membrane. Cell Mol Life Sci 67:3927–3946
187. Holub BJ, Kuksis A, Thompson W (1970) Molecular species of mono-, di-, and triphosphoinositides of bovine brain. J Lipid Res 11:558–564
188. Inoue M, Murase S, Okuyama H (1984) Acyl coenzyme a: phospholipid acyltransferases in porcine platelets discriminate between omega-3 and omega-6 unsaturated fatty acids. Arch Biochem Biophys 231:29–37
189. Hou TY, Monk JM, Fan YY, Barhoumi R, Chen YQ, Rivera GM, McMurray DN, Chapkin RS (2012) n-3 polyunsaturated fatty acids suppress phosphatidylinositol 4,5-bisphosphate-dependent actin remodelling during CD4+ T-cell activation. Biochem J 443:27–37
190. Jolly CA, Jiang YH, Chapkin RS, McMurray DN (1997) Dietary (n-3) polyunsaturated fatty acids suppress murine lymphoproliferation, interleukin-2 secretion, and the formation of diacylglycerol and ceramide. J Nutr 127:37–43
191. Dustin ML, Depoil D (2011) New insights into the T cell synapse from single molecule techniques. Nat Rev Immunol 11:672–684

192. Tybulewicz VL, Henderson RB (2009) Rho family GTPases and their regulators in lymphocytes. Nat Rev Immunol 9:630–644
193. Webb Y, Hermida-Matsumoto L, Resh MD (2000) Inhibition of protein palmitoylation, raft localization, and T cell signaling by 2-bromopalmitate and polyunsaturated fatty acids. J Biol Chem 275:261–270
194. Liang X, Nazarian A, Erdjument-Bromage H, Bornmann W, Tempst P, Resh MD (2001) Heterogeneous fatty acylation of Src family kinases with polyunsaturated fatty acids regulates raft localization and signal transduction. J Biol Chem 276:30987–30994
195. Zeyda M, Staffler G, Horejsi V, Waldhausl W, Stulnig TM (2002) LAT displacement from lipid rafts as a molecular mechanism for the inhibition of T cell signaling by polyunsaturated fatty acids. J Biol Chem 277:28418–28423
196. Zeyda M, Szekeres AB, Saemann MD, Geyeregger R, Stockinger H, Zlabinger GJ, Waldhausl W, Stulnig TM (2003) Suppression of T cell signaling by polyunsaturated fatty acids: selectivity in inhibition of mitogen-activated protein kinase and nuclear factor activation. J Immunol 170:6033–6039
197. Barr VA, Balagopalan L, Barda-Saad M, Polishchuk R, Boukari H, Bunnell SC, Bernot KM, Toda Y, Nossal R, Samelson LE (2006) T-cell antigen receptor-induced signaling complexes: internalization via a cholesterol-dependent endocytic pathway. Traffic 7:1143–1162
198. Chichili GR, Westmuckett AD, Rodgers W (2010) T cell signal regulation by the actin cytoskeleton. J Biol Chem 285:14737–14746
199. Hashimoto-Tane A, Yokosuka T, Ishihara C, Sakuma M, Kobayashi W, Saito T (2010) T-cell receptor microclusters critical for T-cell activation are formed independently of lipid raft clustering. Mol Cell Biol 30:3421–3429
200. Eibert SM, Lee KH, Pipkorn R, Sester U, Wabnitz GH, Giese T, Meuer SC, Samstag Y (2004) Cofilin peptide homologs interfere with immunological synapse formation and T cell activation. Proc Natl Acad Sci USA 101:1957–1962
201. Faure S, Salazar-Fontana LI, Semichon M, Tybulewicz VL, Bismuth G, Trautmann A, Germain RN, Delon J (2004) ERM proteins regulate cytoskeleton relaxation promoting T cell-APC conjugation. Nat Immunol 5:272–279
202. Huang Y, Burkhardt JK (2007) T-cell-receptor-dependent actin regulatory mechanisms. J Cell Sci 120:723–730
203. Nolz JC, Medeiros RB, Mitchell JS, Zhu P, Freedman BD, Shimizu Y, Billadeau DD (2007) WAVE2 regulates high-affinity integrin binding by recruiting vinculin and talin to the immunological synapse. Mol Cell Biol 27:5986–6000
204. Padrick SB, Cheng HC, Ismail AM, Panchal SC, Doolittle LK, Kim S, Skehan BM, Umetani J, Brautigam CA, Leong JM, Rosen MK (2008) Hierarchical regulation of WASP/WAVE proteins. Mol Cell 32:426–438
205. Geyeregger R, Zeyda M, Zlabinger GJ, Waldhausl W, Stulnig TM (2005) Polyunsaturated fatty acids interfere with formation of the immunological synapse. J Leukoc Biol 77:680–688
206. Quann EJ, Liu X, Altan-Bonnet G, Huse M (2011) A cascade of protein kinase C isozymes promotes cytoskeletal polarization in T cells. Nat Immunol 12:647–654
207. Quintana A, Schwindling C, Wenning AS, Becherer U, Rettig J, Schwarz EC, Hoth M (2007) T cell activation requires mitochondrial translocation to the immunological synapse. Proc Natl Acad Sci USA 104:14418–14423
208. Schwindling C, Quintana A, Krause E, Hoth M (2010) Mitochondria positioning controls local calcium influx in T cells. J Immunol 184:184–190
209. Quintana A, Pasche M, Junker C, Al-Ansary D, Rieger H, Kummerow C, Nunez L, Villalobos C, Meraner P, Becherer U, Rettig J, Niemeyer BA, Hoth M (2011) Calcium microdomains at the immunological synapse: how ORAI channels, mitochondria and calcium pumps generate local calcium signals for efficient T-cell activation. EMBO J 30:3895–3912
210. Muller MR, Rao A (2010) NFAT, immunity and cancer: a transcription factor comes of age. Nat Rev Immunol 10:645–656

211. Padhan K, Varma R (2010) Immunological synapse: a multi-protein signalling cellular apparatus for controlling gene expression. Immunology 129:322–328
212. Tuosto L (2011) NF-kappaB family of transcription factors: biochemical players of CD28 co-stimulation. Immunol Lett 135:1–9
213. Waite JC, Skokos D (2012) Th17 response and inflammatory autoimmune diseases. Int J Inflam 2012:819467
214. Balkwill F, Coussens LM (2004) Cancer: an inflammatory link. Nature 431:405–406
215. Miyahara Y, Odunsi K, Chen W, Peng G, Matsuzaki J, Wang RF (2008) Generation and regulation of human CD4+ IL-17-producing T cells in ovarian cancer. Proc Natl Acad Sci USA 105:15505–15510
216. Su X, Ye J, Hsueh EC, Zhang Y, Hoft DF, Peng G (2010) Tumor microenvironments direct the recruitment and expansion of human Th17 cells. J Immunol 184:1630–1641
217. Kryczek I, Wei S, Zou L, Altuwaijri S, Szeliga W, Kolls J, Chang A, Zou W (2007) Cutting edge: Th17 and regulatory T cell dynamics and the regulation by IL-2 in the tumor microenvironment. J Immunol 178:6730–6733
218. Sfanos KS, Bruno TC, Maris CH, Xu L, Thoburn CJ, DeMarzo AM, Meeker AK, Isaacs WB, Drake CG (2008) Phenotypic analysis of prostate-infiltrating lymphocytes reveals TH17 and Treg skewing. Clin Cancer Res 14:3254–3261
219. Zhang B, Rong G, Wei H, Zhang M, Bi J, Ma L, Xue X, Wei G, Liu X, Fang G (2008) The prevalence of Th17 cells in patients with gastric cancer. Biochem Biophys Res Commun 374:533–537
220. Kato T, Furumoto H, Ogura T, Onishi Y, Irahara M, Yamano S, Kamada M, Aono T (2001) Expression of IL-17 mRNA in ovarian cancer. Biochem Biophys Res Commun 282:735–738
221. Steiner GE, Newman ME, Paikl D, Stix U, Memaran-Dagda N, Lee C, Marberger MJ (2003) Expression and function of pro-inflammatory interleukin IL-17 and IL-17 receptor in normal, benign hyperplastic, and malignant prostate. Prostate 56:171–182
222. Numasaki M, Fukushi J, Ono M, Narula SK, Zavodny PJ, Kudo T, Robbins PD, Tahara H, Lotze MT (2003) Interleukin-17 promotes angiogenesis and tumor growth. Blood 101:2620–2627
223. Tesmer LA, Lundy SK, Sarkar S, Fox DA (2008) Th17 cells in human disease. Immunol Rev 223:87–113
224. Maniati E, Soper R, Hagemann T (2010) Up for mischief? IL-17/Th17 in the tumour microenvironment. Oncogene 29:5653–5662
225. Le Gouvello S, Bastuji-Garin S, Aloulou N, Mansour H, Chaumette MT, Berrehar F, Seikour A, Charachon A, Karoui M, Leroy K, Farcet JP, Sobhani I (2008) High prevalence of Foxp3 and IL17 in MMR-proficient colorectal carcinomas. Gut 57:772–779
226. Liu J, Duan Y, Cheng X, Chen X, Xie W, Long H, Lin Z, Zhu B (2011) IL-17 is associated with poor prognosis and promotes angiogenesis via stimulating VEGF production of cancer cells in colorectal carcinoma. Biochem Biophys Res Commun 407:348–354
227. Tosolini M, Kirilovsky A, Mlecnik B, Fredriksen T, Mauger S, Bindea G, Berger A, Bruneval P, Fridman WH, Pages F, Galon J (2011) Clinical impact of different classes of infiltrating T cytotoxic and helper cells (Th1, th2, treg, th17) in patients with colorectal cancer. Cancer Res 71:1263–1271
228. Chae WJ, Gibson TF, Zelterman D, Hao L, Henegariu O, Bothwell AL (2010) Ablation of IL-17A abrogates progression of spontaneous intestinal tumorigenesis. Proc Natl Acad Sci USA 107:5540–5544
229. Nam JS, Terabe M, Kang MJ, Chae H, Voong N, Yang YA, Laurence A, Michalowska A, Mamura M, Lonning S, Berzofsky JA, Wakefield LM (2008) Transforming growth factor beta subverts the immune system into directly promoting tumor growth through interleukin-17. Cancer Res 68:3915–3923
230. Wu S, Rhee KJ, Albesiano E, Rabizadeh S, Wu X, Yen HR, Huso DL, Brancati FL, Wick E, McAllister F, Housseau F, Pardoll DM, Sears CL (2009) A human colonic commensal promotes colon tumorigenesis via activation of T helper type 17T cell responses. Nat Med 15:1016–1022

231. He D, Li H, Yusuf N, Elmets CA, Li J, Mountz JD, Xu H (2010) IL-17 promotes tumor development through the induction of tumor promoting microenvironments at tumor sites and myeloid-derived suppressor cells. J Immunol 184:2281–2288
232. Wang L, Yi T, Kortylewski M, Pardoll DM, Zeng D, Yu H (2009) IL-17 can promote tumor growth through an IL-6-Stat3 signaling pathway. J Exp Med 206:1457–1464
233. Hyun YS, Han DS, Lee AR, Eun CS, Youn J, Kim HY (2012) Role of IL-17A in the development of colitis-associated cancer. Carcinogenesis 33:931–936
234. Durant L, Watford WT, Ramos HL, Laurence A, Vahedi G, Wei L, Takahashi H, Sun HW, Kanno Y, Powrie F, O'Shea JJ (2010) Diverse targets of the transcription factor STAT3 contribute to T cell pathogenicity and homeostasis. Immunity 32:605–615
235. Kortylewski M, Xin H, Kujawski M, Lee H, Liu Y, Harris T, Drake C, Pardoll D, Yu H (2009) Regulation of the IL-23 and IL-12 balance by Stat3 signaling in the tumor microenvironment. Cancer Cell 15:114–123
236. Maloy KJ, Kullberg MC (2008) IL-23 and Th17 cytokines in intestinal homeostasis. Mucosal Immunol 1:339–349
237. Langowski JL, Zhang X, Wu L, Mattson JD, Chen T, Smith K, Basham B, McClanahan T, Kastelein RA, Oft M (2006) IL-23 promotes tumour incidence and growth. Nature 442:461–465
238. Fiaschi T, Chiarugi P (2012) Oxidative stress, tumor microenvironment, and metabolic reprogramming: a diabolic liaison. Int J Cell Biol 2012:762825
239. Babior BM (1999) NADPH oxidase: an update. Blood 93:1464–1476
240. Chiarugi P, Fiaschi T (2007) Redox signalling in anchorage-dependent cell growth. Cell Signal 19:672–682
241. Storz P (2005) Reactive oxygen species in tumor progression. Front Biosci 10:1881–1896
242. Chapkin RS, Hong MY, Fan YY, Davidson LA, Sanders LM, Henderson CE, Barhoumi R, Burghardt RC, Turner ND, Lupton JR (2002) Dietary n-3 PUFA alter colonocyte mitochondrial membrane composition and function. Lipids 37:193–199
243. Hong MY, Chapkin RS, Barhoumi R, Burghardt RC, Turner ND, Henderson CE, Sanders LM, Fan YY, Davidson LA, Murphy ME, Spinka CM, Carroll RJ, Lupton JR (2002) Fish oil increases mitochondrial phospholipid unsaturation, upregulating reactive oxygen species and apoptosis in rat colonocytes. Carcinogenesis 23:1919–1925
244. Costantini P, Jacotot E, Decaudin D, Kroemer G (2000) Mitochondrion as a novel target of anti-cancer chemotherapy. J Natl Cancer Inst 92:1042–1053
245. Ng Y, Barhoumi R, Tjalkens RB, Fan YY, Kolar S, Wang N, Lupton JR, Chapkin RS (2005) The role of docosahexaenoic acid in mediating mitochondrial membrane lipid oxidation and apoptosis in colonocytes. Carcinogenesis 26:1914–1921
246. Fan YY, Ran Q, Toyokuni S, Okazaki Y, Callaway ES, Lupton JR, Chapkin RS (2011) Dietary fish oil promotes colonic apoptosis and mitochondrial proton leak in oxidatively stressed mice. Cancer Prev Res 4:1267–1274
247. Kolar S, Barhoumi R, Jones CK, Wesley J, Lupton JR, Fan YY, Chapkin RS (2011) Interactive effects of fatty acid and butyrate-induced mitochondrial Ca(2)(+) loading and apoptosis in colonocytes. Cancer 117:5294–5303
248. Kolar SS, Barhoumi R, Callaway ES, Fan YY, Wang N, Lupton JR, Chapkin RS (2007) Synergy between docosahexaenoic acid and butyrate elicits p53-independent apoptosis via mitochondrial Ca(2+) accumulation in colonocytes. Am J Physiol Gastrointest Liver Physiol 293:G935–G943
249. Crim KC, Sanders LM, Hong MY, Taddeo SS, Turner ND, Chapkin RS, Lupton JR (2008) Upregulation of p21Waf1/Cip1 expression in vivo by butyrate administration can be chemoprotective or chemopromotive depending on the lipid component of the diet. Carcinogenesis 29:1415–1420
250. Sanders LM, Henderson CE, Hong MY, Barhoumi R, Burghardt RC, Wang N, Spinka CM, Carroll RJ, Turner ND, Chapkin RS, Lupton JR (2004) An increase in reactive oxygen species by dietary fish oil coupled with the attenuation of antioxidant defenses by dietary pectin enhances rat colonocyte apoptosis. J Nutr 134:3233–3238

251. Saw CL, Huang Y, Kong AN (2010) Synergistic anti-inflammatory effects of low doses of curcumin in combination with polyunsaturated fatty acids: docosahexaenoic acid or eicosapentaenoic acid. Biochem Pharmacol 79:421–430
252. Kim W, Fan YY, Smith R, Patil B, Jayaprakasha GK, McMurray DN, Chapkin RS (2009) Dietary curcumin and limonin suppress CD4+ T-cell proliferation and interleukin-2 production in mice. J Nutr 139:1042–1048
253. Simopoulos AP (1991) Omega-3 fatty acids in health and disease and in growth and development. Am J Clin Nutr 54:438–463
254. Blasbalg TL, Hibbeln JR, Ramsden CE, Majchrzak SF, Rawlings RR (2011) Changes in consumption of omega-3 and omega-6 fatty acids in the United States during the 20th century. Am J Clin Nutr 93:950–962
255. Russo GL (2009) Dietary n-6 and n-3 polyunsaturated fatty acids: from biochemistry to clinical implications in cardiovascular prevention. Biochem Pharmacol 77:937–946
256. Simopoulos AP (2002) The importance of the ratio of omega-6/omega-3 essential fatty acids. Biomed Pharmacother 56:365–379
257. Colditz GA, Wolin KY, Gehlert S (2012) Applying what we know to accelerate cancer prevention. Sci Transl Med 4:127rv124
258. Moses H III, Dorsey ER, Matheson DH, Thier SO (2005) Financial anatomy of biomedical research. JAMA 294:1333–1342
259. Kaiser J (2012) Profile: Bert Vogelstein. Cancer genetics with an edge. Science 337:282–284

Chapter 13
Natural Products as Anti-inflammatory Agents

Gary Stoner and Li-Shu Wang

Abstract Inflammation is triggered by numerous factors including oxidative stress, environmental pollutants, microbial agents, and physical damage to tissues. Chronic inflammation, characterized by a prevalence of macrophages and lymphocytes in the affected tissues and the overexpression of a host of cellular cytokines, chemokines, and inflammatory enzymes, promotes all stages of cancer development including initiation, promotion, cell transformation, angiogenesis, invasion, and metastasis. This chapter describes some key mechanisms by which naturally occurring dietary compounds, either alone or in combination, reduce the harmful effects of inflammation and the risk for cancer development. The most extensively studied compounds are a series of polyphenols which influence the inflammatory process in multiple ways including their ability to scavenge oxidative radicals, influence carcinogen activation and detoxification, and regulate expression levels of numerous transcription activators and their associated cytokines, chemokines, and inflammatory enzymes. In the past, the inhibitory effects of naturally occurring compounds, especially the polyphenols, have been attributed mainly to their intrinsic antioxidant capacity; however, it is likely that their direct binding to cellular macromolecules and the associated effects on gene transcription and translation, as well as on enzyme activity, may be equally as important.

G. Stoner, Ph.D. (✉) • L.-S. Wang, Ph.D.
Division of Hematology and Oncology, Department of Medicine, Medical College of Wisconsin, 8701 Watertown Plank Road, Milwaukee, WI 53226, USA
e-mail: gstoner@mcw.edu; liswang@mcw.edu

A.J. Dannenberg and N.A. Berger (eds.), *Obesity, Inflammation and Cancer*, Energy Balance and Cancer 7, DOI 10.1007/978-1-4614-6819-6_13,

13.1 Introduction

It has been known for decades that the cellular components of the inflammatory response are critical for their role in the killing of cancer cells and, hopefully, curing the disease. In recent years, however, it has become apparent that inflammation, especially chronic inflammation, is also a significant risk factor for the development of cancer. Chronic inflammation has been associated with the various stages of tumorigenesis including initiation, promotion, progression, invasion, and metastasis [1, 2]. One of the many cancer types in which there is a strong association between chronic inflammation and tumor development is colon cancer. Meta-analysis has shown that patients suffering from inflammatory bowel diseases, such as Crohn's disease and ulcerative colitis, have a 33-fold increase in risk for colon cancer relative to the general population [3]. Inflammatory bowel diseases affect approximately 1–2 of every 1,000 people in developed countries and are on the rise worldwide [4]. Links between inflammation and cancer have also been confirmed in a number of animal models, including cancers of the colon [5], liver [6], pancreas [7], prostate [8], and lung [9].

One mechanism through which chronic inflammation and carcinogenesis are linked is via the activities of reactive oxygen (ROS) and nitrogen (RNS) species (i.e., RONS). In normal tissues and cells, RONS are important for regulating signaling cascades that govern multiple cellular functions. However, in an inflammatory microenvironment, infiltrating leukocytes raise the level of RONS to amounts that exceed the ability of cells to eliminate them and the cells become "stressed." Stress occurs because the activities of antioxidant phase II enzymes such as superoxide dismutase, catalase, glutathione peroxidase, quinone oxidoreductase, epoxide hydrolase, and others are not sufficient to remove the excess RONS. The infiltrating leukocytes that produce RONS include macrophages, monocytes, neutrophils, T and B lymphocytes, mast cells, and natural killer cells [1, 2]. Some of these cell types produce growth factors, cytokines, and chemokines that stimulate cell proliferation, angiogenesis, invasion, and metastasis. These factors also chemo-attract additional lymphocytes, monocytes, and macrophages leading to a chronically activated microenvironment that becomes a relentless source of genetic and histopathologic damage to the surrounding epithelium. Ultimately, this damage can lead to the development of cancer in all of its stages.

Epidemiological studies provide evidence that persons who consume a diet rich in vegetables and fruit have a reduced risk of cancer at multiple organ sites [10–12]. Because fruits and vegetables are major sources of antioxidants, it is thought that the antioxidants are largely responsible for their cancer-preventive effects. This hypothesis is supported by epidemiological data, implicating micronutrient antioxidants in cancer risk reduction, and by experimental data from in vitro and in vivo studies [13, 14]. The antioxidant effects of vegetables and fruits are due, in part, to their content of flavonoids and phenolic acids [15–17]. Some of the polyphenolic compounds with demonstrated cancer-preventive effects include the catechins in green tea, theaflavins in black tea, curcumin in turmeric, resveratrol in red wine, quercetin in apples, and ellagic acid and the anthocyanins in berries. Collectively, these antioxidants exhibit a significant ability to scavenge RONS [15, 18, 19], and this is associated, at least in part, with their inhibitory effects on multiple cell-signaling pathways and inflammatory processes. For example, one of the most extensively investigated polyphenols is curcumin, present in the Indian spice, turmeric. Cell-signaling pathways that are inhibited by curcumin alone include activator protein-1 (AP-1), nuclear factor-kappa B (NF-κB), signal transducer and activator of transcription 3 (STAT3), mitogen-activated protein kinase (MAPK), protein kinase B (Akt), cyclooxygenase-2 (COX-2), 5-lipoxygenase (5-LOX), c-Jun N terminal kinase (JNK), human growth factor receptor 2 (HER2), IκB kinase (IKK), epidermal growth factor receptor (EGFR), B-cell lymphoma 2 (Bcl-2), B-cell lymphoma extra large (Bcl-X_L), poly ADP ribose polymerase (PARP), and proapoptotic caspases [20]. Curcumin and other anticancer compounds are also effective in reducing the activities of inflammatory cytokines and chemokines as well as in influencing the trafficking of leukocytes that participate in the inflammatory process.

In the remaining portion of this chapter, we summarize some of the known molecular mechanisms by which naturally occurring compounds, foods, and food extracts elicit anti-inflammatory effects. We refer to whole foods and food extracts as "mixtures" of bioactive compounds. Mechanisms not discussed in this chapter include the effects of naturally occurring compounds on inflammation-associated phase II enzyme induction and angiogenesis; these mechanisms will be discussed elsewhere in this series.

13.2 Inflammation-Related Molecular Targets of Natural Products

13.2.1 Transcription Activators

Transcription activators (factors) are proteins that bind to specific DNA sequences, thereby controlling the transcription of genetic information from DNA to mRNA. Transcription factors perform this function alone or with other proteins in a complex by promoting or blocking the recruitment of RNA polymerase to specific genes. Abnormal expression of these factors can lead to many of the hallmarks of cancer including increased proliferation, reduced apoptosis, malignant conversion, inflammation, angiogenesis, tissue invasion, and metastasis. A large number of transcription factors have been identified in the past few decades and several of these are associated with inflammatory processes and cancer [1, 20]. The following section describes two of the most investigated transcription factors in inflammation and tumor development and mechanisms by which they are influenced by dietary agents. For a more complete discussion on this topic, the reader is referred to the review of Aggarwal and Shishodia [20].

13.2.1.1 Nuclear Factor-Kappa B

NF-κB is a protein complex that is present in nearly all animal cell types [21, 22]. It is important in regulating cellular responses to various stimuli because it belongs to a family of "rapid acting" primary transcription factors, i.e., transcription factors present in cells in an inactive state that do not require new protein synthesis to be activated. Known inducers of NF-κB activity include reactive oxygen species, inflammatory factors such as tumor necrosis factor-alpha (TNFα) and interleukin 1-beta (IL-1β), bacterial endotoxins, chemical carcinogens and toxins, ultraviolet light, and X-rays [20]. Activation of the NF-κB complex is initiated by signal-induced degradation of I kappa B (IκB) proteins. This occurs primarily through activation of a kinase termed IKK. With the degradation of IκB, the NF-κB complex translocates from the cytoplasm to the nucleus where, ultimately, it induces the expression of an ever-increasing number of target genes, including genes that suppress apoptosis (Bcl-2 and Bcl-XL), induce proliferation (cyclin D1, c-myc) and angiogenesis (vascular endothelial growth factor [VEGF]), and promote tissue invasion and metastasis (matrix metalloproteases [MMP]) and inflammation (COX-2, 5-LOX, inducible nitric oxide synthase [iNOS], TNF-α, NADPH oxidase [NOX], and others). Many of these activated target genes are crucial to the development of cancer at multiple tissue sites.

Because of its central role in the control of several cancer-associated cellular functions, numerous investigations have been conducted to identify dietary inhibitors of NF-κB [20]. Among those identified include resveratrol [21], curcumin [22], ursolic acid [23], lycopene [24], apigenin [25], silymarin [26], ellagic acid [27], cyanidin-3-glucoside [28], caffeic acid phenethyl ester (CAPE) [29], epigallocatechin-3-gallate (EGCG) [30], [6]-gingerol [31], S-allyl cysteine [32], flavopiridol [33], benzyl and phenylethyl isothiocyanates [34, 35], indole-3-carbinol [36], sulforaphane [37], quercetin [38], anethole [39], and alpha-lipoic acid [40]. In addition to pure compounds, various "mixtures" of naturally occurring compounds, including soy isoflavones (genistein, daidzein, and glycitein) [41], black tea extract (BTE) [42], lyophilized whole black raspberries (BRBs) and strawberries [43, 44], anthocyanin extracts of black raspberries [43], ginger extract [45], grape seed proanthocyanidins [46, 47], and cranberry proanthocyanidins [48, 49], have been shown to exhibit inhibitory effects on NF-κB activation in vitro and in vivo.

Aggarwal and Shishodia [20] discussed mechanisms by which naturally occurring compounds inhibit NF-κB activation. These include the blocking of one or more steps in the NF-κB signaling pathway, the translocation of NF-κB into the nucleus, binding of the NF-κB dimers to DNA, or interactions with the basal transcriptional machinery. Caffeine, for example, is a potent inhibitor of ultraviolet (UV)-induced NF-κB activation by its inhibitory action on protein kinase C (PKC), resulting in the inhibition of p38-induced activation of the MAPK pathway [50]. Resveratrol suppresses tumor necrosis factor (TNF)-induced phosphorylation and nuclear translocation of the p65 subunit of NF-κB and NF-κB reporter gene transcription [21]. Diosgenin suppresses TNF-induced activation of IKK leading to inhibition of TNF-dependent phosphorylation and degradation of IκBα and nuclear translocation of the p65 subunit of NF-κB [51]. CAPE suppresses NF-κB activation by interfering with the binding of the p50-p65 complex to DNA [29]. BTE inhibits the nuclear translocation of NF-κB/p65 and the DNA-binding activity of NF-κB [42]. Cranberry proanthocyanidins inhibit the production of MMP by reducing the phosphorylation of at least five intracellular kinases and NF-κB p65 activity [48]. These represent only a few mechanisms by which dietary factors inhibit NFκB activation, and ultimately, these actions reduce the expression of numerous NF-κB-related proteins such as COX-2, iNOS, TNF, and various cytokines and chemokines associated with inflammation.

13.2.1.2 Activator Protein-1

Activator protein-1 (AP-1) is a transcriptional regulator composed of members of the Fos and Jun families of DNA-binding proteins. AP-1 regulates gene expression in response to a variety of stimuli including growth factors, cytokines, environmental stresses such as UV radiation, and bacterial and viral infections. AP-1 activation is linked to increased cell proliferation, cell transformation, innate immune response, and inflammation. AP-1 appears to enhance cell proliferation by activating cyclin D1 and repressing suppressor genes such as p53, p16, and p21cipl/wafl [52]. Importantly, AP-1 promotes the transition of tumor cells from an epithelial to a mesenchymal morphology which is one of the early steps in metastasis. Both AP-1 and NF-κB are inducible by many of the same stimuli, and in mouse epidermal JB-6 cells, both are required for maintaining the transformed phenotype [53]. With respect to inflammation, many of the cytokine genes and other factors that drive inflammation are regulated cooperatively by a transcription factor complex consisting of AP-1 and nuclear factor of activated T cells (NFAT). AP-1/NFAT-dependent gene regulation has been demonstrated for the proinflammatory cytokines: interleukins 1, 2, and 6 (IL-1, IL-2, IL-6), TNFα, TNFβ, interferon gamma (IFNγ), Fas ligand (FasL), CD40 ligand (CD40L), and granulocyte-macrophage colony-stimulating factor (GM-CSF) [54]. Overexpression of AP-1, therefore, contributes significantly to the inflammatory process.

Several dietary factors have been shown to suppress AP-1 activation and function. These include resveratrol [55], curcumin [56], EGCG from green tea [57], capsaicin [58], oleandrin [59], alpha-mangostin [60], theaflavin-3,3′-digallate [61], anthocyanidins (delphinidin, petunidin, cyanidin) [62], and chlorogenic acid [63]. Mixtures of naturally occurring compounds that inhibit AP-1 include blackberry extracts [64], whole lyophilized BRBs [65] and BRB extracts [66], and apple peel extract [67]. As for NF-κB, the mechanisms though which these agents influence AP-1 activation and function are varied. Resveratrol inhibited 12-*O*-tetradecanoylphorbol-13-acetate (TPA) and UV-induced AP-1 activation in cultured HeLa cells by diminishing PKC and various protein tyrosine kinases upstream of the MAPK pathway [55]. Delphinidin and blackberry extracts inhibited TPA- or UV-induced activation of AP-1 in JB-6 mouse epidermal cells by reducing phosphorylation of protein kinases in the JNK and extracellular signal-related protein kinase (ERK) pathways [64].

EGCG and theaflavin-3,3-digallate inhibited TPA-induced transformation of JB-6 mouse epidermal cells and NIH3T3 cells, respectively, by inhibiting TPA-induced PKC and by reducing the binding of AP-1 to DNA [57, 61]. The remaining dietary compounds and mixtures influence the activation and functions of AP-1 in similar ways leading to reduced cell proliferation, inflammation, and the risk for cancer development.

13.2.2 Tumor Necrosis Factor

TNF, originally discovered in 1968 by Kolb and Granger [68], is a key inflammatory cytokine produced chiefly by activated macrophages and monocytes. The primary role of TNF is the regulation of different immune cell types. A local increase in TNF in tissues will cause the cardinal signs of inflammation to occur, i.e., heat, edema, redness, pain, and loss of function. Dysregulation of TNF has been implicated in a variety of human diseases including inflammatory bowel disease [69] and cancer [70]. The induction of proinflammatory genes by TNF is primarily due to the ability of TNF to activate NF-κB in multiple different cell types [20]. As stated above, NF-κB activation leads to the expression of inflammation-associated COX-2, iNOS, and various chemokines and cytokines. TNF promotes cell proliferation, in part, by activation of JNK and p38 genes in the MAPK cascade, thereby serving as a growth factor for multiple tumor types.

Due to its pivotal role in inflammation and in mediating tumor development, there has been a vigorous search for agents that suppress TNF activity. Some of the known dietary inhibitors of TNF production act by inhibiting NF-κB activation and the subsequent release of TNF from cells. These include resveratrol [71, 72], quercetin [71, 72], EGCG [73], curcumin [74], capsaicin [75], and nordihydroguaiaretic acid [76]. Other dietary compounds inhibit TNF production from monocytes and macrophages [77]. Among these are the terpenoid, cynaropicrin [78], silymarin [79], epicatechin [80], narginin [81], and lignans [82], to name a few. These agents function by either inhibiting the expression of a series of kinase enzymes or exhibiting inhibitory effects on cyclic AMP. Several long-chain fatty acids, including docosahexaenoic acid (DHA) and eicosahexaenoic acid (EHA) in fish oil, have been shown to suppress the serum level of TNF in vivo [83]. Numerous dietary phenols have been shown to inhibit TNF-induced cytotoxicity in vitro. For example, a series of catechols were found to be effective in reducing TNF-induced cytotoxicity, presumably due to their iron-chelating activity and subsequent inhibition of lipoxygenase enzymes [84]. The flavonols, kaempherol, quercetin, myricetin, morin, and rutin also inhibit TNF cytotoxicity, and the C-3 free hydroxyl group on these compounds appears to play a pivotal role in their protective effect. In contrast to the protective effect of flavonols, no protective activity was found for the flavones (chrysin, apigenin, and luteolin), and in fact, apigenin and chrysin which possess one (C-4′) hydroxyl group or none on the B-ring, respectively, enhanced TNF cytotoxicity [85]. These results indicate that relatively minor changes in the structure of phenolic compounds can have a profound effect on their ability to influence TNF cytotoxicity, an important consideration in choosing agents for clinical trials.

In summary, several target sites have been identified for affecting TNF production and the ensuing inflammation associated with TNF. Of these targets, NF-κB is the most interesting as it regulates TNF production and, in turn, TNF regulates NF-κB reciprocally to produce its biological effects. There are very few natural products that inhibit TNF at the posttranslational level. For a more extensive discussion of natural inhibitors of TNF production, secretion, and function, the reader is referred to the review by Habtemariam [77].

13.2.3 Cyclooxygenase-2

Cyclooxygenase (*COX*), or prostaglandin H synthase, is an enzyme responsible for the conversion of arachidonic acid (AA) into the prostanoids: prostaglandins, prostacyclin, and thromboxane [86]. Overproduction of prostanoids can result in increased cell proliferation, inflammation, and angiogenesis and reduced apoptosis. There are two major isoforms of *COX*: COX-1 and COX-2. COX-1 is a constitutive enzyme found in most mammalian cells, whereas COX-2 is undetectable in most normal tissues. COX-2 can be induced in tissues by various toxins and carcinogens, tumor promoters, proinflammatory cytokines including TNF, and growth factors and is found in abundance in activated macrophages and in other cells at sites of inflammation. Depending upon the inducer and the cell type, several transcription factors including AP-1, NF-κB, and STAT-3 are capable of stimulating COX-2 transcription [87, 88]. Importantly, COX-2 is overexpressed in nearly all premalignant lesions and tumors of the skin, head and neck, esophagus, stomach, colon, breast, lung, pancreas, liver, prostate, bladder, cervix, and uterus [20]. In view of its ubiquitous presence in premalignant lesions and in tumors of many types, COX-2 is an important target for cancer chemoprevention and therapy.

As might be expected, the dietary factors mentioned above that suppress the expression of NF-κB and AP-1 have the potential of inhibiting COX-2 expression. A partial list of these includes luteolin [89], resveratrol [90], curcumin [91], genistein [92], and EGCG [93]. In addition, several studies have demonstrated the ability of mixtures of naturally occurring compounds to suppress the expression of COX-2. Our laboratory observed the ability of diets containing 5 % and 10 % lyophilized BRBs to suppress the expression of NF-κB in carcinogen-treated rat esophagus, and this correlated with inhibition of COX-2 expression, reduced prostaglandin E_2 levels in esophageal tissues, and about a 55 % reduction in tumorigenesis [43]. Montrose et al. [94] evaluated the effects of dietary intervention of 10 % lyophilized BRBs on disease severity in an experimental mouse model of ulcerative colitis. C57BL/6J mice were administered with 3 % dextran sodium sulfate (DSS) in drinking water for 7 days along with either control diet or a 10 % lyophilized BRB diet. Dietary BRBs reduced DSS-induced injury to the colonic epithelium and this was associated with a significant reduction in COX-2 protein levels in the colon and PGE_2

levels in plasma. BRB treatment also suppressed mRNA levels of the proinflammatory cytokines TNF-α and IL-1β in colon tissues. These effects on COX-2, TNF-α, and IL-1β were correlated with reduced mRNA levels of phosphorylated IκBα (PIκBα) in the colon which might be expected to result in an inhibition of NF-κB activity. Other mixtures of naturally occurring compounds that have been shown to suppress the expression of COX-2 include cranberry and blueberry extracts [95, 96], pomegranate juice [97], green tea extracts [98], grape seed proanthocyanidins [99], and soy isoflavones [100]. Because these dietary compounds and mixtures are predominately polyphenols, it is likely that their antioxidant effects play a significant role in their ability to inhibit COX-2 expression and COX-2-associated inflammation.

Mixtures of naturally occurring compounds have also been shown to suppress COX-2 expression in human tissues in vivo. In a phase 1b clinical trial, Shumway et al. [101] observed that the topical application of a 10 % black raspberry gel to oral dysplastic lesions (0.5 g 4 times a day for 6 weeks) resulted in an approximate 41 % reduction in histopathologic grade of the lesions and about a 50 % reduction in loss of heterozygosity (LOH) at three tumor suppressor gene loci. In a companion article, Mallery et al. [102] reported significant reductions in COX-2 and iNOS protein levels in the surface epithelium of the berry gel-treated oral lesions by quantitative immunohistochemistry, but only the reduction in COX-2 was significant. More recently, Chen et al. [44] reported that the oral administration of lyophilized strawberries (60 g total/day) in a slurry of water to Chinese subjects at high risk for the development of esophageal squamous cell carcinoma reduced the histologic grade of dysplastic esophageal lesions in 29 (80.6 %) ($p<0.0001$) of 36 subjects. This observation was associated with reduced protein expression levels of iNOS by 79.5 % ($p<0.001$), COX-2 by 62.9 % ($p<0.001$), pNF-κB-p65 by 62.6 % ($p<0.001$), and phospho-S6 (pS6) by 73.2 %. Measurement of pS6 by Western blot is used to assess the activity of the mammalian target of rapamycin (mTOR) gene which plays a central role in cell proliferation [103]. For a more extensive discussion of the mechanisms of inhibition of COX-2 by naturally occurring compounds, the reader is referred to the review of Aggarwal and Shishodia [20].

13.2.4 Lipoxygenase

Lipoxygenases (LOX) are the key enzymes for the conversion of AA to biologically active leukotrienes. There are three types of LOX: 5-LOX, 12-LOX, and 15-LOX. Several studies suggest a link between 5-LOX and cancer development in animals and humans. 5-LOX catalyzes the oxidation of AA at the 5 position to yield 5-hydroperoxyeicosatetraenoic acid (5-HPETE) and then converts 5-HPETE to leukotriene A_4 (LTA_4). Depending upon the cell type, LTA_4 is converted to a series of additional leukotrienes which are released from leukocytes and mast cells to provoke inflammation. LTB_4, in particular, has potent chemokinetic and chemotactic activity towards leukocytes and is considered a potent mediator of inflammation. Both of the AA metabolizing enzymes, COX-2 and 5-LOX, are commonly overexpressed in tumors, and blocking both enzymes simultaneously is considered to be a promising approach to treat inflammatory diseases including cancer.

The plant kingdom is a valuable source for 5-LOX and dual 5-LOX/COX inhibitors. Importantly, natural inhibitors of 5-LOX also tend to be effective in reducing the expression levels of COX. Schneider and Bucar [104] provide a list of over 180 different plant extracts and natural compounds isolated from different plant species that exhibit inhibitory effects against 5-LOX and/or 5-LOX/COX in cultured leukocytes and macrophages. Some of the more commonly known inhibitors were found to be linoleic, oleic, and palmitic acids; silibinin; β-sitosterol; luteolin; quercetin; rosmarinic acid; allicin; resveratrol; ferulic acid; EGCG; gingerol; and curcumin. The IC_{50} values for most of these compounds were between 1 and 50 μM although several were effective at doses below 1 μM. Thus, it is apparent that many of these inhibitors are likely to exhibit inhibitory effects in vivo at doses that are achievable pharmacologically.

13.2.5 Inducible Nitric Oxide Synthase

Nitric oxide synthases are a family of three enzymes that catalyze the production of nitric oxide (NO) from L-arginine. NO plays an important role as a cellular signaling molecule as well as a cytotoxic or regulatory molecule of the innate immune response [105]. NO is synthesized for short periods of time (seconds to minutes) following activation of constitutively expressed endothelial NO synthase (eNOS) or neuronal NO synthase (nNOS). In contrast, inducible NO synthase (iNOS) is expressed after cell activation only and it produces NO for relatively long periods of time (hours to days). The principal inducers of iNOS in many different cell types are the proinflammatory cytokines, TNF-α, IL-1β, and IFN-γ, and lipopolysaccharides (LPS) [106], and the expression of iNOS is dependent, in part, upon activation of the NF-κB signaling pathway and the Janus kinase/signal transducer and activator of transcription (JAK-STAT) pathway [107]. iNOS-derived NO plays a critical role in the pathophysiology of several human diseases including Crohn's disease, ulcerative colitis and cancer. iNOS is expressed in a variety of human tumor types including esophagus, lung, breast, bladder, prostate, colorectal, and melanoma [20].

Several investigations have been conducted to identify naturally occurring inhibitors of iNOS expression in vitro and in vivo. Among the individual compounds that have been shown to inhibit iNOS expression in vitro are the polyphenols, EGCG [108], [6]-gingerol [109], resveratrol [110], curcumin [111], indole-3-carbinol [112], quercetin and rutin [113], and chlorogenic acid [114]. This inhibition most likely occurs through the suppression of NF-κB and/or other signaling pathways (JAK-STAT, MAPK, etc.). For example, Lin and Lin [108] reported that EGCG and other tea polyphenols inhibited iNOS mRNA and protein expression in LPS-activated murine macrophages through suppression of the binding of NF-κB to the iNOS promoter, thereby inhibiting the induction of iNOS. Resveratrol was found to inhibit iNOS expression in beta-amyloid-treated C6 glioma cells by downregulation of

NF-κB [110]. Tedeschi et al. [115] reported that green tea inhibits cytokine-induced iNOS expression in human lung adenocarcinoma A549 cells and human colon DLD-1 cells by downregulation of the DNA-binding activity of STAT-1α. Some of these compounds are also effective in inhibiting iNOS expression in vivo. For example, chlorogenic acid protected mice against LPS-induced lung injury by markedly downregulating iNOS expression, myeloperoxidase activity, and the migration of polymorphonuclear neutrophils into bronchiolar lavage fluid [114].

Numerous mixtures of naturally occurring compounds have also been shown to be effective in downregulating the expression of iNOS, usually in combination with COX-2. Kim et al. [116] evaluated a series of plant extracts for their ability to inhibit NO generation in murine macrophages stimulated with LPS and INF-α. Extracts from avocado, basil, Chinese mustard, mitsuba, and red turnip were particularly effective in inhibiting NO production. The roots of *Rhododendron mucronulatum* (RM) have been used in Asian traditional medicine for the treatment of dysuria and fever for centuries [117]. An alcohol extract of RM, found to contain numerous polyphenolic compounds, was highly effective in inhibiting protein levels of iNOS and COX-2 in LPS-stimulated HaCaT cells. Chitosomes (liposomes in which chitosan is bound to soy lecithin) "loaded with" cranberry proanthocyanidins were found to be effective in inhibiting iNOS and COX-2 expression in murine macrophages activated with LPS [118]. Anthocyanin-rich extracts from the acai berry were found to be potent inhibitors of iNOS and COX-2 expression in mouse brain microglial cells, and this correlated with reductions in p38-MAPK, TNF-α, and NF-κB [119]. Our laboratory observed an inhibition of iNOS mRNA and protein expression in carcinogen-treated rat esophageal tissues by diets containing either 5 % whole BRB powder or an anthocyanin-rich fraction of BRBs [43, 65]. These effects of BRBs and their component anthocyanins were correlated with downregulation of NF-κB and COX-2. We stated above that the oral administration of strawberry powder (60 g/day) in a slurry of water for 6 months to Chinese subjects at high risk for esophageal squamous cell carcinoma resulted in about an 80 % regression of mildly dysplastic lesions [44]. This observation was associated with significant reductions in protein expression levels of iNOS and COX-2 which may have been due to downregulation of the phosphorylated form of NF-κB-p65. Collectively, these observations indicate that iNOS is an important target for dietary intervention in the treatment of inflammation and in the prevention of cancer.

13.2.6 Proinflammatory Cytokines

The tumor microenvironment contains innate immune cells (including macrophages, neutrophils, mast cells, myeloid-derived suppressor cells, dendritic cells, and natural killer cells) and adaptive immune cells (T and B lymphocytes) in addition to the cancer cells and their surrounding stroma. These cells communicate with each other by direct contact or by cytokine and chemokine production that act in autocrine and paracrine manners to control tumor growth. The most abundant immune cells within the tumor microenvironment are the tumor-associated macrophages (TAMs) and T lymphocytes. TAMs are one of the most important players in inflammation and cancer as they promote tumor growth and are obligatory for angiogenesis, invasion, and metastasis [120]. TAMs are classified into M1 and M2 types, analogous to Th1 and Th2 T lymphocytes. M1 macrophages, when activated by various stimuli such as INFγ and LPS, express high levels of the proinflammatory cytokines TNF-α, IL-1β, IL-6, IL-12, and IL-23 as well as iNOS [1]. In contrast, M2 macrophages, induced by IL-4, IL-10, and IL-13, downregulate IL-12 expression and show increased expression of the anti-inflammatory cytokine, IL-10. The tumor-promoting cytokines are the M1 cytokines (TNF-α, IL-1β, IL-6, IL-12, or IL-23), whereas IL-10, the M2 cytokine, has been shown to be tumor suppressive [121].

The proinflammatory cytokines activate various transcription factors such as NF-κB, AP-1, and STAT3 in premalignant cells to induce genes that stimulate cell proliferation, survival, and tumor development. As discussed above, TNF-α activates both NF-κB and AP-1 transcription factors, but in the skin, its tumor-promoting effects are mediated by AP-1 [122]. STAT3 activation in cancer cells is dependent upon a number of growth factors and cytokines, including IL-6, IL-11, and epidermal growth factor (EGF) [123]. The development of specific tumor types tends to be associated with individual cytokines [2]. For example, autocrine production of IL-1β promotes growth of pancreatic carcinoma cell lines [124]. IL-6 acts as a paracrine growth factor for colorectal cancer, non-Hodgkin's lymphoma, renal cell carcinoma, and bladder cancer [125–128]. IL-8 has been detected in multiple cancer types and its expression in human melanomas and ovarian cancers correlates with their metastatic potential [129].

Given the importance of the proinflammatory cytokines in the various stages of tumor development, numerous studies have been conducted to identify naturally occurring compounds and mixtures that inhibit cytokine expression. Typically, cytokine production is induced in cultured cells of various types by treatment with TNF-α, INFγ, LPS, or other inducers, and the individual compounds or mixtures are evaluated for their ability to reduce cytokine expression. Among the numerous compounds that downregulate IL-6 expression in vitro include resveratrol [130], curcumin [131], ellagic acid [132], genistein [133], lycopene [134], capsaicin [135], and EGCG [136]. To varying extents, these compounds also downregulate the expression of IL-12 and IL-1α or IL-1β in cultured cells. Downregulation of these cytokines is often associated with inhibition of MAPKs such as ERK1/2, p38, or JNK as well as the transcription factors, NF-κB, AP-1, or STAT3.

In vivo, green tea polyphenols (GTP) have been tested for inhibition of TPA-induced cytokine expression in mouse skin and in transgenic mouse models that overexpress specific cytokines [136, 137]. Topical application of GTP to the skin of mice previously treated with TPA resulted in reduced expression of TPA-induced IL1-α in the skin [136]. Administration of GTP in the drinking water to TNF-α transgenic mice which are susceptible to idiopathic pulmonary fibrosis led to reduced expressions of TNF-α and IL-6 [137]. In addition to GTP, other mixtures shown to inhibit cytokine expression include grape seed proanthocyanidins, an ethanol extract of pomegranate, tart cherry anthocyanins, and black tea theaflavins.

Grape seed proanthocyanidins inhibited IL-17-induced IL-6 production in cultured human pulmonary epithelial cells by reducing the expressions of MAPK and NF-κB [138]. Pomegranate extract inhibited TNF-α induced production of IL-6 and NO in cultured human osteoblasts [139]. Tart cherry anthocyanins showed an additive effect along with atorvastatin in inhibiting LPS-induced IL-6 production by adipose stem cells [140]. Black tea theaflavins inhibited TPA-induced inflammation in the ears of CD-1 mice, and this correlated with reduced levels of IL-1β, IL-6, PGE_2, and leukotriene B_4 suggesting that the anti-inflammatory activity of the theaflavins may have been due to their ability to inhibit AA metabolism via lipoxygenase and COX pathways [141]. These results suggest that inflammatory cytokines are desirable targets for the anti-inflammatory effects of naturally occurring compounds and mixtures. In addition, cytokines may be excellent biomarkers of effect when measured both in tissues and in serum/plasma.

13.3 Summary and Conclusions

Inflammation is a very complex process, initiated by a host of injurious factors, and characterized by leukocyte infiltration into the affected tissues and the aberrant expression of multiple genes that drive the inflammatory process. The conventional wisdom is that the injurious agents induce one or more transcription activators (e.g., Nf-κB, AP-1, STAT3) and these activators, in turn, cause the overexpression of inflammatory enzymes (e.g., COX-2, iNOS, 5-LOX), cytokines, and chemokines (e.g., TNF, IL-1β, IL-6, IL-12) (Fig. 13.1). Cytokines and chemokines then chemo-attract additional inflammatory leukocytes into the tissues leading ultimately to a chronic state of inflammation and persistent tissue damage. One of the

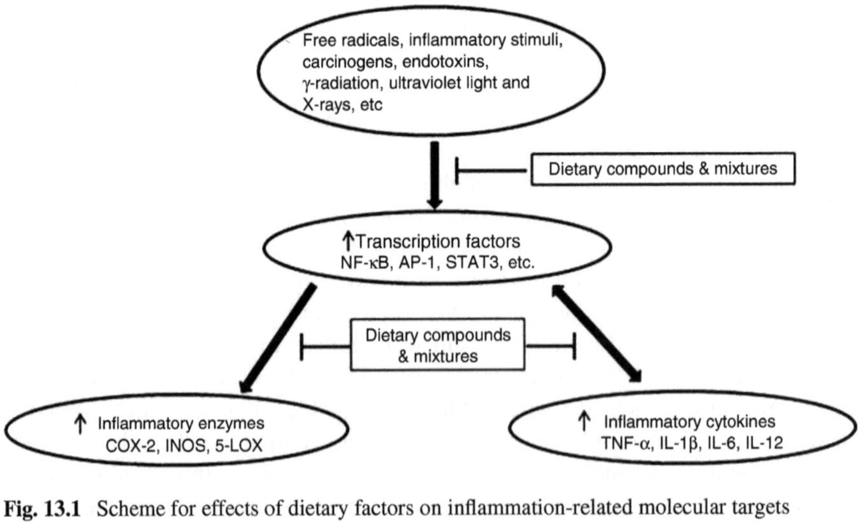

Fig. 13.1 Scheme for effects of dietary factors on inflammation-related molecular targets

outcomes of chronic inflammation is cancer, and indeed, inflammation is associated with essentially all stages of tumor development, i.e., initiation, promotion, malignant conversion, angiogenesis, invasion, and metastasis. Given this, there has been an active search in the past few decades for both synthetic and naturally occurring inhibitors of the inflammatory process to reduce the risk for cancer development.

This chapter lists individual dietary compounds that influence molecular targets involved in the inflammatory process. It is clear that some of the most effective dietary inhibitors of carcinogenesis (e.g., resveratrol, curcumin, EGCG, isothiocyanates, triterpenoids, silymarin, [6]-gingerol) positively influence multiple molecular targets associated with inflammation, providing additional evidence of the close relationship between inflammation and carcinogenesis. The wide-ranging effects of these inhibitors on inflammation targets may be due, at least in part, to the fact that many of them are polyphenols that exhibit antioxidant activity and might be expected to inhibit multiple RONS-driven signaling pathways. Interestingly, most of these agents are not well absorbed and, as such, can be consumed for long periods without producing harmful side effects. In many cases, it is not clear whether the parent compounds themselves or one or more of their metabolites are responsible for the observed biological effects. In that regard, it is important to determine the role, if any, of the enteric microbiome in the metabolism of these compounds. It is also essential to examine the pharmacokinetics of uptake and distribution of these compounds into blood and tissues to provide clues as to how they might be formulated for optimal delivery to target tissues. Finally, studies should be undertaken to determine if the compounds themselves, or their metabolites, interact directly with promoter sequences of inflammatory genes to influence gene expression and/or bind directly to inflammatory proteins (e.g., COX-2, iNOS,5-LOX, TNF, IL-6) to influence their activities.

Whole lyophilized foods and food extracts also exhibit a wide range of inhibitory effects on the expression of inflammation-associated genes. The removal of water from foods by lyophilization and the grinding of dried foods into a powder can result in a significant concentration of their bioactive constituents. For example, we reported that the active constituents in black raspberries are concentrated tenfold by the lyophilization process and that BRB powder, in different formulations, is chemopreventive for oral, esophageal, and colon cancers in rodents and in humans [43, 44, 94, 101, 102, 142, 143]. The inhibitory effects of berry powder on inflammation-associated genes are an important component of the overall chemopreventive effects of BRBs in rodents [143] and, likely, in humans. For example, we reported that the oral administration of BRB powder at a total dose of 60 g/day to colon cancer patients for periods of 1–9 weeks resulted in reduced plasma levels of GM-CSF and IL-8 [144]. Similarly, in obese individuals given a high-fat, high-caloric meal, the oral administration of 45 g/day of BRB powder for 4 days resulted in a significant reduction in plasma levels of IL-6 [145]. It is likely that the other foods and food extracts mentioned in this chapter have similar effects on inflammatory biomarkers suggesting that the routine consumption of anti-inflammatory agents in the human diet in the form of food concentrates and extracts could well reduce the overall risk of humans to cancer.

References

1. Grivennikov SI, Greten FR, Karin M (2010) Immunity, inflammation, and cancer. Cell 140:883–899
2. Aggarwal BB, Shishodia S, Sandur SK, Pandey MK, Sethi G (2006) Inflammation and cancer: how hot is the link? Biochem Pharmacol 72:1605–1621
3. Canavan C, Abrams KR, Mayberry J (2006) Meta-analysis: colorectal and small bowel cancer risk in patients with Crohn's disease. Aliment Pharmacol Ther 23:1097–1104
4. Molodecky NA, Soon IS, Rabi DM, Ghali WA, Ferris M, Chernoff G et al (2012) Increasing incidence and prevalence of the inflammatory bowel diseases with time, based on systematic review. Gastroenterology 142:46–54.e42
5. Clapper ML, Cooper HS, Chang W-CL (2007) Dextran sulfate sodium-induced colitis-associated neoplasia: a promising model for the development of chemopreventive interventions. Acta Pharmacol Sin 28:1450–1459
6. Rogers AB, Houghton J (2009) Helicobacter-based mouse models of digestive system carcinogenesis. Methods Mol Biol 511:267–295
7. Satake K, Mukai R, Kato Y, Umeyama K (1986) Effects of cerulein on the normal pancreas and on experimental pancreatic carcinoma in the Syrian golden hamster. Pancreas 1:246–253
8. Vykhovanets EV, Resnick MI, MacLennan GT, Gupta S (2007) Experimental rodent models of prostatitis: limitations and potential. Prostate Cancer Prostatic Dis 10:15–29
9. McKenna IM, Waalkes MP, Chen LC, Gordon T (1997) Comparison of inflammatory lung responses in Wistar rats and C57 and DBA mice following acute exposure to cadmium oxide fumes. Toxicol Appl Pharmacol 146:196–206
10. Steinmetz KA, Potter JD (1991) Vegetables, fruit, and cancer. II. Mechanisms. Cancer Causes Control 2:427–442
11. Block G, Patterson B, Subar A (1992) Fruit, vegetables, and cancer prevention: a review of the epidemiological evidence. Nutr Cancer 18:1–29
12. Willett WC (1995) Diet, nutrition, and avoidable cancer. Environ Health Perspect 103:165–170
13. Sun J, Chu Y-F, Wu X, Liu RH (2002) Antioxidant and antiproliferative activities of common fruits. J Agric Food Chem 50:7449–7454
14. Chu Y-F, Sun J, Wu X, Liu RH (2002) Antioxidant and antiproliferative activities of common vegetables. J Agric Food Chem 50:6910–6916
15. Kähkönen MP, Hopia AI, Heinonen M (2001) Berry phenolics and their antioxidant activity. J Agric Food Chem 49:4076–4082
16. Stoner GD (2004) Cancer chemoprevention: promising cancer chemoprevention agents. In: Kelloff GJ, Hawk ET, Sigman CC (eds) Chemoprevention by fruit phenolic compounds, vol 1. Humana Press, New Jersey, pp 419–435
17. Stoner GD (2003) Bioprocesses and biotechnology for functional foods and nutraceuticals. In: Neeser JR, German JB (eds) Foods and food components in the prevention of cancer; Marcel Dekker, New York, chapter 17, pp 331–373
18. Wang SY, Jiao H (2000) Scavenging capacity of berry crops on superoxide radicals, hydrogen peroxide, hydroxyl radicals, and singlet oxygen. J Agric Food Chem 48:5677–5684
19. Leonard SS, Cutler D, Ding M, Vallyathan V, Castranova V, Shi X (2002) Antioxidant properties of fruit and vegetable juices: more to the story than ascorbic acid. Ann Clin Lab Sci 32:193–200
20. Aggarwal BB, Shishodia S (2006) Molecular targets of dietary agents for prevention and therapy of cancer. Biochem Pharmacol 71:1397–1421
21. Manna SK, Mukhopadhyay A, Aggarwal BB (2000) Resveratrol suppresses TNF-induced activation of nuclear transcription factors NF-κB, activator protein-1, and apoptosis: potential role of reactive oxygen intermediates and lipid peroxidation. J Immunol 164:6509–6519
22. Singh S, Aggarwal BB (1995) Activation of transcription factor NF-κB is suppressed by curcumin (diferuloylmethane). J Biol Chem 270:24995–25000

23. Shishodia S, Majumdar S, Banerjee S, Aggarwal BB (2003) Ursolic acid inhibits nuclear factor-κB activation induced by carcinogenic agents through suppression of IκBα kinase and p65 phosphorylation. Cancer Res 63:4375–4383
24. Kim G-Y, Kim J-H, Ahn S-C, Lee H-J, Moon D-O, Lee C-M et al (2004) Lycopene suppresses the lipopolysaccharide-induced phenotypic and functional maturation of murine dendritic cells through inhibition of mitogen-activated protein kinases and nuclear factor-κB. Immunology 113:203–211
25. Shukla S, Gupta S (2004) Suppression of constitutive and tumor necrosis factor α-induced nuclear factor (NF)-κB activation and induction of apoptosis by apigenin in human prostate carcinoma PC-3 cells: correlation with down-regulation of NF-κB-responsive genes. Clin Cancer Res 10:3169–3178
26. Agarwal R (2000) Cell signaling and regulators of cell cycle as molecular targets for prostate cancer prevention by dietary agents. Biochem Pharmacol 60:1051–1059
27. Edderkaoui M, Odinokova I, Ohno I, Gukovsky I, Go VL, Pandol SJ et al (2008) Ellagic acid induces apoptosis through inhibition of nuclear factor kappa B in pancreatic cancer cells. World J Gastroenterol 14:3672–3680
28. Huang C, Zhang D, Li J, Tong Q, Stoner GD (2007) Differential inhibition of UV-induced activation of NF-κB and AP-1 by extracts from black raspberries, strawberries, and blueberries. Nutr Cancer 58:205–212
29. Orban Z, Mitsiades N, Burke TR Jr, Tsokos M, Chrousos GP (2000) Caffeic acid phenethyl ester induces leukocyte apoptosis, modulates nuclear factor-kappa B and suppresses acute inflammation. Neuroimmunomodulation 7:99–105
30. Yang F, Oz HS, Barve S, de Villiers WJS, McClain CJ, Varilek GW (2001) The green tea polyphenol (–)-epigallocatechin-3-gallate blocks nuclear factor-κB activation by inhibiting IκB kinase activity in the intestinal epithelial cell line IEC-6. Mol Pharmacol 60:528–533
31. Kim SO, Chun KS, Kundu JK, Surh YJ (2004) Inhibitory effects of [6]-gingerol on PMA-induced COX-2 expression and activation of NF-kappaB and p38 MAPK in mouse skin. Biofactors 21:27–31
32. Geng Z, Rong Y, Lau BHS (1997) S-Allyl cysteine inhibits activation of nuclear factor kappa B in human T cells. Free Radic Biol Med 23:345–350
33. Takada Y, Aggarwal BB (2004) Flavopiridol inhibits NF-κB activation induced by various carcinogens and inflammatory agents through inhibition of IκBα kinase and p65 phosphorylation. J Biol Chem 279:4750–4759
34. Murakami A, Song M, Ohigashi H (2007) Phenethyl isothiocyanate suppresses receptor activator of NF-kappaB ligand (RANKL)-induced osteoclastogenesis by blocking activation of ERK1/2 and p38 MAPK in RAW264.7. Biofactors 30:1–11
35. Lai KC, Huang AC, Hsu SC, Kuo CL, Yang JS, Wu SH et al (2010) Benzyl isothiocyanate (BITC) inhibits migration and invasion of human colon cancer HT29 cells by inhibiting matrix metalloproteinase-2/-9 and urokinase plasminogen (uPA) through PKC and MAPK signaling pathway. J Agric Food Chem 58:2935–2942
36. Takada Y, Andreeff M, Aggarwal BB (2005) Indole-3-carbinol suppresses NF-κB and IκBα kinase activation, causing inhibition of expression of NF-κB-regulated antiapoptotic and metastatic gene products and enhancement of apoptosis in myeloid and leukemia cells. Blood 106:641–649
37. Heiss E, Herhaus C, Klimo K, Bartsch H, Gerhäuser C (2001) Nuclear factor κB is a molecular target for sulforaphane-mediated anti-inflammatory mechanisms. J Biol Chem 276:32008–32015
38. Ruiz PA, Braune A, Hölzlwimmer G, Quintanilla-Fend L, Haller D (2007) Quercetin inhibits TNF-induced NF-κB transcription factor recruitment to proinflammatory gene promoters in murine intestinal epithelial cells. J Nutr 137:1208–1215
39. Chainy GB, Manna SK, Chaturvedi MM, Aggarwal BB (2000) Anethole blocks both early and late cellular responses transduced by tumor necrosis factor: effect on NF-kappaB, AP-1, JNK, MAPKK and apoptosis. Oncogene 19:2943–2950

40. Suzuki YJ, Aggarwal BB, Packer L (1992) α-Lipoic acid is a potent inhibitor of NF-κB activation in human T cells. Biochem Biophys Res Commun 189:1709–1715
41. Valachovicova T, Slivova V, Bergman H, Shuherk J, Sliva D (2004) Soy isoflavones suppress invasiveness of breast cancer cells by the inhibition of NF-kappaB/AP-1-dependent and -independent pathways. Int J Oncol 25:1389–1395
42. Song Y-A, Park Y-L, Kim K-Y, Chung C-Y, Lee G-H, Cho D-H et al (2011) Black tea extract prevents lipopolysaccharide-induced NF-kappaB signaling and attenuates dextran sulfate sodium-induced experimental colitis. BMC Complement Altern Med 11:91
43. Wang LS, Hecht SS, Carmella SG, Yu N, Larue B, Henry C et al (2009) Anthocyanins in black raspberries prevent esophageal tumors in rats. Cancer Prev Res 2:84–93
44. Chen T, Yan F, Qian J, Guo M, Zhang H, Tang X et al (2012) Randomized phase II trial of lyophilized strawberries in patients with dysplastic precancerous lesions of the esophagus. Cancer Prev Res 5:41–50
45. Rhode J, Fogoros S, Zick S, Wahl H, Griffith K, Huang J et al (2007) Ginger inhibits cell growth and modulates angiogenic factors in ovarian cancer cells. BMC Complement Altern Med 7:44
46. Mantena SK, Katiyar SK (2006) Grape seed proanthocyanidins inhibit UV-radiation-induced oxidative stress and activation of MAPK and NF-κB signaling in human epidermal keratinocytes. Free Radic Biol Med 40:1603–1614
47. Sharma SD, Meeran SM, Katiyar SK (2007) Dietary grape seed proanthocyanidins inhibit UVB-induced oxidative stress and activation of mitogen-activated protein kinases and nuclear factor-kappaB signaling in in vivo SKH-1 hairless mice. Mol Cancer Ther 6:995–1005
48. La VD, Howell AB, Grenier D (2009) Cranberry proanthocyanidins inhibit MMP production and activity. J Dent Res 88:627–632
49. Feldman M, Tanabe S, Howell A, Grenier D (2012) Cranberry proanthocyanidins inhibit the adherence properties of Candida albicans and cytokine secretion by oral epithelial cells. BMC Complement Altern Med 12:6
50. Ravi D, Muniyappa H, Das K (2008) Caffeine inhibits UV-mediated NF-κB activation in A2058 melanoma cells: an ATM-PKCδ-p38 MAPK-dependent mechanism. Mol Cell Biochem 308:193–200
51. Shishodia S, Aggarwal BB (2006) Diosgenin inhibits osteoclastogenesis, invasion, and proliferation through the downregulation of Akt, I kappa B kinase activation and NF-kappa B-regulated gene expression. Oncogene 25:1463–1473
52. Eferl R, Wagner EF (2003) AP-1: a double-edged sword in tumorigenesis. Nat Rev Cancer 3:859–868
53. Hsu T-C, Young MR, Cmarik J, Colburn NH (2000) Activator protein 1 (AP-1)- and nuclear factor κB (NF-κB)-dependent transcriptional events in carcinogenesis. Free Radic Biol Med 28:1338–1348
54. Zenz R, Eferl R, Scheinecker C, Redlich K, Smolen J, Schonthaler H et al (2008) Activator protein 1 (Fos/Jun) functions in inflammatory bone and skin disease. Arthritis Res Ther 10:201
55. Yu R, Hebbar V, Kim DW, Mandlekar S, Pezzuto JM, Kong AN (2001) Resveratrol inhibits phorbol ester and UV-induced activator protein 1 activation by interfering with mitogen-activated protein kinase pathways. Mol Pharmacol 60:217–224
56. Han SS, Keum YS, Seo HJ, Surh YJ (2002) Curcumin suppresses activation of NF-kappaB and AP-1 induced by phorbol ester in cultured human promyelocytic leukemia cells. J Biochem Mol Biol 35:337–342
57. Dong Z (2000) Effects of food factors on signal transduction pathways. Biofactors 12:17–28
58. Han SS, Keum Y-S, Seo H-J, Chun K-S, Lee SS, Surh Y-J (2001) Capsaicin suppresses phorbol ester-induced activation of NF-κB/Rel and AP-1 transcription factors in mouse epidermis. Cancer Lett 164:119–126

59. Manna SK, Sah NK, Newman RA, Cisneros A, Aggarwal BB (2000) Oleandrin suppresses activation of nuclear transcription factor-κB, activator protein-1, and c-Jun NH2-terminal kinase. Cancer Res 60:3838–3847
60. Lee Y-B, Ko K-C, Shi M-D, Liao Y-C, Chiang T-A, Wu P-F et al (2010) α-Mangostin, a novel dietary xanthone, suppresses TPA-mediated MMP-2 and MMP-9 expressions through the ERK signaling pathway in MCF-7 human breast adenocarcinoma cells. J Food Sci 75:H13–H23
61. Chen Y-C, Liang Y-C, Lin-Shiau S-Y, Ho C-T, Lin J-K (1999) Inhibition of TPA-induced protein kinase C and transcription activator protein-1 binding activities by theaflavin-3,3′-digallate from black tea in NIH3T3 cells. J Agric Food Chem 47:1416–1421
62. Hou D-X, Kai K, Li J-J, Lin S, Terahara N, Wakamatsu M et al (2004) Anthocyanidins inhibit activator protein 1 activity and cell transformation: structure–activity relationship and molecular mechanisms. Carcinogenesis 25:29–36
63. Feng R, Lu Y, Bowman LL, Qian Y, Castranova V, Ding M (2005) Inhibition of activator protein-1, NF-κB, and MAPKs and induction of phase 2 detoxifying enzyme activity by chlorogenic acid. J Biol Chem 280:27888–27895
64. Feng R, Bowman LL, Lu Y, Leonard SS, Shi X, Jiang B-H et al (2004) Blackberry extracts inhibit activating protein 1 activation and cell transformation by perturbing the mitogenic signaling pathway. Nutr Cancer 50:80–89
65. Chen T, Hwang H, Rose ME, Nines RG, Stoner GD (2006) Chemopreventive properties of black raspberries in N-nitrosomethylbenzylamine-induced rat esophageal tumorigenesis: down-regulation of cyclooxygenase-2, inducible nitric oxide synthase, and c-Jun. Cancer Res 66:2853–2859
66. Huang C, Li J, Song L, Zhang D, Tong Q, Ding M et al (2006) Black raspberry extracts inhibit benzo(a)pyrene diol-epoxide-induced activator protein 1 activation and VEGF transcription by targeting the phosphotidylinositol 3-kinase/Akt pathway. Cancer Res 66:581–587
67. Ding M, Lu Y, Bowman L, Huang C, Leonard S, Wang L et al (2004) Inhibition of AP-1 and neoplastic transformation by fresh apple peel extract. J Biol Chem 279:10670–10676
68. Kolb WP, Granger GA (1968) Lymphocyte in vitro cytotoxicity: characterization of human lymphotoxin. Proc Natl Acad Sci U S A 61:1250–1255
69. Brynskov J, Foegh P, Pedersen G, Ellervik C, Kirkegaard T, Bingham A et al (2002) Tumour necrosis factor α converting enzyme (TACE) activity in the colonic mucosa of patients with inflammatory bowel disease. Gut 51:37–43
70. Locksley RM, Killeen N, Lenardo MJ (2001) The TNF and TNF receptor superfamilies: integrating mammalian biology. Cell 104:487–501
71. Wadsworth TL, Koop DR (1999) Effects of the wine polyphenolics quercetin and resveratrol on pro-inflammatory cytokine expression in RAW 264.7 macrophages. Biochem Pharmacol 57:941–949
72. Kawada N, Seki S, Inoue M, Kuroki T (1998) Effect of antioxidants, resveratrol, quercetin, and N-acetylcysteine, on the functions of cultured rat hepatic stellate cells and Kupffer cells. Hepatology 27:1265–1274
73. Lambert JD, Sang S, Hong J, Yang CS (2010) Anticancer and anti-inflammatory effects of cysteine metabolites of the green tea polyphenol, (−)-epigallocatechin-3-gallate. J Agric Food Chem 58:10016–10019
74. Abe Y, Hashimoto SHU, Horie T (1999) Curcumin inhibition of inflammatory cytokine production by human peripheral blood monocytes and alveolar macrophages. Pharmacol Res 39:41–47
75. Singh S, Natarajan K, Aggarwal BB (1996) Capsaicin (8-methyl-N-vanillyl-6-nonenamide) is a potent inhibitor of nuclear transcription factor-kappa B activation by diverse agents. J Immunol 157:4412–4420
76. Eugui EM, DeLustro B, Rouhafza S, Ilnicka M, Lee SW, Wilhelm R et al (1994) Some antioxidants inhibit, in a co-ordinate fashion, the production of tumor necrosis factor-alpha, IL-beta, and IL-6 by human peripheral blood mononuclear cells. Int Immunol 6:409–422

77. Habtemariam S (2000) Natural inhibitors of tumour necrosis factor-alpha production, secretion and function. Planta Med 66:303–313
78. Natarajan K, Manna SK, Chaturvedi MM, Aggarwal BB (1998) Protein tyrosine kinase inhibitors block tumor necrosis factor-induced activation of nuclear factor-kappaB, degradation of IkappaBalpha, nuclear translocation of p65, and subsequent gene expression. Arch Biochem Biophys 352:59–70
79. Zi X, Mukhtar H, Agarwal R (1997) Novel cancer chemopreventive effects of a flavonoid antioxidant silymarin: inhibition of mRNA expression of an endogenous tumor promoter TNF alpha. Biochem Biophys Res Commun 239:334–339
80. Fujiki H, Suganuma M, Okabe S, Sueoka N, Komori A, Sueoka E et al (1998) Cancer inhibition by green tea. Mutat Res 402:307–310
81. Kawaguchi K, Kikuchi S, Hasegawa H, Maruyama H, Morita H, Kumazawa Y (1999) Suppression of lipopolysaccharide-induced tumor necrosis factor-release and liver injury in mice by naringin. Eur J Pharmacol 368:245–250
82. Cho JY, Park J, Yoo ES, Yoshikawa K, Baik KU, Lee J et al (1998) Inhibitory effect of lignans from the rhizomes of Coptis japonica var. dissecta on tumor necrosis factor-alpha production in lipopolysaccharide-stimulated RAW264.7 cells. Arch Pharm Res 21:12–16
83. Endres S, Ghorbani R, Kelley VE, Georgilis K, Lonnemann G, van der Meer JWM et al (1989) The effect of dietary supplementation with n-3 polyunsaturated fatty acids on the synthesis of interleukin-1 and tumor necrosis factor by mononuclear cells. N Engl J Med 320:265–271
84. O'Donnell VB, Spycher S, Azzi A (1995) Involvement of oxidants and oxidant-generating enzyme(s) in tumour-necrosis-factor-alpha-mediated apoptosis: role for lipoxygenase pathway but not mitochondrial respiratory chain. Biochem J 310:133–141
85. Habtemariam S (1997) Flavonoids as inhibitors or enhancers of the cytotoxicity of tumor necrosis factor-alpha in L-929 tumor cells. J Nat Prod 60:775–778
86. Marnett LJ, Kalgutkar AS (1999) Cyclooxygenase 2 inhibitors: discovery, selectivity and the future. Trends Pharmacol Sci 20:465–469
87. Subbaramaiah K, Dannenberg AJ (2003) Cyclooxygenase 2: a molecular target for cancer prevention and treatment. Trends Pharmacol Sci 24:96–102
88. Kim EJ, Raval AP, Perez-Pinzon MA (2008) Preconditioning mediated by sublethal oxygen-glucose deprivation-induced cyclooxygenase-2 expression via the signal transducers and activators of transcription 3 phosphorylation. J Cereb Blood Flow Metab 28:1329–1340
89. Baumann J, von Bruchhausen F, Wurm G (1980) Flavonoids and related compounds as inhibition of arachidonic acid peroxidation. Prostaglandins 20:627–639
90. Subbaramaiah K, Chung WJ, Michaluart P, Telang N, Tanabe T, Inoue H et al (1998) Resveratrol inhibits cyclooxygenase-2 transcription and activity in phorbol ester-treated human mammary epithelial cells. J Biol Chem 273:21875–21882
91. Plummer SM, Holloway KA, Manson MM, Munks RJ, Kaptein A, Farrow S et al (1999) Inhibition of cyclo-oxygenase 2 expression in colon cells by the chemopreventive agent curcumin involves inhibition of NF-kappaB activation via the NIK/IKK signalling complex. Oncogene 18:6013–6020
92. Mutoh M, Takahashi M, Fukuda K, Matsushima-Hibiya Y, Mutoh H, Sugimura T et al (2000) Suppression of cyclooxygenase-2 promoter-dependent transcriptional activity in colon cancer cells by chemopreventive agents with a resorcin-type structure. Carcinogenesis 21:959–963
93. Gerhäuser C, Klimo K, Heiss E, Neumann I, Gamal-Eldeen A, Knauft J et al (2003) Mechanism-based in vitro screening of potential cancer chemopreventive agents. Mutat Res 523–524:163–172
94. Montrose DC, Horelik NA, Madigan JP, Stoner GD, Wang L-S, Bruno RS et al (2011) Anti-inflammatory effects of freeze-dried black raspberry powder in ulcerative colitis. Carcinogenesis 32:343–350
95. Bodet C, Chandad F, Grenier D (2007) Cranberry components inhibit interleukin-6, interleukin-8, and prostaglandin E2 production by lipopolysaccharide-activated gingival fibroblasts. Eur J Oral Sci 115:64–70

96. Lau FC, Bielinski DF, Joseph JA (2007) Inhibitory effects of blueberry extract on the production of inflammatory mediators in lipopolysaccharide-activated BV2 microglia. J Neurosci Res 85:1010–1017
97. Adams LS, Seeram NP, Aggarwal BB, Takada Y, Sand D, Heber D (2006) Pomegranate juice, total pomegranate ellagitannins, and punicalagin suppress inflammatory cell signaling in colon cancer cells. J Agric Food Chem 54:980–985
98. Kundu JK, Na H-K, Chun K-S, Kim Y-K, Lee SJ, Lee SS et al (2003) Inhibition of phorbol ester-induced COX-2 expression by epigallocatechin gallate in mouse skin and cultured human mammary epithelial cells. J Nutr 133:3805S–3810S
99. Sharma S, Katiyar S (2010) Dietary grape seed proanthocyanidins inhibit UVB-induced cyclooxygenase-2 expression and other inflammatory mediators in UVB-exposed skin and skin tumors of SKH-1 hairless mice. Pharm Res 27:1092–1102
100. Lau TY, Leung LK (2006) Soya isoflavones suppress phorbol 12-myristate 13-acetate-induced COX-2 expression in MCF-7 cells. Br J Nutr 96:169–176
101. Shumway BS, Kresty LA, Larsen PE, Zwick JC, Lu B, Fields HW et al (2008) Effects of a topically applied bioadhesive berry gel on loss of heterozygosity indices in premalignant oral lesions. Clin Cancer Res 14:2421–2430
102. Mallery SR, Zwick JC, Pei P, Tong M, Larsen PE, Shumway BS et al (2008) Topical application of a bioadhesive black raspberry gel modulates gene expression and reduces cyclooxygenase 2 protein in human premalignant oral lesions. Cancer Res 68:4945–4957
103. Sun Q, Chen X, Ma J, Peng H, Wang F, Zha X et al (2011) Mammalian target of rapamycin up-regulation of pyruvate kinase isoenzyme type M2 is critical for aerobic glycolysis and tumor growth. Proc Natl Acad Sci U S A 108:4129–4134
104. Schneider I, Bucar F (2005) Lipoxygenase inhibitors from natural plant sources. Part 1: medicinal plants with inhibitory activity on arachidonate 5-lipoxygenase and 5-lipoxygenase[sol]cyclooxygenase. Phytother Res 19:81–102
105. Kröncke F, Kolb B (1998) Inducible nitric oxide synthase in human diseases. Clin Exp Immunol 113:147–156
106. Bogdan C (2001) Nitric oxide and the regulation of gene expression. Trends Cell Biol 11:66–75
107. Kleinert H, Wallerath T, Fritz G, Ihrig-Biedert I, Rodriguez-Pascual F, Geller DA et al (1998) Cytokine induction of NO synthase II in human DLD-1 cells: roles of the JAK-STAT, AP-1 and NF-κB-signaling pathways. Br J Pharmacol 125:193–201
108. Lin Y-L, Lin J-K (1997) (–)-Epigallocatechin-3-gallate blocks the induction of nitric oxide synthase by down-regulating lipopolysaccharide-induced activity of transcription factor nuclear factor-kappaB. Mol Pharmacol 52:465–472
109. Ippoushi K, Azuma K, Ito H, Horie H, Higashio H (2003) [6]-Gingerol inhibits nitric oxide synthesis in activated J774.1 mouse macrophages and prevents peroxynitrite-induced oxidation and nitration reactions. Life Sci 73:3427–3437
110. Tsai S-H, Lin-Shiau S-Y, Lin J-K (1999) Suppression of nitric oxide synthase and the down-regulation of the activation of NFκB in macrophages by resveratrol. Br J Pharmacol 126:673–680
111. Chan MM-Y, Huang H-I, Fenton MR, Fong D (1998) In vivo inhibition of nitric oxide synthase gene expression by curcumin, a cancer preventive natural product with anti-inflammatory properties. Biochem Pharmacol 55:1955–1962
112. Chen YH, Dai HJ, Chang HP (2003) Suppression of inducible nitric oxide production by indole and isothiocyanate derivatives from Brassica plants in stimulated macrophages. Planta Med 69:696–700
113. Chen Y-C, Shen S-C, Lee W-R, Hou W-C, Yang L-L, Lee TJF (2001) Inhibition of nitric oxide synthase inhibitors and lipopolysaccharide induced inducible NOS and cyclooxygenase-2 gene expressions by rutin, quercetin, and quercetin pentaacetate in RAW 264.7 macrophages. J Cell Biochem 82:537–548
114. Zhang X, Huang H, Yang T, Ye Y, Shan J, Yin Z et al (2010) Chlorogenic acid protects mice against lipopolysaccharide-induced acute lung injury. Injury 41:746–752

115. Tedeschi E, Menegazzi M, Yao Y, Suzuki H, Förstermann U, Kleinert H (2004) Green tea inhibits human inducible nitric-oxide synthase expression by down-regulating signal transducer and activator of transcription-1alpha activation. Mol Pharmacol 65:111–120
116. Kim OK, Murakami A, Nakamura Y, Ohigashi H (1998) Screening of edible Japanese plants for nitric oxide generation inhibitory activities in RAW 264.7 cells. Cancer Lett 125:199–207
117. Choi SE, Park KH, Han BH, Jeong MS, Seo SJ, Lee DI et al (2011) Inhibition of inducible nitric oxide synthase and cyclooxygenase-2 expression by phenolic compounds from roots of rhododendron mucronulatum. Phytother Res 25:1301–1305
118. Madrigal-Carballo S, Rodríguez G, Sibaja M, Reed JD, Vila AO, Molina F (2009) Chitosomes loaded with cranberry proanthocyanidins attenuate the bacterial lipopolysaccharide-induced expression of iNOS and COX-2 in raw 264.7 macrophages. J Liposome Res 19:189–196
119. Poulose SM, Fisher DR, Larson J, Bielinski DF, Rimando AM, Carey AN et al (2012) Anthocyanin-rich açai (Euterpe oleracea Mart.) fruit pulp fractions attenuate inflammatory stress signaling in mouse brain BV-2 microglial cells. J Agric Food Chem 60:1084–1093
120. Condeelis J, Pollard JW (2006) Macrophages: obligate partners for tumor cell migration, invasion, and metastasis. Cell 124:263–266
121. Lin WW, Karin M (2007) A cytokine-mediated link between innate immunity, inflammation, and cancer. J Clin Invest 117:1175–1183
122. Angel P, Imagawa M, Chiu R, Stein B, Imbra RJ, Rahmsdorf HJ et al (1987) Phorbol ester-inducible genes contain a common cis element recognized by a TPA-modulated trans-acting factor. Cell 49:729–739
123. Grivennikov SI, Karin M (2010) Dangerous liaisons: STAT3 and NF-κB collaboration and crosstalk in cancer. Cytokine Growth Factor Rev 21:11–19
124. Arlt A, Vorndamm J, Müerköster S, Yu H, Schmidt WE, Fölsch UR et al (2002) Autocrine production of interleukin 1beta confers constitutive nuclear factor kappaB activity and chemoresistance in pancreatic carcinoma cell lines. Cancer Res 62:910–916
125. Klein B, Zhang X, Jourdan M, Content J, Houssiau F, Aarden L et al (1989) Paracrine rather than autocrine regulation of myeloma-cell growth and differentiation by interleukin-6. Blood 73:517–526
126. Voorzanger N, Touitou R, Garcia E, Delecluse H-J, Rousset F, Joab I et al (1996) Interleukin (IL)-10 and IL-6 are produced in vivo by non-Hodgkin's lymphoma cells and act as cooperative growth factors. Cancer Res 56:5499–5505
127. Okamoto M, Kawamata H, Kawai K, Oyasu R (1995) Enhancement of transformation in vitro of a nontumorigenic rat urothelial cell line by interleukin 6. Cancer Res 55:4581–4585
128. Landi S, Moreno V, Gioia-Patricola L, Guino E, Navarro M, de Oca J et al (2003) Association of common polymorphisms in inflammatory genes interleukin (IL)6, IL8, tumor necrosis factor alpha, NFKB1, and peroxisome proliferator-activated receptor gamma with colorectal cancer. Cancer Res 63:3560–3566
129. Luca M, Huang S, Gershenwald JE, Singh RK, Reich R, Bar-Eli M (1997) Expression of interleukin-8 by human melanoma cells up-regulates MMP-2 activity and increases tumor growth and metastasis. Am J Pathol 151:1105–1113
130. Feng Y-H, Zhu Y-N, Liu J, Ren Y-X, Xu J-Y, Yang Y-F et al (2004) Differential regulation of resveratrol on lipopolysaccharide-stimulated human macrophages with or without IFN-γ pre-priming. Int Immunopharmacol 4:713–720
131. Cho JW, Lee KS, Kim CW (2007) Curcumin attenuates the expression of IL-1beta, IL-6, and TNF-alpha as well as cyclin E in TNF-alpha-treated HaCaT cells; NF-kappaB and MAPKs as potential upstream targets. Int J Mol Med 19:469–474
132. Choi YH, Yan GH (2009) Ellagic Acid attenuates immunoglobulin E-mediated allergic response in mast cells. Biol Pharm Bull 32:1118–1121
133. Gao X, Liu K, Huang F, Zhang D, Guo X, Wang M et al (2013) Positive and negative regulation of insulin action by genistein in the endothelium. J Nutr Biochem 24(1):222–230

134. Kang M, Park K, Seo J, Kim H (2011) Lycopene inhibits IL-6 expression in cerulein-stimulated pancreatic acinar cells. Genes Nutr 6:117–123
135. Bhutani M, Pathak AK, Nair AS, Kunnumakkara AB, Guha S, Sethi G et al (2007) Capsaicin is a novel blocker of constitutive and interleukin-6-inducible STAT3 activation. Clin Cancer Res 13:3024–3032
136. Ahmed S, Marotte H, Kwan K, Ruth JH, Campbell PL, Rabquer BJ et al (2008) Epigallocatechin-3-gallate inhibits IL-6 synthesis and suppresses transsignaling by enhancing soluble gp130 production. Proc Natl Acad Sci U S A 105:14692–14697
137. Sueoka N, Suganuma M, Sueoka E, Okabe S, Matsuyama S, Imai K et al (2001) A new function of green tea: prevention of lifestyle-related diseases. Ann N Y Acad Sci 928:274–280
138. Kim H, Kim J, Song H, Park K, Mun K-C, Ha E (2011) Grape seed proanthocyanidin extract inhibits interleukin-17-induced interleukin-6 production via MAPK pathway in human pulmonary epithelial cells. Naunyn Schmiedebergs Arch Pharmacol 383:555–562
139. Kim YH, Choi EM (2009) Stimulation of osteoblastic differentiation and inhibition of interleukin-6 and nitric oxide in MC3T3-E1 cells by pomegranate ethanol extract. Phytother Res 23:737–739
140. Zhou Z, Nair MG, Claycombe KJ (2012) Synergistic inhibition of interleukin-6 production in adipose stem cells by tart cherry anthocyanins and atorvastatin. Phytomedicine 19:878–881
141. Huang M-T, Liu Y, Ramji D, Lo C-Y, Ghai G, Dushenkov S et al (2006) Inhibitory effects of black tea theaflavin derivatives on 12-O-tetradecanoylphorbol-13-acetate-induced inflammation and arachidonic acid metabolism in mouse ears. Mol Nutr Food Res 50:115–122
142. Wang LS, Arnold M, Huang YW, Sardo C, Seguin C, Martin E et al (2011) Modulation of genetic and epigenetic biomarkers of colorectal cancer in humans by black raspberries: a phase I pilot study. Clin Cancer Res 17:598–610
143. Stoner GD (2009) Foodstuffs for preventing cancer: the preclinical and clinical development of berries. Cancer Prev Res (Phila) 2:187–194
144. Mentor-Marcel RA, Bobe G, Sardo C, Wang LS, Kuo CT, Stoner G et al (2012) Plasma cytokines as potential response indicators to dietary freeze-dried black raspberries in colorectal cancer patients. Nutr Cancer 64:820–825
145. Sardo C, Kitzmiller J, Apseloff G, Harris R, Roe D, Stoner G et al. Reducing postprandial inflammation with black raspberries: a pilot study in older obese and overweight men (submitted for publication)

Chapter 14
Calorie Restriction and Cancer Prevention: Established and Emerging Mechanisms

Stephen D. Hursting, Nikki A. Ford, Sarah M. Dunlap, Marcie J. Hursting, and Laura M. Lashinger

Abstract Calorie restriction (CR) is one of the most potent, broadly acting dietary interventions for inducing weight loss and for inhibiting cancer in experimental models. Translation of the mechanistic lessons learned from research on CR to cancer prevention strategies in humans is important given the high prevalence of excess energy intake, obesity, and metabolic syndrome in many parts of the world and also given the established links between obesity-associated metabolic perturbations and increased risk and/or progression of many types of cancers. This chapter synthesizes findings on the biological mechanisms underlying many of the anticancer effects of CR, with emphasis on the role of inflammatory processes and growth factor signaling (well-established mechanisms) as well as vascular perturbations, autophagy, and sirtuins (emerging mechanisms). These CR-responsive pathways and processes represent targets for translating CR research into effective cancer prevention strategies in humans.

S.D. Hursting, Ph.D. (✉)
Department of Nutritional Sciences, The University of Texas at Austin, Austin, TX, USA
e-mail: shursting@austin.utexas.edu

Department of Molecular Carcinogenesis, The University of Texas-MD Anderson Cancer Center, Smithville, TX, USA

N.A. Ford, Ph.D. • S.M. Dunlap, Ph.D. • L.M. Lashinger, Ph.D.
Department of Nutritional Sciences, The University of Texas at Austin, Austin, TX, USA
e-mail: nikkiannford@gmail.com; lalash@austin.utexas.edu

M.J. Hursting, Ph.D.
Clinical Science Consulting, Austin, TX, USA
e-mail: mhursting@sprintmail.com

A.J. Dannenberg and N.A. Berger (eds.), *Obesity, Inflammation and Cancer*, Energy Balance and Cancer 7, DOI 10.1007/978-1-4614-6819-6_14,

14.1 Introduction

Calorie restriction (CR), a dietary regimen in which subjects (typically test animals) receive a reduced energy diet (typically 20–40 % reduction in total energy intake relative to an unrestricted comparison group), is one of the most potent and broadly acting dietary interventions for preventing or reversing weight gain and inhibiting cancer in experimental tumor models [1]. Recent reports of decreased risk of diabetes and cancer in response to CR in rhesus monkeys [2, 3], and observations that CR decreases inflammatory and endocrine markers associated with increased breast cancer risk in women [4–6], suggest that the beneficial effects of CR on metabolism and chronic disease risk observed in rodent models may extend to nonhuman primates and humans.

Observational epidemiologic studies provide further evidence that CR exerts beneficial effects on longevity and cancer risk in humans [1]. For example, inhabitants of Okinawa, Japan, have for centuries consumed significantly fewer calories than residents of the main Japanese islands and have always had lower death rates from cancer and other chronic diseases than inhabitants of the Japanese mainland [7]. It will be interesting to see if the recent westernization of the diet of Okinawans results in increased cancer rates in future. Another example involves patients with early-onset anorexia nervosa and hence periods of energy restriction; these patients have reduced risk of breast cancer [8]. Furthermore, surveillance data from Norwegian women during World War II showed reduced breast cancer risk later in life in association with acute (<1 year) energy restriction (~50 % reduction in calorie intake without significant changes in diet quality) [9]. However, populations with more severe restriction than experienced in Norway, such as survivors of the 1944 Dutch "Hunger Winter," the Jewish Holocaust, and the Siege of Leningrad, actually displayed higher breast cancer rates [10–12]. This indicates a possible threshold beyond which energy restriction (especially when combined with other stressors) may be cancer promoting. This is particularly true for those born around the time of the severe deprivation and stress, suggesting an important perinatal window of susceptibility to metabolic reprogramming in response to energy restriction [13].

These stressful conditions, particularly the confounding effects of severe physical and psychosocial stress, malnutrition, infection, and other factors associated with war conditions in severely affected countries, make many of these wartime surveillance studies a challenge to interpret. These conditions are also in contrast to the controlled conditions characteristic of most CR studies in animal models that consistently show anticancer effects. CR regimens are often referred to as "CR with optimal nutrition" or "undernutrition without malnutrition," and CR experiments typically involve 20–40 % reductions in total energy relative to ad libitum-fed controls but with adequate nutrition and a controlled physical environment [1]. In rodent models, CR regimens administered throughout life are generally more effective against cancer than CR regimens initiated in adulthood, although both early-onset and adult-onset CR, relative to control diet regimens, are protective against a variety of cancer types [1]. In rhesus monkey studies, CR begun in young adults is more protective against cancer than CR initiated later in life [3].

There are several National Institute of Aging-funded clinical trials underway to address the question of whether the observed health benefits of CR in rodents and nonhuman primates translate to humans. One of these trials, called the Comprehensive Assessment of Long-Term Effects of Reducing Intake of Energy (CALERIE) Study, is evaluating the effects of a 2-year CR regimen (25 % less energy than controls) in healthy, nonobese individuals. Preliminary reports on CALERIE indicate that many of the same metabolic and endocrine changes observed in rodents and monkeys are also occurring in humans in response to CR [14, 15]. These findings are consistent with recent studies in women at high risk for breast cancer showing that inflammatory and growth factor signaling pathways are reduced by total CR or 2 days/week of restricted carbohydrate calories [4–6]. The observed metabolic effects of 2-days/week of restricted carbohydrate calories [4] are of particular interest, since it is likely easier and more sustainable for most people to periodically restrict a single macronutrient, such as carbohydrates, than to chronically restrict total energy.

In this chapter we discuss possible mechanisms underlying the anticancer effects of CR, with emphasis on CR-associated changes in inflammation and growth factor signaling, as well as emerging evidence suggesting that vascular perturbations (including angiogenesis) and modulation of sirtuin and autophagy pathways may also play roles in the effects of CR on tumor development and progression. As summarized in Fig. 14.1, we specifically describe the impact of CR on (a) macrophages, cytokines, and other inflammatory mediators; (b) growth factors, including insulin, insulin-like growth factor (IGF)-1, adipokines, and their downstream signaling pathways; (c) vascular integrity factors, including angiogenic regulators; (d) autophagy regulators; and (e) sirtuin pathway components. We discuss how these multifactorial CR-induced changes combine to suppress tumor development and/or progression. Components of these interrelated pathways offer possible mechanism-based targets for the prevention and control of cancers, particularly the estimated 20–25 % [16] of human cancers related to, or caused by, excess body weight and lack of physical activity.

14.2 Calorie Restriction Decreases Chronic Inflammation

Chronic inflammation is characterized by increased circulating cytokines and chemokines that attract immune cells (such as macrophages that also produce inflammatory mediators) into the local microenvironment [17–19]. The inflammatory cascade is further amplified by the release of inflammatory cytokines such as interleukin (IL)-1β, IL-6, tumor necrosis factor (TNF)-α, and monocyte chemoattractant protein (MCP)-1, primarily from macrophages, into the local and systemic circulation. Adipocytes can enlarge past the point of effective oxygen diffusion, which results in hypoxia and eventually necrosis. Free fatty acids escape the engorged/necrotic adipocytes and deposit in other tissues, which in turn promotes insulin resistance, diabetes (through downregulation of insulin receptors and glucose transporters), hepatic steatosis, and pancreatic steatosis, and also activates signaling molecules involved in

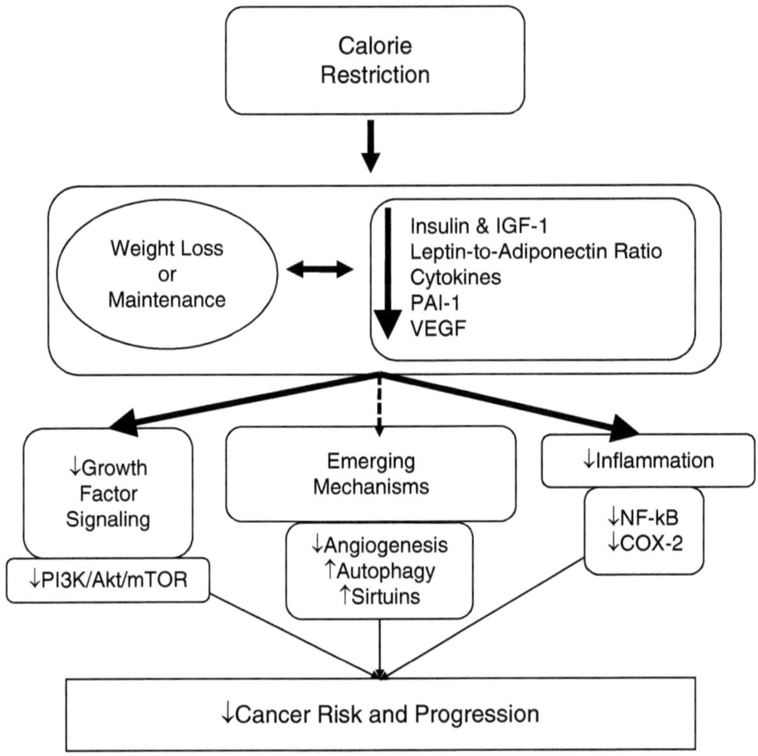

Fig. 14.1 Calorie restriction and cancer prevention: summary of established and emerging mechanisms. An *arrow* preceding text denotes a directional effect (e.g., activity or concentration). *Solid arrow* indicates established effects; *dashed arrows* indicate emerging mechanisms requiring additional study. *IGF-1* insulin-like growth factor-1; *PAI-1* plasminogen activator inhibitor-1; *VEGF* vascular endothelial growth factor; *PI3K* phosphoinositide 3-kinase; *NF-κB* nuclear factor kappa-light-chain-enhancer of activated B-cells; *COX-2* cyclooxygenase-2

epithelial carcinogenesis, such as nuclear factor kappa-light-chain-enhancer of activated B-cells (NF-κB) and cyclooxygenase (COX)-2 [20]. NF-κB is a transcription factor that is activated in response to bacterial and viral stimuli, growth factors, and inflammatory molecules (e.g., TNF-α, IL-6, and IL-1β) and is responsible for inducing gene expression associated with cell proliferation, apoptosis, inflammation, metastasis, and angiogenesis. Activation of NF-κB is a common characteristic of many tumors and is associated with insulin resistance and elevated circulating levels of leptin, insulin, and/or IGF-1 [17, 21, 22].

A connection between chronic inflammation and cancer development was observed 150 years ago when Rudolph Virchow noted an abundance of leukocytes in neoplastic tissue [23]. Inflammation is now considered a hallmark of cancer, and the evidence is accumulating that chronic, "smoldering" inflammation is associated with increased cancer risk [24–26]. Indeed, several tissue-specific inflammatory lesions are established neoplastic precursors for invasive cancer, including inflammatory bowel disease for colon cancer, pancreatitis for pancreatic cancer, dermatitis for certain forms of skin cancer, prostatitis for prostate cancer, and gastritis for gastric cancer [27, 28]. Tumor and preneoplastic microenvironments are composed of multiple cell types including epithelial cells, fibroblasts, mast cells, and cells of the innate and adaptive immune system [29]. As discussed previously, macrophages, which are activated in the obese state, infiltrate tumors and amplify the inflammatory tumor microenvironment, often through NF-κB-dependent production of cytokines and angiogenic factors [29]. COX-2 is another important cancer-related inflammatory mediator that is upregulated in most tumors and catalyzes the synthesis of the potent inflammatory lipid metabolite, prostaglandin E_2. COX-2 expression, an indicator of poor prognosis in multiple cancer types, is increased in response to obesity [30].

CR can prevent much of the chronic inflammation associated with preneoplasia or neoplasia [17, 31–33]. Specifically, CR decreases the number of tumor-associated macrophages (TAMs), NF-κB signaling, circulating and tissue cytokines, and COX-2 expression, in many tissues and tumor types [17, 32, 33]. A major contributor to the proinflammatory tumor environment is the presence of TAMs. The recruitment of TAMs to the tumor microenvironment is largely dependent on the MCP-1, circulating levels of which are consistently decreased by CR [32, 33]. Levels of MCP-1 in tumor tissue are highly correlated with the accumulation of TAMs [34]. Macrophages are capable of polarizing into what is known as an M1, or classically activated cytotoxic macrophage, or an M2, or immunosuppressive macrophage. The cytokines produced by each type of macrophage are what distinguish an M1 from an M2, and tumor tissue typically contains a larger quantity of M2-type macrophages [34]. In addition to producing cytokines and chemokines, TAMs also produce growth factors that enhance proliferation and angiogenesis and contribute to deposition and dissolution of connective tissue [34]. There is also evidence to suggest that NF-κB plays a role in mediating TAM transcriptional programs and, by extension, protumorigenic effects of TAMs [17, 32, 34].

14.3 Calorie Restriction Impacts Growth Signals

14.3.1 Insulin and IGF-1

The peptide hormone insulin is produced by beta cells in the pancreas and released in response to chronic hyperglycemia, which is associated with insulin resistance, aberrant glucose metabolism, chronic inflammation, and the production of other metabolic hormones such as IGF-1, leptin, and adiponectin [35]. Clinical and epidemiologic evidence suggests that elevated levels of circulating insulin or the cleavage product of proinsulin (C-peptide) are associated with increased risk and/or progression of cancers of the breast (pre- and postmenopausal), endometrium, colon, kidney, and pancreas [35, 36]. Insulin exerts tumor-enhancing effects directly via the insulin receptor or via IGF-1 receptor (IGF-1R)/insulin receptor hybrids on preneoplastic and neoplastic cells [37]. High circulating levels of insulin also upregulate hepatic synthesis of IGF-1 essential for growth and development of many tissues, particularly during the prenatal period [35, 37]. In the circulation IGF-1 is typically bound to IGF-binding proteins (IGFBPs) that regulate the amount of free IGF-1 bioavailable to bind to the IGF-1R and elicit growth or survival signaling [35, 37]. Elevated circulating IGF-1 is an established risk factor for many cancer types [38, 39], and in states of obesity/metabolic syndrome, the amount of bioavailable IGF-1 increases, possibly via hyperglycemia-induced suppression of IGFBP synthesis and/or hyperinsulinemia-induced promotion of hepatic growth hormone receptor expression and IGF-1 synthesis [35–38]. In contrast to obesity, CR prevents hyperinsulinemia, enhances insulin sensitivity, increases serum levels and tissue mRNA expression of several IGFBPs, and decreases total and bioavailable circulating IGF-1 levels [1].

The phospatidylinositol-3 kinase (PI3K)/Akt pathway, downstream of both the insulin receptor and IGF-1R, is one of the most commonly activated pathways in epithelial cancers [40]. This pathway integrates intracellular and environmental cues, such as growth factor concentrations and nutrient availability, to regulate cellular survival, proliferation, protein translation, and metabolism. Akt regulates the mammalian target of rapamycin (mTOR) [41], which regulates cell growth, cell proliferation, and survival through downstream mediators. Increased activation of mTOR is common in tumors and many normal tissues from obese and/or diabetic mice, while CR decreases mTOR signaling in these same tumors and normal tissues [42]. mTOR activation is inhibited by increased AMP-activated kinase (AMPK) under low-nutrient conditions [43]. Specific mTOR inhibitors block the tumor-enhancing effects of obesity in mouse models [44, 45].

14.3.2 *Adiponectin, Leptin, and the Leptin-to-Adiponectin Ratio*

Adiponectin is a peptide hormone primarily secreted from visceral white adipose tissue. In contrast to leptin and other adipokines, circulating levels of adiponectin negatively correlate with adiposity and are thus increased by CR and decreased by obesity [46]. Adiponectin functions to counter obesity-related metabolic perturbations, such as insulin resistance and leptin resistance, that impact glucose and fatty acid metabolism, alter insulin responses and increase production of inflammatory cytokines [46]. Thus, possible mechanisms through which adiponectin exerts anti-cancer effects may include increasing insulin sensitivity and decreasing insulin/IGF-1 and mTOR signaling via activation of AMPK [47]. Adiponectin also reduces proinflammatory cytokine expression via inhibition of NF-κB [47, 48].

Leptin is a peptide hormone produced by white adipose tissue, and the leptin receptor is a member of the class I cytokine receptor family that signals through the Janus kinase and signal transducer activator of transcription (JAK/STAT) pathway commonly dysregulated in inflammatory conditions and many cancers [49, 50]. Circulating leptin levels positively correlate with adipose stores and nutritional status and function as an energy sensor to signal the brain to reduce appetite. Leptin has direct effects on peripheral tissues and indirect effects on neuroendocrine regulators of appetite and energy expenditure in the hypothalamus and impacts carcinogenesis, angiogenesis, immune responses, cytokine production, and other biological processes [50]. In the obese state, adipose tissue overproduces leptin, and the brain no longer responds to the signal, resulting in leptin resistance. Insulin, glucocorticoids, TNF-α, and estrogens all stimulate leptin release [50]. CR consistently and robustly decreases systemic leptin levels in a manner dependent on the extent of the adiposity loss [1].

In vitro, animal, and epidemiologic evidence linking adiponectin [51–55] or leptin [56–58] individually to cancer risk is mixed. Intermittent CR suppresses murine mammary tumor incidence in association with decreased leptin-to-adiponectin ratio [51]. Associations between the leptin-to-adiponectin ratio and the metabolic syndrome [59–61] and some cancers [62–64] have also been reported.

14.4 Emerging Mechanisms Underlying the Anticancer Effects of Calorie Restriction

14.4.1 Vascular Effects

Perturbations in the production and/or interactions of several factors that influence key functions of the endothelium, including its roles in regulating angiogenesis, hemostasis, vascular density, inflammation, and vascular-wall integrity, have been linked to cancer. One such vascular-related factor is plasminogen-activated inhibitor (PAI)-1, a serine protease inhibitor produced by endothelial cells, stromal cells, and adipocytes in visceral white adipose tissue [65]. PAI-1, through its inhibition of urokinase-type and tissue-type plasminogen activators, regulates fibrinolysis and integrity of the extracellular matrix [66]. Increased circulating PAI-1 levels, frequently found in obese subjects, are associated with increased risk of atherogenesis and cardiovascular disease, diabetes, and several cancers [65–68]. PAI-1 is also involved in angiogenesis and thus may contribute to obesity-driven tumor cell growth, invasion, and metastasis [68]. Circulating levels of PAI-1 are consistently decreased in response to CR [1], although the mechanistic link between PAI-1 and cancer requires further study.

Another important mediator of vascular integrity is the heparin-binding glycoprotein vascular endothelial growth factor (VEGF) produced by adipocytes and tumor cells. VEGF has mitogenic, angiogenic, and vascular permeability-enhancing activities specific for endothelial cells [69]. The need for nutrients and oxygen triggers tumor cells to produce VEGF, which leads to the formation of new blood vessels to nourish the rapidly growing tumor. VEGF may also facilitate the metastatic spread of tumor cells [70]. Adipocytes communicate with endothelial cells by producing a variety of proangiogenic and vascular permeability-enhancing factors, including VEGF and PAI-1 [71]. In the obese, nontumor setting, these factors stimulate neovascularization in support of the expanding fat mass [71]. Circulating levels of VEGF are increased in obese, relative to lean, humans and animals, and increased tumoral expression of VEGF is associated with poor prognosis in several obesity-related cancers [72–75]. Data to date suggest that CR decreases systemic and tissue VEGF and has antiangiogenic effects in multiple experimental tumor models [73–75].

14.4.2 Autophagy

Autophagy is a cellular degradation pathway involved in the clearance of damaged or unnecessary proteins and organelles. It also provides an alternative source of energy and substrates during periods of restricted dietary intake (such as CR) or metabolic stress to enhance survival. In response to CR, plasma glucose levels (relative to controls) are low, insulin secretion is suppressed, and glucagon is released from the alpha-cells of the pancreas, resulting in increased autophagy in the liver, beta-cells of the pancreas, skeletal muscle, and heart [76, 77]. One of the proposed mechanisms of CR is that under conditions of nutrient limitation, survival is promoted by a shift in metabolic investment from cell replication and growth to maintenance [78]. This tightly regulated process is driven by a group of autophagy-related proteins and is suppressed by the conserved nutrient sensor target of rapamycin (TOR, referred to as mTOR in mammalian species) [79]. CR regulates TOR complex 1, and to a lesser extent TOR complex 2, in many species including flies, worms, yeast, and mammals. TOR complex 1 signaling regulates protein translation and many cellular processes including metabolism and autophagy [79]. Additionally, suppression of nutrient-activated TOR signaling is sufficient to trigger an energy stress response that is coordinated by AMPK, and this metabolic program blunts the growth responses to nutrient availability and promotes autophagy [80].

Several longevity-promoting regimens, including inhibition of TOR with rapamycin, resveratrol, or the natural polyamine spermidine, may require autophagy for their effects [81]. Autophagy activation is essential for clearing cellular damage and preventing disease in normal cells. Tumor cells also utilize autophagy to maintain a favorable metabolic state for daughter cell production, especially under limiting nutrient conditions [82]. However, little is known about what role autophagy plays in CR-mediated effects on tumor development or progression. In particular, the links between autophagy, apoptosis, and energy metabolism in normal vs. cancer cells may provide important insights into the anticancer effects of CR.

14.4.3 Sirtuins

The Sirtuin family of proteins has been implicated in the regulation of endocrine signaling, stress-induced apoptosis, and the metabolic changes associated with energy balance modulation and aging [83–85]. Sirtuins were originally studied in yeast and nematodes, where CR increases lifespan in association with the levels and activity of the Sir2 protein [86–88]. The levels of Sir2, or its mammalian homologue SIRT1, rise in response to CR [84–88]. SIRT1 is an NAD-dependent deacetylase that inhibits stress-induced apoptotic cell death and modulates IGF-1, adiponectin and insulin production, and insulin sensitivity, in some tissues [88–90].

The specific roles of sirtuins in cancer development and/or progression are not yet clear. SIRT1 is upregulated in several tumor types and can inhibit apoptosis and downregulate the expression of tumor suppressor genes (such as p53, which is frequently mutated in many human cancers) to enhance survival of epithelial cancer cells [91–94]. In addition, the SIRT1 activator SRT1720 promotes tumor cell migration and lung metastases in a murine breast cancer model [95]. In contrast, there is also evidence that SIRT1 can act to suppress polyp formation in the APC^{Min} intestinal tumor model [96]. Additionally, in preclinical studies the phytochemical resveratrol activates SIRT1 and reduces cancer development in multiple models [97]. SIRT1 overexpression does not influence the anticancer effects of an every-other-day fasting regimen (a variation of CR) in a p53-deficient mouse model of cancer, suggesting that SIRT1 may have a limited role in the effects of CR on cancer [98]. Given the conflicting data to date regarding the tumor-enhancing, vs. inhibitory, effects of SIRT1 activation and the apparently limited role of SIRT1 in the response to CR, it remains unclear if SIRT1 or other sirtuins represent mechanistic targets for cancer prevention.

14.5 Targets and Strategies for Mimicking the Effects of Calorie Restriction

The identification and development of natural or synthetic agents that mimic some of the protective effects of CR may facilitate new strategies for cancer prevention. Given how difficult it is for many people to adopt a low calorie diet for an extended period, the identification of drugs or other agents that could either complement or even reproduce the anticancer effects of CR without drastic changes in diet and lifestyle is a goal for many pharmaceutical companies. Numerous studies have used microarray analyses to profile the molecular targets responding to CR and other dietary energy balance modulations [99–103]. Most of these studies were focused on understanding CR effects related to aging, and they revealed that the extent to which CR modulates the transcriptome is species specific, tissue specific, and dependent on the duration and intensity of CR. Nonetheless, some emerging patterns from these studies suggest that transcripts involved in inflammation, growth factor signaling (particularly related to the insulin and IGF-1 pathways), oxidative stress, and nutrient metabolism are commonly altered by CR. Application of the emerging field of metabolomics to this question should accelerate the identification of additional targets.

The IGF-1 and Akt/mTOR pathways, and possibly the sirtuin pathway, have emerged as potential key mediators of CR's anticancer effects and are initial targets for possible CR mimetics. Agents or interventions that safely reduce IGF-1, or inhibit one or more components of the signaling pathways downstream of IGF-1 and other growth factors (including Akt and mTOR), even without dietary changes, may provide an effective physiological or pharmacological mimetic of those effects. The hope is that these agents or interventions could be readily adopted by a large proportion of the population, particularly those unable to lose weight and at high risk for cancer or other chronic diseases associated with obesity. Small-molecule inhibitors of IGF-1, antisense inhibitor approaches, anti-IGF-1 antibody therapies, and microRNA-based approaches are under development [104, 105]. In addition, a wide variety of natural agents with demonstrated cancer chemopreventive or chemotherapeutic activity have recently been reported to inhibit the IGF-1 pathway [106].

Pharmacological mTOR inhibitors have emerged as the lead candidates for CR mimetics. Rapamycin treatment extends lifespan and delays cancer in mice, providing additional support for mTOR as a target for mimicking the effects of CR [107]. We have shown that rapamycin or its analogue, RAD001 (everolimus), offsets the obesity-associated increase in growth of mammary or pancreatic tumors [33, 44]. Rapamycin is a potent inhibitor of the mTOR complex 1, but chronic rapamycin exposure has been linked in some studies to disruption of mTOR complex 2 signaling, resulting in impaired glucose tolerance and insulin action [108]. Thus, while mTOR complex 1 inhibition appears to be a good strategy for mimicking many of the anticancer effects of CR, the search for agents that can do so without disrupting mTOR complex 2 signaling is ongoing.

Metformin, a biguanide commonly used to treat type 2 diabetes, is another promising CR mimetic that inhibits mTOR signaling and circumvents the side effects of hyperglycemia/insulin resistance that occur with rapamycin [109]. Metformin inhibits gluconeogenesis through indirect activation of AMPK in the liver and possibly cancer cells and may also exert direct effects on AMPK in cancer cells to decrease mTOR activation [109]. Administration of metformin suppresses tumor development and/or growth in multiple experimental models, including colon, mammary, and hematopoietic cancer models [109]. Epidemiological studies have suggested that type 2 diabetic patients treated with metformin have a lower cancer risk and mortality relative to diabetic patients receiving sulfonylurea, insulin or other therapies [110–112]. A randomized trial is now underway to evaluate the effect of metformin on breast cancer recurrence [113]. Phenformin, another biguanide that has been abandoned for diabetes therapy due to its toxicity from lactic acidosis, is a more potent AMPK inhibitor than metformin and may also have some potential as a CR mimetic for cancer prevention at lower, nontoxic doses [109].

Genetic induction of the Sir2/SIRT1 family of NAD-dependent deacetylases mimics some of the effects of CR [84, 86, 87, 96], although the role of SIRT1 in the anticancer effects of CR is unclear and may be minimal [98]. Sirtuin modulators, including resveratrol and its analogues, and pharmacologic modulators of SIRT1 [97, 114] exert some anticancer activity, although much of this work has been limited to in vitro systems and awaits verification in vivo.

14.6 Conclusions

As summarized in Fig. 14.1, this review considers lessons learned from CR and cancer research to discuss promising molecular targets for cancer prevention, particularly for breaking the obesity-cancer link. Potential targets include components of energy-responsive growth factor and adipokine signaling pathways, inflammatory pathways, vascular regulators, autophagy regulators, and the sirtuin pathway. Clearly, no single pathway accounts for all of the anticancer effects of CR. As with most chronic disease intervention strategies, combination approaches involving lifestyle (including diet and physical activity) and pharmacological interventions that target multiple pathways (and that maximize efficacy and minimize adverse effects) will likely be most successful for preventing cancer. Future studies aimed at further elucidating the mechanisms underlying the anticancer effects of CR and that exploit this mechanistic information to target CR-responsive pathways will facilitate the translation of CR research into effective cancer prevention strategies in humans.

References

1. Hursting SD, Smith SM, Lashinger LM, Harvey AE, Perkins SN (2010) Calories and carcinogenesis: lessons learned from 30 years of calorie restriction research. Carcinogenesis 31:83–89
2. Colman RJ, Anderson RM, Johnson SC, Kastman EK, Kosmatka KJ, Beasley TM, Allison DB, Cruzen C, Simmons HA, Kemnitz JW, Weindruch R (2009) Caloric restriction delays disease onset and mortality in rhesus monkeys. Science 325:201–204
3. Mattson JA, Roth GS, Beasley TM, Tilmon EM, Handy AM, Herbert RL, Longo DL, Allison DB, Young JE, Bryant M, Barnard D, Ward WF, Qi W, Ingram DK, deCabo R (2012) Impact of caloric restriction on health and survival in rhesus monkeys from the NIA study. Nature 489(7415):318–321. doi:10.1038/nature11432
4. Harvie M, Howell A (2012) Energy restriction and the prevention of breast cancer. Proc Nutr Soc 71:263–275
5. Imayama I, Ulrich CM, Alfanao CM, Wang C, Xiao L, Wner MH, Campbell KL, Duggan C, Foster-Schubert KE, Kong A, Mason CE, Wnag CY, Blackburn GL, Bain CE, Thompson HJ, McTiernan A (2012) Effects of a caloric restriction weight loss diet and exercise on inflammatory biomarkers in overweight/obese postmenopausal women: a randomized controlled trial. Cancer Res 72:2314–2326
6. Campbell KL, Foster-Schubert KE, Alfano CM, Wang CC, Wang CY, Duggan CR, Mason C, Imayama I, Kong A, Bain CE, Blackburn GL, Stanczyj FZ, McTiernan A (2012) Reduced-calorie dietary weight loss, exercise and sex hormones in postmenopausal women: a randomized control trial. J Clin Oncol 30:2314–2326
7. Kagawa Y (1978) Impact of westernization on the nutrition of Japanese: changes in physique, cancer, longevity and centenarians. Prev Med 7:205–217
8. Michels KB, Ekbom A (2004) Caloric restriction and incidence of breast cancer. J Am Med Assoc 291:1226–1230
9. Tretli S, Gaard M (1996) Lifestyle changes during adolescence and risk of breast cancer: an ecologic study of the effect of World War II in Norway. Cancer Causes Control 7:507–512
10. Elias SG, Peeters PH, Grobbee DE, van Noord PA (2005) The 1944–1945 Dutch famine and subsequent overall cancer incidence. Cancer Epidemiol Biomarkers Prev 14:1981–1985

11. Keinan-Boker L, Vin-Raviv N, Lipshitz I, Linn S, Barchana M (2009) Cancer incidence in Israeli Jewish survivors of World War II. J Natl Cancer Inst 101:1489–1500
12. Koupil I, Plavinskaja S, Parfenova N, Shestov DB, Danziger PD, Vagero D (2009) Cancer mortality in women and men who survived the siege of Leningrad (1941–1944). Int J Cancer 124:1416–1421
13. Hursting SD, Forman MR (2009) Cancer risk from extreme stressors: lessons from European Jewish survivors of World War II. J Natl Cancer Inst 101:1436–1437
14. Heilbronn LK, de Jonge L, Frisard MI, DeLany JP, Larson-Meyer DE, Rood J, Nguyen T, Martin CK, Volaufova J, Most MM, Greenway FL, Smith SR, Deutsch WA, Williamson DA, Ravussin E (2009) Effect of 6-month calorie restriction on biomarkers of longevity, metabolic adaptation, and oxidative stress in overweight individuals: a randomized controlled trial. JAMA 295:1539–1548
15. Redman LM, Heilbronn LK, Martin CK, de Jonge L, Williamson DA, Delany JP, Ravussin E (2009) Metabolic and behavioral compensations in response to caloric restriction: implications for the maintenance of weight loss. PLoS One 4:e4377
16. Colditz GA, Wollin KY, Gehlert S (2012) Applying what we know to accelerate cancer prevention. Sci Transl Med 127(4):1–9
17. Harvey AE, Lashinger LM, Hursting SD (2011) The growing challenge of obesity and cancer: an inflammatory subject. Ann N Y Acad Sci 1229:45–52
18. Subbaramaiah K, Howe LR, Bhardway P, Du B, Gravaghi C, Yantiss RK, Zhou XK, Blaho VA, Hia T, Yang P, Kopelovich L, Hudis CA, Dannenberg AJ (2011) Obesity is associated with inflammation and elevated aromatase expression in the mouse mammary gland. Cancer Prev Res 4:329–346
19. Olefsky JM, Glass CK (2010) Macrophages, inflammation, and insulin resistance. Annu Rev Physiol 72:219–246
20. O'Rourke RW (2009) Inflammation in obesity-related diseases. Surgery 145:255–259
21. Renehan AG, Roberts DL, Dive C (2008) Obesity and cancer: pathophysiological and biological mechanisms. Arch Physiol Biochem 114:71–83
22. Karin M (2006) Nuclear factor-kappaB in cancer development and progression. Nature 441:431–436
23. Virchow R (1865) Aetiologie der neoplastischen Geschwulst/Pathogenie der neoplastischen Geschwulste [Etiology and pathology of cancerous tumors]. Die Krankhaften Geschwulste. Verlag von August Hirschwald, Berlin, p 57–101
24. Aggarwal BB, Gehlot P (2009) Inflammation and cancer: how friendly is the relationship for patients? Curr Opin Pharmacol 9:351–369
25. Ono M (2008) Molecular links between tumor angiogenesis and inflammation: inflammatory stimuli of macrophages and cancer cells as targets for therapeutic strategy. Cancer Sci 99:1501–1506
26. Del Prete A, Allavena P, Santoro G, Fumarulo R, Corsi MM, Mantovani A (2011) Molecular pathways in cancer-related inflammation. Biochem Med 21:264–275
27. Foltz CJ, Fox JG, Cahill R, Murphy JC, Yan L, Shames B, Schauer DB (1998) Spontaneous inflammatory bowel disease in multiple mutant mouse lines: association with colonization by Helicobacter hepaticus. Helicobacter 3:69–78
28. Coussens LM, Werb Z (2002) Inflammation and cancer. Nature 420:860–867
29. Allavena P, Sica A, Garlanda C, Mantovani A (2008) The Yin-Yang of tumor-associated macrophages in neoplastic progression and immune surveillance. Immunol Rev 222:155–161
30. Koki A, Khan NK, Woerner BM, Dannenberg AJ, Olson L, Seibert K, Edwards D, Hardy M, Isakson P, Masterrer JL (2002) Cyclooxygenase-2 in human pathological disease. Adv Exp Med Biol 507:177–184
31. Perkins SN, Hursting SD, Phang JM, Haines DC (1998) Calorie restriction reduces ulcerative dermatitis and infection-related mortality in p53-deficient and wild-type mice. J Invest Dermatol 111:292–296

32. Harvey A, Lashinger L, Otto G, Malone L, Hursting SD (2012) Decreased systemic insulin-like growth factor-1 in response to calorie restriction modulates tumor growth, NF-kB activation, and inflammation-related gene expression. Mol Carcinog. doi:10.1002/mc.21940
33. Lashinger LM, Malone LM, Brown GW, Daniels EA, Goldberg JA, Otto G, Fischer SM, Hursting SD (2011) Rapamycin partially mimics the anticancer effects of calorie restriction in a murine model of pancreatic cancer. Cancer Prev Res 4:1041–1051
34. Sica A, Schioppa T, Mantovani A, Allavena P (2006) Tumour-associated macrophages are a distinct M2 polarised population promoting tumour progression: potential targets of anticancer therapy. Eur J Cancer 42:717–727
35. Pollak M (2012) The insulin and insulin-like growth factor family in neoplasia: an update. Nat Rev Cancer 12:159–169
36. Hursting SD, Berger NA (2010) Energy balance, host-related factors, and cancer progression. J Clin Oncol 28:4058–4065
37. Gallagher EJ, Fierz Y, Ferguson RD, LeRoith D (2010) The pathway from diabetes and obesity to cancer, on the route to targeted therapy. Endocr Pract 16:864–873
38. Renehan AG, Zwahlen M, Minder C, O'Dwyer ST, Shalet SM, Egger M (2004) Insulin-like growth factor (IGF)-1, IGF binding protein-3, and cancer risk: systematic review and meta-regression analysis. Lancet 363:1346–1353
39. Price AJ, Allen NE, Appleby PN, Crowe FL, Travis RC, Tipper SJ, Overvad K, Gronbæk H, Tjonneland A, Fons Johnsen N, Rinaldi S, Kaaks R, Lukanova A, Boeing H, Aleksandrova K, Trichopoulou A, Trichopoulos D, Andarakis G, Palli D, Krogh V, Tumino R, Sacerdote C, Bueno-de-Mesquita HB, Arguelles MV, Sanchez MJ, Chirlaque MD, Barricarte A, Larranaga N, Gonzalez CA, Stattin P, Johansson M, Khaw KT, Wareham NJ, Gunter MJ, Riboli E, Key TJ (2012) Insulin-like growth factor-1 concentration and risk of prostate cancer: results from the European Prospective Investigation into Cancer and Nutrition. Cancer Epidemiol Biomarkers Prev 21(9):1531–1541. doi:10.1158/1055-9965.EPI-12-0481-T
40. Wong KK, Engelman JA, Cantley LC (2010) Targeting the PI3K signaling pathway in cancer. Curr Opin Genet Dev 20:87–90
41. Memmott RM, Dennis PA (2009) Akt-dependent and -independent mechanisms of mTOR regulation in cancer. Cell Signal 21:656–664
42. Moore T, Beltran LD, Carbajal S, Strom S, Traag J, Hursting SD, DiGiovanni J (2008) Dietary energy balance modulates signaling through the Akt/mammalian target of rapamycin pathway in multiple epithelial tissues. Cancer Prev Res 1:65–76
43. Lindsley JE, Rutter J (2004) Nutrient sensing and metabolic decisions. Comp Biochem Physiol B Biochem Mol Biol 139:543–559
44. De Angel RE, Conti CJ, Wheatley KE, Brenner AJ, deGraffenried LA, Hursting SD (2012) The enhancing effects of obesity on mammary tumor growth and Akt/mTOR pathway activation persist after weight loss and are reversed by Rad001. Mol Carcinog. doi:10.1002/mc.21878
45. Nogueira LM, Dunlap SM, Ford NA, Hursting SD (2012) Calorie restriction and rapamycin inhibit MMTV-Wnt-1 mammary tumor growth in a mouse model of postmenopausal obesity. Endocr Relat Cancer 19:57–68
46. Vaiopoulos AG, Marinou K, Christodoulides C, Koutsilieris M (2012) The role of adiponectin in human vascular physiology. Int J Cardiol 155:188–193
47. Barb D, Williams CJ, Neuwirth AK, Mantzoros CS (2007) Adiponectin in relation to malignancies: a review of existing basic research and clinical evidence. Am J Clin Nutr 86:s858–s866
48. Stofkova A (2009) Leptin and adiponectin: from energy and metabolic dysbalance to inflammation and autoimmunity. Endocr Regul 43:157–168
49. Gautron L, Elmquist JK (2011) Sixteen years and counting: an update on leptin in energy balance. J Clin Invest 121:2087–2093
50. Villanueva EC, Myers MG (2008) Leptin receptor signaling and the regulation of mammalian physiology. Int J Obes 32(suppl 7):S8–S12

51. Rogozina OP, Bonorden MJ, Seppanen CN, Grande JP, Cleary MP (2011) Effect of chronic and intermittent calorie restriction on serum adiponectin and leptin and mammary tumorigenesis. Cancer Prev Res 4:568–581
52. Dalamaga M, Diakopoulos KN, Mantzoros CS (2012) The role of adiponectin in cancer: a review of current evidence. Endocr Rev 33:547–594
53. Grossmann ME, Nkhata KJ, Mizuno NK, Ray A, Cleary MP (2008) Effects of adiponectin on breast cancer cell growth and signaling. Br J Cancer 98:370–379
54. Rzepka-Gorska I, Bedner R, Cymbaluk-Ploska A, Chudecka-Glaz A (2008) Serum adiponectin in relation to endometrial cancer and endometrial hyperplasia with atypia in obese women. Eur J Gynaecol Oncol 29:594–597
55. Tian YF, Chu CH, Wh MH, Chang CL, Yang T, Chou YC, Hsu GC, Yu CP, Yu JC, Sun CA (2007) Anthropometric measures, plasma adiponectin, and breast cancer risk. Endocr Relat Cancer 14:669–677
56. Stattin P, Lukanova A, Biessy C, Soderberg S, Palmqvist R, Kaaks R, Olsson T, Jellum E (2004) Obesity and colon cancer: does leptin provide a link? Int J Cancer 109:149–152
57. Zheng Q, Dunlap SM, Zhu J, Downs-Kelly E, Rich JN, Hursting SD, Berger NA, Reizes O (2011) Leptin deficiency suppresses MMV-Wnt-1 mammary tumor growth in obese mice and abrogates tumor initiating cell survival. Endocr Relat Cancer 18:491–503
58. Fenton JI, Hord NG, Lavigne JA, Perkins SN, Hursting SD (2005) Leptin, insulin-like growth factor-1, and insulin-like growth factor-2 are mitogens in ApcMin/+ but not Apc+/+ colonic epithelial cells. Cancer Epidemiol Biomarkers Prev 14:1646–1652
59. Jung CH, Rhee EJ, Choi JH, Bae JC, Yoo SH, Kim WJ, Park CY, Mok JO, Kim CH, Lee WY, Oh KW, Park SW, Kim SW (2010) The relationship of adiponectin/leptin ratio with homeostasis model assessment insulin resistance index and metabolic syndrome in apparently healthy Korean male adults. Korean Diabetes J 34:237–243
60. Mirza S, Qu HQ, Li Q, Martinez PJ, Rentfro AR, McCormick JB, Fisher-Hoch SP (2011) Adiponectin/leptin ratio and metabolic syndrome in a Mexican American population. Clin Invest Med 34:E290
61. Labruna G, Pasanisi F, Nardelli C, Caso R, Vitale DF, Contaldo F, Sacchetti L (2011) High leptin/adiponectin ratio and serum triglycerides are associated with an "at-risk" phenotype in young severely obese patients. Obesity 19:1492–1496
62. Cleary MP, Ray A, Rogozina OP, Dogan S, Grossman ME (2009) Targeting the adiponectin:leptin ratio for postmenopausal breast cancer prevention. Front Biosci 1:329–357
63. Ashizawa N, Yahata T, Quan J, Adachi S, Yoshihara, Tanaka K (2010) Serum leptin-adiponectin ratio and endometrial cancer risk in postmenopausal female subjects. Gynecol Oncol 119:65–69
64. Chen DC, Chung YF, Yeh YT, Chaung HC, Kuo FC, Fu OY, Chen HY, Hou ME, Yuan SS (2006) Serum adiponectin and leptin levels in Taiwanese breast cancer patients. Cancer Lett 237:109–114
65. Iwaki T, Urano T, Umemura K (2012) PAI-1, progress in understanding the clinical problem and its etiology. Br J Haematol 157:291–298
66. Skurk T, Hauner H (2004) Obesity and impaired fibrinolysis: role of adipose production of plasminogen activator inhibitor-1. Int J Obes Relat Metab Disord 28:1357–1364
67. Carter JC, Church FC (2009) Obesity and breast cancer: the roles of peroxisome proliferator-activated receptor-gamma and plasminogen activator inhibitor-1. PPAR Res 2009: 345320–345333
68. McMahon GA, Petitclerc E, Stefansson S, Smith E, Wong MK, Westwick RJ, Ginsburg D, Brooks PC, Lawrence DA (2001) Plasminogen activator inhibior-1 regulates tumor growth and angiogenesis. J Biol Chem 276:33964–33968
69. Byrne AM, Bouchier-Hayes DJ, Harmey JH (2005) Angiogenic and cell survival functions of vascular endothelial growth factor (VEGF). J Cell Mol Med 9:777–794
70. Saharinen P, Eklund L, Pulkki K, Bono P, Alitalo K (2011) VEGF and angiopoietin signaling in tumor angiogenesis and metastasis. Trends Mol Med 17:347–362

71. Cao Y (2007) Angiogenesis modulates adipogenesis and obesity. J Clin Invest 117: 2362–2368
72. Liu Y, Tamimi RM, Collins LC, Schnitt SJ, Gilmore HL, Connolly JL, Colditz GA (2011) The association between vascular endothelial growth factor expression in invasive breast cancer and survival varies with intrinsic subtypes and use of adjuvant systemic therapy: results from the Nurses' Health Study. Breast Cancer Res Treat 129:175–184
73. Powolny AA, Wang S, Carlton PS, Hoot DR, Clinton SK (2008) Interrelationships between dietary restriction, the IGF-1 axis, and expression of vascular endothelial growth factor by prostate adenocarcinoma in rats. Mol Carcinog 47:458–476
74. Lashinger LM, Malone LM, MacArthur MJ, Goldberg JA, Daniels EA, Pavone A, Colby JK, Smith NC, Perkins SN, Fischer SM, Hursting SD (2011) Genetic reduction of insulin-like growth factor-1 mimics the anticancer effects of calorie restriction on cyclooxygenase-2-driven pancreatic neoplasia. Cancer Prev Res 4:1030–1040
75. Blando J, Moore T, Hursting SD, Jiang G, Saha A, Beltran L, Shen J, Repass J, Strom S, DiGiovanni J (2011) Dietary energy balance modulates prostate cancer progression in Hi-Myc mice. Cancer Prev Res 4:2002–2014
76. Mizushima N, Yamamoto A, Matsui M, Yoshimori T, Ohsumi Y (2004) In vivo analysis of autophagy in response to nutrient starvation in mice expressing a fluorescent autophagosome marker. Mol Biol Cell 15:1101–1111
77. Ezaki J, Matsumoto N, Takeda-Ezaki M, Komatsu M, Takahashi K, Hiraoka Y, Taka H, Fujimura T, Takehana K, Yoshida M, Iwata J, tanida I, Furuya N, Zheng DM, Tada N, Tanaka K, Kominami E, Ueno T (2011) Liver autophagy contributes to the maintenance of blood glucose and amino acid levels. Autophagy 7:727–736
78. Kapahi P, Chen D, Rogers AN, Katewa SD, Li PW, Thomas EL, Kockel L (2010) With TOR, less is more: a key role for the conserved nutrient-sensing pathway in aging. Cell Metab 11:453–465
79. Chang YY, Juhasz G, Goraksha-Hicks P, Arsham AM, Mallin DR, Muller LK, Neufield TP (2009) Nutrient-dependent regulation of autophagy through the target of rapamycin pathway. Biochem Soc Trans 37:232–236
80. Kim J, Kundu M, Viollet B, Guan KL (2011) AMPK and mTOR regulate autophagy through direct phosphorylation of Ulk1. Nat Cell Biol 13:132–141
81. Madeo F, Tavermarakis N, Kroemer G (2010) Can autophagy promote longevity? Nat Cell Biol 12:842–846
82. Guo JY, Chen HY, Matthew R, Fan J, Strohecker AM, Karsli-Uzunbas G, Kamphorst JJ, Chen G, Lemons JM, Karantza V, Coller HA, Dipaola RS, Gelinas C, Rabinowitz JD, White E (2011) Activated ras requires autophagy to maintain oxidative metabolism and tumorigenesis. Genes Dev 25:460–470
83. Yang T, Fu M, Pestell R, Sauve AA (2006) SIRT1 and endocrine signaling. Trends Endocrinol Metab 17:186–191
84. Cohen HY, Miller C, Bitterman KJ, Wall NR, Hekking B, Kessler B, Howitz KT, Gorospe M, de Cabo R, Sinclair DA (2004) Calorie restriction promotes mammalian cell survival by inducing the SIRT1 deacetylase. Science 305:390–392
85. Metoyer CF, Pruitt K (2008) The role of sirtuin proteins in obesity. Pathophysiology 15:103–108
86. Lin SJ, Defossez PA, Guarente L (2000) Requirement of NAD and SIR2 for life-span extension by calorie restriction in Saccharomyces cerevisiae. Science 289:2126–2128
87. Tissenbaum HA, Guarente L (2001) Increased dosage of a sir-2 gene extends lifespan in Caenorhabditis elegans. Nature 410:227–230
88. Bordone L, Cohen D, Robinson A, Motta MC, van Veen E, Czopik A, Steele AD, Crowe H, Marmor S, Luo J, Gu W, Guarente L (2007) SIRT1 transgenic mice show phenotypes resembling calorie restriction. Aging Cell 6:759–767
89. Ramsey KM, Mills KF, Satoh A, Imai S (2008) Age-associated loss of Sirt1-mediated enhancement of glucose-stimulated insulin secretion in beta cell-specific Sirt1-overexpressing (BESTO) mice. Aging Cell 7:78–88

90. Nemoto S, Fergusson MM, Finkel T (2005) SIRT1 functionally interacts with the metabolic regulator and transcriptional coactivator PGC-1{alpha}. J Biol Chem 280:16456–16460
91. Lim CS (2006) SIRT1: tumor promoter or tumor suppressor? Med Hypotheses 67:341–344
92. Ford J, Jiang M, Milner J (2005) Cancer-specific functions of SIRT1 enable human epithelial cancer cell growth and survival. Cancer Res 65:10457–10463
93. Chen WY, Wang DH, Yen RC, Luo J, Gu W, Baylin SB (2005) Tumor suppressor HIC1 directly regulates SIRT1 to modulate p53-dependent DNA-damage responses. Cell 123:437–448
94. Zhao W, Kruse JP, Tang Y, Jung SY, Qin J, Gu W (2008) Negative regulation of the deacetylase SIRT1 by DBC1. Nature 451:587–590
95. Suzuki K, Hayashi R, Ichikawa T, Imanishi S, Yamada T, Inomata M, Miwa T, Matsui S, Usui I, Urakaze M, Matsuya Y, Ogawa H, Sakaurai H, Salki I, Tobe K (2012) SRT 1720, a SIRT1 activator, promotes tumor cell migration and lung tumor metastasis of breast cancer in mice. Oncol Rep 27:1726–1732
96. Firestein R, Blander G, Michan S, Oberdoerffer P, Ogino S, Campbell J, Bhimavarapu A, Luikenhuis S, de Cabo R, Fuchs C, Hahn WC, Guarente LP, Sinclair DA (2008) The SIRT1 deacetylase suppresses intestinal tumorigenesis and colon cancer growth. PLoS One 3:e2020
97. Baur JA, Sinclair DA (2006) Therapeutic potential of resveratrol: the in vivo evidence. Nat Rev Drug Discov 5:493–506
98. Herranz D, Iglesias G, Munoz-Maerin M, Serrano M (2011) Limited role of Sirt1 in cancer protection by dietary restriction. Cell Cycle 10:2215–2217
99. Lee CK, Klopp RG, Weindruch R, Prolla TA (1999) Gene expression profile of aging and its retardation by caloric restriction. Science 285:1390–1393
100. Cao SX, Dhahbi JM, Mote PL, Spindler SR (2001) Genomic profiling of short- and long-term caloric restriction effects in the liver of aging mice. Proc Natl Acad Sci U S A 98: 10630–10635
101. Padovani M, Lavigne JA, Chandramouli GVR, Perkins SN, Barrett JC, Hursting SD, Bennett LM, Berrigan D (2009) Distinct effects of calorie restriction and exercise on mammary gland gene expression in C57BL/6 mice. Cancer Prev Res 2:1076–1087
102. Wheatley KE, Nogueira LM, Perkins SN, Hursting SD (2011) Differential effects of calorie restriction and exercise on the adipose transcriptome in diet-induced obese mice. J Obes 2011:265417
103. Nogueira LM, Lavigne JA, Perkins SN, Chandramoulli GVR, Lui H, Barrett JC, Hursting SD (2012) Dose-dependent effects of calorie restriction on metabolism, gene expression and mammary tumor burden are partially mediated by insulin-like growth factor-1. Cancer Med 1(2):275–288
104. Sachdev D, Yee D (2007) Disrupting insulin-like growth factor signaling as a potential cancer therapy. Mol Cancer Ther 6:1–12
105. McKinsey EL, Parrish JK, Irwin AE, Niemeyer BF, Kern HB, Birks DK, Jedlicka P (2011) A novel oncogenic mechanism in Ewing sarcoma involving IGF pathway targeting by EWS/Fli1-regulated microRNA's. Oncogene 30:4910–4920
106. Gullett NP, Ruhul A, Bayraktar S, Pezzuto JM, Shin DM, Khuri FR, Aggarwal BB, Surh YJ, Kucuk O (2010) Cancer prevention with natural compounds. Semin Oncol 37:258–281
107. Harrison DE, Strong R, Sharp ZD, Nelson JF, Astle CM, Flurkey K, Nadon NL, Wilkinson JE, Frenkel K, Carter CS, Pahor M, Javors MA, Fernandez E, Miller RA (2009) Rapamycin fed late in life extends lifespan in genetically herterogeneous mice. Nature 460:392–395
108. Lamming DW, Ye L, Katajisto P, Goncalves MD, Saitoh M, Stevens DM, Davis JG, Salmon AB, Richardson A, Ahima RS, Guertin DA, Sabatini DM, Bauer JA (2012) Rapamycin-induced insulin resistance is mediated by mTORC2 loss and uncoupled from longevity. Science 335:1638–1643
109. Pollak MN (2012) Investigating metformin for cancer prevention and treatment: the end of the beginning. Cancer Discov 2(9):778–790

110. Dowling RJ, Niraula S, Stambolic V, Goodwin PJ (2012) Metformin in cancer: translational challenges. J Mol Endocrinol 48:R31–R43
111. Decensi A, Puntoni M, Goodwin P, Cazzaniga M, Gennari A, Bonammi B, Gandini S (2010) Metformin and cancer risk in diabetic patients: a systematic review and metaanalysis. Cancer Prev Res 3:1451–1461
112. Currie CJ, Poole CD, Gale EA (2009) The influence of glucose—lowering therapies on cancer risk in type 2 diabetics. Diabetologia 52:1766–1777
113. Goodwin PJ, Stambolic V, Lemieux J, Chen BE, Parulekar WR, Gelmon KA, Hershman DL, Hobday TJ, Ligibel JA, Mayer IA, Pritchard KI, Whelan TJ, Rastogi P, Shepherd LE (2011) Evaluation of metformin in early breast cancer: a modification of the traditional paradigm for clinical testing of antic-cancer agents. Breast Cancer Res Treat 126:215–220
114. Aljada A, Dong L, Mousa SA (2010) Sirtuin-targeting drugs: mechanisms of action and potential therapeutic applications. Curr Opin Investig Drugs 11:1158–1168

Chapter 15
Vascular Targeting of Adipose Tissue

Ahmad Salameh and Mikhail G. Kolonin

Abstract Obesity is a rapidly increasing worldwide threat to health due to its association with cardiovascular disease, diabetes, cancer, and a number of other medical conditions. Considerable efforts are underway to develop drugs for obesity prevention and treatment. Like tumors, white adipose tissue (WAT) overgrowing in obesity depends on functional blood vessels for its expansion and maintenance. Recent findings indicate the apparent vasculogenic role of WAT-derived cells recruited by tumors. Based on these notions, endothelial and perivascular cell populations in WAT have been considered as potential therapy targets in the context of obesity and cancer. In this chapter, we discuss studies aimed at inactivation of WAT vasculature and evaluate it as a prospective approach to treating obesity and its potential implications for cancer and other diseases.

Obesity, defined as a body mass index (BMI) of 30 kg/m^2 or more, is a result of white adipose tissue (WAT) overgrowth [1–4]. Over the past few decades, profound changes in nutrition and lifestyle reaching epidemic proportion in most industrialized countries have led to a sharp increase in the prevalence of obesity and its complications. Decreasing physical activity and increasing food accessibility has resulted in a severe alteration of the balance between intake and expenditure of energy. According to a recent report by the World Health Organization, there are >1 billion overweight adults worldwide, and at least 300 million of these individuals are clinically obese [5].

A. Salameh, Ph.D.
The University of Texas M.D. Anderson Cancer Center, 1515 Holcombe Boulevard, Unit 1347, Room T7.3821, P.O. Box 301439, Houston, TX 77030-1439, USA
e-mail: Ahmad.S.Salameh@uth.tmc.edu

M.G. Kolonin, Ph.D. (✉)
The University of Texas Health Science Center at Houston, 1825 Pressler Street, Room 630-G, Houston, TX 77030, USA
e-mail: Mikhail.G.Kolonin@uth.tmc.edu

A.J. Dannenberg and N.A. Berger (eds.), *Obesity, Inflammation and Cancer*, Energy Balance and Cancer 7, DOI 10.1007/978-1-4614-6819-6_15,

15.1 Obesity as a Health Problem

More than 2,400 years ago, Hippocrates observed: "Sudden death is more common in those who are naturally fat than in the lean." This astute observation has remained incompletely explained for many centuries to come. Obesity is associated with the metabolic syndrome, manifestations of which are low-grade inflammation, dyslipidemia, and type-2 diabetes [6]. The risk of cardiovascular disease is highly elevated in obese individuals [7]. In addition, obesity complicates numerous other medical conditions [8]. Finally, over the past decade, the link between obesity and a number of cancers has been revealed [9, 10]. While originally colorectal and endometrial and breast cancers were identified as obesity linked [11], recent reports have added pancreatic, kidney, and esophageal cancers to the list [12]. As more evidence linking excess BMI and increased cancer risk/progression is accumulating, research is turning its focus to uncover the pathophysiology behind this correlation [10]. The underlying mechanisms discussed elsewhere in this book are complex and incompletely understood [13]. Until now, it has been unclear if excess adipose tissue itself affects cancer progression or if this link is predominantly due to diet and lifestyle of obese individuals [14]. Recent animal studies show that the state of obesity can accelerate tumor growth irrespective of diet [15], which explains clinical observations [16]. Based on the increased progression of cancers surrounded by adipose tissue (e.g., prostate, breast, and uterine) in obese individuals, it has been proposed that adipose tissue has a direct effect on tumor growth [10]. An emerging body of evidence confirms that this cross talk indeed takes place at several levels.

15.2 Adipose Tissue as an Organ

In humans, subcutaneous WAT is present as layers between muscle and dermis, while intraperitoneal WAT is located around the gut, kidneys, and other internal organs [17]. Accumulating evidence indicates that different depots of WAT in the body are regulated independently and have different implications in disease. Specifically, it is overgrowth of intra-abdominal WAT, which encompasses omental and mesenteric (visceral), retroperitoneal, and perigonadal depots, that is associated with inflammation and the metabolic syndrome. Abdominal adiposity is believed to predominantly account for the poor prognosis of obese patients with cancer and cardiovascular disease.

The ability of multicellular organisms to store and release energy on demand has been crucial for their survival throughout evolution. In mammals, this function is primarily executed by adipocytes, the large cells of WAT that store fatty acids esterified to triglycerides in lipid droplets. Efficient management of energy in the body is regulated by intricate communication between multiple organs [18]. In response to signals from the central nervous system (CNS) and the digestive tract, the stored lipids can become mobilized, metabolized, and either released as fatty acids or used to increment thermogenesis by oxidation [19, 20]. As opposed to WAT, brown adipose tissue (BAT) is responsible for energy dissipation in the form of heat [21, 22]. In humans, BAT is clearly present and functional in newborns, whereas adults had been thought to lack BAT until recently. However, recent studies have demonstrated that adults also have BAT, although it is conditionally functional [23].

Adipose tissue is not merely an organ designed to manage energy. It is wired with the immune, cardiovascular, and reproductive systems through the exchange of cells and soluble factors that signal in endocrine and paracrine ways [24, 25]. Mature adipocytes synthesize and secrete numerous enzymes, growth factors, cytokines, and hormones that are involved in overall energy homeostasis [26]. Collectively, factors secreted by cells of adipose tissue are called adipokines [27]. These molecules secreted by adipocytes and other cells of adipose tissue play important roles not only in adipogenesis but also in diverse physiological processes and metabolic pathways in the body including lipid homeostasis and modulation of inflammatory responses. Collectively, adipokines modulate a range of global physiological responses including energy balance, inflammation, angiogenesis (blood vessel formation), hemostasis (regulation of blood coagulation), and blood pressure [28].

15.3 Adipose Tissue Composition

Adipose tissue development and expansion are controlled by the concerted action of extracellular and intracellular signals [4, 29]. Environmental, genetic, and epigenetic stimuli regulate preadipocyte differentiation into lipid-laden adipocytes through a process called adipogenesis (Fig. 15.1). Adipogenic differentiation takes place around the neurovascular bundles. In addition to adipocytes, WAT is composed of other cell types, interactions among which orchestrate obesity progression [30]. As shown in Fig. 15.1, these populations include stromal cells, vascular endothelial cells (EC), perivascular cells (pericytes/mural cells/adventitial cells), and infiltrating blood cells including monocytes/macrophages, lymphocytes, and mast cells [17, 31]. Blood vessels are essential for adipose tissue growth and maintenance as the route for delivery of oxygen, nutrients, and factors modulating differentiation [32]. Another important function of blood vasculature in adipose tissue is serving as a niche for progenitor cells that differentiate into adipocytes [33, 34]. Expansion of WAT in obesity results not only from adipocyte hypertrophy (increase in cell size) but also from hyperplasia (increase in cell number), which relies on progenitor cell proliferation [35]. Adipose progenitors have originally been identified in the stromal/vascular fraction (SVF) of WAT and termed adipose stromal cells (ASC) [36]. Recently it has become apparent that the ASC population appears to contain several distinct subpopulations with different differentiation potentials [37, 38].

The origins of vascular cells in adipose tissue are not completely understood. Adipose tissue is highly vascularized: each adipocyte is encircled by capillaries. Angiogenesis, the process of blood vessel sprouting from existing vasculature well characterized in tumors [39], also has a crucial role in WAT [40]. In the past few

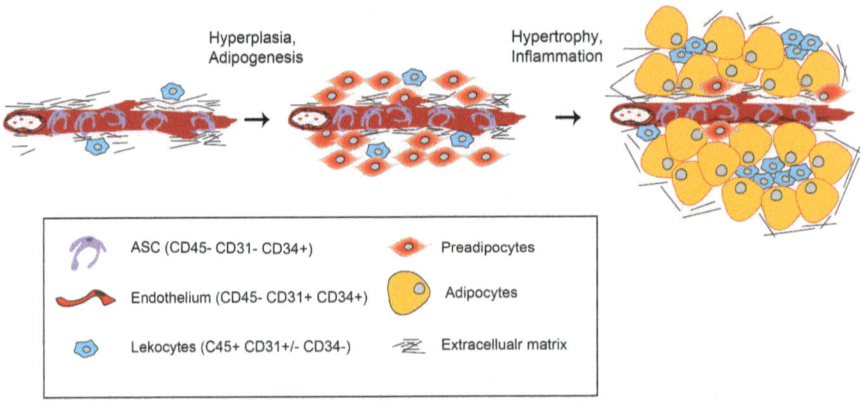

Fig. 15.1 Adipose tissue expansion in obesity. Development of adipose tissue involves proliferation and differentiation of perivascular adipose progenitors residing within the population of adipose stromal cells (ASC) into preadipocytes and finally into lipid-laden adipocytes. Adipogenesis and hypertrophic adipocyte growth are coordinated with angiogenesis and recruitment of leukocytes that are responsible for inflammation onset. Surface markers useful for identification of adipose cell populations are indicated

years, it has become apparent that recruitment of circulating vasculogenic progenitors into blood vessels may also play an important role in adult vascularization [41, 42]. Studies in cancer models have demonstrated that endothelial progenitor cells (EPCs) derived from the bone marrow contribute to neovasculature [41, 43]. In addition, a small percentage of WAT endothelium appears to be derived from transdifferentiating adipocyte progenitors [44]. Therefore, vascular cells of WAT appear to be derived from distinct origins with individual endothelial cell subpopulations distinct in their properties. An important component of adipose vasculature is comprised by adipose progenitor cells residing in the perivascular cell compartment of the blood vessel wall [33, 34].

ASC comprise the majority of cells in the SVF and display multipotency and self-renewal capacity comparable to those reported for mesenchymal stromal cells (MSC) originally characterized in the bone marrow [45]. Mesenchymal progenitors currently occupy the central arena of stem cell biology and regenerative medicine [36, 46]. The ability of MSC to differentiate into cells of mesenchymal lineage, such as adipocytes, osteocytes, and chondrocytes, has coined the term "mesenchymal stem cells" [47]. Although there are reports suggesting that adipocytes can be derived from hematopoietic stem cells [48], there is predominant evidence that bone marrow-derived precursors contribute only to endothelial cells and infiltrating hematopoietic cells but not to mesenchymal stroma and adipocytes [49, 50]. The transition from the adipose progenitors to preadipocytes is still poorly defined, and understanding of intercellular interactions within WAT is incomplete despite recent advancements [51, 52]. In addition to serving as adipocyte progenitors, ASC cooperate with the endothelium and support vascularization [33, 34]. Recent studies show that ASC are expanded in obesity [15]. ASC promote endothelial cell proliferation and blood vessel formation at least in part via trophic effects of secreted vascular endothelial growth factor (VEGF) and other angiogenic molecules [33, 51, 53]. The molecular mechanisms of interactions between adipose EC, ASC, and adipocytes that dictate the commitment of the progenitor cells toward differentiation or mobilization are being investigated [54].

15.4 Adipose Tissue and Cancer

Composition and function of WAT has pointed to the mechanisms through which obesity may be linked with cancer [55]. Inflammatory and endocrine signaling by adipose tissue-derived molecules (adipokines) has been proposed to account for cancer promotion in obesity [10]. Results from animal models, as well as clinical associations, support this hypothesis [28, 56]. In addition, recent studies have shown that cells from adipose tissue are capable of trafficking to tumors, thus enabling paracrine action of adipokines from within the tumor microenvironment [57, 58]. Investigation of the molecular pathways through which adipose cells traffic to tumors and execute their functions is underway [59]. Extracellular matrix (ECM) modulation, immune system suppression, and direct effects on malignant cell survival and proliferation have been pointed to as potential activities of systemic and locally produced adipokines within the tumor [55]. However, accumulating evidence indicates that cells from WAT promote tumor growth to a large extent through supporting tumor vasculature [57, 58]. The ability of WAT-derived cells to stimulate tumor neovascularization has been supported by a number of recent studies [58–60]. When transplanted into host animals, adipose endothelial cells can directly incorporate into the lumen of tumor vasculature, whereas ASC acquire perivascular localization in tumors [58]. Recent comparison of circulating cell populations in lean and obese mice revealed that cancer and obesity result in ASC exodus into the peripheral blood [15]. Clinical studies also demonstrate association of obesity and cancer with increased circulation of endothelial cells and ASC in patients, suggesting that endogenous WAT serves as a source of cells contributing to tumor vasculature [61–63]. The link between cancer relapse and lipotransfer procedures is consistent with this conclusion [64]. Recent studies by our group have provided evidence that ASC traffic from endogenous WAT to tumors where they become incorporated into blood vessels as pericytes [15]. Other groups have confirmed our findings [60, 65, 66]. In addition, we found that, upon recruitment by tumors, ASC can differentiate into adipocytes in an obesity-dependent manner, thus identifying an origin of intratumoral adipocytes previously implicated in cancer progression [67–69]. Extending this evidence, we found that increased tumor vascular patency is associated with elevated proliferation of neighboring malignant cells. Taken together, these findings indicate that cells recruited from endogenous adipose tissue can be recruited by tumors to potentiate the supportive properties of the tumor microenvironment. This previously overlooked phenomenon appears to at least partially account for the association between obesity and cancer.

15.5 Approaches to Obesity Treatment

Physical activity and diet are the logical measures against obesity; however, modern lifestyle complicates their implementation. Surgical interventions to reduce adipose tissue mass, such as gastric banding, are effective and prevent the risk of obesity-associated diseases [70]. However these highly invasive practices are typically applied to severe obesity cases. Pharmacological treatments for obesity available today are inadequate. Those approved by the FDA achieve ~5 % weight loss per year even when combined with diet- and exercise-based behavioral adjustments [71, 72]. The majority of anti-obesity agents tested clinically target the CNS to alter neuronal signals regulating appetite and the gastrointestinal tract to alter nutrient adsorption. There are two new recently approved drugs, Belviq and Qsymia, that act through the brain to promote satiety; however, their marginal efficacy and concerns over their safety limit enthusiasm [73]. Another drug approved in conjunction with reduced fat diet is Orlistat (Alli, Xenical), which inhibits gastrointestinal lipases and hence blocks lipid digestion. Side effects, such as incontinence and reduced vitamin absorption, have limited its popularity. Gut hormones, such as glucagon-like peptide and ghrelin, provide hope as prospective anti-obesity biologics [74]; however, the need for new approaches to obesity prevention and treatment is clearly pressing.

Directly modulating the content or function of adipose tissue is a potential alternative approach to treat obesity and the related diseases. A number of groups have focused on developing anti-obesity therapies through targeting the mature adipocytes and modulation of their differentiation [75]. An attractive idea related to this pursuit is converting energy-storing WAT to energy-burning BAT. In addition to brown adipocytes derived from common skeletal muscle progenitors, brown-like (beige or "brite") adipocytes can arise from white adipocyte progenitors (ASC) both in cell culture and in vivo [76]. This conversion is driven by sympathetic nervous system stimuli, such as cold temperature, and signal transduction cascades triggered by activation of β3-adrenergic receptors in WAT. In mice, expansion of these brown adipocytes within WAT can lead to virtually all adipose depots becoming BAT-like at the expense of WAT [77]. Targeting adipokines, the bioactive products of adipocytes, has also been pursued as a prospective therapeutic approach [18]. However, current pharmaceutical approaches to adipose tissue targeting remain at preclinical stages. Thus, while many potential adipocyte targets have been identified, they are yet to be translated into drugs.

15.6 Adipose Vasculature Targeting as a New Approach to Obesity Treatment

Blood vessels supply oxygen, nutrients, growth factors, and progenitor cells to all organs. Inhibition of neovascularization has arisen as a powerful approach to controlling tissue expansion. Over the past decade, vasculature targeting as a cancer

Table 15.1 Anti-obesity vascular-targeting agents in animal models

Molecule	Targeted tissue	Experimental model	Body mass	References
VEGF	Angiogenic vessels	Inhibitor administration Overexpression	↓ ↓	[30, 88]
VEGFR	Angiogenic vessels	Inhibitor administration	↓	[85]
PlGF	Angiogenic vessels	Inhibitor administration Genetic deficiency	↓ ↑	[90]
SPARC	Angiogenic vessels	Genetic deficiency	↑	[92]
Leptin	Angiogenic vessels	Administration Genetic deficiency	↓ ↑	[96] [97]
Plasminogen	Angiogenic vessels	Administration	↓	[90] [138]
Tiplaxtinin		Genetic deficiency	↑	[139]
Angiostatin	Angiogenic vessels	Administration	↓	[81]
Endostatin	Angiogenic vessels	Administration	↓	[81]
TNP-470 CKD-732	Angiogenic vessels	Administration	↓	[81]
MMP inhibitors: Galardin, Bay-129566, Ro20-2653	Angiogenic vessels	Inhibitor administration	↓	[40, 140, 141]
Thalidomide	Angiogenic vessels	Administration	↓	[81]
Adiponectin	Angiogenic vessels	Administration Genetic deficiency	↓ ↑	[83]
Curcumin (polyphenol)	Angiogenic vessels	Administration	↓	[104]
Ob-x and fumagillin	Angiogenic vessels	Administration	↓	[101, 102]
EGCG (catechin in tea)	Angiogenic vessels	Administration	↓	[40, 142]
Adipotide	Mature vessels	Administration	↓	[117]

EGCG epigallocatechin-3-gallate; *MMPs* metalloproteinases; *PLGF* placental growth factor; *VEGFR* vascular endothelial growth factor receptor

therapy has evolved resulting in clinically approved drugs that have considerably improved the clinical outcome [78, 79]. Like tumors, WAT is a highly dynamic organ, growth and maintenance of which requires continuous remodeling of the capillary networks [32]. Emerging evidence shows that modulators of angiogenesis affect the expansion of WAT mass by regulating the development of adipose vasculature [31]. Pharmaceutical manipulation of adipose tissue neovascularization by angiogenic stimulators and inhibitors might therefore offer novel therapeutic options for the treatment of obesity and related metabolic disorders. Further, recent advances in the development of agents aimed to disrupt the mature vascular components (namely, endothelial and perivascular) represent an alternative direction in anti-obesity drug design. Table 15.1 summarizes studies reporting changes in WAT vasculature, most of which have been performed in animal models. A detailed discussion of some of the individual molecular mechanisms probed in these publications is provided below.

15.6.1 Targeting Immature WAT Vasculature with Angiogenesis Inhibitors

Recently, it has been appreciated that vasculature is critical for WAT development as the gatekeeper of blood access to adipocytes [4, 30, 80]. The pioneering studies from the Folkman group have demonstrated that inhibition of angiogenesis has the potential to prevent not only tumor growth but also obesity onset [39]. Both angiogenesis and vasculogenesis (progenitor cell recruitment from remote organs) appear to play an important role in WAT vascularization. Because WAT growth-underlying obesity proceeds into adulthood, it has been proposed that adipogenesis and angiogenesis feed onto each other in obesity [81, 82]. VEGFs and their receptors (VEGFRs) comprise the system highly active in adipose tissue [80, 83–86]. VEGF-A is a major angiogenic factor that controls proliferation, migration, and permeability of the endothelium [53, 87]. Administration of anti-VEGF antibodies inhibiting adipose angiogenesis affects adipocyte differentiation and systemic lipid metabolism, indicating VEGF as a key factor linking the interplay between angiogenesis and adipogenesis in WAT with physiological responses in the rest of the body [30, 88]. Interestingly, a recent report from the Scherer group shows that consequences of VEGF modulation in WAT are context dependent, highlighting the complexity of adipose tissue biology and emphasizing the point that broad implications of WAT vasculature targeting are yet to be fully understood [88].

A close recent connection between VEGF-B and endothelial fatty acid uptake [89] also indicates the direct dependence of adipogenesis on angiogenesis.

Placental growth factor (PlGF), a homolog of VEGF, enhances angiogenesis in pathological conditions. Loss of PlGF impairs angiogenesis in the ischemic retina, limb, and heart, in wounded skin, and in tumors, without affecting physiological angiogenesis. Administration of a PlGF-neutralizing monoclonal antibody inhibits angiogenesis and fat pad formation, indicating its function in WAT [90]. Another component of the system, matricellular protein SPARC (secreted protein, acidic, and rich in cysteine), also known as osteonectin, binds to VEGF-A, impairs VEGFR-1 activation, and inhibits FGF-2, resulting in inhibition of endothelial cell proliferation. SPARC is produced by adipose tissue and its expression is upregulated in obesity [91]. SPARC-deficient mice on high-fat diet develop larger fat pads as compared to wild-type mice [92]. The apelin/APJ signaling system [93] and angiopoietins along with their receptors, Tie1 and Tie2, are likely to play a role in WAT angiogenesis [94]. Many other components directly or indirectly regulating angiogenesis have been identified in adipose tissue. These include leptin, neuropilin-1, fibroblast growth factors (FGF-2), thrombospondin-1, adiponectin, tissue factor, tumor necrosis factor (TNF-α), transforming growth factor beta (TGF-β), insulin-like growth factor-1 (IGF-1), hepatocyte growth factor (HGF), and various components of the ECM [40, 95]. Leptin, adiponectin, and ECM-related mechanisms have drawn particular attention in respect to the link between obesity and cancer.

Leptin is a hormone/cytokine/growth factor secreted by adipocytes, and its inactivation (in ob/ob mice) or loss of its receptor (in db/db mice) causes morbid obesity due to the role of this system in appetite control [96]. Leptin can, however, induce angiogenesis, suggesting that it might contribute positively to the development of adipose tissue [97]. In genetically obese human individuals carrying homozygous mutations in the leptin gene, administration of recombinant leptin successfully reduces body weight; however, the majority of obese humans have unusually high levels of circulating leptin and are leptin resistant [71, 98]. The potential implication of the leptin role in angiogenesis, which could contribute to obesity in these patients, warrants further investigation. Interestingly, while obese cancer patients have increased levels of leptin in circulation, the opposite trend is observed for adiponectin [56]. One of the functions proposed for adiponectin is inhibition of angiogenesis, which may be implicated in both obesity and cancer [99]. However, both antiangiogenic and pro-angiogenic functions have been reported for adiponectin.

Angiogenesis is mechanistically linked with ECM remodeling [100]. Proteolysis is required not only for the expansion of the adipocyte basement membrane in hypertrophic WAT, but also for cell migration during the development of blood vessels and peripheral nerves in the growing adipose tissue. Thus, both adipogenesis and angiogenesis require proteolytic activity, which is mainly regulated via the plasminogen/plasmin (fibrinolytic) and matrix metalloproteinase (MMP) systems. Recently, some evidence has emerged that proteins of the ADAM (A Disintegrin And Metalloproteinase) and ADAMTS (ADAM with TSP motif) families may also be implicated. The named proteinases are collectively able to cleave a wide variety of substrates, including ECM components, other proteinases and their inhibitors, and matrix receptors, thus fine-tuning adipose tissue remodeling. MMPs and plasmin can also release, activate, or degrade several growth factors and cytokines that play major roles in angiogenesis and are implicated in obesity. Several nutritionally induced obesity models in transgenic mice have been used to study the role of the fibrinolytic system in the development of obesity. Mice deficient in plasminogen, the substrate for both plasminogen activators, showed reduced fat accumulation associated with reduced differentiation of progenitor cells [101]. Indeed, tiplaxtinin, designed as a synthetic inhibitor of PAI-1, reduces body weight in WT mice kept on HFD [90].

A number of preclinical studies using angiogenesis inhibitors for obesity intervention, with many reviewed elsewhere [40], are summarized in Table 15.1. In the pioneering study, treatment with angiostatin (kringle 1–4 domains of plasminogen), endostatin (a C-terminal fragment of collagen XVIII), TNP-470, Bay-129566 (a MMP inhibitor), and thalidomide was tested in mouse models. It has been demonstrated that inhibition of angiogenesis is potent enough to impede WAT expansion even in genetically obese leptin-deficient mice [81]. In addition, agents blocking VEGFR-2 have been shown to prevent the development of obesity in genetic mouse models and studies based on high-fat diets [85]. Inhibition of angiogenesis in the growing adipose tissue associated with a reduction in vascular density and endothelial cell apoptosis caused a decrease in the body weight of obese mice [81, 85, 102, 103]. Although agents such as TNP-470 and its analog CKD-732 are considered to be selective angiogenesis inhibitors, many have off-target effects such as food aversion, which may partially account for weight loss [31]. Therefore, it remains to be determined to which extent WAT is affected through specific targeting of endothelial cells, as opposed to affecting systemic energy balance.

A number of recent reports confirm the initial observation that disrupting WAT neovascularization can prevent the onset of obesity in both genetic and diet-induced obesity models (Table 15.1). For example, curcumin, the major polyphenol in turmeric spice, has been shown to suppress WAT accumulation through its effects on angiogenesis and adipogenesis [104]. Similar results have been obtained with other herbal and synthetic antiangiogenic compounds, such as Ob-x, EGCG (catechin in tea), and fumagillin [105, 106]. However, for most studies, it remains questionable to which extent the effect of these agents directly on the vasculature, as opposed to their effect on other cells in WAT or other organs, affects food intake and adipose tissue levels. Notably, while mice reduce their body weight while on agents such as TNP-470, they regain weight when off schedule [81]. These treatment-regulated cycles can occur several times without causing resistance or other obvious side effects. These findings suggest that antiangiogenic agents can be used repeatedly for the treatment of obesity without encountering drug resistance, which has been observed for cancer [107].

It should indeed be kept in mind that the angiogenic circuits involved in adipose tissue development also are critical to many other biological processes. Although animal treatment with antiangiogenic agents has consistently resulted in inhibition of WAT accumulation, there is much preclinical work to be done before this approach can be considered clinically feasible. Although generally mild, adverse effects of angiogenesis inhibition have been reported in clinical trials [108]. Off-target effects of antiangiogenesis drugs illustrate the fact that neovascularization does occur, although at levels much lower than in development, outside pathological tissues in adulthood [39]. New blood vessel formation generally accompanies tissue remodeling that takes place either upon injury or during certain normal physiological processes. For example, because VEGF/VEGFR signaling is not WAT specific, its inhibition is expected to cause systemic effects. Indeed, VEGF inhibition can result in off-site proteinuria, hypertension, and internal bleeding [109]. Moreover, studies in mouse models show that chronic blockade of certain endothelial pathways can backfire with pathological activation of endothelial cells, perturbed organ physiology, and even vascular tumorigenesis, further emphasizing safety concerns [110]. Little is known regarding the signaling between the angiogenic components of adipose tissues and the CNS as well as other organs [111, 112]. Also, because cell death in WAT results in the recruitment of leukocytes resolving the tissue damage [113], the resulting local and systemic activation in secretion of inflammatory cell-derived cytokines, such as TNF-α and several interleukins (e.g., IL-6 and IL-8), may have systemic implications. A better understanding of the WAT targeting consequences is instrumental in the development of therapeutic approaches.

15.6.2 Targeting "Mature" WAT Vasculature

Studies on tumor models illustrate the notion that vascular disruption can serve as an effective complementary tool to constrain tissue expansion [114, 115]. It has been proposed that the mature vasculature of WAT could be similarly targeted and that depleting the supply of nutrients and oxygen essential for the maintenance of adipocytes could result in obesity reversal after its onset. Based on the notion that tumor vasculature features differential expression of markers [116], our group had proposed the existence of cell surface molecules selectively upregulated in adult WAT blood vessels that could be therapeutically targeted in pathologically expanded WAT. In a proof-of-principle study aiming to identify and use WAT vascular targets for experimental obesity therapy, we screened a combinatorial library in mice for peptides systemically homing to WAT. As a result, we isolated a peptide with the sequence CKGGRAKDC that selectively accumulated in WAT [117]. The uptake of CKGGRAKDC by adipose endothelial cells suggested that the internalizing receptor could serve as a target of therapies directed to WAT. By using WAT membrane protein extracts, we biochemically isolated the vascular receptor of the CKGGRAKDC peptide and identified it as prohibitin (Phb). Follow-up studies have validated the specificity of the CKGGRAKDC peptide for Phb for WAT endothelium cells using nanocarriers [118]. We used a mouse diet-induced obesity model to test in vivo the capacity of the CKGGRAKDC peptide to deliver cytotoxic (proapoptotic) peptide KLAKLAKKLAKLAK to WAT. Daily subcutaneous injections of the CKGGRAKDC-KLAKLAKKLAKLAK fusion peptide (Adipotide) caused rapid obesity reversal [117]. Another study reproduced these results in the rat model [111] and suggested that reduced food consumption in WAT-targeted animals explains the lack of apparent lipodystrophic effects. These observations uncover a previously unappreciated cross talk between the status of WAT vasculature and central regulation of food intake [111]. Recently, Adipotide, which was validated in three monkey models [119], has been shown to have antidiabetic effects [120], and the first-in-human Adipotide clinical trial is ongoing.

In follow-up studies, we identified annexin A2 (annexin II, Anx2) as a Phb-binding protein mimicked by the CKGGRAKDC peptide [121]. Consistent with Phb/Anx2 interaction in WAT endothelium, Anx2 has been previously shown to have a pro-angiogenic function [122]. In addition, a function of Phb in controlling adipogenesis has been uncovered [123]. Our recent report shows that Phb and Anx2 form a complex in lipid rafts isolated from the cell membranes of WAT cells and identify the domains mediating Phb/Anx2 interaction [121]. Currently, our group is characterizing the function of Phb/Anx2 in adipose tissue in order to better establish this complex as a prospective anti-obesity therapy target. Delivery of other therapeutic agents to this molecular "zip code" is also being pursued [118].

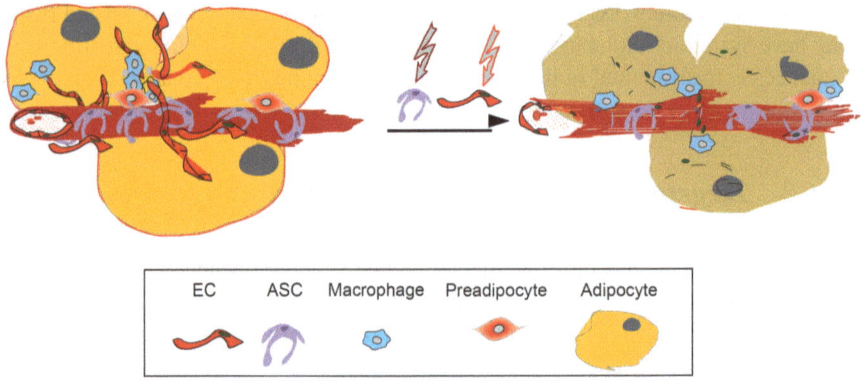

Fig. 15.2 Vascular targeting of adipose tissue as an anti-obesity approach. Targeted depletion of adipose endothelium, the integral component of the vasculature delivering nutrients and oxygen to adipocytes, results in quick tissue resorption in rodent models. Cytoablation of pro-angiogenic adipocyte progenitors (ASC), in parallel with the endothelium, could represent a potential complimentary approach to prevention of adipose tissue regrowth upon discontinuation of anti-vascular therapy

15.6.3 Targeting Perivascular Adipose Cells

Despite the short-term effectiveness in experimental animal models, obesity relapses due to WAT regrowth upon the cessation of treatment with endothelium-targeting agents [117]. Although proliferating endothelial cells serve as building blocks for growing vessels, this process relies on the trophic contribution and mechanical support from stromal cells [33, 54]. Indeed, while endothelial cells and adipocytes undergo cell death under ischemic conditions, ASC are resistant to hypoxia [124]. Therefore, resistance to vascular-targeting agents appears to result from the pro-angiogenic action of surviving ASC that quickly rewire the tissue. We propose that in the future, ASC targeting in parallel with endothelial cell targeting in WAT might provide an approach to long-term control of WAT mass (Fig. 15.2). Accumulating body of evidence indicates that WAT contributes to the pool of progenitor cells mobilized in obesity and cancer [62, 63] and serving as perivascular cells in tumors [15, 58]. Therefore, targeting WAT-derived cells in circulation or upon homing to cancer sites could offer a new direct approach to complementary cancer therapy.

Studies from our research team have contributed to the understanding of how blood vessel formation might be coordinated with ASC differentiation and migration by interaction between SPARC and β1 integrin [125]. SPARC-integrin binding also plays a role in pericyte recruitment during cancer progression [126]. Approaches to SPARC targeting could, therefore, have a two-pronged effect. Another potentially interesting target is aminopeptidase N (CD13), which plays a role in tumor angiogenesis and is also expressed by both endothelial and stromal adipose [127]. Recently, by combinatorial phage display approaches [54, 128, 129], a new decorin isoform (delta-DCN) as marker of ASC with progenitor capacity and a peptide specifically binding to delta-DCN have been identified [31]. Derivatives of this peptide could potentially be developed for directed delivery of therapy depleting ASC serving as perivascular adipocyte progenitors in WAT and possibly directly in tumors. Approaches to targeting mature adipocytes are also being developed [130].

15.7 Unresolved Issues in Adipose Vascular Targeting

It remains to be determined whether targeting individual adipose tissue depots will be possible and if sparing specific depots may be beneficial. Visceral WAT is susceptible to inflammation in severely obese individuals as a result of insufficient oxygenation of grossly enlarged adipocytes, which is the underlying cause of lipotoxicity and the associated pathological consequences [131]. By contrast, subcutaneous WAT, typically remaining sufficiently vascularized, has the potential to benefit metabolism by improving glucose homeostasis and increasing energy consumption [132]. The content and properties of cells differ between visceral and subcutaneous WAT. A recent comparison of angiogenesis occurring in human visceral and subcutaneous WAT did not detect significant differences [95]; however, separately assessing the physiological consequences of targeting vasculature in distinct depots might be important. Although Adipotide appears to not discriminate between WAT depots [117], future studies may uncover agents useful for targeting of vasculature selectively in a WAT depot of interest. Another important issue is the potential importance of vessel types affected by treatment. For example, at this point it is unclear whether Adipotide has a preference toward arterial or venous vasculature and whether vessel size is a predeterminant of peptide binding and effect. In addition, the unclear role of lymphatic vasculature in WAT has to be taken into consideration. Clearly, WAT contains well-vascularized lymph nodes; however, lymphatic capillaries in WAT have not been reported [133].

It is perhaps even more important to foresee the potential effects of targeted therapies on BAT vasculature. A number of results from the rodent models indicate that BAT has a protective effect against the pathological consequences of obesity. The significance of discovering BAT in adults [134] lies in possible new approaches to treatment of obesity and of the associated disorders [52]. Because angiogenesis accompanies BAT formation [135], inhibiting adipose vascularization in a nonspecific manner is expected to have a negative impact on BAT and, therefore, on energy expenditure, potentially defeating the purpose of the treatment. Thus, development of approaches to targeting specifically blood vessels of WAT might be critical. The apparent dependence of adipocyte physiology on the status of the vasculature raises the possibility that vascular-targeting agents could be designed to convert WAT into BAT, rather than destroying tissue altogether, as a more physiological and safer anti-obesity treatment [31]. Development of pharmacological approaches to activate proliferation, vascularization, and/or metabolism of the existing residual BAT could, in theory, tilt the WAT/BAT balance and be used to treat obesity.

15.8 Adipose Vascular Targeting in Cancer Patients?

According to some studies, reduction of WAT mass has protective effects against cancer and other obesity-linked diseases [70], although this has not been consistently confirmed [9]. It is also possible that reduction of adiposity will benefit obese patients who have been diagnosed with cancer. As new vascular-targeting agents are characterized, new options to treat obesity and cancer become available. However, since the composition and function of adipose tissue are complex and multifactorial, it is questionable whether every anti-obesity strategy will benefit patients [8]. Future therapeutic strategies must take into account the biological role of adipose tissue and the well-described pathologies associated with its deficiency [136, 137]. Specifically, the risk of lipotoxicity is noteworthy. WAT lipid storage defects can lead to increased circulating fatty acids that become ectopically deposited in muscle, liver, and other tissues, leading to insulin resistance and metabolic disease, as seen in genetic or pharmacologically induced lipodystrophies [31]. Therefore, aggressive therapies directly or indirectly affecting adipocytes, in the absence of decreased caloric intake, could result in lipotoxicity, which also applies to vascular-targeting approaches. Lack of apparent steatosis in response to the Adipotide is likely due to decreased food consumption and the metronomic regimen of the reported treatments [117]. The extent to which the anorexigenic effects of this and other experimental compounds targeted at vasculature are responsible for weight loss is not easy to assess. But it is certain that antiangiogenesis drugs, to be used for amelioration of obesity without pathological consequences, must be accompanied by concomitant reduction in food intake. The fine balance between WAT and BAT is another important issue to take into consideration in designing vascular-targeting therapies.

It should be noted that adipose tissue targeting might not be an ideal option in particular for cancer patients [8]. Patients undergoing conventional cancer therapies, with chemotherapy or radiotherapy being the most common, tend to experience nausea, upset stomach, and loss of appetite, which results in nutrient deprivation and body weakening. Therefore, the potential adverse side effects of future anti-obesity drugs are a serious issue for cancer patients. An additional concern is the possibility of WAT cell mobilization as a result of certain approaches to weight loss and the hijacking of these cells by tumors. For example, ablation of adipose endothelium and the resulting WAT resorption could represent the setting when intact ASC and infiltrating macrophages become “homeless” and thereby encouraged to execute their trophic functions as components of tumor microenvironment. Although purely speculative at this point, these potential pitfalls are to be tested in animal models.

15.9 Summary

Systemic deregulation of angiogenesis is a hallmark of obesity-associated pathologies including cardiovascular disease, diabetes, and cancer. Therefore, it is possible that targeting neovasculature not only in WAT but also at other sites of ectopic angiogenesis may have a combined therapeutic benefit. Considering the body of evidence demonstrating a link between obesity and cancer, it is clear that new approaches to modulate WAT content and activity could be beneficial for patients. There is an apparent need for noninvasive treatment of obesity compatible with the health status of cancer patients. There may be a big advantage to minimizing the effects of anti-obesity treatment outside of WAT in order to prevent adverse effects on CNS and gut physiology. New obesity therapeutic prototypes that enable WAT-specific targeting hold a potential promise. However, recent findings indicate that systemic and even WAT-specific inhibition of vascular function could be dangerous. Therefore, careful analysis of preclinical models will be necessary to establish whether vascular targeting does indeed represent a viable approach for the treatment of obesity, in particular in the cancer setting. As the field of vascular targeting and its application to obesity and cancer treatment is evolving, a number of questions remain to be answered. Continued efforts in unraveling the mechanisms through which obesity supports cancer development and progression might open other new avenues for therapy.

References

1. Kras KM et al (1999) Adipocyte development is dependent upon stem cell recruitment and proliferation of preadipocytes. Obes Res 7(5):491–497
2. Sun K, Kusminski CM, Scherer PE (2011) Adipose tissue remodeling and obesity. J Clin Invest 121(6):2094–2101
3. Ducharme NA, Bickel PE (2008) Lipid droplets in lipogenesis and lipolysis. Endocrinology 149(3):942–949
4. Rosen ED, Spiegelman BM (2006) Adipocytes as regulators of energy balance and glucose homeostasis. Nature 444(7121):847–853
5. Wisse BE, Kim F, Schwartz MW (2007) Physiology. An integrative view of obesity. Science 318(5852):928–929
6. Cowey S, Hardy RW (2006) The metabolic syndrome: a high-risk state for cancer? Am J Pathol 169(5):1505–1522
7. Flegal KM et al (2007) Cause-specific excess deaths associated with underweight, overweight, and obesity. JAMA 298(17):2028–2037
8. Sirin O, Kolonin MG (2012) Treatment of obesity as a potential complementary approach to cancer therapy. Drug Discov Today, Online ahead of print
9. Kant P, Hull MA (2011) Excess body weight and obesity—the link with gastrointestinal and hepatobiliary cancer. Nat Rev Gastroenterol Hepatol 8(4):224–238
10. Roberts DL, Dive C, Renehan AG (2010) Biological mechanisms linking obesity and cancer risk: new perspectives. Annu Rev Med 61:301–316
11. Calle EE, Kaaks R (2004) Overweight, obesity and cancer: epidemiological evidence and proposed mechanisms. Nat Rev Cancer 4(8):579–591
12. Eheman C et al (2012) Annual report to the nation on the status of cancer, 1975–2008, featuring cancers associated with excess weight and lack of sufficient physical activity. Cancer 118(9):2338–2366
13. Vona-Davis L, Howard-McNatt M, Rose DP (2007) Adiposity, type 2 diabetes and the metabolic syndrome in breast cancer. Obes Rev 8(5):395–408
14. Hursting SD et al (2007) The obesity-cancer link: lessons learned from a fatless mouse. Cancer Res 67(6):2391–2393

15. Zhang Y et al (2012) Stromal progenitor cells from endogenous adipose tissue contribute to pericytes and adipocytes that populate the tumor microenvironment. Cancer Res 72(20):5198–5208
16. Pierce J, Natarajan L, Caan BJ (2007) Influence of a diet very high in vegetables, fruit, and fiber and low in fat on prognosis following treatment for breast cancer: the Women's Healthy Eating and Living (WHEL) randomized trial. JAMA 298(3):289–298
17. Ailhaud G, Grimaldi P, Negrel R (1992) Cellular and molecular aspects of adipose tissue development. Annu Rev Nutr 12:207–233
18. Klein J et al (2006) Adipose tissue as source and target for novel therapies. Trends Endocrinol Metab 17(1):26–32
19. Lafontan M (2005) Fat cells: afferent and efferent messages define new approaches to treat obesity. Annu Rev Pharmacol Toxicol 45:119–146
20. Klaus S (2004) Adipose tissue as a regulator of energy balance. Curr Drug Targets 5(3):241–250
21. Tam CS, Lecoultre V, Ravussin E (2012) Brown adipose tissue: mechanisms and potential therapeutic targets. Circulation 125(22):2782–2791
22. Cannon B, Nedergaard J (2004) Brown adipose tissue: function and physiological significance. Physiol Rev 84(1):277–359
23. Nedergaard J, Bengtsson T, Cannon B (2007) Unexpected evidence for active brown adipose tissue in adult humans. Am J Physiol Endocrinol Metab 293(2):444–452
24. Park J, Euhus DM, Scherer PE (2011) Paracrine and endocrine effects of adipose tissue on cancer development and progression. Endocr Rev 32(4):550–570
25. Kershaw EE, Flier JS (2004) Adipose tissue as an endocrine organ. J Clin Endocrinol Metab 89(6):2548–2556
26. Paz-Filho G et al (2011) Associations between adipokines and obesity-related cancer. Front Biosci 16:1634–1650
27. Trayhurn P, Wood IS (2004) Adipokines: inflammation and the pleiotropic role of white adipose tissue. Br J Nutr 92(3):347–355
28. Khandekar MJ, Cohen P, Spiegelman BM (2011) Molecular mechanisms of cancer development in obesity. Nat Rev Cancer 11(12):886–895
29. Hausman DB et al (2001) The biology of white adipocyte proliferation. Obes Rev 2(4):239–254
30. Nishimura S et al (2007) Adipogenesis in obesity requires close interplay between differentiating adipocytes, stromal cells, and blood vessels. Diabetes 56(6):1517–1526
31. Daquinag AC, Zhang Y, Kolonin MG (2011) Vascular targeting of adipose tissue as an anti-obesity approach. Trends Pharmacol Sci 32(5):300–307
32. Cao Y (2007) Angiogenesis modulates adipogenesis and obesity. J Clin Invest 117(9):2362–2368
33. Traktuev D et al (2008) A population of multipotent CD34-positive adipose stromal cells share pericyte and mesenchymal surface markers, reside in a periendothelial location, and stabilize endothelial networks. Circ Res 102(1):77–85
34. Tang W et al (2008) White fat progenitor cells reside in the adipose vasculature. Science 322(5901):583–586
35. Spalding KL et al (2008) Dynamics of fat cell turnover in humans. Nature 453(7196): 783–787
36. Gimble JM, Katz AJ, Bunnell BA (2007) Adipose-derived stem cells for regenerative medicine. Circ Res 100(9):1249–1260
37. Daquinag AC et al (2011) An isoform of decorin is a resistin receptor on the surface of adipose progenitor cells. Cell Stem Cell 9(1):74–86
38. Lee NK et al (2011) Identification of a novel peptide ligand targeting visceral adipose tissue via transdermal route by in vivo phage display. J Drug Target 19(9):805–813
39. Folkman J (2006) Angiogenesis. Annu Rev Med 57:1–18
40. Cao Y (2010) Adipose tissue angiogenesis as a therapeutic target for obesity and metabolic diseases. Nat Rev Drug Discov 9(2):107–115

41. Bertolini F et al (2006) The multifaceted circulating endothelial cell in cancer: towards marker and target identification. Nat Rev Cancer 6(11):835–845
42. Tilki D et al (2009) Emerging biology of vascular wall progenitor cells in health and disease. Trends Mol Med 15(11):501–509
43. Wels J et al (2008) Migratory neighbors and distant invaders: tumor-associated niche cells. Genes Dev 22(5):559–574
44. Gupta RK et al (2012) Zfp423 expression identifies committed preadipocytes and localizes to adipose endothelial and perivascular cells. Cell Metab 15(2):230–239
45. Bianco P, Robey PG, Simmons PJ (2008) Mesenchymal stem cells: revisiting history, concepts, and assays. Cell Stem Cell 2(4):313–319
46. Caplan AI, Correa D (2011) The MSC: an injury drugstore. Cell Stem Cell 9(1):11–15
47. Prockop DJ (1997) Marrow stromal cells as stem cells for nonhematopoietic tissues. Science 276(5309):71–74
48. Sera Y et al (2009) Hematopoietic stem cell origin of adipocytes. Exp Hematol 37(9):1108–1120
49. Tomiyama K et al (2008) Characterization of transplanted green fluorescent protein+ bone marrow cells into adipose tissue. Stem Cells 26(2):330–338
50. Koh YJ et al (2007) Bone marrow-derived circulating progenitor cells fail to transdifferentiate into adipocytes in adult adipose tissues in mice. J Clin Invest 117(12):3684–3695
51. Rodeheffer MS, Birsoy K, Friedman JM (2008) Identification of white adipocyte progenitor cells in vivo. Cell 135(2):240–249
52. Zeve D, Tang W, Graff J (2009) Fighting fat with fat: the expanding field of adipose stem cells. Cell Stem Cell 5(5):472–481
53. Yancopoulos GD et al (2000) Vascular-specific growth factors and blood vessel formation. Nature 407(6801):242–248
54. Kolonin MG et al (2012) Alternative origins of stroma in normal organs and disease. Stem Cell Res 8(2):312–323
55. Zhang Y, Bellows CF, Kolonin MG (2010) Adipose tissue-derived progenitor cells and cancer. World J Stem Cells 2(5):103–113
56. Grossmann ME et al (2010) Obesity and breast cancer: status of leptin and adiponectin in pathological processes. Cancer Metastasis Rev 29(4):641–653
57. Zhang J et al (2012) Automated analysis of investigational near-infrared fluorescence lymphatic imaging in humans. Biomed Opt Express 3(7):1713–1723
58. Zhang Y et al (2009) White adipose tissue cells are recruited by experimental tumors and promote cancer progression in mouse models. Cancer Res 69(12):5259–5266
59. Klopp AH et al (2012) Omental adipose tissue-derived stromal cells promote vascularization and growth of endometrial tumors. Clin Cancer Res 18(3):771–782
60. Kidd S et al (2012) Origins of the tumor microenvironment: quantitative assessment of adipose-derived and bone marrow-derived stroma. PLoS One 7(2):e30563
61. Mancuso P et al (2011) Circulating perivascular progenitors: a target of PDGFR inhibition. Int J Cancer 129(6):1344–1350
62. Bellows CF et al (2011) Circulation of progenitor cells in obese and lean colorectal cancer patients. Cancer Epidemiol Biomarkers Prev 20(11):2461–2468
63. Bellows CF et al (2011) Influence of BMI on level of circulating progenitor cells. Obesity 19(8):1722–1726
64. Bertolini F et al (2012) Adipose tissue cells, lipotransfer and cancer: a challenge for scientists, oncologists and surgeons. Biochim Biophys Acta 1826:209–214
65. Lin G et al (2010) Effects of transplantation of adipose tissue-derived stem cells on prostate tumor. Prostate 70(10):1066–1073
66. Martin-Padura I et al (2012) The white adipose tissue used in lipotransfer procedures is a rich reservoir of CD34+ progenitors able to promote cancer progression. Cancer Res 72(1):325–334
67. Nieman KM et al (2011) Adipocytes promote ovarian cancer metastasis and provide energy for rapid tumor growth. Nat Med 17(11):1498–1503

68. Dirat B et al (2011) Cancer-associated adipocytes exhibit an activated phenotype and contribute to breast cancer invasion. Cancer Res 71(7):2455–2465
69. Zyromski NJ et al (2009) Obesity potentiates the growth and dissemination of pancreatic cancer. Surgery 146(2):258–263
70. Adams TD et al (2007) Long-term mortality after gastric bypass surgery. N Engl J Med 357(8):753–761
71. Friedman JM (2009) Obesity: causes and control of excess body fat. Nature 459(7245):340–342
72. Bray GA, Greenway FL (2007) Pharmacological treatment of the overweight patient. Pharmacol Rev 59(2):151–184
73. Dolgin E (2012) A history of drugs on the weight list. Nat Med 18(6):843
74. Tharakan G, Tan T, Bloom S (2010) Emerging therapies in the treatment of 'diabesity': beyond GLP-1. Trends Pharmacol Sci 32(1):8–15
75. Nawrocki AR, Scherer PE (2005) Keynote review: the adipocyte as a drug discovery target. Drug Discov Today 10(18):1219–1230
76. Cinti S (2009) Transdifferentiation properties of adipocytes in the adipose organ. Am J Physiol Endocrinol Metab 297(5):977–986
77. Wu J et al (2012) Beige adipocytes are a distinct type of thermogenic fat cell in mouse and human. Cell 150(2):366–376
78. Carmeliet P, Jain RK (2000) Angiogenesis in cancer and other diseases. Nature 407(6801):249–257
79. Ferrara N, Kerbel RS (2005) Angiogenesis as a therapeutic target. Nature 438(7070):967–974
80. Hausman GJ, Richardson RL (2004) Adipose tissue angiogenesis. J Anim Sci 82(3):925–934
81. Rupnick MA et al (2002) Adipose tissue mass can be regulated through the vasculature. Proc Natl Acad Sci U S A 99(16):10730–10735
82. Christiaens V, Lijnen HR (2010) Angiogenesis and development of adipose tissue. Mol Cell Endocrinol 318(1–2):2–9
83. Brakenhielm E, Cao Y (2008) Angiogenesis in adipose tissue. Methods Mol Biol 456:65–81
84. Gealekman O et al (2008) Enhanced angiogenesis in obesity and in response to PPARgamma activators through adipocyte VEGF and ANGPTL4 production. Am J Physiol Endocrinol Metab 295(5):1056–1064
85. Fukumura D et al (2003) Paracrine regulation of angiogenesis and adipocyte differentiation during in vivo adipogenesis. Circ Res 93(9):88–97
86. Voros G et al (2005) Modulation of angiogenesis during adipose tissue development in murine models of obesity. Endocrinology 146(10):4545–4554
87. Carmeliet P (2005) Angiogenesis in life, disease and medicine. Nature 438(7070):932–936
88. Sun K et al (2012) Dichotomous effects of VEGF-A on adipose tissue dysfunction. Proc Natl Acad Sci U S A 109(15):5874–5879
89. Hagberg CE et al (2010) Vascular endothelial growth factor B controls endothelial fatty acid uptake. Nature 464(7290):917–921
90. Lijnen HR et al (2006) Impaired adipose tissue development in mice with inactivation of placental growth factor function. Diabetes 55(10):2698–2704
91. Tartare-Deckert S, Chavey C, Monthouel MN, Gautier N, Van Obberghen E (2001) The matricellular protein SPARC/osteonectin as a newly identified factor up-regulated in obesity. J Biol Chem 276:2231–22237
92. Bradshaw AD et al (2003) SPARC-null mice display abnormalities in the dermis characterized by decreased collagen fibril diameter and reduced tensile strength. J Invest Dermatol 120(6):949–955
93. Kunduzova O et al (2008) Apelin/APJ signaling system: a potential link between adipose tissue and endothelial angiogenic processes. FASEB J 22(12):4146–4153
94. Oike Y, Tabata M (2009) Angiopoietin-like proteins—potential therapeutic targets for metabolic syndrome and cardiovascular disease. Circ J 73(12):2192–2197

95. Ledoux S et al (2008) Angiogenesis associated with visceral and subcutaneous adipose tissue in severe human obesity. Diabetes 57(12):3247–3257
96. Friedman JM, Halaas JL (1998) Leptin and the regulation of body weight in mammals. Nature 395(6704):763–770
97. Sierra-Honigmann MR et al (1998) Biological action of leptin as an angiogenic factor. Science 281(5383):1683–1686
98. Licinio J et al (2004) Phenotypic effects of leptin replacement on morbid obesity, diabetes mellitus, hypogonadism, and behavior in leptin-deficient adults. Proc Natl Acad Sci U S A 101(13):4531–4536
99. Delort L et al (2012) New insights into anticarcinogenic properties of adiponectin: a potential therapeutic approach in breast cancer? Vitam Horm 90:397–417
100. Crandall DL, Hausman GJ, Kral JG (1997) A review of the microcirculation of adipose tissue: anatomic, metabolic, and angiogenic perspectives. Microcirculation 4(2):211–232
101. Hoover-Plow J, Ellis J, Yuen L (2002) In vivo plasminogen deficiency reduces fat accumulation. Thromb Haemost 87(6):1011–1019
102. Brakenhielm E et al (2004) Angiogenesis inhibitor, TNP-470, prevents diet-induced and genetic obesity in mice. Circ Res 94(12):1579–1588
103. Tam J et al (2009) Blockade of VEGFR2 and not VEGFR1 can limit diet-induced fat tissue expansion: role of local versus bone marrow-derived endothelial cells. PLoS One 4(3):e4974
104. Ejaz A et al (2009) Curcumin inhibits adipogenesis in 3T3-L1 adipocytes and angiogenesis and obesity in C57/BL mice. J Nutr 139(5):919–925
105. Kim MY et al (2010) The anti-angiogenic herbal composition Ob-X inhibits adipose tissue growth in obese mice. Int J Obes (Lond) 34(5):820–830
106. Lijnen HR, Frederix L, Van Hoef B (2010) Fumagillin reduces adipose tissue formation in murine models of nutritionally induced obesity. Obesity (Silver Spring) 18(12):2241–2246
107. Bergers G, Hanahan D (2008) Modes of resistance to anti-angiogenic therapy. Nat Rev Cancer 8(8):592–603
108. Kappers MH et al (2009) Cardiovascular and renal toxicity during angiogenesis inhibition: clinical and mechanistic aspects. J Hypertens 27(12):2297–2309
109. Schmidinger M, Bellmunt J (2010) Plethora of agents, plethora of targets, plethora of side effects in metastatic renal cell carcinoma. Cancer Treat Rev 36(5):416–424
110. Yan M et al (2010) Chronic DLL4 blockade induces vascular neoplasms. Nature 463(7282):E6–E7
111. Kim DH, Woods SC, Seeley RJ (2010) Peptide designed to elicit apoptosis in adipose tissue endothelium reduces food intake and body weight. Diabetes 59(4):907–915
112. Woods SC, Seeley RJ (2000) Adiposity signals and the control of energy homeostasis. Nutrition 16(10):894–902
113. Pajvani UB et al (2005) Fat apoptosis through targeted activation of caspase 8: a new mouse model of inducible and reversible lipoatrophy. Nat Med 11(7):797–803
114. Carmeliet P (2000) Mechanisms of angiogenesis and arteriogenesis. Nat Med 6(4):389–395
115. Kerbel RS (2008) Tumor angiogenesis. N Engl J Med 358(19):2039–2049
116. Kolonin MG (2009) Tissue-specific targeting based on markers expressed outside endothelial cells. Adv Genet 67:61–102
117. Kolonin MG et al (2004) Reversal of obesity by targeted ablation of adipose tissue. Nat Med 10(6):625–632
118. Hossen MN et al (2010) Ligand-based targeted delivery of a peptide modified nanocarrier to endothelial cells in adipose tissue. J Control Release 147(2):261–268
119. Barnhart KF et al (2011) A peptidomimetic targeting white fat causes weight loss and improved insulin resistance in obese monkeys. Sci Transl Med 3(108):108–112
120. Kim DH et al (2012) Rapid and weight-independent improvement of glucose tolerance induced by a peptide designed to elicit apoptosis in adipose tissue endothelium. Diabetes 61(9):2299–2310
121. Staquicini FI et al (2011) Vascular ligand-receptor mapping by direct combinatorial selection in cancer patients. Proc Natl Acad Sci U S A 108(46):18637–18642

122. Ling Q et al (2004) Annexin II regulates fibrin homeostasis and neoangiogenesis in vivo. J Clin Invest 113(1):38–48
123. Liu D et al (2012) Mitochondrial dysfunction and adipogenic reduction by prohibitin silencing in 3T3-L1 cells. PLoS One 7(3):e34315
124. Suga H et al (2010) Adipose tissue remodeling under ischemia: death of adipocytes and activation of stem/progenitor cells. Plast Reconstr Surg 126(6):1911–1923
125. Nie J et al (2008) Combinatorial peptides identify $\alpha 5\beta 1$ integrin as a receptor for the matricellular protein SPARC on adipose stromal cells. Stem Cells 26(10):2735–2745
126. Rivera LB, Brekken RA (2011) SPARC promotes pericyte recruitment via inhibition of endoglin-dependent TGF-beta1 activity. J Cell Biol 193(7):1305–1319
127. Guzman-Rojas L et al (2012) Cooperative effects of aminopeptidase N (CD13) expressed by nonmalignant and cancer cells within the tumor microenvironment. Proc Natl Acad Sci U S A 109(5):1637–1642
128. Arap W et al (2002) Targeting the prostate for destruction through a vascular address. Proc Natl Acad Sci U S A 99(3):1527–1531
129. Sergeeva A et al (2006) Display technologies: application for the discovery of drug and gene delivery agents. Adv Drug Deliv Rev 58(15):1622–1654
130. Liu J et al (2012) Selection of aptamers specific for adipose tissue. PLoS One 7(5):e37789
131. Freedland ES (2004) Role of a critical visceral adipose tissue threshold (CVATT) in metabolic syndrome: implications for controlling dietary carbohydrates: a review. Nutr Metab 1(1):12
132. Tran TT, Kahn CR (2010) Transplantation of adipose tissue and stem cells: role in metabolism and disease. Nat Rev Endocrinol 6(4):195–213
133. Rutkowski JM, Davis KE, Scherer PE (2009) Mechanisms of obesity and related pathologies: the macro- and microcirculation of adipose tissue. FEBS J 276(20):5738–5746
134. Paidisetty S, Blodgett TM (2009) Brown fat: atypical locations and appearances encountered in PET/CT. Am J Roentgenol 193(2):359–366
135. Xue Y et al (2009) Hypoxia-independent angiogenesis in adipose tissues during cold acclimation. Cell Metab 9(1):99–109
136. Reitman ML et al (1999) Transgenic mice lacking white fat: models for understanding human lipoatrophic diabetes. Ann N Y Acad Sci 892:289–296
137. Moitra J et al (1998) Life without white fat: a transgenic mouse. Genes Dev 12(20): 3168–3181
138. Lijnen HR (2008) Angiogenesis and obesity. Cardiovasc Res 78(2):286–293
139. Crandall DL et al (2006) Modulation of adipose tissue development by pharmacological inhibition of PAI-1. Arterioscler Thromb Vasc Biol 26(10):2209–2215
140. Demeulemeester D, Collen D, Lijnen HR (2005) Effect of matrix metalloproteinase inhibition on adipose tissue development. Biochem Biophys Res Commun 329:105–110
141. Bourlier V et al (2005) Protease inhibitor treatments reveal specific involvement of matrix metalloproteinase-9 in human adipocyte differentiation. J Pharmacol Exp Ther 312(3): 1272–1279
142. Shin JH, Shin DW, Noh M (2009) Interleukin-17A inhibits adipocyte differentiation in human mesenchymal stem cells and regulates pro-inflammatory responses in adipocytes. Biochem Pharmacol 77:1835–1844

Chapter 16
Anti-inflammatory Effects of Exercise

Michael Gleeson

Abstract Chronic low-level inflammation is associated with obesity and increased incidence of cancer and chronic disease states. Regular exercise reduces the risk of some cancers as well as chronic metabolic and cardiorespiratory diseases, in part because exercise exerts anti-inflammatory effects. The anti-inflammatory effects of regular exercise may be mediated via both a reduction in visceral fat mass (with a subsequent decreased release of adipokines) and the induction of an anti-inflammatory environment with each bout of exercise via the release of IL-6 from working skeletal muscle, reduced expression of TLRs on monocytes/macrophages, inhibition of monocyte/macrophage infiltration into adipose tissue, phenotypic switching of macrophages within adipose tissue, a reduction in the circulating numbers of pro-inflammatory monocytes and an increase in the circulating numbers of regulatory T cells.

16.1 Exercise in Cancer Prevention and Therapy

There is consensus that exercise training protects against some types of cancers and can have a beneficial role in therapy for cancer patients. Exercise training enhances aspects of anti-tumour immunity and reduces inflammatory mediators. However, the data linking immunological and inflammatory mechanisms, physical activity and cancer risk reduction remains tentative. Determining if regular physical activity reduces cancer risk through immunological mechanisms is of public health relevance and could lead to tailored and novel exercise prescriptions. Comprehensive

M. Gleeson, Ph.D. (✉)
School of Sport, Exercise and Health Sciences, Loughborough University,
Loughborough, Leicestershire, LE11 3TU, UK
e-mail: m.gleeson@lboro.ac.uk

A.J. Dannenberg and N.A. Berger (eds.), *Obesity, Inflammation and Cancer*,
Energy Balance and Cancer 7, DOI 10.1007/978-1-4614-6819-6_16,

reviews by the International Agency for Research on Cancer [8] and the World Cancer Research Fund [104] identified an independent protective effect of physical activity on colon and postmenopausal breast cancer risk. Evidence is also mounting for risk reduction by physical activity for endometrial, lung and pancreatic cancers.

Physical activity has a therapeutic effect in cancer patients by reducing cancer recurrence, enhancing health outcomes and increasing survival. Women who exercised moderately prior to [28] and after a breast cancer diagnosis had significant improvements in overall and disease-specific survival and quality of life compared to sedentary counterparts [82, 100]. Similarly, studies with colorectal [55] and prostate [45, 75] cancer survivors have suggested that mortality and disease progression are approximately 50 % lower in physically active than sedentary individuals.

Numerous cohort studies have shown a link between obesity and cancer incidence overall and for selected cancer sites including endometrial, breast, colon and oesophageal adenocarcinoma [4]. There is also data showing that individuals who lose weight and maintain the loss have a reduced cancer incidence and mortality, providing some hope that weight loss in obese individuals may help them prevent cancer. Several possible mechanisms have been suggested to explain the association of obesity with increased risk of certain cancers: (1) Fat tissue produces excess amounts of oestrogen, high levels of which have been associated with the risk of breast, endometrial and some other cancers; (2) Obese people often have increased circulating levels of insulin and insulin-like growth factor-1, which may promote the development of certain tumours; (3) Fat cells produce adipokines that may stimulate or inhibit cell growth. For example, leptin, which is more abundant in obese people, seems to promote cell proliferation, whereas adiponectin, which is less abundant in obese people, may have antiproliferative effects; (4) Fat cells may also have direct and indirect effects on other tumour growth regulators, including mammalian target of rapamycin (mTOR) and AMP-activated protein kinase; (5) Obese people often have chronic low-level inflammation, which has been associated with increased cancer risk. Other possible mechanisms include altered immune responses, effects on intracellular signalling pathways involving nuclear factor kappa beta and oxidative stress.

Currently, the biological mechanisms relating exercise and cancer are not well understood. Potential mediators include prevention of obesity, reduction in body weight and/or adiposity, decreases in reproductive hormone levels, altered growth factor milieu, enhanced antioxidant defence mechanisms and changes in immune function including reduced inflammation and enhanced anti-tumour immunity. The relative contribution of these mechanisms in specific cancer types remains unknown. With respect to the hypothesis that exercise induces alterations in immune mediators, more is known about exercise-induced changes in inflammatory mediators than about changes in specific anti-tumour mechanisms.

The association between chronic inflammation and cancer is well established [14]. Human cross-sectional studies demonstrate an inverse relationship between regular physical activity and inflammatory biomarkers including C-reactive protein

(CRP), tumour necrosis factor-alpha (TNF-α) and interleukin-6 (IL-6) [42, 70]. Reductions in CRP levels with exercise training have been reported [42]. Although exercise may reduce inflammatory biomarkers, clinical trials indicate variable outcomes, with an effect of exercise on CRP in some but not all studies [74]. Less work has been done with IL-6 in humans, but again there are conflicting results [101]. There can be little doubt that regular physical activity is beneficial in preventing some cancers, as well as in decreasing recurrence, increasing survival and improving quality of life for cancer patients. Animal studies (e.g. [17, 18, 39]) indicate that multiple biological pathways may be involved including a reduction in inflammation and an enhancement of anti-tumour immunity. To date, neither of these mechanisms has been studied in adequate detail to fully understand their role in cancer prevention and therapy with respect to exercise.

The focus of this chapter is to explain the various mechanisms by which exercise exerts its anti-inflammatory effects. Additional research is needed to determine which inflammatory mediators and anti-tumour immune mechanisms are most sensitive to exercise and the dose, duration and frequency of exercise needed to achieve the most potent anti-inflammatory or anti-tumour effects.

16.2 The Links Between Sedentary Behaviour, Chronic Inflammation and Chronic Disease

The prevalence of obesity continues to rise worldwide and is being accompanied by proportional increases in a host of other medical conditions associated with derangements of immunometabolism [50] such as type 2 diabetes (T2DM), cardiovascular diseases, chronic obstructive pulmonary disease, dementia, depression and cancer. Inflammation appears to be aetiologically linked to the pathogenesis of all these conditions [65], and the development of a chronic low-grade inflammatory state (as indicated by elevated levels of circulating inflammation markers such as IL-6, TNF-α and CRP) has been established as a predictor of risk for several of them [73]. Importantly, physical inactivity and sedentary behaviour increase the risk of all these conditions [36, 69, 99]. An inactive lifestyle leads to the accumulation of visceral fat and consequently the activation of a network of inflammatory pathways that results in inflammation in adipose tissue, increased release of adipokines (peptides and proteins including some cytokines that are secreted from white adipose tissue) and the development of a low-grade systemic inflammatory state [65]. Chronic inflammation promotes the development of insulin resistance, atherosclerosis, neurodegeneration and tumour growth and subsequently the development of several diseases associated with physical inactivity (Fig. 16.1). Exercise has anti-inflammatory effects, and therefore, in the long term, regular physical activity can protect against the development of these chronic diseases as well as having other benefits for health, functional capacity and quality of life [36, 69] which are summarised in Table 16.1. Furthermore, exercise can be used as a treatment for (or to

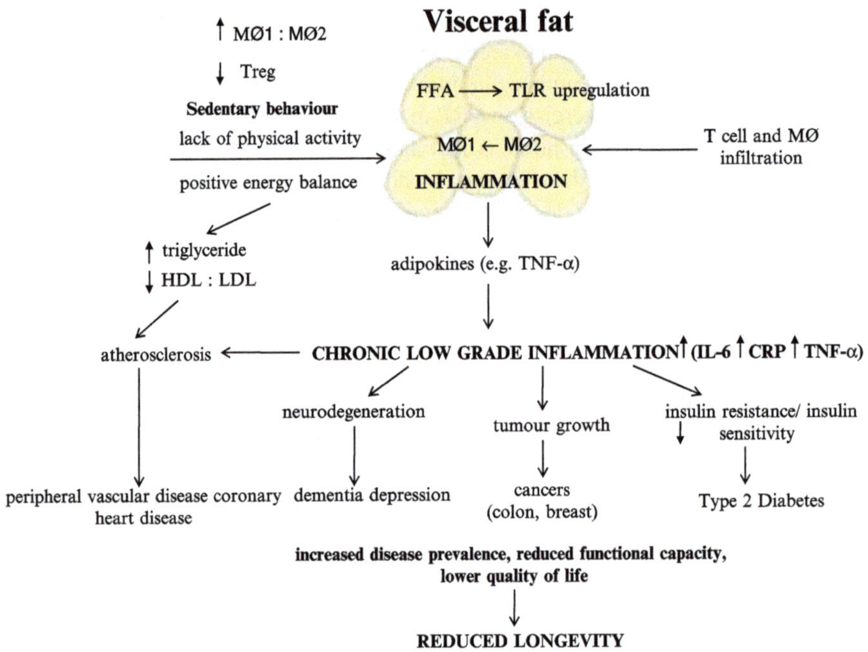

Fig. 16.1 Physical inactivity and positive energy balance lead to an accumulation of visceral fat which becomes infiltrated by pro-inflammatory macrophages and T cells. The pro-inflammatory M1 macrophage phenotype predominates and inflamed adipose tissue releases adipokines and TNF-α that lead to a state of persistent low-grade systemic inflammation. This promotes the development of insulin resistance, tumour growth, neurodegeneration and atherosclerosis. The latter is exacerbated by the deleterious changes in the blood lipid profile associated with a lack of physical activity

ameliorate the symptoms of) many of these conditions which is increasingly being promoted as the concept that "exercise is medicine".

Obviously, exercise increases energy expenditure and burns off some of the body fat that would otherwise accumulate when individuals eat more dietary energy than they need. In that simple sense, exercise reduces the risk of developing obesity and excessive adiposity. Regular exercise also imbues cardiovascular health benefits by improving the blood lipid profile by decreasing the concentration of plasma triglycerides and small LDL particles and by increasing the concentration of protective HDL cholesterol [46]. These beneficial alterations in plasma lipids are presumed to limit the development of atherosclerosis. However, the protective effect of a physically active lifestyle against chronic inflammation associated diseases may, to some extent, be ascribed to an anti-inflammatory effect of exercise [42]. This may be mediated not only via a reduction in visceral fat mass (with a subsequent decreased production and release of adipokines) but also by induction of an anti-inflammatory environment with each bout of exercise [70]. The remainder of this chapter will explain the possible mechanisms by which exercise exerts its anti-inflammatory effect.

Table 16.1 Summary of the interaction between physical activity and major diseases assessing evidence that exercise may (a) lower disease risk and (b) have therapeutic value in treating disease (see Pedersen and Saltin [69] and Hardman and Stensel [36] for further detail)

Disease	Evidence that physical activity may lower disease risk and/or have therapeutic value in treating disease
Cancer	High levels of PA are associated with lower risk of colon and breast cancer. PA may lower cancer risk by systemic (reduced body fat and insulin levels, enhanced immune function) and site-specific (reduced sex steroid hormone levels for breast cancer, decreased bowel transit time for colon cancer) mechanisms. Some observational and RCT evidence supports a therapeutic role for PA in preserving mobility and function in cancer patients
CHD	A large body of epidemiological evidence demonstrates that high levels of PA and PF are associated with a lower risk of developing CHD. RCTs show that regular PA can favourably modify CHD risk factors including (but not limited to) dyslipidaemia, hypertension and obesity. RCTs also show that PA improves survival in CHD patients
Stroke	Evidence that high levels of PA and PF reduce the risk of stroke is suggestive but not as compelling as that for CHD. RCTs show that PA can lower, but not necessarily normalise, blood pressure in hypertensive individuals
T2D	Observational epidemiological evidence consistently demonstrates an association between high levels of PA/PF and a reduced risk of developing T2D. RCTs show that lifestyle intervention (diet and PA) can lower body mass, improve glucose tolerance and reduce the risk of developing T2D in high-risk patients. In patients with T2D, high levels of PA and PF are associated with a reduced risk of CHD and all-cause mortality
Dementia	Observational epidemiological studies indicate that higher levels of PA are associated with a lower risk of cognitive decline and dementia in older adults. Some limited evidence is available from RCTs to suggest that PA induces modest improvements in cognition in people who are at increased risk of dementia/Alzheimer's disease
Other	There is some evidence from observational and intervention studies to support a role for PA for enhancing physical function and improving quality of life in those suffering from chronic heart failure, chronic obstructive pulmonary disease, depression, intermittent claudication, osteoarthritis and osteoporosis

CHD coronary heart disease; *PA* physical activity; *PF* physical fitness; *RCT* randomised controlled trial; *T2D* type 2 diabetes mellitus

16.3 Anti-inflammatory Effects of Exercise

The anti-inflammatory effects of exercise have mostly been ascribed to two possible mechanisms: (1) increased production and release of anti-inflammatory cytokines from contracting skeletal muscle [70] and (2) reduced expression of toll-like receptors (TLRs) on monocytes and macrophages [26] with subsequent inhibition of downstream responses such as pro-inflammatory cytokine production, antigen presentation and costimulatory molecule expression [33]. However, the anti-inflammatory effects of exercise arise not only from these two mechanisms but also other effects of exercise that recently have been established such as the inhibition of monocyte/macrophage infiltration into adipose tissue [43], phenotypic switching of macrophages within adipose tissue [43], a reduction in the circulating numbers of pro-inflammatory monocytes [95] and an increase in the circulating numbers of IL-10-secreting regulatory T cells [98, 107]. The major focus of this part of the chapter is to explain these various mechanisms.

16.3.1 Reduction in Visceral Fat Mass

The accumulation of body fat, particularly in the abdomen, liver and muscles, is associated with increased all-cause mortality [71], the development of T2DM [5], cardiovascular disease [35], dementia [102] and several cancers [105]. The production of pro-inflammatory adipokines is increased with adipose tissue expansion whereas the amounts of anti-inflammatory cytokines produced are reduced. This leads to the development of a state of persistent system low-grade inflammation [108]. Regular exercise can reduce waist circumference and cause considerable reductions in abdominal/visceral fat, even in the absence of any loss of body weight, in both men and women regardless of age [77]. Therefore, increased physical activity can bring about a reduction in systemic inflammation [108] via a reduction in pro-inflammatory adipokine secretion as a direct result of lowering the amount of fat stored in abdominal depots.

16.3.2 Release of IL-6 from Contracting Muscle

One of the earliest reports on the effects of exercise on cytokines observed an elevation in the circulating levels of several cytokines following the completion of a marathon [60]. Some years later, several reports on the effects of exercise on cytokines began to appear in the scientific literature—the impetus behind this increase was probably the development of sensitive, specific, commercially available, assays for the detection of a large number of cytokines. One of the earliest and most consistent findings has been that of an elevation in the circulating level of IL-6 following prolonged strenuous exercise. In an important study, Nehlsen-Cannarella et al. [58] demonstrated that the plasma IL-6 concentration was dramatically increased following 2.5 h of high-intensity running. Furthermore, it was shown that when subjects consumed a carbohydrate beverage during exercise, the increase in the circulating IL-6 concentration was decreased compared with subjects who consumed a placebo. While these studies provide no mechanistic insight into the source of the exercise-induced increase in the circulating IL-6 concentration or its biological purpose, this study acted as a stimulus for subsequent investigations into the effects of exercise on cytokines. Indeed, many subsequent studies have demonstrated an increase in the circulating concentration of several cytokines following prolonged strenuous exercise. Increases in the circulating concentrations of both pro-inflammatory cytokines (e.g. IL-1β, TNF-α) and anti-inflammatory cytokines

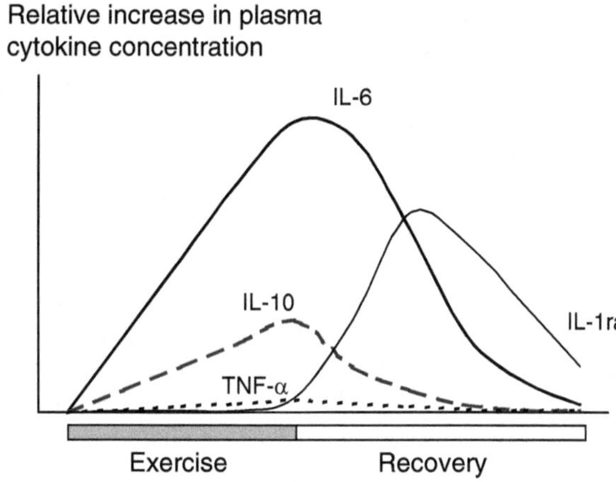

Fig. 16.2 The cytokine response to prolonged strenuous exercise (modified from Febbraio and Pedersen [21])

(e.g. IL-6 and IL-10) [62, 63], cytokine inhibitors (e.g. IL-1 receptor antagonist and soluble TNF receptors) [62], chemokines (e.g. IL-8, macrophage inflammatory protein (MIP) and monocyte chemotactic protein-1 (MCP-1)) [64, 91, 92] and colony-stimulating factors [92] have been reported following endurance exercise. However, the finding of an increase in the circulating IL-6 concentration following prolonged exercise is the most marked and consistent exercise-induced response of any cytokine so far examined (see Fig. 16.2).

At rest, approximately 30 % of circulating IL-6 arises from the adipose tissue but only about 10 % of this can be attributed to the adipocytes with the remainder coming mostly from adipose tissue resident macrophages. Other sources of circulating IL-6 at rest include blood leukocytes (predominantly monocytes), the brain and the liver. An early study indicated that circulating monocytes were unlikely to be the source of the exercise-induced increase in the plasma IL-6 concentration [97] as 1 h of strenuous exercise caused no changes in the amount of IL-6 mRNA detected in peripheral blood mononuclear cells (PBMCs) despite an elevation in the plasma IL-6 level. This finding was later confirmed by Starkie et al. [84] who demonstrated that monocyte intracellular IL-6 protein expression was unchanged following a bout of prolonged strenuous exercise; importantly, Starkie et al. [85] also demonstrated that exercise had no significant effects on TNF-α or IL-1β production from monocytes. Several possible sites of origin where suggested for the exercise-induced increase in the circulating level of IL-6, with contracting skeletal muscle receiving the most attention following the observation by Steensberg et al. [89] that IL-6 released from the exercising leg could account for the rise in the circulating IL-6 concentration during prolonged exercise.

Although initial studies supported the hypothesis that an increase in the plasma level of IL-6 was related to exercise-induced muscle damage [12], it soon became apparent that muscle damage per se was only a minor contributor to the exercise-induced rise in the circulating level of IL-6. Firstly, although many of the initial studies examining the effects of exercise on cytokines used running as an exercise model, several studies have also examined the cytokine response to prolonged bicycle exercise which induces only a minimal degree of muscle damage (if any) and consequently does not trigger an inflammatory response. However, cycling exercise does result in a considerable elevation (typically 3–10-fold higher than resting values) in the circulating IL-6 concentration comparable to that of running at the same mode-specific relative intensity [86]. However, perhaps the strongest evidence that muscle damage is not a prerequisite for an increase in the systemic IL-6 concentration in response to exercise came from Croisier et al. [15]. In this study, subjects performed two bouts of eccentric muscle contractions separated by a period of 3 weeks. After the initial exercise bout, the expected elevation in serum myoglobin (a marker of muscle damage) and delayed onset muscle soreness was observed, in addition to a rise in the circulating IL-6 concentration. Importantly, it is well known that following a period of recovery from an initial bout of muscle damaging exercise, a second exercise bout identical to the first causes a much lower level of muscle damage. Therefore, and as expected, the second exercise bout resulted in minimal increases in serum myoglobin and muscle soreness, yet the increase in the circulating IL-6 concentration was very similar to that observed in response to the initial bout of exercise. These studies provide compelling evidence that the increase in the circulating IL-6 concentration following exercise is not primarily related to muscle damage.

During and following exercise of sufficient load, the active skeletal muscle markedly increases both cellular and circulating levels of IL-6 [66]. With prolonged exercise (>2.5 h), plasma IL-6 levels can increase over 100-fold although more modest increases are reported with shorter duration exercise [24]. Increases have also been noted using intermittent exercise protocols of relatively short duration [54]. With exercise, the increase in circulating IL-6 is transient, normally returning to resting levels within 1 h after exercise. The plasma IL-6 concentration increases exponentially with exercise duration [89], and a major stimulus of its synthesis and release appears to be a fall in muscle glycogen content ([44, 67]; Fig. 16.3). Increases in intracellular calcium and increased formation of reactive oxygen species are also capable of activating transcription factors known to regulate IL-6 synthesis [24].

IL-6 appears responsible for the subsequent rise in circulating levels of the anti-inflammatory cytokines IL-10 and IL-1 receptor antagonist (IL-1ra) and also stimulates the release of cortisol from the adrenal glands [88]. The causal role of IL-6 in stimulating IL-10, IL-1ra and cortisol secretion is substantiated by the observation that intravenous infusion of IL-6 totally mimics the acute anti-inflammatory effects of a bout of exercise both with regard to elevations of plasma IL-10, IL-1ra and cortisol [88] and with regard to suppression of endotoxin-stimulated increases in TNF-α levels [83]. These actions and possible associated health effects are summarised in Fig. 16.4. The IL-6 half-life is prolonged by combining with the soluble IL-6 receptor (sIL-6R), and it is this complex which is crucial in determining the biological activity of IL-6. The expression of IL-6R in tissues is limited to hepatocytes, leukocytes and adipocytes [76] with a relatively low expression in resting skeletal muscle [44]. Exercise training increases the expression of IL-6R on the muscle membrane [44] removing some of the dependency on the circulating sIL-6 receptor.

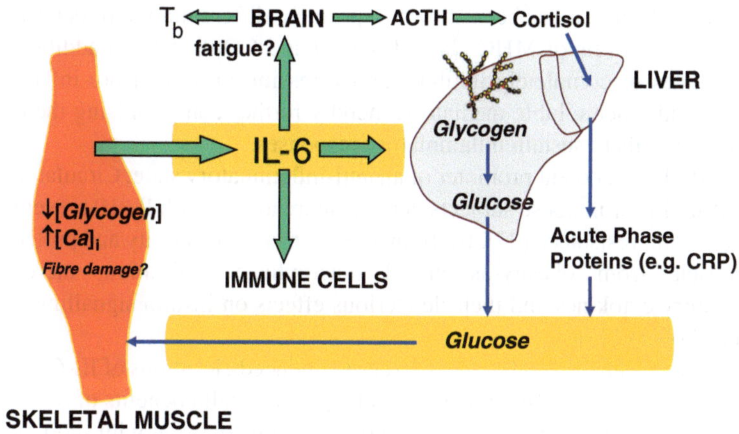

Fig. 16.3 Rises in intracellular calcium ion concentration with contraction and glycogen depletion stimulate the production of IL-6 by the working muscles. IL-6 is then released from the muscle resulting in an elevation in the systemic IL-6 concentration. IL-6 also stimulates liver glycogen breakdown in the liver, lipolysis in adipose tissue and the secretion of ACTH, cortisol, IL-10, IL1ra and CRP and may also be involved in the development of central fatigue

Fig. 16.4 Biological role of muscle-derived IL-6 during exercise. *Full lines* represent acute effects of IL-6 infusion or exercise-induced elevations of circulating IL-6. *Broken lines* represent possible health effects of IL-6 (modified from Steensberg [87])

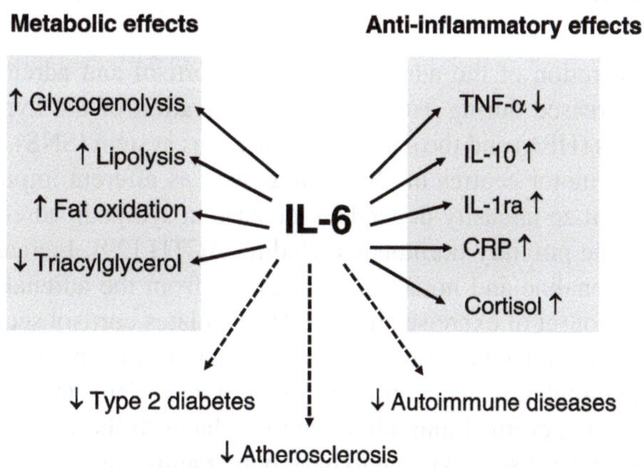

IL-1ra is secreted mainly by monocytes and macrophages and inhibits the pro-inflammatory actions of IL-1β [27]. IL-10 is known to be produced primarily by regulatory T cells and monocytes but also by Th2 cells, macrophages, dendritic cells, B cells, CD8+ T cells, Th1 cells and Th17 cells [52]. Irrespective of the cellular source, the principal role of IL-10 appears to be containment and suppression of inflammatory responses so as to downregulate adaptive immune effector responses [57] and minimise tissue damage in response to microbial challenges. Accordingly, IL-10 induces downregulation of MHC antigens, the intercellular adhesion molecule-1 and the costimulatory molecules CD80 and CD86 on antigen presenting cells, and it has been shown to promote differentiation of dendritic cells expressing low levels of MHC class II, CD80 and CD86 [52]. In addition, IL-10 downregulates or completely inhibits the expression of several pro-inflammatory cytokines and other soluble mediators, thereby further compromising the capacity of effector T cells to sustain inflammatory responses.

Thus, IL-10 is a potent promoter of an anti-inflammatory state. Circulating levels of IL-10 are lower in obese subjects, and acute treatment with IL-10 prevents lipid-induced insulin resistance [37]. IL-10 increases insulin sensitivity and protects skeletal muscle from obesity-associated macrophage infiltration, increases in inflammatory cytokines and their deleterious effects on insulin signalling and glucose metabolism [37].

A limitation of the hypothesis that exercise-induced elevations of IL-6 are mostly responsible for the anti-inflammatory and long-term health benefits of regular exercise is that substantial increases in circulating IL-6 do not occur with short durations of low-/moderate-intensity exercise [24] despite the known health benefits (e.g. reduced risk of heart disease) associated with only very moderate increases in physical activity above that of a totally sedentary lifestyle [56].

16.3.3 Increased Levels of Circulating Cortisol and Adrenaline

Secretion of the adrenal hormones cortisol and adrenaline into the circulation is increased during exercise due to activation of the hypothalamic-pituitary-adrenal axis (HPA) and the sympathetic nervous system (SNS), respectively. Impulses from the motor centres in the brain as well as afferent impulses from working muscles elicit an intensity-dependent increase in sympathoadrenal activity and in release of some pituitary hormones including ACTH [29]. Increased SNS activity stimulates adrenaline and noradrenaline release from the adrenal medulla within seconds of the onset of exercise and ACTH stimulates cortisol secretion from the adrenal cortex within a matter of minutes. These hormonal responses usually precede the rises in circulating concentrations of cytokines. Thus, the magnitude of the elevations in plasma cortisol and adrenaline is related to the intensity and duration of exercise [29]. Cortisol is known to have potent anti-inflammatory effects [16], and catecholamines downregulate the lipopolysaccharide (LPS)-induced production of TNF-α, IL-6 and IL-1β by immune cells [7]. Cortisol secretion is also augmented by the aforementioned rise in circulating IL-6 [88]. Thus, hormones, myokines and cytokines all contribute to the anti-inflammatory effect of exercise (Fig. 16.5).

16.3.4 Inhibition of Macrophage Infiltration into Adipose Tissue

The expression of pro-inflammatory cytokines, chemokines and cell adhesion molecules in adipose tissue is increased in obese mice [38]. Macrophages and also

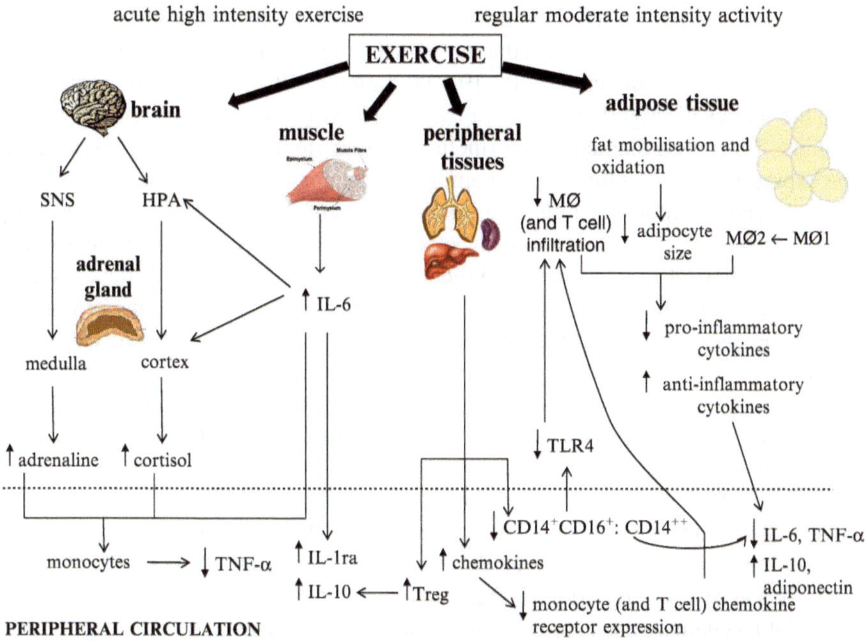

Fig. 16.5 Potential mechanisms contributing to the anti-inflammatory effects of exercise. Activation of the HPA and sympathetic nervous system (SNS) leads to the release of cortisol and adrenaline from the adrenal cortex and medulla, respectively. These hormones inhibit the release of TNF-α by monocytes. IL-6 produced by contracting skeletal muscle also downregulates the production of TNF-α by monocytes and may stimulate further cortisol release. Acute elevations in IL-6 stimulate release of IL-1 receptor antagonist (IL-1ra) from monocytes and macrophages, thus increasing the circulating concentrations of this anti-inflammatory cytokine. Exercise training mobilises T_{Reg} cells, a major source of the anti-inflammatory cytokine IL-10, and decreases the proportion of inflammatory (CD14+CD16+) monocytes, compared with classical (CD14++) monocytes. Following exercise, CD14++ monocytes express less TLR4 and thereby induce a reduced inflammatory response, marked by lower levels of pro-inflammatory cytokines and reduced adipose tissue infiltration. Exercise also increases plasma concentrations of key inflammatory immune cell chemokines; repeated elevations of such chemokines may lead to a downregulation of their cellular receptors, resulting in reduced tissue infiltration. A reduction in adipose tissue mass and adipocyte size, along with reduced macrophage infiltration, and a switch from a M1 to M2 phenotype all may contribute to a reduction in pro-inflammatory cytokine release from adipose tissue, such as IL-6 and TNF-α, and an increase in anti-inflammatory cytokine release, such as adiponectin and IL-10

T cells, which infiltrate adipose tissue in obesity, are known to regulate the inflammatory state of adipose tissue [41]. Thus, the migration of PBMCs towards sites of inflammation, including adipose tissue and damaged vascular endothelium, is central to the development of sustained tissue inflammation [109]. It is thought that the size of the adipocytes triggers macrophage infiltration rather than overall obesity and that recruitment of the macrophages may be stimulated by the chemokines, MCP-1 and MIP [11].

Exercise might limit the movement of PBMCs into inflamed adipose tissue in a similar manner to its effect of reducing PBMC migration towards a virus-infected human bronchial epithelial cell line [9]. Migration of PBMCs from the circulation into the tissues is a tightly regulated process involving a gradient of release of chemokines from the inflamed tissue (including from immune cells residing within), the expression of complimentary chemokine receptors on PBMCs and the expression of adhesion molecules on both immune and endothelial cells. Acute bouts of exercise reduce T cell migration towards the supernatants of human rhinovirus-infected human BEAS-2B airway epithelial cells in a manner that is independent of any involvement of adhesion molecules or exercise-induced elevations of cortisol or catecholamines [9]. However, it is known that acute exercise stress results in the release of chemokines into the circulation from multiple sources, and sustained exposure of PBMCs to physiological concentrations of chemokines including MCP-1 results in chemokine receptor internalisation [48]. This is thought to serve as a negative feedback mechanism to reduce migration and thereby terminate accumulation of PBMCs in inflamed tissue. It is therefore possible that an active lifestyle creates an environment of repeated short-lasting elevations in plasma chemokines that act over time to downregulate expression of their receptors on PBMCs and restrict migration of these cells towards adipose tissue. However, this potential mechanism needs to be explored further in humans.

Whether exercise acts to inhibit the release of chemokines from human adipose tissue and in this way reduce macrophage infiltration is not clear. Certainly, evidence from murine studies has shown that obese mice deficient in the macrophage chemokines MCP-1 and CXCL14 do not exhibit inflammatory responses such as macrophage infiltration, increased tissue expression of IL-6 and insulin resistance [41]. However, while exercise training reduces macrophage infiltration into adipose tissue in obese mice it has little effect on adipose tissue MCP-1 and CXCL14 expression [43], perhaps suggesting that these chemokines are not key to the exercise-induced restriction of macrophage infiltration of adipose tissue. Recent findings in humans also do not report any independent effect of 12 weeks of exercise training on adipose tissue expression of MCP-1 (or MIP, TNF-α and IL-6), despite falls in circulating concentrations of MCP-1 with exercise. Diet-induced weight loss or weight loss in combination with exercise was associated with non-significant falls in adipokine mRNA expression in adipose tissue, and significant decreases in CD14 expression in adipose tissue were only found with diet alone.

In mice, training is reported to decrease the tissue expression of intercellular adhesion molecule-1 (ICAM-1) [43], the expression of which is known to be increased in obesity in humans [10]. Furthermore, antagonism of ICAM-1 in obese mice prevents macrophage infiltration into adipose tissue [13], and circulating ICAM-1 levels were reduced by 6 months of progressive aerobic exercise training in patients with T2DM without changes in body mass and waist circumference [110]. Obviously, further studies in humans are required to ascertain the role of exercise training on ICAM-1 in adipose tissue, but ICAM-I might also play a role in the exercise-induced reduction of macrophage infiltration into adipose tissue.

Macrophage activation has been defined into two separate polarisation states, M1 and M2 [49]. The M1 macrophage produces TNF-α, IL-6 and nitric oxide, while the M2 macrophage produces anti-inflammatory cytokines and arginase. Therefore, M1 macrophages induce a chronic inflammatory state, and M2 macrophages subdue inflammation in adipose tissue. The inflammation state of adipose tissue also appears to be associated with a preferential recruitment of M1 macrophages and/or a phenotypic switch of macrophage polarisation in adipose tissue towards the M1 phenotype. Therefore, it is possible that the attenuated inflammatory state of adipose tissue associated with chronic exercise training occurs by both suppression of macrophage infiltration and acceleration of phenotypic switching from M1 to M2 macrophages. A recent study in mice fed with a high-fat diet to induce obesity provided some evidence that exercise training induces the phenotypic switching from M1 to M2 macrophages in adipose tissue as well as inhibiting M1 macrophage infiltration into adipose tissue [43], although studies in humans are lacking.

16.3.5 Down-Regulation of Monocyte/Macrophage TLR Expression

TLRs are highly conserved transmembrane proteins that play an important role in the detection and recognition of microbial pathogens, and they can also be activated by endogenous danger signals of tissue damage such as heat shock proteins [40]. The key product of TLR signalling in antigen presenting cells is the production of pro-inflammatory cytokines and proteins, and thus, the TLR pathway plays an important role in mediating whole body inflammation [93]. Following a prolonged bout of strenuous exercise, the expression of TLRs 1, 2 and 4 on monocytes is decreased for at least several hours ([33, 47, 61]; Fig. 16.6). Prolonged exercise also results in a decreased induction of MHCII (Fig. 16.7), costimulatory molecules CD80 and CD86 and cytokines following stimulation with known TLR ligands [47]. Whether this reduction in cell-surface expression of TLRs is due to a down-regulation of TLR gene expression, a shedding of TLRs from the cell surface or re-internalisation by the cell remains to be established.

Evidence is now emerging that TLRs may be involved in the link between a sedentary lifestyle, inflammation and disease. Exercise training studies and cross-sectional comparisons between physically active and inactive subjects have shown a reduced inflammatory response of blood monocytes to endotoxin stimulation in vitro and a lowered TLR4 expression at both the cell surface and mRNA level in physically active individuals ([26, 33]; Fig. 16.8) which is associated with decreased inflammatory cytokine production [90] and has been shown to occur in both young and elderly adults (Fig. 16.9). Over the long term, a decrease in TLR expression may represent a beneficial effect for health because it decreases the inflammatory capacity of leukocytes, thus altering whole body chronic inflammation and possibly reducing the risk of developing chronic disease. The precise physiological stimulus

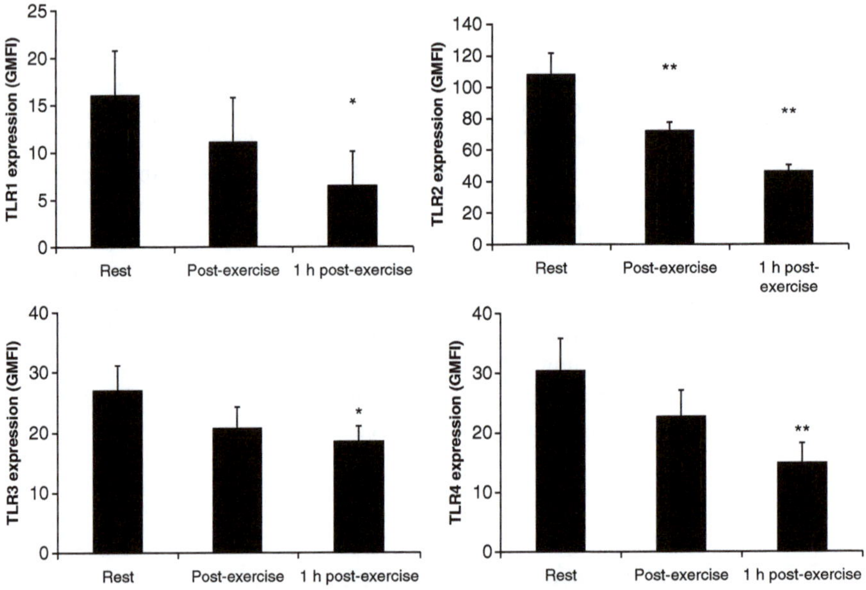

Fig. 16.6 Effect of 2.5-h cycling at $60\,\%VO_2$ max in temperate (20 °C) conditions on monocyte toll-like receptor (TLR expression) (data from Oliveira and Gleeson [61])

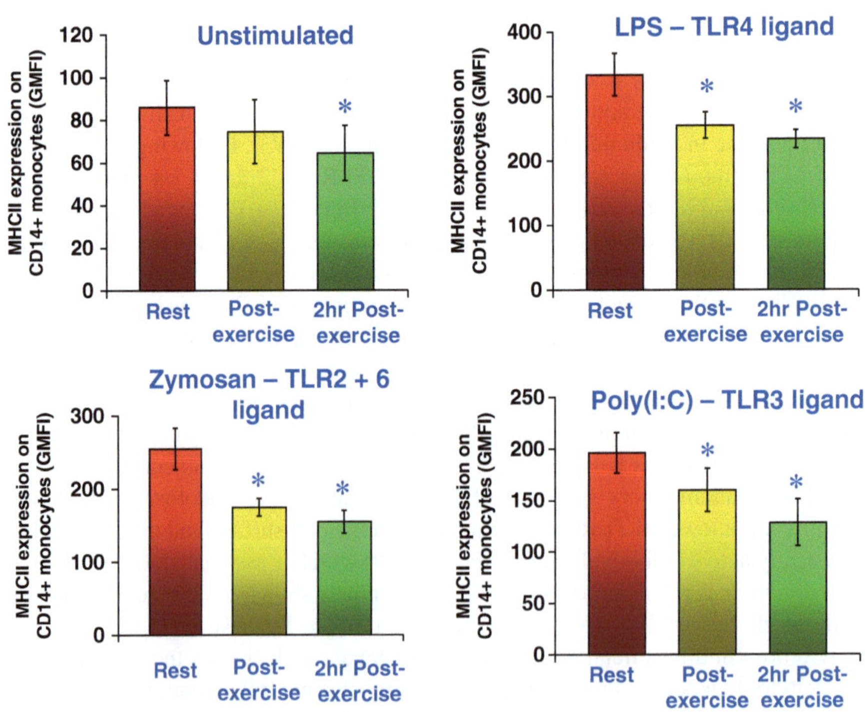

Fig. 16.7 Effects of exercise on TLR activation by TLR ligands—upregulation of MHCII. *Asterisk* indicates significantly different from pre-exercise (data from Lancaster et al. [47])

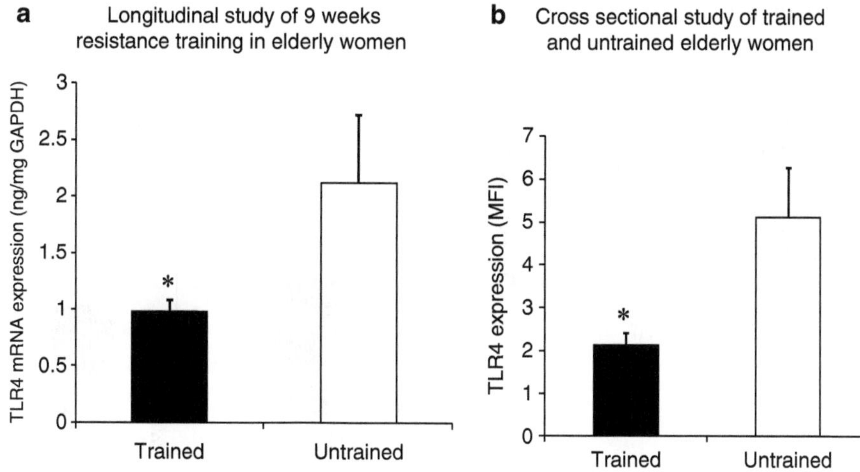

Fig. 16.8 Effect of exercise training and TLR4-mRNA and TLR4 expression. (**a**) A longitudinal study of 9 weeks resistance training in elderly women (data from Flynn et al. [25]) and (**b**) a cross-sectional study of trained and untrained elderly women (data from McFarlin et al. [53])

Fig. 16.9 Effects of moderate exercise training and age on monocyte TLR4 expression. Data are for old and young subjects before and after a 12-week training intervention in the inactive groups or maintained activity level in the active groups (modified from Stewart et al. [90])

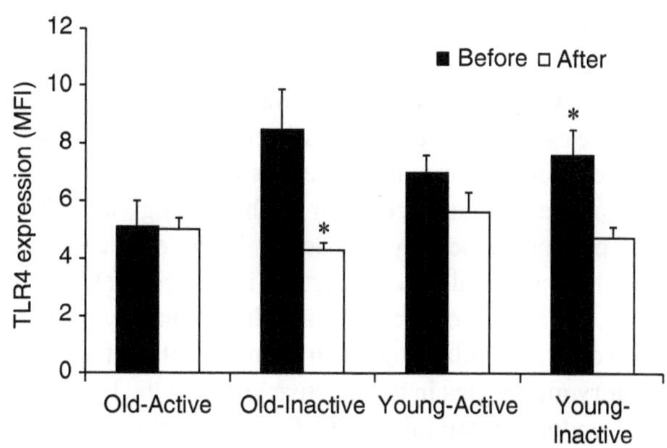

mediating an exercise-induced decrease in cell-surface TLR expression is not known; however, several possible signals have been implicated including anti-inflammatory cytokines, stress hormones and heat shock proteins [33].

16.3.6 Reduced Numbers of Pro-inflammatory Monocytes in Blood

There are two main populations of monocytes, classical (CD14++) and inflammatory (CD14+CD16+), that differentially express cell-surface TLR4, with the

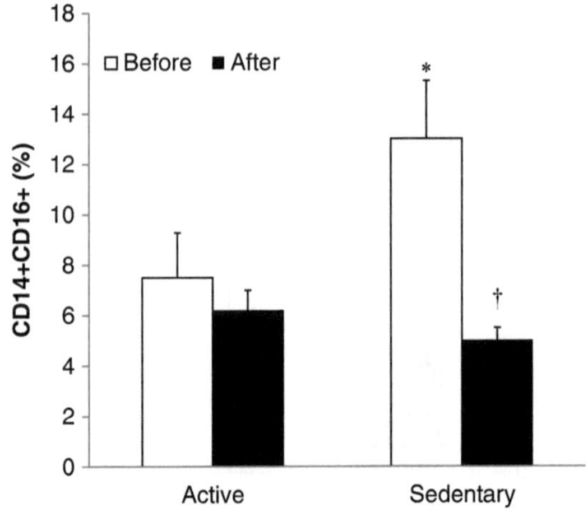

Fig. 16.10 Inflammatory monocyte subset (CD14+CD16+) percentage in physically active and sedentary subjects before and after 12 weeks of exercise training (for sedentary subjects) or continued physical activity (for active subjects). *Baseline group effect ($P<0.05$). **Significant group × time effect ($P<0.05$) (modified from Timmerman et al. [95])

inflammatory monocytes expressing 2.5 times more cell-surface TLR4 [81]. Despite constituting only 10 % of the total monocyte population, inflammatory monocytes contribute significantly to the inflammatory potential of the monocyte pool as a whole [6]. The circulating, inflammatory monocyte percentage is elevated in patients with rheumatoid arthritis [2], cardiovascular disease [79] and T2DM [34], and it has been suggested that inflammatory monocytes play a significant role in the pathogenesis of several diseases linked to inflammation and obesity [72]. Transient increases in the inflammatory monocyte percentage after a single, acute bout of intense exercise have been observed [80] followed by a rapid return to baseline during recovery, but regular exercise appears to reduce the proportion of inflammatory monocytes in the circulation in the resting state. For example, a cross-sectional comparison of healthy physically inactive elderly men and women with an age-matched physically active group indicated that sedentary people have a twofold higher percentage of circulating, inflammatory monocytes [95]. Furthermore, just 12 weeks of regular exercise training significantly reduced the percentage of inflammatory monocytes in the inactive group to the level of the active group (Fig. 16.10), and endotoxin-stimulated TNF-α production was reduced significantly after the training intervention. Based on previous reports that glucocorticoid therapy selectively depletes CD14+CD16+ monocytes [23], it is interesting to speculate that exercise-induced transient spikes in cortisol may have played a role in reducing CD14+CD16+ monocytes with exercise training.

16.3.7 Increased Circulating Numbers of Regulatory T Cells

CD4+CD25+ regulatory T cells specifically express the gene encoding forkhead/winged helix transcription factor (Foxp3) [78] and suppress immune responses via cell contact-dependent mechanisms. Studies show that the depletion of these cells causes autoimmune disease and enhances the immune response to foreign antigens [22]. Interestingly, a recent study showed that a 12-week programme of tai chi chuan exercise induced a significant increase in regulatory T cells [107]. Production of the regulatory T cell mediators transforming growth factor α (TGF-α) and IL-10 in response to specific antigen stimulation (varicella zoster virus) was also significantly increased after this exercise programme. Furthermore, a study of patients with T2DM showed that regular tai chi chuan exercise altered the Th1/Th2/Treg balance by increasing Foxp3 but not TGF-α expression [106].

In a study that used a running mouse model, the responses of circulating regulatory T cells to moderate and high-intensity exercise training were examined. Only the high-intensity training resulted in increases in regulatory T cell numbers and activation and was also associated with reduced pro-inflammatory and increased anti-inflammatory cytokine expression [98]. Intriguingly, the logical conclusion from these findings is that high-intensity exercise training might be more beneficial than moderate intensity in reducing risk of chronic cardiovascular and metabolic diseases via its anti-inflammatory effects. This notion is supported by another recent study that showed that the combination of high-intensity aerobic plus resistance exercise training, in addition to daily physical activity, is required to achieve a significant anti-inflammatory effect in T2DM patients [3].

16.3.8 Other Factors

During acute exercise, there is also a marked increase in growth hormone, prolactin, heat shock proteins, and other factors that have immunomodulatory effects [68]. Taken together, it appears that each bout of exercise induces an anti-inflammatory environment. Various mechanisms can contribute to this (Fig. 16.5), and it seems likely that their relative importance will vary dependent on the frequency, intensity and duration of the exercise performed. Intuitively, we might expect IL-6 to assume greater relative importance when the exercise is prolonged and glycogen-depleting whereas catecholamine-mediated effects are likely to assume greater importance with shorter duration, high-intensity exercise.

16.4 Exercise Is Medicine

In view of the anti-inflammatory effects of exercise described above and the role of inflammation in the pathogenesis of disease, it is not surprising that exercise is now considered a prophylactic for preventing several major diseases as well as an effective therapy for many conditions/diseases (Table 16.1). Perhaps the strongest evidence for the role of exercise in disease prevention comes from randomised controlled trials evaluating the effectiveness of lifestyle intervention in preventing T2DM (for a review, see [30]). These studies have demonstrated conclusively that lifestyle intervention (combined diet and exercise) is effective in preventing T2DM in groups of individuals who are at high risk of the disease by virtue of having impaired fasting glucose and/or impaired glucose tolerance as well as being overweight/obese. A limitation of these studies is that they did not isolate the independent effects of exercise and diet in preventing T2DM, but the effectiveness of exercise is supported by the finding in the Finnish Diabetes Prevention Study [96] that among those in the intervention group who did not reach the goal of losing 5 % of their initial body mass but who achieved the goal of exercising for more than 4 h per week, the odds ratio of diabetes was 0.2 (i.e. 80 % lower) than in intervention participants who remained sedentary. Thus, although more needs to be learnt about the role of exercise in preventing T2DM, it is clear that exercise makes a valuable contribution to an overall lifestyle package for preventing this disease.

In addition, exercise appears to have major benefits for the treatment of T2DM. The findings of one non-randomised study (the Malmö feasibility study) showed that 54 % of participants with early stage T2DM were in remission by the end of a 5-year diet and exercise intervention [20]. Moreover, prospective observational studies indicate that high levels of physical activity and/or physical fitness are effective in reducing the risk of cardiovascular disease and/or all-cause mortality and there is evidence implicating inflammation in the pathogenesis of T2DM [50], and it is therefore likely that the therapeutic benefits of exercise for those with T2DM are due, at least in part, to the well-established anti-inflammatory effects of regular exercise [42].

Aside from its role in preventing and treating T2DM, there is a good evidence that exercise is effective in preventing several other major diseases particularly cardiovascular disease [94], breast cancer [19] and colon cancer [103], and there is some evidence to support a role of exercise in preventing dementia [1]. Moreover, while exercise should not be considered a panacea there is evidence to support a role for exercise as a therapy for many diseases/conditions beyond those mentioned above including chronic obstructive pulmonary disease, chronic kidney disease, asthma and osteoporosis [69].

16.5 The Elite Athlete Paradox

Although regular moderate-intensity exercise is associated with a reduced incidence of upper respiratory tract infection (URTI) compared with a completely sedentary state [51, 59], the long hours of hard training that elite athletes undertake appears to make them more susceptible to URTIs [31]. This is also likely attributable to the anti-inflammatory effects of exercise inducing a degree of immunodepression. An increased risk of minor infections may be the (small) price to be paid for the long-term health benefits of regular exercise at high dosage.

A recent murine study indicated that intensive exercise training results in an increased anti-inflammatory cytokine (IL-10) response to antigen exposure [98], and a study on human endurance athletes revealed that those who were illness-prone during a 4-month period of winter training had fourfold higher IL-10 production by antigen-stimulated whole blood culture compared with athletes who remained illness-free during the same period [32]. There is now extensive evidence from both murine and human studies that IL-10 production usually imposes some limits on the effectiveness of antipathogen immune responses, especially innate immunity and adaptive Th1 responses. These studies suggest that very high training loads induce a large enough anti-inflammatory state to increase the risk of picking up minor infections.

16.6 Conclusions

Regular exercise reduces the risk of some cancers as well as chronic metabolic and cardiorespiratory diseases, in part because exercise exerts anti-inflammatory effects. The anti-inflammatory effects of regular exercise may be mediated via both a reduction in visceral fat mass and the induction of an anti-inflammatory environment with each bout of exercise. Various mechanisms may contribute to the anti-inflammatory effects of exercise including increased release of IL-6 from working skeletal muscle, increased release of cortisol and adrenaline from the adrenal glands, reduced expression of TLRs on monocytes/macrophages, inhibition of monocyte/macrophage infiltration into adipose tissue, phenotypic switching of macrophages within adipose tissue, a reduction in the circulating numbers of pro-inflammatory monocytes and an increase the circulating numbers of regulatory T cells. At present, we do not know what the relative importance of the different anti-inflammatory mechanisms that have thus far been identified are, though it seems likely that this will depend on the modes, frequencies, intensities and durations of exercise performed. The anti-inflammatory effects of exercise are also likely to be responsible for depressed immunity that makes the elite athlete more susceptible to common infections.

References

1. Abbott RD, White LR, Ross GW, Masaki KH, Curb JD, Petrovitch H (2004) Walking and dementia in physically capable elderly men. JAMA 292:1447–1453
2. Baeten D, Boots AM, Steenbakkers PG, Elewaut D, Bos E, Verheijden GF, Berheijden G, Miltenburg AM, Rijnders AW, Veys EM, De Keyser F (2000) Human cartilage gp39+, CD16+ monocytes in peripheral blood and synovium: correlation with joint destruction in rheumatoid arthritis. Arthritis Rheum 43:1233–1243
3. Balducci S, Zanuso S, Nicolucci A, De Feo P, Cavallo S, Cardelli P, Fallucca S, Alessi E, Fallucca F, Pugliese G, Italian Diabetes Exercise Study (IDES) Investigators (2010) Effect of an intensive exercise intervention strategy on modifiable cardiovascular risk factors in subjects with type 2 diabetes mellitus: a randomized controlled trial: the Italian Diabetes and Exercise Study (IDES). Arch Intern Med 170:1794–1803
4. Basen-Engquist K, Chang M (2011) Obesity and cancer risk: recent review and evidence. Curr Oncol Rep 13:71–76

5. Bays HE (2009) "Sick fat," metabolic disease, and atherosclerosis. Am J Med 122:S26–S37
6. Belge KU, Dayyani F, Horelt A, Siedlar M, Frankenberger M, Grankenberger B, EspeviK T, Ziegler-Heitbrock L (2002) The proinflammatory CD14+CD16+DR++ monocytes are a major source of TNF. J Immunol 168:3536–3542
7. Bergmann M, Gornikiewicz A, Sautner T, Waldmann E, Weber T, Mittlböck M, Roth E, Függer R (1999) Attenuation of catecholamine-induced immunosuppression in whole blood from patients with sepsis. Shock 12:421–427
8. Bianchini F, Kaaks R, Vainio H (2002) Weight control and physical activity in cancer prevention. Obes Rev 3:5–8
9. Bishop NC, Walker GJ, Gleeson M, Wallace FA, Hewitt CRA (2009) Human T lymphocyte migration towards the supernatants of human rhinovirus infected airway epithelial cells: influence of exercise and carbohydrate intake. Exerc Immunol Rev 15:42–59
10. Bosanská L, Michalský D, Lacinová Z, Dostálová I, Bártlová M, Haluzíková D, Matoulek M, Kasalický M, Haluzík M (2010) The influence of obesity and different fat depots on adipose tissue gene expression and protein levels of cell adhesion molecules. Physiol Res 59:79–88
11. Bruun JM, Lihn AS, Pedersen SB, Richelsen B (2005) Monocyte chemoattractant protein-1 release is higher in visceral than subcutaneous human adipose tissue (AT): implication of macrophages resident in the AT. J Clin Endocrinol Metab 90:2282–2289
12. Bruunsgaard H, Galbo H, Halkjaer-Kristensen J, Johansen TL, MacLean DA, Pedersen BK (1997) Exercise-induced increase in serum interleukin-6 in humans is related to muscle damage. J Physiol 499:833–841
13. Chow FY, Nikolic-Paterson DJ, Ozols E, Atkins RC, Tesch GH (2005) Intercellular adhesion molecule-1 deficiency is protective against nephropathy in type 2 diabetic db/db mice. J Am Soc Nephrol 16:1711–1722
14. Coussens LM, Werb Z (2002) Inflammation and cancer. Nature 420:860–867
15. Croisier JL, Camus G, Venneman I, Deby-Dupont G, Juchmès-Ferir A, Lamy M, Crielaard JM, Deby C, Duchateau J (1999) Effects of training on exercise-induced muscle damage and interleukin 6 production. Muscle Nerve 22:208–212
16. Cupps TR, Fauci AS (1982) Corticosteroid-mediated immunoregulation in man. Immunol Rev 65:133–155
17. Davis JM, Kohut ML, Jackson DA, Colbert LH, Mayer EP, Ghaffar A (1998) Exercise effects on lung tumor metastases and in vitro alveolar macrophage antitumor cytotoxicity. Am J Physiol 274:R1454–R1459
18. Demarzo MM, Martins LV, Fernandes CR, Herrero FA, Perez SE, Turatti A, Garcia SB (2008) Exercise reduces inflammation and cell proliferation in rat colon carcinogenesis. Med Sci Sports Exerc 40:618–621
19. Eliassen HA, Hankinson SE, Rosner B, Holmes MD, Willet WC (2010) Physical activity and risk of breast cancer among postmenopausal women. Arch Intern Med 170:1758–1764
20. Eriksson KF, Lindgärde F (1991) Prevention of type 2 (non-insulin-dependent) diabetes mellitus by diet and physical exercise. The 6-year Malmö feasibility study. Diabetologia 34:891–898
21. Febbraio MA, Pedersen BK (2002) Muscle-derived interleukin-6: mechanisms for activation and possible biological roles. FASEB J 16:1335–1347
22. Fernandez MA, Puttur FK, Wang YM, Howden W, Alexander SI, Jones CA (2008) T regulatory cells contribute to the attenuated primary CD8+ and CD4+ T cell responses to herpes simplex virus type 2 in neonatal mice. J Immunol 180:1556–1564
23. Fingerle-Rowson G, Angstwurm M, Andreesen R, Ziegler-Heitbrock HW (1998) Selective depletion of CD14+CD16+ monocytes by glucocorticoid therapy. Clin Exp Immunol 112:501–506
24. Fischer CP (2006) Interleukin-6 in acute exercise and training: what is the biological relevance? Exerc Immunol Rev 12:6–33
25. Flynn MG, McFarlin BK, Phillips MD, Stewart LK, Timmerman KL (2003) Toll-like receptor 4 and CD14 mRNA expression are lower in resistive exercise-trained elderly women. J Appl Physiol 95:1833–1842
26. Flynn MG, McFarlin BK (2006) Toll-like receptor 4: link to the anti-inflammatory effects of exercise? Exerc Sport Sci Rev 34:176–181

27. Freeman BD, Buchman TG (2001) Interleukin-1 receptor antagonist as therapy for inflammatory disorders. Expert Opin Biol Ther 1:301–308
28. Friedenreich CM, Gregory J, Kopciuk KA, Mackey JR, Courneya KS (2009) Prospective cohort study of lifetime physical activity and breast cancer survival. Int J Cancer 124:1954–1962
29. Galbo H (1983) Hormonal and metabolic adaptation to exercise. Thieme-Stratton, New York
30. Gill JMR, Cooper AR (2008) Physical activity and prevention of type 2 diabetes mellitus. Sports Med 38:807–824
31. Gleeson M (2005) Immune function in sport and exercise. Elsevier, Edinburgh
32. Gleeson M, Bishop NC, Oliveira M, McCauley T, Tauler P, Muhamad AS (2012) Respiratory infection risk in athletes: association with antigen-stimulated IL-10 production and salivary IgA secretion. Scand J Med Sci Sports 22:410–417
33. Gleeson M, McFarlin BK, Flynn MG (2006) Exercise and toll-like receptors. Exerc Immunol Rev 12:34–53
34. Guilietti A, van Etten E, Overbergh L, Stoffels K, Bouillon R, Mathieu C (2007) Monocytes from type 2 diabetic patients have a pro-inflammatory profile. 1,25-Dihydroxyvitamin D(3) works as anti-inflammatory. Diabetes Res Clin Pract 77:47–57
35. Haffner SM (2007) Abdominal adiposity and cardiometabolic risk: do we have all the answers? Am J Med 120:S10–S16
36. Hardman AE, Stensel DJ (2009) Physical activity and health: the evidence explained, 2nd edn. Routledge, Oxford, pp 120–121
37. Hong EG, Ko HJ, Cho YR, Kim HJ, Ma Z, Yu TY, Friedline RH, Kurt-Jones E, Finberg R, Fischer MA, Granger EL, Norbury CC, Hauschka SD, Philbrick WM, Lee CG, Elias JA, Kim JK (2009) Interleukin-10 prevents diet-induced insulin resistance by attenuating macrophage and cytokine response in skeletal muscle. Diabetes 58:2525–2535
38. Jiao P, Chen Q, Shah S, Du J, Tao B, Tzameli I, Yan W, Xu H (2009) Obesity-related upregulation of monocyte chemotactic factors in adipocytes: involvement of nuclear factor-kappaB and c-Jun NH2-terminal kinase pathways. Diabetes 58:104–115
39. Ju J, Nolan B, Cheh M, Bose M, Lin Y, Wagner GC, Yang CS (2008) Voluntary exercise inhibits intestinal tumorigenesis in Apc(Min/+) mice and azoxymethane/dextran sulfate sodium-treated mice. BMC Cancer 8:316
40. Kaisho T, Akira S (2006) Toll-like receptor function and signalling. J Allergy Clin Immunol 117:979–987
41. Kanda H, Tateya S, Tamori Y, Kotani K, Hiasa K, Kitazawa R, Kitazawa S, Miyachi H, Maeda S, Egashira K, Kasuga M (2006) MCP-1 contributes to macrophage infiltration into adipose tissue, insulin resistance, and hepatic steatosis in obesity. J Clin Invest 116:1494–1505
42. Kasapis C, Thompson PD (2005) The effects of physical activity on serum C-reactive protein and inflammatory markers: a systematic review. J Am Coll Cardiol 45:1563–1569
43. Kawanishi N, Yano H, Yokogawa Y, Suzuki K (2010) Exercise training inhibits inflammation in adipose tissue via both suppression of macrophage infiltration and acceleration of phenotypic switching from M1 to M2 macrophages in high-fat-diet-induced obese mice. Exerc Immunol Rev 16:105–118
44. Keller C, Steensberg A, Hansen AK, Fischer CP, Plomgaard P, Pedersen BK (2005) Effect of exercise, training, and glycogen availability on IL-6 receptor expression in human skeletal muscle. J Appl Physiol 99:2075–2079
45. Kenfield SA, Stampfer MJ, Giovannucci EL, Chan JM (2011) Physical activity and survival after prostate cancer diagnosis in the health professionals follow-up study. J Clin Oncol 29:726–732
46. Kraus WE, Houmard JA, Duscha BD, Knetzger KJ, Wharton MB, McCartney JS, Bales CW, Henes S, Samsa GP, Otvos JD, Kulkarni KR, Slentz CA (2002) Effects of the amount and intensity of exercise on plasma lipoproteins. N Engl J Med 347:1483–1492
47. Lancaster GI, Khan Q, Drysdale P, Wallace F, Jeukendrup AE, Drayson MT, Gleeson M (2005) The physiological regulation of toll-like receptor expression and function in humans. J Physiol 563:945–955

48. Maffei M, Funicello M, Vottari T, Gamucci O, Costa M, Lisi S, Viegi A, Ciampi O, Bardi G, Vitti P, Pinchera A, Santini F (2009) The obesity and inflammatory marker haptoglobin attracts monocytes via interaction with chemokine (C-C motif) receptor 2 (CCR2). BMC Biol 17:87
49. Martinez FO, Sica A, Mantovani A, Locati M (2008) Macrophage activation and polarization. Front Biosci 13:453–461
50. Mathis D, Shoelson S (2011) Immunometabolism: an emerging frontier. Nat Rev Immunol 11:81
51. Matthews CE, Ockene IS, Freedson PS, Rosal MC, Merriam PA, Hebert JR (2002) Moderate to vigorous physical activity and risk of upper-respiratory tract infection. Med Sci Sports Exerc 34:1242–1248
52. Maynard CL, Weaver CT (2008) Diversity in the contribution of IL-10 to cell-mediated immune regulation. Immunol Rev 226:219–233
53. McFarlin BK, Flynn MG, Campbell WW, Stewart LK, Timmerman KL (2004) TLR4 is lower in resistance-trained older women and related to inflammatory cytokines. Med Sci Sports Exerc 36:1876–1883
54. Meckel Y, Eliakim A, Seraev M, Zaldivar F, Cooper DM, Sagiv M, Nemet D (2009) The effect of a brief sprint interval exercise on growth factors and inflammatory mediators. J Strength Cond Res 23:225–230
55. Meyerhardt JA, Heseltine D, Niedzwiecki D, Hollis D, Saltz LB, Mayer RJ, Thomas J, Nelson H, Whittom R, Hantel A, Schilsky RL, Fuchs CS (2006) Impact of physical activity on cancer recurrence and survival in patients with stage III colon cancer: findings from CALGB 89803. J Clin Oncol 24:3535–3541
56. Miyashita M, Burns SF, Stensel DJ (2008) Accumulating short bouts of brisk walking reduces postprandial plasma triacylglycerol concentrations and resting blood pressure in healthy young men. Am J Clin Nutr 88:1225–1231
57. Moore KW, de Waal MR, Coffman RL, O'Garra A (2001) Interleukin-10 and the interleukin-10 receptor. Annu Rev Immunol 19:683–765
58. Nehlsen-Cannarella SL, Fagoaga OR, Nieman DC, Henson DA, Butterworth DE, Schmitt RL, Bailey EM, Warren BJ, Utter A, Davis JM (1997) Carbohydrate and the cytokine response to 2.5 h of running. J Appl Physiol 82:1662–1667
59. Nieman DC, Henson DA, Austin MD, Sha W (2011) Upper respiratory tract infection is reduced in physically fit and active adults. Br J Sports Med 45:987–992
60. Northoff H, Berg A (1991) Immunologic mediators as parameters of the reaction to strenuous exercise. Int J Sports Med 12:S9–S15
61. Oliveira M, Gleeson M (2010) The influence of prolonged cycling on monocyte Toll-like receptor 2 and 4 expression in healthy men. Eur J Appl Physiol 109:251–257
62. Ostrowski K, Hermann C, Bangash A, Schjerling P, Nielsen JN, Pedersen BK (1998) A trauma-like elevation of plasma cytokines in humans in response to treadmill running. J Physiol 513:889–894
63. Ostrowski K, Rohde T, Asp S, Schjerling P, Pedersen BK (1999) Pro- and anti-inflammatory cytokine balance in strenuous exercise in humans. J Physiol 515:287–291
64. Ostrowski K, Rohde T, Asp S, Schjerling P, Pedersen BK (2001) Chemokines are elevated in plasma after strenuous exercise in humans. Eur J Appl Physiol 84:244–245
65. Ouchi N, Parker JL, Lugus JJ, Walsk K (2011) Adipokines in inflammation and metabolic disease. Nat Rev Immunol 11:85–97
66. Pedersen BK, Edward F (2009) Adolph distinguished lecture: muscle as an endocrine organ: IL-6 and other myokines. J Appl Physiol 107:1006–1014
67. Pedersen BK, Fischer CP (2007) Beneficial health effects of exercise–the role of IL-6 as a myokine. Trends Pharmacol Sci 28:152–156
68. Pedersen BK, Hoffman-Goetz L (2000) Exercise and the immune system: regulation, integration, and adaptation. Physiol Rev 80:1055–1081
69. Pedersen BK, Saltin B (2006) Evidence for prescribing exercise as therapy in chronic disease. Scand J Med Sci Sports 16:S5–S65
70. Petersen AM, Pedersen BK (2005) The anti-inflammatory effect of exercise. J Appl Physiol 98:1154–1162

71. Pischon T, Boeing H, Hoffmann K, Bergmann M, Schulze MB, Overvad K, van der Schouw YT, Spencer E, Moons KG, Tjønneland A, Halkjaer J, Jensen MK, Stegger J, Clavel-Chapelon F, Boutron-Ruault MC, Chajes V, Linseisen J, Kaaks R, Trichopoulou A, Trichopoulos D, Bamia C, Sieri S, Palli D, Tumino R, Vineis P, Panico S, Peeters PH, May AM, Bueno-de-Mesquita HB, van Duijnhoven FJ, Hallmans G, Weinehall L, Manjer J, Hedblad B, Lund E, Agudo A, Arriola L, Barricarte A, Navarro C, Martinez C, Quirós JR, Key T, Bingham S, Khaw KT, Boffetta P, Jenab M, Ferrari P, Riboli E (2008) General and abdominal adiposity and risk of death in Europe. N Engl J Med 359:2105–2120
72. Poitou C, Dalmas E, Renovato M, Benhamo V, Hajduch F, Abdennour M, Kahn JF, Veyrie N, Rizkalla S, Fridman WH, Sautès-Fridman C, Clément K, Cremer I (2011) CD14dimCD16+ and CD14+CD16+ monocytes in obesity and during weight loss: relationships with fat mass and subclinical atherosclerosis. Arterioscler Thromb Vasc Biol 31:2322–2330
73. Pradhan AD, Manson JE, Rifai N, Buring JE, Ridker PM (2001) C-reactive protein, interleukin 6, and risk of developing type 2 diabetes mellitus. JAMA 286:327–334
74. Puglisi MJ, Fernandez ML (2008) Modulation of C-reactive protein, tumor necrosis factor-alpha, and adiponectin by diet, exercise, and weight loss. J Nutr 138:2293–2296
75. Richman EL, Kenfield SA, Stampfer MJ, Paciorek A, Carroll PR, Chan JM (2011) Physical activity after diagnosis and risk of prostate cancer progression: data from the cancer of the prostate strategic urological research endeavor. Cancer Res 71:3889–3895
76. Rose-John S, Scheller J, Elson G, Jones SA (2006) IL-6 biology is co-ordinated by membrane bound and soluble receptors: role in inflammation and cancer. J Leukoc Biol 80:227–236
77. Ross R, Bradshaw AJ (2009) The future of obesity reduction: beyond weight loss. Nat Rev Endocrinol 5:319–325
78. Sakaguchi S (2005) Naturally arising Foxp3-expressing CD25+CD4+ regulatory T cells in immunological tolerance to self and non-self. Nat Immunol 6:345–352
79. Schlitt A, Heine GH, Blankenberg S, Espinola-Klein C, Dopheide JF, Bickel C, Lackner KJ, Iz M, Meyer J, Darius H, Rupprecht HJ (2004) CD14+CD16+ monocytes in coronary artery disease and their relationship to serum TNF-alpha levels. Thromb Haemost 92:419–424
80. Simpson RJ, McFarlin BK, McSporran C, Spielmann G, ó Hartaigh B, Guy K (2009) Toll-like receptor expression on classic and pro-inflammatory blood monocytes after acute exercise in humans. Brain Behav Immun 23:232–239
81. Skinner NA, MacIsaac CM, Hamilton JA, Visvanathan K (2005) Regulation of Toll-like receptor (TLR)2 and TLR4 on CD14dimCD16+ monocytes in response to sepsis-related antigens. Clin Exp Immunol 141:270–278
82. Speck RM, Courneya KS, Masse LC, Duval S, Schmitz KH (2010) An update of controlled physical activity trials in cancer survivors: a systematic review and meta-analysis. J Cancer Surviv 4:87–100
83. Starkie R, Ostrowski SR, Jauffred S, Febbraio M, Pedersen BK (2003) Exercise and IL-6 infusion inhibit endotoxin-induced TNF-alpha production in humans. FASEB J 17:884–886
84. Starkie RL, Angus DJ, Rolland J, Hargreaves M, Febbraio MA (2000) Effect of prolonged, submaximal exercise and carbohydrate ingestion on monocyte intracellular cytokine production in humans. J Physiol 528:647–655
85. Starkie RL, Arkinstall MJ, Koukoulas I, Hawley JA, Febbraio MA (2001) Carbohydrate ingestion attenuates the increase in plasma interleukin-6, but not skeletal muscle interleukin-6 mRNA, during exercise in humans. J Physiol 533:585–591
86. Starkie RL, Rolland J, Angus DJ, Anderson MJ, Febbraio MA (2001) Circulating monocytes are not the source of elevations in plasma IL-6 and TNF-alpha levels after prolonged running. Am J Physiol 280:C769–C774
87. Steensberg A (2003) The role of IL-6 in exercise-induced immune changes and metabolism. Exerc Immunol Rev 9:40–47
88. Steensberg A, Fischer CP, Keller C, Moller K, Pedersen BK (2003) IL-6 enhances plasma IL-1ra, IL-10, and cortisol in humans. Am J Physiol Endocrinol Metab 285:E433–E437

89. Steensberg A, van Hall G, Osada T, Sacchetti M, Saltin B, Pedersen BK (2000) Production of interleukin-6 in contracting human skeletal muscles can account for the exercise-induced increase in plasma interleukin-6. J Physiol 529:237–242
90. Stewart LK, Flynn MG, Campbell WW, Craig BA, Robinson JP, McFarlin BK, Timmerman KL, Coen PM, Felker J, Talbert E (2005) Influence of exercise training and age on CD14+ cell surface expression of toll-like receptor 2 and 4. Brain Behav Immun 19:389–397
91. Suzuki K, Nakaji S, Yamada M, Totsuka M, Sato K, Sugawara K (2002) Systemic inflammatory response to exhaustive exercise. Cytokine kinetics. Exerc Immunol Rev 8:6–48
92. Suzuki K, Nakaji S, Yamada M, Liu Q, Kurakake S, Okamura N, Kumae T, Umeda T, Sugawara K (2003) Impact of a competitive marathon race on systemic cytokine and neutrophil responses. Med Sci Sports Exerc 35:348–355
93. Takeda K, Kaisho T, Akira SS (2003) Toll-like receptors. Annu Rev Immunol 21:335–376
94. Tanasescu M, Leitzmann MF, Rimm EB, Willett WC, Stampfer MJ, Hu FB (2002) Exercise type and intensity in relation to coronary heart disease in men. JAMA 288:1994–2000
95. Timmerman KL, Flynn MG, Coen PM, Markofski MM, Pence BD (2008) Exercise training-induced lowering of inflammatory (CD14+CD16+) monocytes: a role in the anti-inflammatory influence of exercise? J Leukoc Biol 84:1271–1278
96. Tuomilehto J, Lindstrom J, Eriksson JG, Valle TT, Hamalainen H, Ilanne-Parikka P, Keinanen-Kiukaanniemi S, Laakso M, Louheranta A, Rastas M, Salminen V, Uusitupa M (2001) Prevention of type 2 diabetes mellitus by changes in lifestyle among subjects with impaired glucose tolerance. N Engl J Med 344:1343–1350
97. Ullum H, Haahour PM, Diamant M, Palmø J, Halkjaer-Kristensen J, Pedersen BK (1994) Bicycle exercise enhances plasma IL-6 but does not change IL-1 alpha, IL-1 beta, IL-6, or TNF-alpha pre-mRNA in BMNC. J Appl Physiol 77:93–97
98. Wang J, Song H, Tang X, Yang Y, Niu Y, Ma Y (2012) Effect of exercise training intensity on murine T-regulatory cells and vaccination response. Scand J Med Sci Sports 22:643–652
99. Warren TY, Barry V, Hooker SP, Sui X, Church TS, Blair SN (2010) Sedentary behaviors increase risk of cardiovascular disease mortality in men. Med Sci Sports Exerc 42:879–885
100. West-Wright CN, Henderson KD, Sullivan-Halley J, Ursin G, Deapen D, Neuhausen S, Reynolds P, Chang E, Ma H, Bernstein L (2009) Long-term and recent recreational physical activity and survival after breast cancer: the California Teachers Study. Cancer Epidemiol Biomarkers Prev 18:2851–2859
101. Wetmore CM, Ulrich CM (2004) Mechanisms associating physical activity with cancer incidence: exercise and immune function. CRC Taylor Francis, Boca Raton
102. Whitmer RA, Gustafson DR, Barrett-Connor E, Haan MN, Gunderson EP, Yaffe K (2008) Central obesity and increased risk of dementia more than three decades later. Neurology 71:1057–1064
103. Wolin KY, Yan Y, Colditz GA (2011) Physical activity and risk of colon adenoma: a meta-analysis. Br J Cancer 104:882–885
104. World Cancer Research Fund (2007) Food, nutrition and the prevention of cancer: a global perspective. WCRF, London
105. Xue F, Michels KB (2007) Diabetes, metabolic syndrome, and breast cancer: a review of the current evidence. Am J Clin Nutr 86:S823–S835
106. Yeh SH, Chuang H, Lin LW, Hsiao CY, Wang PW, Liu RT, Yang KD (2009) Regular Tai Chi Chuan exercise improves T cell helper function of patients with type 2 diabetes mellitus with an increase in T-bet transcription factor and IL-12 production. Br J Sports Med 43:845–850
107. Yeh SH, Chuang H, Lin LW, Hsiao CY, Eng HL (2006) Regular tai chi chuan exercise enhances functional mobility and CD4CD25 regulatory T cells. Br J Sports Med 40:239–243
108. Yudkin JS (2007) Inflammation, obesity, and the metabolic syndrome. Horm Metab Res 39:707–709
109. Zeyda M, Huber J, Prager G, Stulnig TM (2011) Inflammation correlates with markers of T-cell subsets including regulatory T cells in adipose tissue from obese patients. Obesity 19:743–748
110. Zoppini G, Targher G, Zamboni C, Venturi C, Cacciatori V, Moghetti P, Muggeo M (2006) Effects of moderate-intensity exercise training on plasma biomarkers of inflammation and endothelial dysfunction in older patients with type 2 diabetes. Nutr Metab Cardiovasc Dis 16:543–549

Index

A.J. Dannenberg and N.A. Berger (eds.), *Obesity, Inflammation and Cancer*, Energy Balance and Cancer 7, DOI 10.1007/978-1-4614-6819-6,

D